Analytical Techniques and Applications of Evapotranspiration

Analytical Techniques and Applications of Evapotranspiration

Edited by **Elizabeth Lamb**

Callisto Reference

New York

Published by Callisto Reference,
106 Park Avenue, Suite 200,
New York, NY 10016, USA
www.callistoreference.com

Analytical Techniques and Applications of Evapotranspiration
Edited by Elizabeth Lamb

International Standard Book Number: 978-1-63239-070-7 (Hardback)

Printed in China.

Contents

Preface

The book includes an analysis of the assessed measurement methods of evapotranspiration. It includes the related operations to the optimization of water usage in the field of agriculture and to the study of ecosystems. Also, various measuring methods are discussed with comprehensive explanations and examples. Some of them are methods at the leaf, agro-ecosystem and plant levels. For optimized water usage in agricultural field, definite measurements on transpiration requirements of crops and various cultivars, as well as conclusions of various irrigation methods and experiments in semi-arid areas for open field, greenhouse and potted grown plants are discussed in the book. Apart from this, some other topics discussed within the book are the features of ET of crops in saline surroundings, effects of ET on groundwater quality in xeric surroundings and the application of ET to climatic classification. The book presents an overview for both researchers and students who wish to know more about these subjects.

Various studies have approached the subject by analyzing it with a single perspective, but the present book provides diverse methodologies and techniques to address this field. This book contains theories and applications needed for understanding the subject from different perspectives. The aim is to keep the readers informed about the progresses in the field; therefore, the contributions were carefully examined to compile novel researches by specialists from across the globe.

Indeed, the job of the editor is the most crucial and challenging in compiling all chapters into a single book. In the end, I would extend my sincere thanks to the chapter authors for their profound work. I am also thankful for the support provided by my family and colleagues during the compilation of this book.

Editor

Part 1

Measuring Techniques for the Spatial and Temporal Characterisation of the ET

1

Spatial and Temporal Variation in Evapotranspiration

Jerry L. Hatfield and John H. Prueger
National Laboratory for Agriculture and the Environment
United States of America

1. Introduction

Evapotranspiration represents the combined loss of soil water from the earth's surface to the atmosphere through evaporation of water from the soil or plant surfaces and transpiration via stomates of the plant. In agricultural production systems these two losses of water represent a major component of the water balance of the crop. If we examine evapotranspiration over time throughout a growing season of a crop then the fractions of evaporation and transpiration will not remain constant. When there is a small plant partially covering the soil then the energy impinging on the soil surface will be used to evaporate water from the soil surface; however, as the crop develops and completely covers the soil then transpiration becomes the dominant process. There is a spatial and temporal aspect to evapotranspiration which exists but is often ignored in our consideration of the dynamics of water loss from the earth's surface.

One of the major questions which exists is how uniform is evapotranspiration over a given production field or over a landscape because of the limited amount of information on the spatial variation of evapotranspiration. There have been a limited number of research studies on the spatial variation in evapotranspiration. Many of these studies utilize remote sensing data as shown by Zhang et al. (2010) in which they developed a spatial-temporal evapotranspiration model for the Hebei Plain in China. They found the temporal variation in evapotranspiration was due to crop growth and the irrigation regime while spatial variation was caused by the type of crop being grown. An aspect of evapotranspiration is the use of reference pan evaporation to provide a surrogate for the atmospheric evaporation and the results from a study by Zhang et al. (2009) showed spatial variation was induced by changes in the driving variables, e.g., windspeed, solar radiation, or temperature. Variations in these parameters would be expected to create spatial differences in evapotranspiration from crop surfaces. Spatial and temporal variation in crop reference evapotranspiration has been studied by Zhang et al. (2010) across a river basin in China and observed the spatial variation in reference evapotranspiration was low in the cool months (January to April) and large in the warm months (May to August). The driving variable inducing the spatial variation in the warm months was most closely related to variation in the available energy among locations.

Li et al. (2006) evaluated the combination of remote sensing data combined with surface energy balance to evaluate the spatial variation in evapotranspiration and found the mean values of evapotranspiration were similar across a range of spatial scales. However, the

standard deviation decreased with higher spatial resolution and when the increased above 480 m, there was a loss of spatial structure in the evapotranspiration maps. Using a Raman lidar system, Eichinger et al. (2006) observed large spatial variation in evapotranspiration in corn (*Zea mays* L.) and soybean (*Glycine max* (L.) Merr.) linked with small elevation differences within the fields. These observations would suggest spatial structure has different scales and there are few studies which have attempted to evaluate spatial variation and the underlying causes. Mo et al. (2004) used a simulation model to evaluate evapotranspiration and found spatial variation was closely related to spatial patterns in precipitation and leaf area of the crop. This is similar to observations by Hatfield et al. (2007) from an experiment in central Iowa in which they observed that spatial variation in energy and carbon fluxes among different corn and soybean fields could be attributed to three factors. These factors were presence of cumulus clouds in the afternoon, variation in precipitation amounts across a watershed, and differences in the soil water availability in the soil profile. These studies demonstrate that there is spatial and temporal variation present in evapotranspiration from agricultural surfaces.
Evapotranspiration is a process controlled by the available energy, gradient of water vapor, availability of water for evaporation, and the gradient of windspeed as the transport process. The linkages among these parameters can be more easily seen in an expanded mathematical description of the latent heat flux (λE) given as

$$\lambda E = \frac{\left(\rho\lambda m/_{P}\right)(e_s - e_a)}{(r_c + r_{av})} \quad [1]$$

where λ is the latent heat of vaporization (J kg^{-1}), ρ the density of air (kg m^{-3}), m the ratio of molecular weight of water vapor to than of air (0.622), P the barometric pressure (kPa), e_s the saturation vapor pressure, e_a the actual vapor pressure of the air immediately above the surface, r_c the canopy resistance for water vapor transfer (s m^{-1}), and r_{av} the aerodynamic resistance for water vapor transfer (s m^{-1}). There has been much written about the linkages among these parameters; however, for a surface, evapotranspiration must be placed in context of the surface energy balance so that the balance of energy is expressed as

$$R_n - G - H = \lambda E \quad [2]$$

where R_n is the net radiation at the surface (J m^{-2} s^{-1}), G the soil heat flux (J m^{-2} s^{-1}), and H the sensible heat flux (J m^{-2} s^{-1}). It is the combination of the various factors which gives rise to the potential spatial and temporal variation in evapotranspiration. For example, the annual variation in solar radiation causes the amount of energy available for evapotranspiration to vary in a predictable way throughout the year. Farmer et al. (2003) found that climate and landscape were the two critical affecting the soil water balance. Kustas and Albertson (2003) observed spatial variation across the landscapes and proposed that our understanding of the critical knowledge gaps affecting spatial and temporal variation in evapotranspiration is lacking.
Measurements of energy balance components and estimates of evapotranspiration from Eq. 1 or 2 are often conducted over a single site within a production field or a landscape. The assumption from this measurement is that these values represent that particular surface with sufficient accuracy from which we derive an understanding of the dynamics of the surface. There are few studies in the literature which have directly measured evapotranspiration within a field to quantify the spatial variation and the factors which

create variation. The studies mentioned above have used remote sensing imagery as a surrogate for the energy balance and their results show there is spatial variation at relatively small scales; however, these scales are still often larger than areas within a production field. We have been addressing the problem of quantifying the spatial and temporal variation in evapotranspiration through a series of related studies across corn and soybean fields in central Iowa. These studies provide us insights into how crop management interacts with the landscape to induce variation in evapotranspiration.

2. Methodological approach

2.1 Energy balance measurements

The experimental site for these studies is located in central Iowa in a production field typical of the area on large (30-35 ha) fields located at 41.967° N, 93.695° W on a Clarion-Nicollet-Webster Soil Association using micrometeorological measurements of H_2O vapor and CO_2 exchanges above the canopy using an energy approach described by Hatfield et al., (2007). The energy balance approach used in these studies combines fast response of CO_2 and H_2O vapor signals with sonic anemometers, net radiation components, soil heat flux, and surface temperature. The use of this approach requires a large area to meet the fetch requirements and data have been collected at this site since 1998 where the data capture rate for these systems is greater than 95% (Hernandez-Ramirez et al., 2009).

Turbulent fluxes of sensible and latent heat (H & LE) and CO_2 were measured using the eddy covariance (EC). Each EC system is comprised of a three-dimensional sonic anemometer (CSAT3 Campbell Scientific Inc. Logan, UT[1]) and a fast response water vapor (H_2O) and CO_2 density open path infrared gas analyzer (IRGA) (LI7500 LICOR Inc., Lincoln, NE). In both the corn and soybean fields, EC instrument height is maintained on the 10 m towers at approximately 2 h (where h = canopy height in m) above the surface. The sampling frequency for the EC systems was 20 Hz with all of the high frequency data directly transmitted to the laboratory.

Ancillary instrumentation on each tower includes a 4-component net radiometer (R_n) (CNR-1 Kipp & Zonen Inc., Saskatoon, Sask.), soil heat flux plates (G) (REBS HFT-3) Cu-Co Type T soil thermocouples, two high precision infrared radiometric temperature sensors (IRT 15° fov) (Apogee Instruments Inc., Logan, UT) and an air temperature/ relative humidity (T_a) (RH) sensor (Vaisala HMP-35, Campbell Scientific Inc. Logan UT). The R_n, air temperature/humidity and one IRT (45° angle of view) sensor are mounted 4.5 m above ground level (AGL). The second IRT sensor is located 0.15 m AGL with a nadir view providing continuous radiometric temperatures of the soil surface. Four soil heat flux plates are placed 0.06 m below the soil, two within the plant row and two within the inter-row space. Pairs of soil thermocouples are placed 0.02 and 0.04 m below the surface and above each soil heat flux plate. Soil water content in the top 0.1m at each site will be measured with Delta-T Theta Probes (Dynamax Houston TX) and together with soil temperature data used to compute the storage component of the soil heat flux. The sampling frequency for the ancillary instrumentation is 0.1 Hz (10 s) with measured values stored as 10 min averages.

2.2 Field scale studies

To evaluate the impact of management on evapotranspiration, production sized fields have been used as experimental units because of the need to quantify the effects of N management on crop growth and yield and water use across a series of soil types. Fields range in size from 32 to 96 ha and are located in the Clarion-Nicollet-Webster Soil Association in central Iowa within the Walnut Creek watershed. This 5,400 ha watershed has been used for extensive research on environmental quality in relation to farming practices as described by Hatfield et al. (1999). Nitrogen management practices have varied across each year in response to the observations obtained from these experiments. The goal of these experiments has been to quantify the interactions of water and N across soil types with different N management practices. The most intensive studies have been conducted within a 60 ha field divided into two fields in a corn-soybean rotation with the primary emphasis on the corn portion of the rotation. The corn hybrid grown in these studies was Pioneer 33P67[1] for the duration of the study. The management practices placed different N rates in the field in large strips of 10 ha so the field was divided into no more than three strips in any one year. Within the field plant sampling, energy balance and crop yield plots were located within a given soil type. In this field, the predominant soils are Clarion, Canisteo, and Webster soils. Within each soil type and N management practice a plot area were identified and marked with GPS coordinates in order to locate the exact area among growing seasons.

Nitrogen management practices have been similar from 1997 through 2001. Nitrogen rates applied in 1997 and 1998 using a starter application at planting of 56 kg ha^{-1} only with the second treatment having the N starter rate and the sidedress rate determined by the Late Spring Nitrate Test (LSNT). The third treatment was the starter plus a rate to represent a non-limiting N rate of an additional 168 kg ha^{-1}. In 1999, 2000, and 2001 N application was modified to further refine rates based on leaf chlorophyll measurements and soil tests obtained from the 1997 and 1998 experiments. The rates applied were 56, 112, 168, or 232 kg N ha^{-1} to different soils, planting rates, and plant population densities (75,000 and 85,000 plants ha^{-1}). In 2000 and 2001, N was applied as either anhydrous ammonia in the fall or liquid urea anhydrous (UAN) in the spring at planting with a sidedress application. These applications were applied with production scale equipment to the field. Soil N concentrations were measured prior to spring operations, after planting, and at the end of the growing season after harvest to a soil depth of 1.5 m using a 5 cm core. Cores were subdivided into depth increments to estimate the N availability throughout the root zone at each of the sampling times. Sample position was recorded with a GPS unit to ensure accurate location of each subsequent sample.

2.3 Watershed scale studies

A watershed scale was conducted in the Walnut Creek Watershed in central Iowa located 5 km south of Ames, Iowa (41°75′ N, 93°41′W) as part of an ongoing long-term monitoring effort to assess interactions of crop water use, CO_2 uptake, and yield as a function of nitrogen management for corn and soybeans. Walnut Creek Watershed is a 5100 ha watershed of intensive corn and soybean production fields ranging in size from 40-160 ha. These two crops occupy approximately 85% of the land area in the watershed. The topography of the watershed and surrounding areas are characterized by flat to gently rolling terrain with elevations in the watershed ranging from 265 – 363 m with the lowest elevations situated on the eastern end of the watershed where the Walnut Creek drains. Details of production, tillage and nutrient management systems within the watershed are described in Hatfield et al. (1999).

[1] Mention of trade names or commercial products in this article is solely for the purpose of providing specific information and does not imply recommendation or endorsement by the U.S. Department of Agriculture.

To most extensive and intensive experiment was conducted in 2002 as part of a remote sensing soil moisture experiment (SMEX02) being was conducted across the Walnut Creek Watershed. This study provided the opportunity to place 12 eddy covariance (EC) stations across the watershed to measure and evaluate the spatial and temporal variation among fluxes across typical corn and soybean production fields in the Upper Midwest region. These stations were in operation during the intensive measurement period of the remote sensing campaign (Kustas et al., 2003) and continued to record measurements until late August 2002. These sites were distributed across the Walnut Creek watershed as shown in Fig 1 and sites 10 and 11 represent in the intensive field sites for the experiments conducted since 1998 on combinations of nitrogen management and water across soils types described above. For each site in the field the soil type was extracted from the soil map from Boone or Story County, Iowa. Eddy covariance sites were located in a range of soil types typical of central Iowa and in most fields the location represented over 0.20 of the total area in the field. The primary difference among the soils was the soil water holding capacity in the upper 1 m of the soil profile (Table 1). This provided an excellent opportunity to not only measure and evaluate differences in turbulent fluxes between corn and soybeans but also the spatial and temporal variability of turbulent flux exchange of CO_2 and H_2O across the agricultural landscape. The full details of the SMCAEX study are described in Kustas et al. (2005).

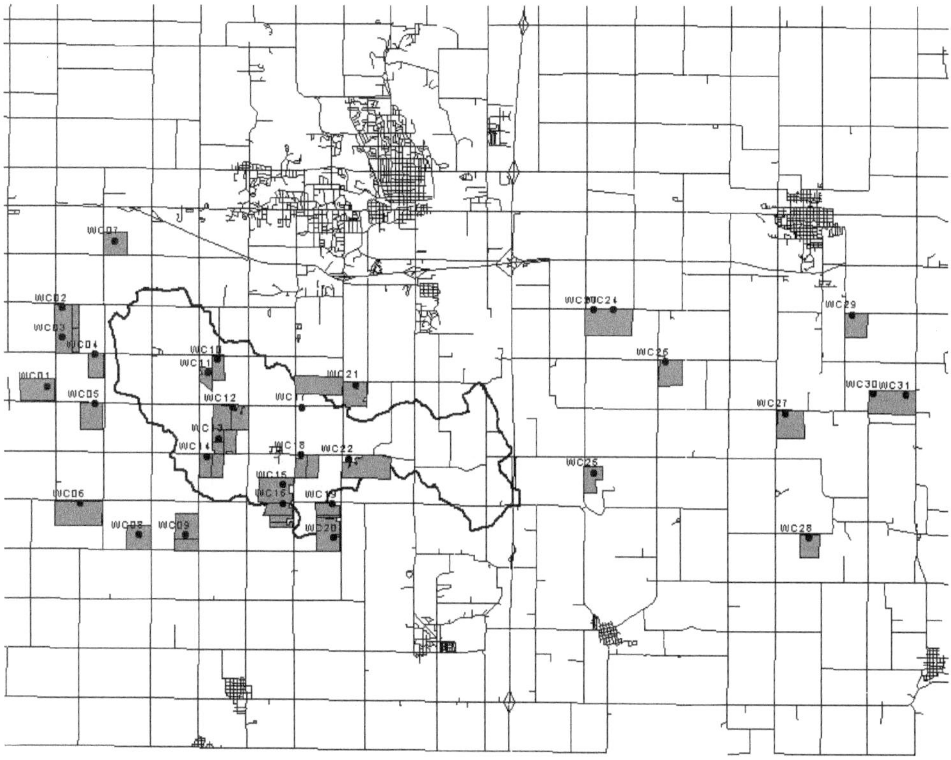

Fig. 1. Distribution of the energy balance and evapotranspiration measurement sites across Walnut Creek watershed in 2002.

Site	Crop	Soil Type	Fraction of Field	Soil Water Holding Capacity (mm for upper 1 m)
3	Soybean	Clarion, fine-loamy, mixed, mesic Typic Hapludolls	0.30	212
6	Corn	Clarion, fine-loamy, mixed, mesic Typic Hapludolls	0.24	212
10	Corn	Nicollet, Fine-loamy, mixed, mesic Aquic Hapludolls	0.16	220
11	Soybean	Harps, Fine-loamy, mesic Typic Calciaquolls	0.18	221
13	Soybean	Harps, Fine-loamy, mesic Typic Calciaquolls	0.12	221
14	Soybean	Clarion, fine-loamy, mixed, mesic Typic Hapludolls	0.24	212
25	Corn	Spillville, Fine-loamy, mixed, mesic Cumulic Hapludolls	0.41	214
33	Corn	Nicollet, Fine-loamy, mixed, mesic Aquic Hapludolls	0.10	220
151	Corn	Clarion, fine-loamy, mixed, mesic Typic Hapludolls	0.34	212
152	Corn	Canisteo, Fine-loamy, mixed (calcareous), mesic Typic Haplaquolls	0.33	209
161	Soybean	Clarion, fine-loamy, mixed, mesic Typic Hapludolls	0.35	212
162	Soybean	Clarion, fine-loamy, mixed, mesic Typic Hapludolls	0.35	212

Table 1.

3. Observations across scales

3.1 Temporal variation among years

Variation among years for evapotranspiration in rainfed areas is dependent upon the amount of precipitation stored within the soil profile. If there is adequate storage capacity, then annual variation in evapotranspiration will more dependent upon the available energy than upon the amount of available water. In areas with soils with limited soil water holding capacity then a more direct relationship will be evident. Across central Iowa, which would be typical of the Corn Belt, there is large annual variation in evapotranspiration as evidenced in the data from 1998 (Fig.2), 1999 (Fig. 3), and 2000 (Fig. 4).

Two important details are evident from these three years which represent fairly typical years in central Iowa. First, there is little evapotranspiration occurring the winter months and fall as evidenced by the relatively small cumulative values during these intervals. Evapotranspiration does not begin to become significant portion of the energy balance (Eq. 2) until about DOY 100 and begins to diminish after DOY 300 (Figs. 2, 3 and 4). These seasonal patterns are consistent among years with very similar times in which evapotranspiration values begin to increase in the spring and decrease in the fall. Second, cumulative values of evapotranspiration are relatively smooth compared to precipitation values, which occur in infrequent storms, not every day, and throughout the year. Third, annual total values of evapotranspiration are more similar among years than are annual precipitation totals. As an example, total evapotranspiration for 1998 was 476 mm, 1999 - 500 mm, and 2000 - 433 mm while total precipitation for 1998 was 933, for 1999 - 743, and for 2000 - 454 mm. Temporal variation in evapotranspiration among years will be dependent upon the energy available and at the annual time scale there are minor differences among years.

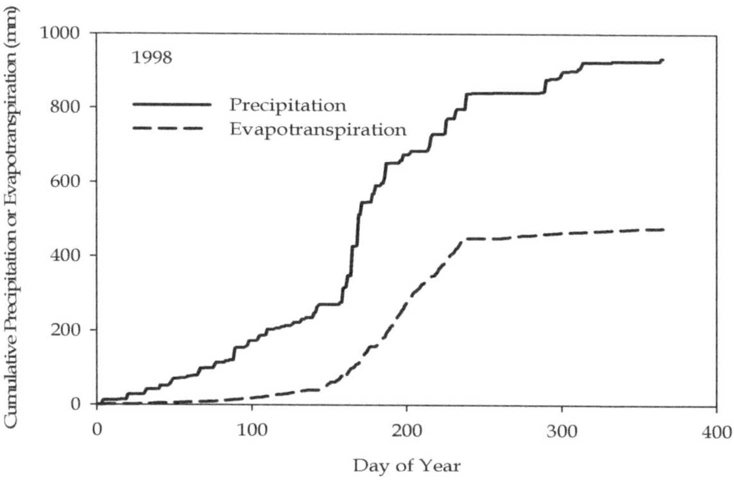

Fig. 2. Annual cumulative precipitation and evapotranspiration for a corn production field for Central Iowa in 1998.

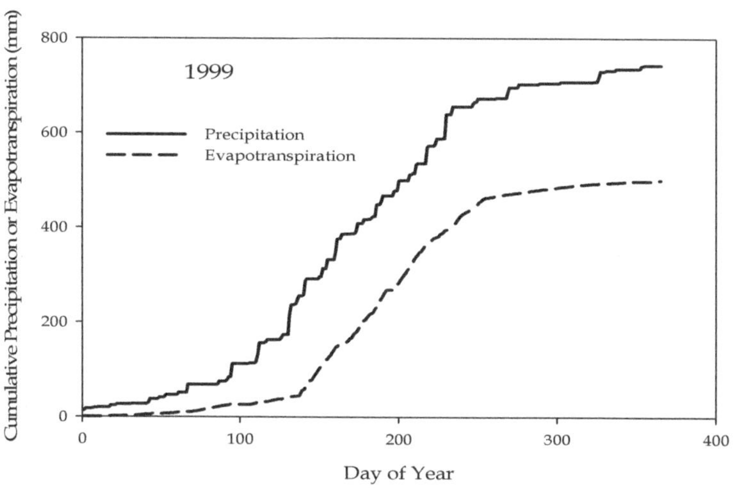

Fig. 3. Annual cumulative precipitation and evapotranspiration for a corn production field for Central Iowa in 1999.

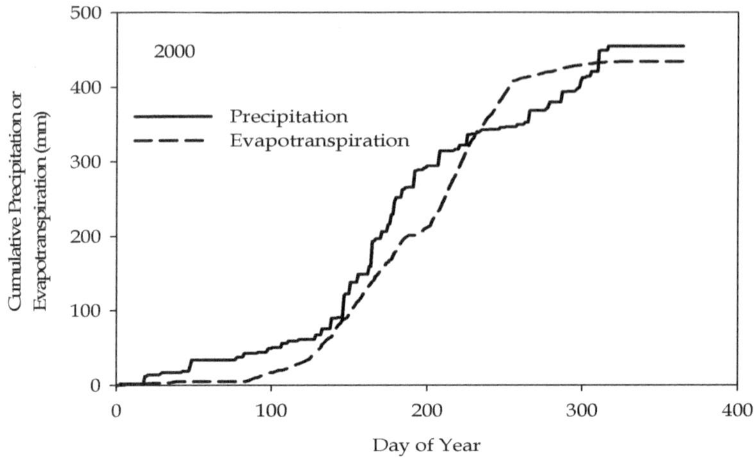

Fig. 4. Annual cumulative precipitation and evapotranspiration for a corn production field for Central Iowa in 2000.

3.2 Spatial variation within production fields

Spatial variation of evapotranspiration within fields is more significant than often thought based on the results shown in Figs 2, 3, and 4. It is assumed that evapotranspiration across a field would be relatively consistent because the energy balance components would be consistent. We have examined this aspect across both corn and soybean fields and found there is a large spatial variation induced by soil water holding capacity. An example of this variation is shown in Fig. 5 for evapotranspiration from a corn crop in an Okoboji soil compared to a Clarion soil and a Nicollet soil. The Okoboji soil is a high organic matter soil (soil organic matter of 7-9%) compared to a Nicollet soil (soil organic matter of 3-5%) and a Clarion soil (soil organic matter of 1-2%). These soils represent three different positions on the landscape with the Clarion soil being the upper part of the landscape in the Clarion-Nicollet-Webster soil association while the Okoboji soils are the lower part of the landscape and often considered to be poorly drained soils while the Nicollet soil is about midway on the slope.

There are large differences in the seasonal totals among these three soils (Fig. 5). The seasonal totals for the Okoboji and Nicollet soils are quite similar at 575 and 522 mm, respectively while the Clarion soil has an annual total of 310 mm. There are differences among the patterns of evapotranspiration throughout the year for the three soils. These types of patterns are not uncommon based on our multiple years of measurements across this field in which we have measured evapotranspiration in different soils. In this field, the evapotranspiration from the Okoboji soil begins slower at the beginning of the season because the tillage practice leaves this area with crop residue which decreases soil water evaporation rates and also the plant growth tends to be slower in this area of the field. In the Nicollet soils, there is more soil water evaporation and earlier plant growth because these areas of the field show an increased rate of growth because of the more favorable growth conditions. In contrast, the Clarion soils behave similar to the Nicollet soils in the early season but then at as the crop grows there is insufficient soil water to maintain the water supply and evapotranspiration becomes limited. This is a common occurrence in these fields and we often observe evapotranspiration totals in the Clarion soils at least half of the soils with the higher water holding capacity. These areas of the field also exhibit water deficits throughout the growing season because the soil is unable to supply the water required to meet the atmospheric demand and the canopy resistance term (Eq. 1) is much higher in these plants than in other soils within the field. There is a spatial variation of evapotranspiration within a field induced by the soil water holding capacity and this will influence the ability of the plant to be able to extract water to meet atmospheric demand.

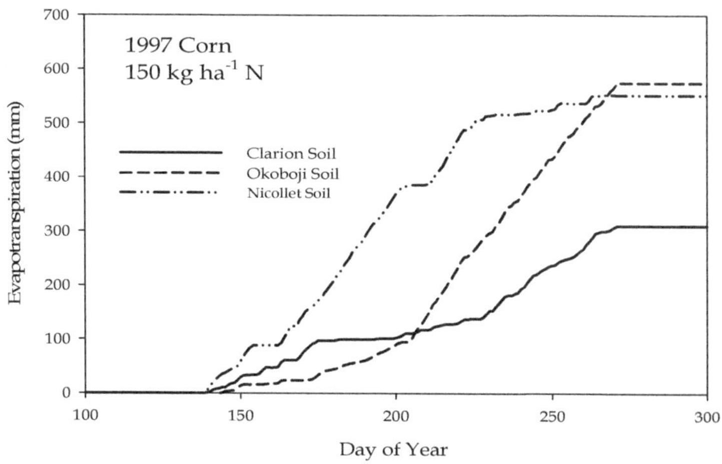

Fig. 5. Seasonal cumulative evapotranspiration from a corn crop grown in an Okoboji and Clarion soil within an individual field during 1997.

Spatial variation of evapotranspiration within a field can be affected by the effect of soil management practices on the crop growth patterns. We have investigated the interactions of nitrogen management with evapotranspiration across production fields. These seasonal totals are confined to the growing season because of the minimal amount of evapotranspiration during the other times of the year as shown in the earlier section. Nitrogen management interacted with soil type in the seasonal evapotranspiration totals reflective of the effect of nitrogen on growth in the different soils. In this study, we compared water use and crop growth in a Webster and Clarion soil. The Webster soil is similar to the Nicollet soil with soil organic matter contents of 3-5%. In this study, the seasonal evapotranspiration totals for both the fall and spring nitrogen application rates showed differences among soils with the Clarion soil having less evapotranspiration than the Webster soils (Fig. 6). There is an interesting effect of nitrogen application rates in this study because the application of 200 kg ha^{-1} on the Clarion soil actually reduced evapotranspiration compared to the 100 kg ha^{-1} rate (Fig. 6). We have observed this response in different years because the low water holding capacity soils cannot supply adequate water for evapotranspiration and there is actually a reduction in plant growth from the excess nitrogen applied. The reverse effect is found in the Webster soil where there is no difference in evapotranspiration rates until late in the growing season when the additional nitrogen from the 200 kg ha^{-1} rate is able to sustain growth and maintain evapotranspiration rates compared to the 100 kg ha^{-1} rate (Fig. 6).

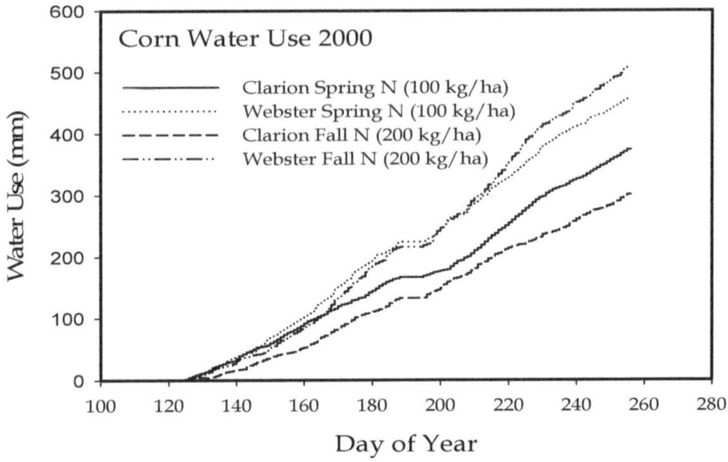

Fig. 6. Seasonal cumulative evapotranspiration values for corn in central Iowa from two different soils in 2000 with different nitrogen rates and application times.

Spatial variation patterns within a field have often been assumed to be minimal; however, these differences are larger than expected because of the differences in soil water holding capacity. The seasonal evapotranspiration patterns represent the combined effects of soils and management and these differences will affect the ability of a crop to endure water stress during the growing season. In rainfed environments, it is critical for precipitation events to maintain the soil water supply at an optimum level and if there is a limitation in the ability of the soil to store water and meet the evapotranspiration rate then crops will undergo water deficit stress.

3.3 Spatial variation among production fields

There have been few studies which have attempted to quantify the differences in evapotranspiration rates among fields. The primary reason is the expense of the array of equipment and the labor requirements to establish this observational network. As part of the SMEX2002 experiment described by Kustas et al. (2003) we were able to establish a network of energy balance stations and eddy correlation equipment across Walnut Creek watershed in central Iowa as shown in Fig. 1. The details of the study have been reported by Hatfield et al. (2007) and they observed variability among fields was due to three factors. Within a day, differences in the energy balance components and evapotranspiration was caused by the presence of cumulus clouds. Clouds are not evenly distributed across the watershed and differentially shade one area of the watershed more than another. These effects do not persist from one day to the next because the presence of clouds over a given field changes among days. However, these effects do induce evapotranspiration differences among fields.

The second factor which caused differences among fields was the spatial variation in precipitation events across the watershed. In temperate climates it is not unusual for convective rainfall amounts to be variable across space and this changes the amount of water available for evaporation. The scale of differences induced by variable precipitation is difficult to assess and across a small area (10 km^2) there could large differences in evapotranspiration. These differences may occur as a result of increased soil water evaporation from the soil surface when the plants are small because of the exposed soil. These differences caused by differential rainfall would be expected to diminish as the crop canopy develops because the amount of exposed soil would decrease and evapotranspiration would be dominated by transpiration from the canopy.

The third factor which caused a difference in the spatial variation in evapotranspiration is related to the soil water holding capacity as shown in Table 1. Across the different sites for the experiment in 2002, Hatfield et al. (2007) observed differences among sites as shown in Figs. 7 and 8. These differences were large for the short-term observations in this study.

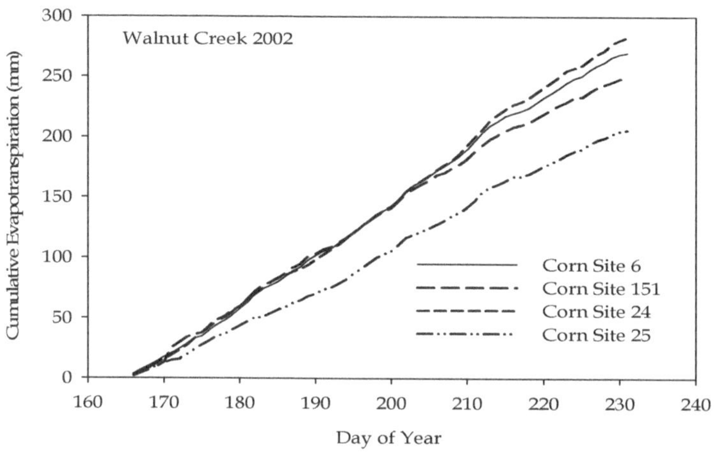

Fig. 7. Cumulative evapotranspiration across four corn fields with detailed measurements in Walnut Creek watershed in 2002.

These observations reveal important components of factors which induce spatial variation in evapotranspiration. For the four corn fields, there was a significant difference in the cumulative evapotranspiration for field 25 compared to the other fields (Fig. 7). In addition to the measurements being made in the soil with a lower water holding capacity, this field also had less rainfall during this portion of the growing season. These differences occurred early in the season and persisted throughout the period of measurements. This is in contrast to the other three fields in which there were similar evapotranspiration values until late in the growing season in which soil water holding capacity became the dominant factor. This degree of differential response would be expected if the energy input and rainfall amounts were the same but the storage factor changed.

In the soybean fields, there was little difference in the early season evapotranspiration among field and the differences among fields began to appear when the growth of the plant achieved full cover and water use rates were at their peak (Fig. 8). Separation among the fields was due to the soil water holding capacity of the field in which measurements were being made. The differences among fields were as large as 25-30 mm which is significant in terms of crop water use requirements and crop growth.

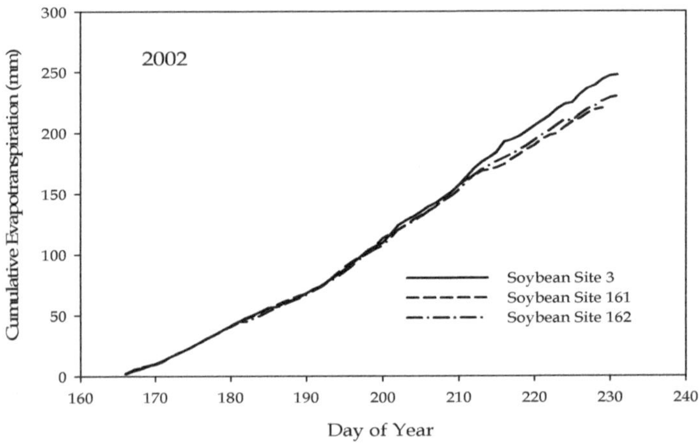

Fig. 8. Cumulative evapotranspiration across four soybean fields with detailed measurements in Walnut Creek watershed in 2002.

In both the corn and soybean observations, there are some notes of caution in terms of understanding spatial and temporal variation in evapotranspiration. Spatial variation of evapotranspiration is a result of a combination of factors and care must be exercised in the placement of energy balance and evapotranspiration equipment within fields and across landscapes in order to capture information from sites representative of the area. These differences can be controlled; however, rainfall patterns and cumulus cloud formation on the shorter time intervals cannot be controlled but should be measured to ensure proper comparisons among sites can be conducted. Overall, the spatial variation in evapotranspiration is due to a complex set of interactions affected the evapotranspiration at a given site. One of the overlooked factors is the soil water holding capacity and the depth of the water extraction caused by differences in rooting depth. These are often considered to be small; however, in our observations these factors can account for 100-200 mm of seasonal water use differences among sites. These differences coupled with spatial variation in rainfall during the growing season can lead to even greater differences among sites. In temperate regions, the spatial pattern of rainfall is a random event while the spatial variation in soil characteristics is a fixed position on the landscape causing the exact seasonal pattern of evapotranspiration for a given year to be a combination of the soil and weather patterns. Understanding the factors causing spatial variation in evapotranspiration will lead to improved capabilities for water management in cropping systems.

4. References

Eichinger, W.E., Cooper, D.I., Hipps, L.E., Kustas, W.P., Neale, C.M.N. & Prueger, J.H. (2006). Spatial and temporal variation in evapotranspiration using Raman lidar. Adv. Water Res. 29: 369-381.

Farmer, D., Sivapalan, M. & Jothityangkoon, C. (2003). Climate, soil, and vegetation controls upon the variability of water balance in temperate and semiarid landscapes. Water Resource Res. 39:1035, doi:10.1029/2001WR00003238.

Hatfield, J.L., Prueger, J.H. & Kustas, W.P. (2007). Spatial and temporal variation of energy and carbon fluxes in Central Iowa. Agron. J. 99:285-296.

Hatfield J.L., Jaynes, D.B., Burkart, M.R., Cambardella, C.A., Moorman, T.B., Prueger, J.H. & Smith, M.A. (1999). Water Quality in Walnut Creek Watershed: Setting and Farming Practices. J. Environ. Qual. 28:11-24.

Hernandez-Ramirez, G., Hatfield, J.L., Parkin, T.B., Prueger, J.H. & Sauer, T. J. (2010). Energy balance and turbulent flux partitioning in a corn-soybean rotation in the Midwestern U.S. Theor. Appl. Climatol. 100:79-92.

Kustas, W.P. & Albertson, J.D. (2003). Effects of surface temperature on land atmosphere exchange: A case study from Monsoon 90. Water Resource Res. 39:1159, doi:10.1029/2001WR001226.

Kustas, W.P., Hatfield, J.L. & Prueger, J.H. (2005). The Soil Moisture Atmosphere Coupling Experiment (SMACEX): Background, Hydrometeorological Conditions and Preliminary Findings. J. Hydrometeorol. 6:791-804.

Kustas, W.P., Prueger, J.H., Hatfield, J.L., MacPherson,J.I., Wolde, M., Neale, C.M.U., Eichinger, W.E., Cooper, D.I., Norman, J.M. & Anderson, M. (2003). An overview of the Soil-Moisture-Atmospheric-Coupling-Experiment (SMACEX) in central Iowa. American Meteorological Society, 17th Conference on Hydrology, Long Beach, CA Feb. 09-12, pp 1-5.

Li, Z.Q., Yu, G.R., Li, Q.K., Fu, Y.L & Li, Y.N. (2006). Effect of spatial variation on areal evapotranspiration in Haibei, Tiber plateau, China. Int J. Remote Sens. 27:3487-3498.

Mo, X.G., Liu, S.X., Lin, Z.H. & Zhao, W.M. (2004). Simulating temporal and spatial variation in evapotranspiration over the Lushi Basin. J. Hydrology. 285:125-142.

Zhang, S.W., Yei, Y.P., Li, H.J. & Wang, Z. (2010). Temporal-spatial variation in crop evapotranspiration in Hebei Plain, China. J. Food Agric. Environ. 8:672-677.

Zhang, X.Q., Ren, Y., Yin, Z.Y., Lin, Z.Y, & Zheng, D. (2009). Spatial and temporal variation patterns of reference evapotranspiration across the Qingahi-Tibetan Plateau during 1971-2004. J Geophysical Res. Atmos. 114: D15105.

Zhang, X.T., Kang, S.K., Zhang, L. & Liu, J.Q. (2010). Spatial variation of climatology monthly crop reference evapotranspiration and sensitivity coefficients in Shiyang river basin of northwest China. Agric. Water Mgt. 97:1506-1510.

2

Is It Worthy to Apply Different Methods to Determine Latent Heat Fluxes? - A Study Case Over a Peach Orchard

F. Castellví
Dept. Environmental and Soil sciences,University of Lleida, Lleida, Spain

1. Introduction

Knowledge of evapotranspiration, *ET*, is crucial for hydrological and micrometeorological studies including model calibration, because *ET* links two fundamental equations; the surface energy balance and the water balance. However, regardless the use of direct and indirect methods, measuring *ET* is difficult due to a number of limitations and/or shortcomings involved in methodologies and instrumentation. For irrigation planning, *ET* has traditionally been estimated using crop coefficient, Kc, values, where the actual evapotranspiration, *ETa*, is determined as, *ETa* = Kc x *ETo* [*ETo* is the reference evapotranspiration]. *ETo* is estimated from weather or climate data using semi-empirical equations (Doorenbos and Pruitt, 1977; Allen et al., 1998; Allen et al., 2005). Weighing lysimeters have been traditionally used to measure *ETa* and estimate the Kc = *ETc*/*ETo*, where *ETc* = *ETa* assuming little or no loss of evaporation due to stress. However, acquisition and maintenance of a large weighing lysimeter is expensive (Scott et al., 2005), and they are not transportable. Therefore, it is difficult to assess Kc over different agricultural areas because Kc evaluation is constrained to a unique crop per season, to a given climate and to a site-specific crop management. It is of interest to mention that it is often difficult to achieve uniformity within and outside of the lysimeter for tree orchards because of the size of the plants, and that under windy conditions lysimeters may not be reliable. For many agricultural institutions, the eddy covariance, EC, method offers an alternative to lysimeters, however, the method is somewhat complex, instrumentation is stringent and gas analyzers are expensive. Less costly micrometeorological methods, based on similarity theory (e.g., the Bowen ratio -energy balance method, aerodynamic method, etc.) have been investigated for use in agriculture (Meyers and Baldocchi, 2005; Drexler et al., 2004; Brustaert, 1988), but neither method has been widely adopted likely due to shortcomings under some climate features. It is not straightforward to measure accurate local humidity gradients and similarity based methods operate within the inertial sub-layer. Therefore, its application is often constrained to non-shallow inversions and to extensive fields to meet fetch requirements. Thus, full season monitoring over a growing crop with a relatively tall canopy requires sensor mounting at high levels, and the lack of fetch often limits application of similarity-based methods. To better interpret the measurements obtained using direct micrometeorological methods, including the EC method, it is highly recommended to take the measurements required to evaluate the closure of the surface energy balance equation as an indirect test of latent heat flux, *LE*, reliability (Brutsaert, 1988; Twine et al., 2000; Wilson et al., 2002; Foken et al, 2006; Castellví et al., 2008). The Penman-Monteith, PM, equation involves the aerodynamic resistance, but also the canopy resistance which makes PM equation difficult to apply for estimating *LE* in other than grazed surfaces not short of water (Allen et al., 1996). Sap flow measurements provide a method to measure transpiration, but despite it is not generally applicable, evaporation is not accounted which limits its application to irrigation system having little or no surface wetting. Soil moisture monitoring is widely used to estimate *ET*, but water movement between soil layers (even when the soil is unsaturated) is not easy to model, and soil moisture sensors measure water content within a small volume and the site measurement is often not representative of the entire field. Hence, due to the difficulty to obtain accurate *ET* (or *LE*) time series, even when affordability to direct measurement (i.e., using lysimeters and the EC method) is not a problem, simultaneous application of alternative methods or approaches may be especially useful when they do not share shortcomings. However, it is desirable to apply simple and affordable methods if they are proven reliable. An alternative method may help to perform quality control including gap filling. For *LE* estimation, the micrometeorological approach known as the residual method has been widely used (Brutsaert, 1988; Twine et al., 2000)

$$LE = Rn - G - H \tag{1}$$

Where, Rn is the net radiation, G is the soil heat flux and H is the sensible heat flux. In Eq. (1) the total rate of energy storage and other additional averaged energy sources (or sinks) including advective terms were neglected. However, Eq. (1) appears to hold for most agricultural surfaces having adequate fetch (Oncley et al., 2007), including tree orchards (Teixeira et al., 2008). When the available net surface energy, (Rn - G) is available, Eq. (1) combined with a method to estimate H allows for LE estimation. The objective of this study is to compare the hourly and daily series of LE determined over a peach orchard using a large weighing lysimeter, LE_{Lys}, and from Eq. (1) estimating H using the EC method, LE_{EC}, and the Surface renewal, SR, method, LE_{SR} (Paw U et al., 1995; Castellví, 2004; Paw U et al., 2005). The SR approach was selected as an alternative method because the instruments required are affordable, robust, easy to maintain, transportable, sense much larger area than a lysimeter, and it is expected reliable under windy conditions.

For this study, the SR model used combines SR analysis, mixing-length theory for momentum (Harman and Finnigan, 2007) and mixing-layer analogy (Raupach et al., 1996; Graefe, 2004). Therefore, it operates close to the canopy top which avoids installation of tall mast as required for other micrometeorological methods. To preserve stationarity, hourly and daily *LE* estimates (Eq. 1) were determined integrating the half-hourly values of the measured available net surface energy and sensible heat flux. The *LE* estimates were compared with measurements from a weighing lysimeter located nearby.

2. Theory

SR analysis [pioneered by Paw U et al. (1995)] combines SR theory [pioneered by Higbie (1935)] and the analysis of the scalar time trace (a subject of research during decades that still is of major interest). The SR theory (originally developed to investigate interfacial heat transfer between a liquid and a gas) assumes that heat transport occurs when a fluid parcel travelling at a given height in the bulk of the flow (above the interface) descends and takes contact with the heated surface for a period during which the parcel is heated. There is an unsteady diffusion transport during the contact until the parcel is, by continuity, replaced or renewed by other parcel coming from aloft. The detection of sequential sweeps and ejections of parcels from the interface is crucial because the heating take place in between. Paw U et al. (1995) applied the SR concept over land surfaces to investigate scalar transfer between sources located at the surface and the atmospheric surface layer. That is, consider an air parcel travelling at a given height above the canopy. Due to shear stress, SR analysis assumes that at some instant the parcel moves down into the canopy and remains in contact with the canopy elements for a period after it is ejected upwards and replaced by another parcel sweeping in from aloft. During the connect period the parcel has been enriched (depleted) of scalar due to the exchange between the air and sources (sinks). Each sweep and ejection is an injection of scalar from the sources into the bulk of the atmosphere which is characterized by a given amount of scalar released during the renewal period. The total flux of a scalar is, therefore, driven by the continuous renewal (i.e., replacement of air parcels) across the canopy top averaged for a given period (typically, half-hour). For sensible heat flux, when high-frequency air temperature measurements are taken at a given height z, the renewal process can be visually inferred in the time trace as a rather regular low-frequency ramp-like (asymmetric triangle shape) pattern. Paw U et al. (1995) presented a diagram of the SR process (Fig. 1a) and abstracted an ideal scheme for a ramp-like event in the air temperature trace (scheme 1 in Fig. 1a). Chen et al. (1997a) presented a slightly different ramp model (scheme 2 in Fig. 1a) that neglects the quiescent period but includes a micro-front period instead of an instantaneous ejection. Whatever the scheme, the amplitude, A, and period, τ, are the ramp dimensions that characterize an injection of sensible heat flux. The eddy responsible of such injections of scalar (i.e., the coherent structure) explains most of the total flux determined at the measurement height using the EC method (Hongyan et al., 2004). Analysis of time series, therefore, consists on identification of coherent structures to extract their ramp dimensions as shown in Fig. 1b. A method based on structure functions for determining the mean ramp dimensions sequentially observed in a trace (typically half-hour) is described in Appendix A.

Based on the scalar conservation equation for a planar homogeneous turbulent flow, assuming that the air parcel is uniformly heated by sources (sinks) below the measurement height, z, with no heat lost from the parcel top while it remains in contact with the canopy, the mean sensible heat flux density can be estimated as (Paw U et al., 1995)

$$H = \rho C_p (\alpha z) \frac{A}{\tau} \tag{2}$$

where ρ and C_p are the density and specific heat at constant pressure and, z, is the top of the air parcel which represents a volume per unit area. The parameter α is included to correct for the assumptions invoked.

When measurements are taken above the canopy, based on the one dimensional diffusion equation, the following relationship to estimate α over the averaging period was derived (Castellví, 2004)

$$\alpha = \begin{cases} \left[\dfrac{k}{\pi} \dfrac{(z-d)}{z^2} \tau u_* \phi_h^{-1}(\varsigma) \right]^{1/2} & z > z^* \\ \left[\dfrac{k}{\pi} \dfrac{(z^*-d)}{z^2} \tau u_* \phi_h^{-1}(\varsigma) \right]^{1/2} & h < z \le z^* \end{cases} \tag{3}$$

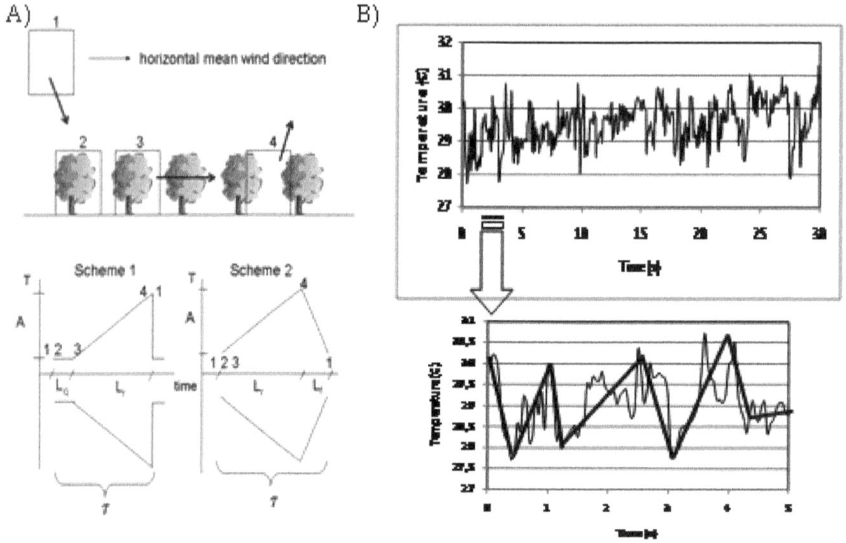

Fig. 1. A) Diagram of the renewal motion. The time course of the air temperature within the air parcel for the different positions is idealized in two ramp models. Scheme 1 assumes a quiescent period and a sharp instantaneous drop in temperature. Scheme 2 neglects the quiescent period and assumes a finite micro-front. L_r, L_q and L_f denote the warming, quiescent and micro-front periods, respectively. A is the ramp amplitude and τ is the total ramp duration. B) Air temperature measured at 10 Hz versus time observed under unstable conditions over a peach tree during 30 s. Few ramps fitted (by eye) are shown for the first 5 s.

Equations (2) and (3) provide a method exempt of α calibration for estimating sensible heat flux (Snyder et al., 1996; Castellví et al., 2008). In Eq. (3), d is the zero-plane displacement, z^* is the roughness layer depth (height from the ground to the base of the inertial sub-layer), h is the canopy height, u_* is the friction velocity, k = 0.4 is the Von Kármán constant, $\phi_h(\zeta)$ is the stability function for heat transfer described below in Eq. (5), and ζ is an stability parameter defined as, $(z\text{-}d)/L_O$, with L_O being the Obukov length defined as

$$L_O = -\frac{u_*^{\,3}}{k\,g\left(\overline{w'T_v'}\right)} T_v \tag{4}$$

where g is the acceleration due to gravity and T_v is the virtual temperature, which can be replaced with T in dry environments. Likely, the most widely used formulation for $\phi_h(\zeta)$ is (Högström, 1988; Foken, 2006)

$$\phi_h(\varsigma) = \begin{cases} (0.95 + 7.8\varsigma) & 0 \le \varsigma \le 1 \\ 0.95\,(1 - 11.6\varsigma)^{-1/2} & -2 \le \varsigma \le 0 \end{cases} \tag{5}$$

It is of interest to mention that other expressions covering a wider range of stability conditions are available (Kader and Perepelkin, 1989).

2.1 Procedure to estimate the sensible heat flux

To solve Eq. (2), other equations providing estimates for z^*, d and u_* are required. Therefore, an iterative procedure to solve the sensible heat flux must be implemented. Certainly, to achieve the full potential of SR analysis, the input required should be constrained to measurements taken at a single height. A model described in Castellví and Snyder (2009) which is based on a mixing-length theory for momentum (Harman and Finnigan, 2007) combined with mixing-layer analogy (Raupach et al., 1996) was used for half-hourly z^* estimates as a function of the canopy architecture (canopy height, h, leaf area index, LAI, and the crown thickness relative to the canopy height, f), drag coefficient at leave scale, c_d, and the turbulent intensity near the canopy top, I_u ($=\frac{\sigma_u}{u_h}$; where σ_u is the turbulent standard deviation of the horizontal wind speed, u, and u_h is the mean u at the canopy top). It is described in Appendix B. Appendix C shows the procedure used to solve H. Accordingly, z^* estimation taking measurements taken at one height may be obtained using the following expression

$$z^* = (c+1)h \quad \text{where} \quad c = \frac{h}{(h-d)}\frac{(fI_u)^2}{(c_d LAI)} \approx 4.45 I_u^2 \tag{6}$$

3. The field experiment and main climate features

The experiment was carried out at the University of California Kearney Research and Extension Center in Parlier (CA) from August 2nd to October 16th 2007. The peach orchard (Crimson Lady) architecture consisted on mature trees, 3.95 m tall, with a dense crown from 0.75 to the canopy top, 4 m distance between trunks in a row and 4.5 m between rows. The leaf area index was $LAI \approx 3.0$ and the ratio of the depth of foliage (m) to the canopy height was $f \approx 0.95$. The orchard was mainly surrounded by grapevines and bare soil. The lysimeter consisted of an underground chamber that houses a balance-beam weighing system (Fred Lourence, Precision Lysimeters, Red Bluff, CA) which is (2m x 4m x 2m deep) containing two trees with similar spatial separation of those in the orchard. Weight changes were logged hourly and the error in latent heat flux was 0.05 mm h^{-1} (Scott et al., 2005). The lysimeter fetch in the prevailing wind direction was 240 m. As a rule of thumb, fetch requirements are estimated according to the ratio 1:100 (i.e., the adjusted surface sublayer grows 1 m for each 100 m distance to the leading edge in the wind direction) and the basis of the inertial sublayer is located of about two-three times the canopy height (Brutsaert, 1988; Kaimal and Finnigan, 1994). The latter prevents the use of traditional micrometeorological methods to estimate turbulent surface fluxes close to the lysimeter, especially those that require measurement of gradients. Latent heat flux, Eq. (1), was evaluated half-hourly nearby the lysimeter. To prevent potential differences due to influence of local advection, instrumentation above the canopy was deployed 40 m apart in the cross mean streamwise direction having same fetch. Net radiation was measured using a net radiometer (REBS, Inc Q7.1, Bellevue, Wa) placed at 6.0 m. The soil heat flux was measured as described in Fuchs and Tanner (1967) and it was obtained as the average of three measurements to account for spatial variability. The three wind speed components and sonic temperature were measured using a sonic anemometer (81000RE, RM Young, USA). It was deployed at 5.5 m to ensure measurements within the adjusted surface layer and to avoid the region of the flow with maximum absorption of momentum. The raw sonic data was recorded at 10 Hz using a data logger (CR1000, Campbell Sci., USA). The protocol used as a reference for comparison described in Mauder et al. (2007) was applied using the TK2 package [free distributed by the University of Bayreuth (Mauder and Foken, 2004)], to determine half-hourly means, variances and covariances. Half-hourly fluxes were used to evaluate the hourly fluxes in Eq. (1). Because the latent heat flux estimates were determined as a residual of the surface energy balance (Eq. 1), the work of expansion of moist air parcels under constant pressure were included in the sensible heat flux. Therefore, the covariance of the vertical wind speed with the sonic temperature, which is close to the virtual temperature, was used because it mostly accounts for the work of expansion of the moist air parcel (Paw U et al., 2000). In practice, when air temperature is measured the work of expansion can be estimated as 0.076 times the latent heat flux (Paw U et al., 2000).

Climate features. Clear skies and no rainfall were observed during the experiment which is the typical climate features in the San Joaquin valley. The following half-hourly means and standard deviations, respectively, were observed; 0.8 m s^{-1} and 0.5 m s^{-1} for the horizontal wind speed, 22.4 C and 7.1 C for sonic air temperature, 0.21 m s^{-1} and 0.15 m s^{-1} for the friction velocity, and 24 W m^{-2} and 70 W m^{-2} for H determined using the EC system; 106 W m^{-2} and 138 W m^{-2} for LE measured using the lysimeter; and 129 W m^{-2} and 196 W m^{-2} for (Rn- G). The number of half-hourly samples collected for unstable and stable cases was 1692 and 1904, respectively. The larger dataset for stable cases was due to the formation of a capping inversion during the afternoon due to regional advection of sensible heat flux in the San Joaquin Valley. Typically, the surface boundary layer becomes near neutral and stable from about two-three hours after noon until sunrise.

4. Results and discussion

To prevent stationarity, sensible and latent heat fluxes using the EC method and SR analysis were estimated half-hourly. The hourly and daily flux estimates were determined by integrating the half-hourly values and were denoted using the corresponding subscripts EC and SR. According to Eq. (1), LE_{EC} and LE_{SR} were determined as; Rn - G - H_{EC} and Rn - G - H_{SR}, respectively. To be consistent with the lysimeter (the reference), the fluxes were expressed in mm. Table 1 shows the linear regression analysis (slope, a, intercept, b, determination coefficient, R^2) and the root mean square error, $Rmse$, to compare the hourly H_{SR} versus H_{EC} for unstable and stable cases and for all the data. It is of interest to mention that, despite in micrometeorology the EC method is considered a reference it operates best when deployed well above the canopy top (i.e., in the inertial sublayer) because large eddies are easier to sample. Thus, likely the EC method slightly underestimated the actual sensible heat flux. Regardless, realistic H estimates must be well correlated with H_{EC} because the EC system directly measures the turbulence. Table 1 shows that, regardless of the stability cases, the slopes were rather close to one, the correlations were high, and the intercepts and the $Rmse$ were small. The performance between H_{SR} and H_{EC} is shown in Fig 2A and, subsequently, LE_{SR} and LE_{EC} were similar (not shown) regardless of the integration period (hourly and daily).

Figure 2B compares the hourly LE_{EC} versus LE_{Lys}. Visually, differences observed between LE_{EC} and LE_{Lys} are in the order of H (Fig. 2A). One may presume that integration of errors due to missing energy terms in the surface energy balance, errors in measuring the available net surface energy, deployment of the EC system to close to the canopy, natural spatial variability in fluxes over a heterogeneous surface enhanced from trees within a lysimeter and differences in footprints may explain most of the scatter observed in Fig. 2B. However, Fig. 3 shows the course of daily LE_{EC}, LE_{SR} and LE_{Lys}, and it is observed that time series did

Case:	*a*	*b* (mm)	R^2	*Rmse* (mm)
Unstable	0.97	0.00	0.93	0.035
Stable	1.07	-0.01	0.89	0.02
All data	1.03	-0.01	0.96	0.02

Table 1. Hourly H_{SR} versus H_{EC} for unstable and stable cases and for all the data.

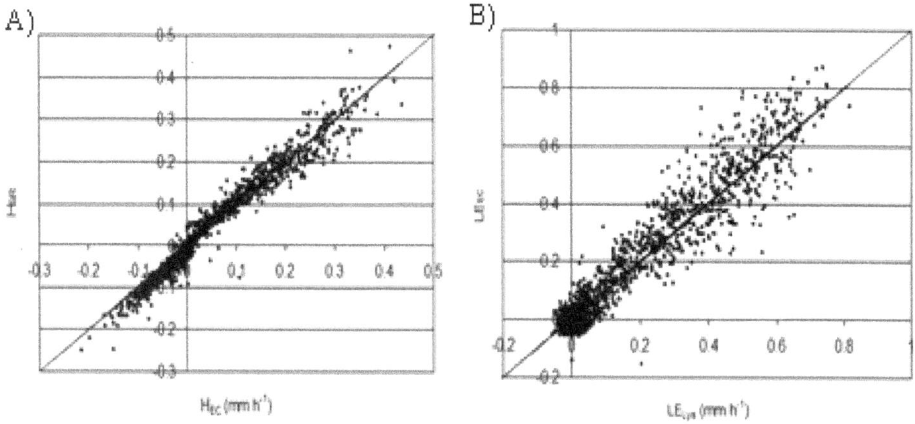

Fig. 2. Hourly (A) H_{SR} versus H_{EC}, and (B) LE_{EC} versus LE_{Lys}.

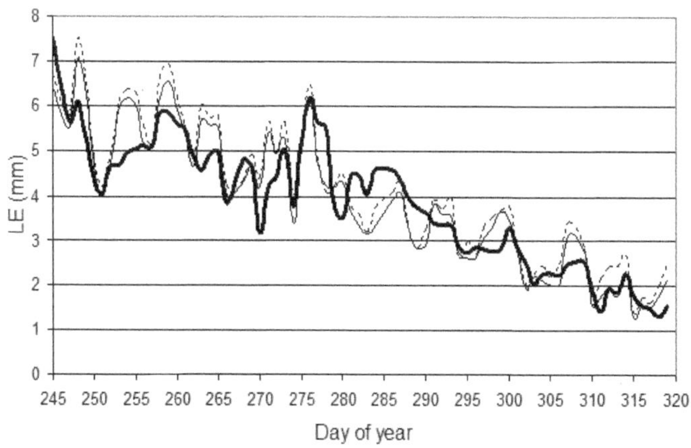

Fig. 3. Daily evapotranspiration LE_{Lys} (thick solid line), LE_{EC} (thin solid line) and LE_{SR} (dashed line).

not matched. Partially, it could be expected as a result of the scatter shown in Fig. 2B (i.e., errors did not balanced at daily scale). In Fig. 3 it is shown that during the first month, LE_{EC} and LE_{SR} were higher than LE_{Lys}. During the period from day of the year 273 to 290, LE_{EC}

and LE_{SR} fluctuated at a rate smaller than LE_{Lys}. After looking at the lysimeter management reports, the different time course shown in Fig. 3 obeyed to irrigation problems. Blockage of the drip irrigation system led to insufficient water inside the lysimeter during the first month, so the ET from the lysimeter did not increase with evaporative demand like the remainder of the orchard, which had adequate soil moisture. As a consequence, after day of year 273 the lysimeter was flood irrigated to return the soil to well-watered conditions. Thus, the deficit irrigation due to the system problem explains the lower LE_{Lys} during the first month and flooding of the lysimeter explains the higher LE_{Lys} after day 273. For the campaign, the total ET estimated by LE_{EC} was 4.8% lower than LE_{SR}, and both were close to the total LE_{Lys} (i.e., within 2%). Partly, this issue can be explained because the extra amount of water applied to restore the lysimeter was evaluated by the difference in the irrigation observed in and out of the lysimeter.

Discussion. Half hourly sensible heat flux estimates, H_{SR}, determined by the simple relationship, $(z^*-h) = 2h$, were found to overestimate H_{EC} by 55%. Other experiments carried out under windy conditions to estimate H over olive and nectarine orchards show that SR analysis performed close to the EC method taking z^* as constant (Castellví and Martinez-Cob, 2005; Castellví et al., 2006a). Therefore, appears that at sites were light winds are often observed application of SR analysis requires estimation of z^*. Though not directly comparable because measurements were taken in the inertial sublayer and fetch was large, a study carried out over short drip irrigated grass (0.10 - 0.15 m tall) under similar weather features, comparison of hourly LE_{EC} and LE_{SR} versus LE_{Lys} showed excellent agreement (Castellví and Snyder, 2010). The latter was as a consequence that, despite over irrigated surfaces $(Rn - G)$ explains much of LE, adding in Eq. (1) the H contribution the correlation of the hourly LE_{EC} and LE_{SR} versus LE_{Lys} was significantly improved.

5. Summary and concluding remarks

An experiment was carried out in a peach orchard to study the reliability of the latent heat flux estimates using the EC method and SR analysis for estimating sensible heat flux in conjunction with the surface energy balance equation. Therefore, it is assumed that the net surface energy $(Rn - G)$ is available. The fetch was limited, the site is influenced by regional advection of sensible heat flux, and light winds occur during most of the day. As a consequence, measurements were taken close to the canopy and the roughness sub-layer depth was estimated on a half-hourly basis to better estimate the α parameter required in SR analysis. A model based on a mixing-length theory for momentum combined with mixing-layer analogy allowed z^* estimates from some characteristic canopy parameters and the turbulent intensity measured near the canopy top. To test the reliability of the z^* estimates, detailed profiles of the wind speed and air temperature are needed. They were unavailable, regardless, determination of z^* appears difficult due to limited fetch. It is shown the potential of SR analysis to estimate the sensible heat flux. It was found that SR analysis was highly correlated with the EC method. Latent heat fluxes estimated as, $LE = Rn - G - H$, were compared with lysimeter data, and the comparison clearly showed the impact of irrigation management on the lysimeter ET. Further research is required covering long campaigns, however, this study suggests that SR analysis for estimating LE using the residual method can be taken in consideration as methods that may help to perform quality control of LE time series.

6. Acknowledgements

The author thanks to R.L Snyder (Univ. of Calif., Davis, CA), N.Rambo and S. Ewert (Dept. of Water Res., San Joaquin District, Fresno, CA), S. Johnson (Univ. of Calif., Kearney Agric. Res. and Ext. Center, Parlier, CA), J. Ayars (USDA ARS, Fresno, CA) and A. Russo (Univ. of Catania, IT) who carried out the field experiment, and to Asun, Carla and Tania for their help in using various facilities at Lleida (Spain). This work was supported by a grant (#4600004549) from the California Department of Water Resources (Sacramento, CA), Ministerio de Ciencia y Innovación and TRANSCLA project (CGL2009-12797-C03-01) of Spain.

7. Appendix A: Determination of the mean ramp dimensions using structure functions

It is of interest to mention that the technique based on structure functions of different order to determine the ramp dimensions is objective (i.e., there is not need to implement filters as for other methods), all the measured data is used, and assumes ramps in a sequence. Based on ramp model shown in scheme 1, Van Atta (1977) pioneered this technique that assumes a statistical independency between the coherent structure and the smallest eddies. The ramp model shown in scheme 2 is described. It accounts for a micro-front period. Therefore it is more realistic than the model shown in scheme 1 (Chen et al., 1997a). A structure function of order n is defined as

$$S^n(r) = \frac{1}{m-j}\sum_{i=1+j}^{m}\left(T_i - T_{i-j}\right)^n$$

where m is the number of data points in the 30-minute interval measured at frequency (f) in Hz, n is the power of the function, j is a sample lag between data points corresponding to a time lag ($r = j/f$), and T_i is the ith temperature sample.

Ramp dimensions A, τ and t_f can be determined by fitting the 2nd, 3th and 5th order structure functions to the following equations for different time lags

If, $0 \le r \le t_f$;

$$\frac{S^n_{(r)}}{r} = (-1)^n \frac{A^n}{\tau}\left[\left(\frac{r}{t_f}\right)^{n-1}\left(1+\frac{t_f - r}{\tau - t_f}\right) - \frac{(n-1)}{(n+1)}\left(\frac{r}{t_f}\right)^2\left(1+\frac{t_f - r}{\tau - t_f}\right)\right]$$

If, $t_f < r \le (\tau - t_f)$

$$\frac{S^n_{(r)}}{r} = (-1)^n \frac{A^n}{\tau}\left(\frac{(\tau - r)^n + (-1)^n(\tau - r)r^{n-1}}{\left(\tau - t_f\right)^n} - \frac{(n-1)t_f\tau^n f_n}{(n+1)r\left(\tau - t_f\right)^n}\right)$$

If, $(\tau - t_f) < r \le \tau$;

$$\frac{S^n_{(r)}}{r} = \frac{A^n}{\tau}\left(\frac{(\tau - r)^n}{r t_f^{n-1}} - \frac{(n-1)\left(\tau - t_f\right)^{n+1}}{(n+1) r t_f^n}\right)$$

$$\text{with } f_n = \begin{cases} 1 & n=2 \\ 1-2\left(\frac{r}{\tau}\right) & n=3 \\ 1-\frac{9}{2}\left(\frac{r}{\tau}\right)+\frac{15}{2}\left(\frac{r}{\tau}\right)^2-5\left(\frac{r}{\tau}\right)^3 & n=5 \end{cases}$$

If t_f is neglected in the above expressions, the resulting expression for $0 < r \leq \tau$ can be used to determine the ramp dimensions A and τ according to scheme 1 (Van Atta, 1977). The mean ramp amplitude is determined by solving the following expression for the real roots:

$$A^3 + pA + q = 0$$

Where $p = 10S^2(r) - \frac{S^5(r)}{S^3(r)}$ and $q = 10S^3(r)$. For $r \ll Lr$, the following expression for the inverse ramp frequency and ramp amplitude holds, Eq. (A1):

$$A^3 = -\frac{S^3(r)}{r}\tau$$

Therefore, several r values can be used to linearize A.1. According to Chen et al. (1997a), the shortest time lag to be used for linearization, r_{1G}, is that produces the first global maximum of $S^3(r)/r$ because A.1 does not hold for $r < r_{1G}$. Thus, the initial time lag, r_{ini}, used to linearize A.1, was $r_{ini}=r_{1G}$. To estimate the maximum time lag, r_{end} to be used for linearization so that $r \ll Lr$, the second global maximum of $S^3(r)/r$ was determined. Based on the third order structure function for , $t_f < r \leq (\tau - t_f)$, the second global maximum occurs at a time lag, r_{2G}, giving $r_{2G} \approx 3/4\ \tau$. According to Qiu et al. (1995), $Lq \approx 0.25\ \tau$ in scheme 1, and therefore, $r_{2G} \approx Lr$. The last r used to linearize A.1 was determined as 2% of r_{2G} to insure $r_{end} \ll Lr$.

8. Appendix B: Estimating the roughness layer depth

Turbulence in the roughness layer is characterized by the presence of distinct coherent structures generated near the canopy top (Finnigan and Shaw 2000; Shaw et al. 2006). Therefore, similarity does not apply within the roughness layer (Kaimal and Finnigan, 1994). As a result many of the characteristic properties of the roughness sub-layer differ from those of the inertial sub-layer and resemble more those of a plane mixing layer [i.e., the mixing-layer analogy reported by Raupach et al. (1996)]. In the following, to simplify, it is assumed that the canopy is homogeneous and dense at an extent that is capable to absorb all the momentum transferred into the canopy by coherent structures. Therefore, because the location of the ground is irrelevant to the dynamics involved, the natural choice of z-axis origin is the location where the physical processes change. That is, the z-axis origin is set at the canopy top which is at height h. Accordingly, $(z^* - h)$ defines the sub-layer that extends from the canopy top to the bottom of the inertial sub-layer. Within the canopy, the horizontal mean wind speed, u, and shear reach a maximum at the canopy top. They are attenuated within the canopy at a rate determined by the drag exerted by the canopy elements. Attenuation is mainly important in the upper part of the canopy and the vertical profile of the mean wind speed follows an exponential decay which can be expressed as (Kaimal and Finnigan, 1994)

$$u_{(z)} = u_h \, e^{\left(\nu z / h\right)} \tag{B.1}$$

Where u_h is the mean wind speed at the canopy top and ν is the extinction coefficient. By invoking a mixing-length model, Eq. (B.1) may be explained as follows. Using the mixing length for momentum, l_m, the mean shear can be expressed as

$$\frac{\partial u_{(z)}}{\partial z} = \frac{u_*}{l_m} = \beta \frac{u_h}{l_m} \tag{B.2}$$

Based on (B.1) and (B.2), coefficients β ($= \frac{u_*}{u_h}$) and ν are related by, $\nu = \beta \frac{h}{l_m}$.

The wind speed at the canopy top is a relevant variable in the mixing-layer analogy because the gradient of the shear stress is constant above the canopy, whereas it is drastically extinguished within the canopy. Therefore, for convenience, the right term in (B.2) includes the wind speed at the canopy top which allows using observed empirical relationships (i.e., scales). According to the time averaged equation for momentum, τ, within the canopy (Raupach and Tom, 1981; Kaimal and Finnigan, 1994) which on the basis of (B.2) can be expressed as

$$\frac{du_{(z)}}{dt} = 0 = -\frac{1}{\rho}\frac{\partial \tau}{\partial z} - \frac{D_{(z)}}{\rho} \equiv \frac{\partial \left(lm \frac{\partial u}{\partial z} \right)^2}{\partial z} - c_d \, a_{(z)} \, u^2 \tag{B.3}$$

where ρ is the air density, $D(z)$ is the drag on the canopy, c_d is the drag coefficient at the leaf scale and a is the leaf area per unit volume. Assuming l_m is a constant, the following expression is derived from (B.1), (B.2) and (B.3):

$$lm = 2\beta^3 \left(c_d a_{(z)} \right)^{-1} \tag{B.4}$$

Therefore, l_m scales with $(c_d \, a_{(z)})^{-1}$ which is a vertical length related with the capability of the coherent structure to penetrate into the canopy. According to mixed-layer analogy, the following scale holds (Raupach et al., 1996)

$$(z^* - h_c) = b \frac{u_h}{\left(\frac{\partial u_{(z)}}{\partial z} \right)_{z=0}} \tag{B.5}$$

There is evidence that b varies within the range 2 to 3 (Graefe, 2004). Combining (B.2), (B.4) and (B.5) the roughness sub-layer may be estimated as

$$z^* = h + \frac{2b\beta^2}{c_d \, a_{(z \approx 0)}} \tag{B.6}$$

Over contrasting surfaces, the following expression to estimate a at the upper part of the canopy holds (Graefe, 2004)

$$a_{(z\approx 0)} = \frac{(h_c - d)}{h_c^2}\frac{LAI}{f^2} \tag{B.7}$$

where LAI is the leaf area index and f the crown thickness relative to the canopy height. When measurements are taken close to the top of the canopy top, the relationship $u^* \approx 0.5\sigma_u$ holds (Kaimal and Finnigan, 1994), where σ_u is the mean u turbulent standard deviation. Thus, we arrive to the approximation: $2b\beta^2 \approx \left(\frac{\sigma_u}{u_h}\right)^2$. Therefore, for a given canopy and atmospheric surface layer stability condition, (B.6) can be estimated as a linear relationship of the canopy height

$$(z^* - h) = ch \quad \text{with} \quad c = \frac{h}{(h-d)}\frac{(fI_u)^2}{(c_d LAI)} \tag{B.8}$$

where I_u $(=\frac{\sigma_u}{u_h})$ is the turbulent intensity. Variation of the roughness sub-layer depth due to stability conditions in (B.8) are mainly accounted through I_u, and likely through c_d and d.

9. Appendix C: Determination of the sensible heat flux

Estimation of z^* (B.8) presumes that measurements are taken at z=h. However, in practice instrumentation is deployed slightly higher to avoid potential shortcomings and damage. In general, depending on the stability conditions, the measurement height may fall within the roughness or in the inertial sublayer. Therefore, the correct expression to estimate α must be selected, and the friction velocity may be estimated as follows (Kaimal and Finnigan, 1994)

$$u_* = \begin{cases} \dfrac{ku_r}{\left(\ln\dfrac{(z_r - d)}{z_0} - \Psi_m(\varsigma)\right)} & z > z^* \\ 0.5\sigma_u & z < z^* \end{cases} \tag{C.1}$$

where, u_r is the wind speed at reference height z_r, and $\Psi_m(\zeta)$ is the integrated Businger-Dyer relationship for momentum (Paulson, 1970)

$$\Psi_m(\varsigma) = \begin{cases} 2\ln(0.5(1+x)) + \ln(0.5(1+x^2)) - 2\arctan(x) + 0.5\pi & \varsigma \le 0 \\ -4.7\varsigma & \varsigma > 0 \end{cases} \tag{C.2}$$

where, x is, $x=(1-16\zeta)^{1/4}$. It is presumed that the wind profile is unavailable. Therefore, for measurements taken in the inertial sub-layer over dense canopies the zero-plane displacement and the aerodynamic surface roughness length, z_o, can be estimated as a portion of the canopy height, h, d=0.67 h and z_o=0.12 h (Brutsaert, 1988; Wieringa, 1993). The stability parameter and the ramp amplitude for temperature have different sign (Fig.1).

Therefore, after determining the ramp dimensions (Appendix A), the appropriate expressions for $\phi_h(\zeta)$ and $\Psi_m(\zeta)$ are known. Estimation of the mean drag coefficient at the leaf scale is a compromise. It depends on the shape and orientation of the leaves and, in principle, it may depend on the velocity field within the canopy through the Reynolds number though such relationship is not still clear (Brutsaert, 1988). Numerical adjustment to produce the best agreement between models of transfer of momentum within the canopy and observations is used to determine c_d values (Ionue, 1981; Pingtong and Takahashi, 2000; Mohan and Tiwari, 2004). Because c_d values are mostly indirectly determined under neutral conditions, if possible, it is best to carry out a short campaign for adjustment. Regardless of the stability conditions, the c_d was set to, c_d=0.2, which is a typical value (Graefe, 2004). It is of interest to mention that because the zero-plane displacement, roughness length and the friction velocity were estimated, the c_d value obtained accounts for such uncertainties. A recent study carried out over orange trees covering a year of measurements shows that a short dataset of about two weeks is enough to adjust B.8 by trial and error to fit H_{SR} and H_{EC}. Therefore, the SR model appears robust. Next, the sensible heat flux, ζ, and u_* are solved simultaneously by iteration. To start the iteration procedure, neutral conditions are assumed for the actual atmospheric surface layer. This gives an approximation of the actual friction velocity (case z > z*), parameter α and H. The latter values allow for the first approximation of ζ, and the process is iterated until convergence is achieved.

10. References

Allen, R.G., Pruitt, W.O., Businger, J.A., Fritschen, L.J., Jensen, M.E., Quinn, F.H., 1996. Evaporation and Transpiration. Hydrology Handbook (2nd edition). R.J. Heggen (overall editing), T.P. Wootton, C.B. Cecilio, L.C. Fowler, S.L. Hui (Task Committee Members). 125-252. ASCE Manual and Reports on Engineering Practice No. 28. American Society of Civil Engineers. New York, NY, USA.

Allen, R.G., Pereira, L.S., Raes, D., and Smith, M., 1998. Crop evapotranspiration: Guidelines for computing crop water requirements. Irrigation and Drainage Paper no. 56, FAO of United Nations, Rome. 300 p.

Allen, R.G., Walter, I.A., Elliott, R.L., Howell, T.A., Itenfisu, D., Jensen, M.E. and Snyder, R.L. 2005. The ASCE Standardized Reference Evapotranspiration Equation. Amer. Soc. of Civil Eng. Reston, Virginia. 192p.

Brutsaert, W., 1988. Evaporation into the atmosphere, D.Reidel P.C., Dordrecht, Holland.

Castellví, F., 2004. Combining surface renewal analysis and similarity theory: A new approach for estimating sensible heat flux. Water Resour. Res. 40, W05201. DOI:10.1029/2003WR002677.

Castellví, F., Martinez-Cob., A., 2005. Estimating sensible heat flux using surface renewal analysis and the variance method. A study case over olive trees at Sastago (NE, Spain). Water Resour. Res. 41, W09422, doi:10.1029/2005WR004035.

Castellví, F., Snyder, R.L., Baldocchi, D.D., Martinez-Cob, A., 2006a. A comparison of new and existing equations for estimating sensible heat flux using surface renewal and similarity concepts. Water Resour. Res. 42, W08406, doi:10.1029/2005WR004642.

Castellví, F., Snyder, R.L., Baldocchi, D.D., 2008. Surface energy-balance closure over rangeland grass using the eddy covariance method and surface renewal analysis. Agric. For. Meteorol. 148, 1147-1160.

Castellví, F., Snyder, R.L., 2009. Sensible heat flux estimates using surface renewal analysis: A study case over a peach oprchard. Agric. For. Meteorol. 149, 1397-1402.

Castellví, F., Snyder, R. L., 2010. A comparison between latent heat fluxes over grass using a weighing lysimeter and surface renewal analysis. Journal of Hydrology 381, 213-220.

Chen, W., Novak, M.D., Black, T.A., Lee, X., 1997a. Coherent eddies and temperature structure functions for three contrasting surfaces. Part I: Ramp model with finite micro-front time. Bound. Layer Meteorol. 84: 99-123.

Doorenbos J., Pruitt, W.O., 1977. Crop water requirements. FAO Irrigation and Drainage Paper No 24, FAO, Rome.

Drexler, J.Z., Snyder, R.L., Spano, D., Paw U, K.T., 2004. A review of models and micrometeorological methods used to estimate wetland evapotranspiration. Hydrological Processes 18: 2071-2101.

Finnigan, J.J, Shaw, R.H., 2000. A wind tunnel study of air flow in waving-wheat: An EOF analysis of the structure of the large-eddy motion. Bound. Layer Meteorol. 96, 211-255.

Foken, T., 2006. 50 years of the Monin-Obukhov similarity theory. Bound. Layer Meteorol. 119, 431-447.

Fuchs, M., Tanner, C.B., 1967. Evaporation from a drying soil. J. Appl. Meteorol. 6, 852–857.

Graefe, J., 2004. Roughness layer corrections with emphasis on SVAT model applications. Agric. For. Meteorol. 124, 237–251.

Harman, I.N., Finnigan, J.J., 2007. A simple unified theory for flow in the canopy and roughness sublayer. Bound. Layer Meteorol. 123, 339–363.

Higbie, R.,1935. The rate of absorption of a pure gas into a still liquid during short periods of exposure, Trans. Am. Inst. Che. Engr. 31, 355-388.

Högström, U., 1988. Non-dimensional wind and temperature profiles in the atmospheric surface layer: A re-evaluation. Bound. Layer Meteorol. 42, 55-78.

Hongyan, C., Jiayi, C., Hu, F., Qingcun, Z., 2004. The coherent structure of water vapour transfer in the unstable atmospheric surface layer. Boundary-Layer Meteorology 111, 534-552.

Inoue, K., 1981. A model study of microstructure of wind turbulence of canopy flow. Bull. Nat. Inst. Agric. Sci. A 27, 69-89.

Kader BA, Perepelkin VG. 1989. Effect of the unstable stratification on wind and temperature profiles in the surface layer. Izv, Akad. Nauk SSSR., Fiz.Atmos. Oheana 25: 787-795.

Kaimal, J.C, Finnigan, J.J., 1994. Atmospheric Boundary Layer Flows. Oxford Univ Press.

Katul, G.G., Hsieh, C., Oren, R., Ellsworth, D., Philips, N., 1996. Latent and sensible heat flux predictions from a uniform pine forest using surface renewal and flux variance methods. Bound. Layer Meteorol. 80, 249-282.

Mauder, M., Foken, T., 2004. Documentation and instruction manual of the eddy covariance software package TK2. Universität Bayreuth, Abt.Mikrometeorologie, Arbeitsergebnisse 26-44. http://www.geo.unibayreuth.de/mikrometeorologie

Mauder, M., Oncley, S.P., Vogt, R., Weidinger, T., Ribeiro, L., Bernhofer, C., Foken, T., Kosiek, W., De Bruin H.A. R., Liu, H., 2007. The energy balance experiment EBEX-2000. Part II: Intercomparison of eddy-covariance sensors and post-field data processing methods. Boundary-Layer Meteorology 123, DOI: 10.1007/s10546-006-9139-4.

Meyers, T.P., Baldocchi, D.D., 2005. Current micrometeorological flux methodologies with applications in agriculture Micrometeorology in Agricultural Systems, Agronomy Monograph 47, Chapter 16, ASA-CSSA-SSSA Publishers, Madison (Wi, USA).

Mohan, M., Tiwari, M.K., 2004. Study of momentum transfers within a vegetation canopy. Proc. Indian Acad. Sci. (Earth Planet. Sci.) 113 (1), 67-72.

Oncley, S.P., Foken, T., Vogt, R., Kohsiek, W., DeBruin, H.A.R., Bernhofer, C., Christen, A., Van Gorsen, E., Grantz, D., Feigenwinter, C., Lehner, I., Liebethal, C., Liu, H., Mauder, M., Pitacco, A., Ribeiro, L., Weidinger, T., 2007. The energy balance experiment EBEX-2000. Part I: overview and energy balance. Bound. Layer Meteorol. 123, Doi: 10.1007/s10546-007-9161-1.

Paulson, C.A., 1970. The mathematical representation of wind speed and temperature profiles in the unstable atmospheric surface layer. Journal of Climate and Applied Meteorol. 9, 857-861.

Paw U, K. T., Qiu, J., Su, H.B., Watanabe, T., Brunet, Y., 1995. Surface renewal analysis: a new method to obtain scalar fluxes without velocity data. Agric. For. Meteorol. 74, 119-137.

Paw U, K. T., Baldocchi, D.D., Meyers, T.P., Wilson, K.B., 2000. Correction of eddy-covariance measurements incorporating both advective effects and density fluxes. Bound. Layer Meteorol. 97, 487-511.

Paw U, K.T., Snyder, R.L., Spano, D., Su, H.B., 2005. Surface renewal estimates of scalar exchanges, Micrometeorology in Agricultural Systems, Agronomy Monograph 47, Chapter 20, ASA-CSSA-SSSA Publishers, Madison (Wi, USA).

Pingtong, Z., Takahashi, H., 2000. A first-order closure model for the wind flow within and above vegetation canopies. Agricultural and Forest Meteorology 103, 301-313.

Qiu, J., Paw U, K.T., Shaw, R.H., 1995. Pseudo-Wavelet analysis of turbulence patterns in three vegetations layers. Bound. Layer Meteorol. 72, 177–204.

Raupach, M.R., Thom, A.S., 1981. Turbulence in and above plant canopies. Annu. Rev. Fluid Mech. 13, 97–129.

Raupach, M.R., Finnigan, J.J., Brunet, Y., 1996. Coherent Eddies in Vegetation canopies: The Mixing-Layer Analogy. Bound. Layer Meteorol. 78, 351-382.

Shaw, R.H., Finnigan, J.J., Patton, E.G., Fitzmaurice, L., 2006. Eddy structure near the plant interface. Proceedings of the 17th symposium on boundary layers and turbulence. AMS, San Diego, CA, J2.1.

Scott, R., Williams, L.E., Ayars, J.E., Trout,T.J., 2005. Weighing lysimeters aid study of water relations in tree and vine crops. California Agriculture, 59 (2), 133-136.

Snyder, R.L., Spano, D., Paw U, K.T., 1996. Surface renewal analysis for sensible and latent heat flux density. Boundary-Layer Meteorology 77, 249-266.

Teixeira, A.H.C., Bastiaanssen, W.G.M., Moura, M.S.BSoares, J.M., Ahmad, M.D., Bos, M.G., 2008. Energy and water balance measurements for water productivity analysis in irrigated mango trees, Northeast Brazil, Agric. For. Meteorol. 148, 1527-1537.

Twine, T.E., Kustas, W.P., Norman, J.M., Cook, D.R., Houser, P.R., Meyers, T. P., Prueger, J.H., Starks, P.J., Wesely, M.L., 2000. Correcting eddy-covariance flux underestimates over a grassland. Agric. For. Meteorol. 103 (3), 279-300.

Van Atta, C.W., 1977. Effect of coherent structures on structure functions of temperature in the atmospheric boundary layer, Arch. of Mech. 29, 161-171.

Wieringa, J., 1993. Representative roughness parameters for homogeneous terrain, Bound. Layer Meteorol. 63, 323-363.

Wilson, K., Goldstein, A., Falge, E., Aubinet, M., Baldocchi, D.D., Berbigier, P., Bernhofer, C., Ceulemans, R., Dolman, H., Field, C., Grelle, A., Ibrom, A., Law, B.E., Kowalski, A.,Meyers, T., Moncrieff, J., Monson, R., Oechel, W., Tenhumen, J., Valentini, R., Verma, S., 2002. Energy balance closure at FLUXNET sites, Agric. For. Meteorol. 113, 223-143.

3

Evapotranspiration Estimation Using Micrometeorological Techniques

Simona Consoli
University of Catania, Department of Agri-food and Environmental Systems Management, Hydraulic Section Catania
Italy

1. Introduction

Accurate evapotranspiration, ET, data are crucial for irrigation management projects, especially in drought prone regions. Evapotranspiration rates can be estimated by micrometeorological methods and the energy balance equation, soil depletion techniques, mass exchange methods, or by using weighting lysimeters. These methods usually are expensive, difficult to operate, and some of them present problems for measurements in heterogeneous vegetation. Therefore, the search for accurate methods for estimating ET fluxes using low-cost, transportable and robust instrumentations is a subject of interest.

The eddy covariance (EC) method is the commonly used micrometeorological technique providing direct measurements of latent heat flux (or evapotranspiration). It adopts a sonic anemometer to measure high-frequency vertical wind speed fluctuations about the mean and an infrared gas analyzer to measure high frequency water concentration fluctuations. These fluctuations are paired to determine the mean covariance of the wind speed and humidity fluctuations about the mean to directly estimate latent heat flux (LE). In the EC method, the sensible heat flux is also estimated using the covariance of the fluctuation in vertical wind speed and variations in temperature about their means. While the preferred method for measuring turbulent fluxes is the eddy covariance (EC) method, the lack of closure is unresolved and a full guidance on experimental set up and raw data processing is still unavailable.

Other energy balance approaches, such as the Bowen ratio and aerodynamic methods, have a sound theoretical basis and can be highly accurate for some surfaces under acceptable conditions.

Biometeorological measurements and theory identified large, organized eddies embedded in turbulent flow, called "coherent structures" as the entities which exchange water vapour, heat, and other scalars between the atmosphere and plant communities.

Based on these studies, a new method for estimating scalar fluxes called "Surface Renewal (SR)" was proposed by Paw U and Brunet (1991). Surface Renewal (SR) theory in conjunction with the analysis of the observed ramp-like patterns in the scalar traces provides an advantageous method for estimating the surface flux density of a scalar. The method was tested with air temperature data recorded over various crop canopies. Results of the studies (Snyder et al., 1996; Spano et al., 1997; Consoli et al., 2006; Castellvì et al., 2008) have demonstrated good SR performance in terms of flux densities estimation, well correlated with EC measurements. The approach has the advantages to (i) require as input only the measurement of scalar trace; (ii) involve lower costs for the experiment set-up, with respect to the EC method; (iii) operate in either the roughness or inertial sub-layers; (iv) avoid levelling, shadowing and high fetch requirements. Snyder et al. (1996) and Spano et al. (1997), however, have indicated the SR method currently requires an appropriate calibration factor, depending on the surface being measured.

The goal of this chapter is to evaluate the suitability of simple, low-cost methods (i.e. Surface Renewal, aerodynamic methods) to determine sensible heat flux (H) for use with net radiation and soil heat flux density to estimate latent heat flux density or evapotranspiration (LE, or ET). To evaluate the potential for using Surface Renewal analysis to determine H and LE over heterogeneous canopies, high frequency temperature measurements were taken over citrus orchards under semi-arid climatic conditions in Sicily (Italy).

2. Theoretical background

2.1 Energy balance

Biometeorology traditionally depends from the first law of thermodynamics (conservation of energy). In physics, the conservation of energy law says that the "total amount of energy in any isolated system remains constant and can't be recreated, although it may change forms"; for example friction turns kinetic energy into thermal energy.

For a thermodynamic system the first law of thermodynamics may be stated as:

$$\delta Q = dU + \delta W \tag{1}$$

Where δQ is the amount of energy added to the system, δW is the amount of energy lost by the system due to the work done by the system on its surroundings and dU is the increase in the internal energy of the system.

Energy flux can take on several forms. The most obvious energy flux is radiated from all the bodies and absorbed from the sun. Energy may also flow from an object into the ground, and it is called ground heat flux (G). Any object in contact with the air will usually transfer some energy to the air or vice versa, this is called sensible heat flux (H). Evaporation or condensation may occur on a surface, both represent a mass flux to or from the surface. The sign conventions of energy budgets are not completely standardized. Usually the radiation entering a surface is considered positive, and radiation leaving a surface is considered negative. All other energy fluxes have the reverse sign convention. For example sensible heat moving from the air into a surface is considered negative, and sensible heat flux moving from the surface to the air is defined as positive.

$$Rn = LE + H + G + Misc. \tag{2}$$

Where:

- *Rn is the net radiation [W/m^2];*
- *LE is the latent heat flux density [W/m^2];*
- *H is the sensible heat flux density [W/m^2];*
- *G is the soil heat flux density [W/m^2];*
- *Misc. is a miscellaneous of flux densities due to the biochemical processes; it's a very small part and generally it can be neglected.*

It's customary to use energy flux densities [W/m^2] rather than energy flow [W], so that the analysis of a surface is not specific to the particular surface area being considered.

2.1.1 Sensible heat flux

Heat transfer from a surface to the surrounding atmosphere must take place by either molecular transfer, turbulent transfer, or both processes.

Considering the case of temperature distribution in the air adjacent to a certain surface (Fig. 1). A molecule randomly travels from height (z) to $(z+\ell)$, where (ℓ) is the typical distance that a molecule must travel before it collides with another molecule, transferring its thermal kinetic energy. The molecule is assumed to move with the speed of sound (approximately 300 m s^{-1} near earth's surface). When this molecule from height (z) travels to $(z+\ell)$, it collides and transfers its heat such that a molecule at height $(z+\ell)$ now has the original temperature of the first molecule (temperature expresses the kinetic energy of the molecules).

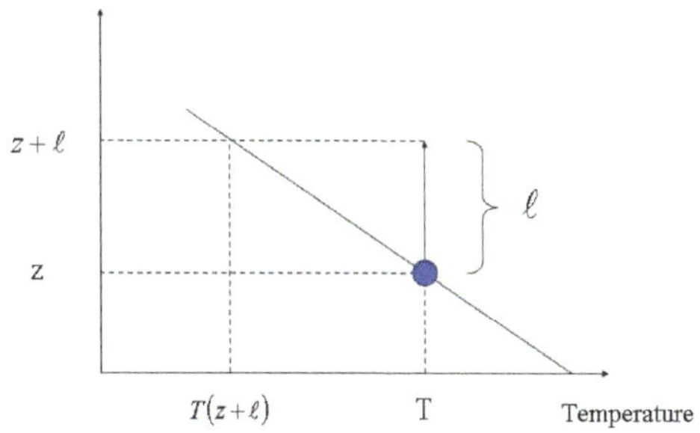

Fig. 1. Mean free path of the air molecule.

The heat transferred per molecule is:

$$h_t = mC_p\left(T_{(z)} - T_{(z+\ell)}\right) \tag{3}$$

Where m is the mass of the molecule and C_p is the specific heat per unit mass.
By using the first-order Taylor approximation:

$$T_{(z+\ell)} = T_{(z)} + \left(\frac{\partial T}{\partial z}\right) + \ell \tag{4}$$

The heat transferred per molecule becomes:

$$mC_p\left(T_{(z)} - \left(T_{(z)} + \frac{\partial T}{\partial z}\ell\right)\right) = -mC_p\frac{\partial T}{\partial z}\ell \tag{5}$$

To find the heat transferred per unit surface area, it's necessary to know the number of molecules per unit area moving away from height (z) to $(z+\ell)$. One may define the number of molecules per unit volume, N. At any time, approximately one-third of the molecules will have components in the z direction; the other two-thirds will have components in the x and y directions. The number density considered active in this process is then N/3.

The flux density F, defined as the number of molecules crossing a unit area perpendicular to z axis in a unit time, is calculated by multiplying N/3 the average speed of the molecules, c:

$$F = \frac{Nc}{3} \tag{6}$$

The heat flux density is the molecular flux density F times the heat carried by each molecule: $-mC_p \frac{\partial T}{\partial z} \ell$:

$$H = -(F)\left(mC_p \frac{\partial T}{\partial z} \ell\right) = -\left(\frac{Nc}{3}\right)\left(mC_p \frac{\partial T}{\partial z} \ell\right) \tag{7}$$

N·m is the mass of a molecule times the number of molecules per unit volume = mass/volume = density (ρ), so eq. (7) becomes:

$$H = \frac{-c\ell}{3} \rho C_p \frac{\partial T}{\partial z} \tag{8}$$

We can define the thermal diffusivity k as $c\ell/3$ [m^2 s^{-1}], thus resulting the equation for sensible heat:

$$H = -k\rho C_p \frac{\partial T}{\partial z} \tag{9}$$

The previous equation for H computation requires the measurement of the gradient of temperature (or concentration) at very small distances from the surface (Stull, 1988).

2.1.2 Latent heat flux

Latent energy caused by evaporation is the major form of cooling for many organisms, including most plants.

The term latent energy flux density describes the energy used in the transfer of water vapour molecules from one phase to another. The energy is used to create a phase change resulting in a mass transfer.

The mass flux of water vapour is needed to determine the latent energy flux. The mass flux density [Kg m^{-2} s] times the energy gained (or lost) per kg of water evaporated, gives the energy flux [J m^{-2} s]. The latent heat of vaporization for water is 2.5 x 10^6 [J kg^{-1}] at 0 °C and decreases to 2.406 x 10^6 [J kg^{-1}] at 40 °C.

It's possible to model the mass flux density of water vapour [kg m^{-2} s^{-1}], "E", in the same manner as for the flux of sensible heat. The gradient theory, derived and based on molecular turbulent transfer is (Stull, 1988):

$$E = -K_w \frac{\partial \rho_v}{\partial z} \tag{10}$$

K_w is the turbulent vapour transport coefficient [m^2 s^{-1}], dependent on stability and turbulent conditions; ρ_v is the absolute humidity [kg m^{-3}]. The absolute humidity is defined as the mass of water vapour per unit volume air.

$$E = \frac{\left[\rho_v(0) - \rho_v(z)\right]}{r_w} \tag{11}$$

where $\rho_v(0)$ is the absolute humidity at the surface and $\rho_v(z)$ is the absolute humidity of the air at height z, r_w is the resistance to water vapour flux [s m^{-1}].
The vapour pressure of water is more used than the specific humidity. The conversion from absolute humidity [kg m^{-3}] to vapour pressure "e" [Pa]:

$$\rho_v = 0.622\frac{e}{P}\rho \tag{12}$$

To obtain the latent energy flux density [J m^{-2} s^{-1}], it is necessary to multiply the evaporative flux density by the latent heat of vaporization, L; e(0) is the vapour pressure of water at the surface, e(z) is the vapour pressure at height z, P is the atmospheric pressure and ρ is the air density [kg m^{-3}] (about 1.2 kg m^{-3}).

$$(L)(E) = \frac{L\rho 0.622}{P} \cdot \frac{\left[e(0) - e(z)\right]}{r_w} \tag{13}$$

If:

$$\gamma \equiv \frac{PC_p}{0.622L} \tag{14}$$

Evaporation from organisms is termed "transpiration", and must be combined with the evaporation term to arrive at the total latent energy transfer from a surface. The combined term for plants is called "Evapotranspiration". For the case of leaves, transpiration takes place in the "sub-stomatal cavity", where vapour pressure "$e_s(T_l)$" is assumed to be saturated at leaf temperature T_l (Stull, 1988).
The resistance term includes two terms, a term for the resistance through the stomata and cuticle "r_s", and a term for the generalized transfer of water vapour through the atmosphere "r_w":

$$LE = \frac{\rho C_p}{\gamma}\frac{\left[e_s(T_l) - e(z)\right]}{(r_w + r_s)} \tag{15}$$

By expressing the saturated vapour pressure in terms of a first-order Taylor approximation "around" the air temperature:

$$e_s(T_l) \approx e_s(T(z)) + \Delta(T_l - T(z)) \tag{16}$$

Where:

$$\Delta \equiv \frac{\partial e}{\partial T} \tag{17}$$

By using the energy budget equation

$$(Rn - G) = H + LE \tag{18}$$

With sensible heat H estimated from the bulk transfer equation:

$$H = \rho C_p \frac{\left(T_l - T(z)\right)}{r_h} \tag{19}$$

r_h is the resistance to heat transfer, also called "aerodynamic resistance". Then:

$$(Rn - G) - LE = H = \rho C_p \frac{\left(T_l - T(z)\right)}{r_h} \tag{20}$$

$$T_l - T(z) = \frac{r_h}{\rho C_p}(Rn - G - LE) \tag{21}$$

Finally it's possible to collect the LE terms to obtain the "Penman-Monteith" equation (Allen et al., 1998):

$$LE = \frac{(Rn - G) + \frac{\rho C_p}{\Delta r_h}\left(e_s\left(T(z)\right) - e(z)\right)}{1 + \frac{\gamma}{\Delta}\frac{r_w + r_s}{r_h}} \tag{22}$$

The Penman-Monteith equation for evapotranspiration is formed by assuming that the stomatal resistance "r_s" is equal to zero. This is true for surfaces that do not have stomatal-type structures that are fully open due to extremely wet conditions. An estimation of evapotranspiration when r_s=0 is called "potential evapotranspiration".

3. Micrometeorological techniques

3.1 Introduction

The surface-atmosphere exchange of scalars (heat, water vapour, carbon dioxide etc.), and vectors (momentum) has been measured and estimated using a variety of techniques. These include eddy-covariance (Swinbank 1955), the eddy-accumulation method (Desjardins, 1972), and gradient/micrometeorological methods (Pruitt, 1963). All of these methods require data logging capable of recording at a range of acceptable frequencies.

In the Eddy Covariance method, high frequency logging (10 Hz or higher) is needed for the 3-D velocity field and the scalar of interest. This, generally, requires sonic anemometer, which is relatively complicated and expensive. The scalar must also be sampled with costly, complex sensors at similar frequencies.

For temperature, however, either the sonic-based temperature (Paw U et al., 2000) or simple, inexpensive temperature sensors such as thermocouples are used. In the eddy accumulation methods, and the relaxed eddy-accumulation, sampling of the scalar is conditionally based on the vertical velocity signal. This reduces the necessity for high-frequency scalar measurements.

In micrometeorological methods, which depend on the measurements of the gradients, such as Bowen Ratio Energy Budget (BREB), advective, or flux-gradient methods, calibrated and carefully matched sensors are needed. In addition, other requirements and limitations exist for each of these methods.

To obtain the sensible heat flux density, the BREB requires measurements of the ground heat flux, biomass storage, and net radiation, and two pairs of high resolution matched temperature and humidity measurements. Under some conditions, small errors in the matched pair measurements can result in large errors of sensible and latent heat flux density. Also, large errors in latent heat flux density can occur near sunrise and sunset when the Bowen Ratio $\beta = H/\lambda E \approx -1.0$, and the surface is assumed to be horizontally homogeneous, resulting in only vertical transport.

For horizontal and vertical advective-gradient methods several sets of matched sensors pairs are needed. Vertical flux-gradient methods require at least one pair of matched scalar sensors and several wind speed measurements to obtain estimates of the vertical turbulent transport coefficients. Again, horizontal homogeneity is assumed in BREB, and errors in the matched sensors or departures from similarity, may result in potential flux estimation errors.

Tillman (1972) first reported the use of high frequency scalar data (temperature variance data) to determine good estimates of sensible heat flux density (H) under unstable conditions. Weaver (1990) used temperature variance data, similarity theory, and calibration coefficients, that vary depending on the surface and energy balance, to make reasonable estimates of H over semi-arid grass lands and brush under unstable conditions. In 1991 Paw U et al., proposed a new high frequency temperature method (surface renewal) that provides estimates of H regardless of the stability conditions and without the need for temperature profile and wind-speed data. The concept of Surface Renewal was originally developed in the chemical engineering literature, and it is considered an abstraction and simplification of the "transilient" paradigm elucidated by Stull (1984; 1988) for the atmosphere.

3.2 Surface renewal theory

3.2.1 Coherent structures

Coherent structures are characterized by repeated temporal and spatial patterns of the velocity and scalar field. Near a surface, where the fluid is assumed to reach zero velocity, a shear zone must be created if the fluid is moving with respect to the surface. It's theorized that the coherent structures are created by the shear. The atmospheric coherent structures are analogous to those found in theoretical and laboratory flows called "plane mixing layers", defined as flows initially starting out as two distinct layers with different mean speeds, both in parallel directions and in contact with each other at a plane boundary (Paw U et al., 1992). Coherent structures are observed to have ejections and sweeps as common features. In an ejection, the near surface fluid rises upward into the fluid. This is associated with a sweep where fluid farther from the surface descends to the near-surface boundary layer. In an ejection, the fluid has lower horizontal velocity and is moving upward with positive vertical velocity. In a sweep, faster moving fluid descends rapidly in a gust (Stull, 1988).

For flows near typical terrestrial ecosystem, plants frequently have large relatively vertical extent into the atmosphere. The momentum drag created by plant structures slows the air, creating the analogy to plane mixing-layer. When a coherent structure is formed associated with the mean shear near the plant canopy top, it consists of linkage of a sweep with at least one ejection. If the sensible heat flux is positive during the sweep phase, a short-lived rapid horizontal-moving parcel of air gusts into the canopy, bringing cooler air down to the plant elements. In the more quiescent ejection phase, slow horizontal-moving air moves upwards. The ejections are weak near the canopy top (Gao et al. 1989), so the air resides in the canopy sufficiently long to show some heating. This is manifested by a slow temperature rise in the temperature trace with time, and it is terminated by the next gust phase, which causes a sudden temperature drop. The resulting temperature trace exhibits a "ramp" pattern (Gao et al., 1989; Paw U et al. 1992). Under stable conditions, the pattern is reversed, with a slow temperature drop as the air in the canopy is cooled by the canopy elements, and a sudden temperature rise as a gust brings down the warmer air from above the canopy (Gao et al., 1989; Paw U et al. 1992). Such patterns have been found in the surface layer of the atmosphere and near vegetation also by other researchers and are sometimes associated with gravity (buoyancy) waves.

Van Atta (1977) identified the relationship between structure functions, turbulence and ramp patterns. The structure function, which has been used extensively in turbulence data analysis, is identified as:

$$S^n(r) = \frac{1}{m}\sum_{i=1}^{i=m-j}\left[T(i+j)-T(i)\right]^n \tag{23}$$

Where: m is the number of data points, n is the order of the structure function, j is the sample lag, and i is the data point number. Depending on the sample frequency (f), the sample lag corresponds to a time lag (r) in seconds. For example, if f=8 Hz and j=2, r=2/8=0.25 s.

Van Atta (1977) evidenced that ramps were regular patterns of fixed geometry that had instantaneous terminations for unstable condition, the slow temperature rise would be terminated by an instantaneous temperature drop (Fig. 2a), and for stable conditions, the slow temperature drop would be terminated by an instantaneous temperature rise (Fig 2b). This resulted in a fixed set of probabilities for the structure function values, crucial for the analysis of ramp geometry.

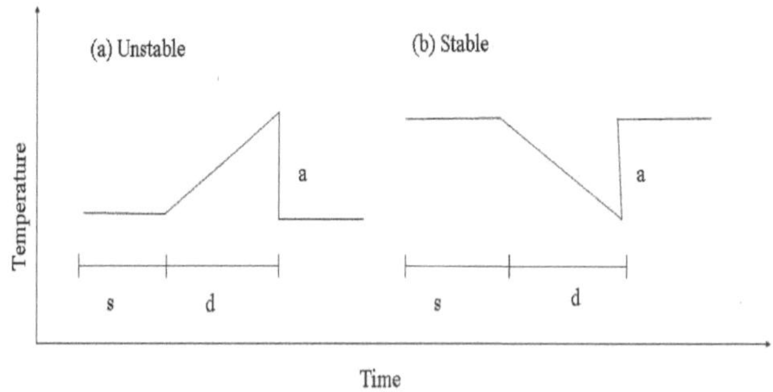

Fig. 2. Hypothetical temperature ramp for (a) unstabe and (b) stable atmospheric conditions

The fixed probabilities are used to determine the ramp dimensions, a, the amplitude of the ramp pattern, s, the spacing between the sudden temperature drop (or rise) and the beginning of the gradual temperature rise (or drop), and, the duration of the gradual temperature rise or fall, d. According to the probability analysis the higher order moments of the structure functions are related to the above ramp dimensions in the following manner:

$$S^2 = \frac{a^2 r}{(d+s)}\left[1-\frac{1}{3}\left(\frac{r}{d}\right)^2\right] \tag{24}$$

$$S^3 = -\frac{a^3 r}{(d+s)}\left[1-\frac{3}{2}\left(\frac{r}{d}\right)+\frac{1}{2}\left(\frac{r}{d}\right)^3\right] \tag{25}$$

$$S^5 = -\frac{a^5 r}{(d+s)}\left[1-\frac{5}{2}\left(\frac{r}{d}\right)+\frac{10}{3}\left(\frac{r}{d}\right)^2-\frac{5}{2}\left(\frac{r}{d}\right)^2-\frac{2}{3}\left(\frac{r}{d}\right)^5\right] \tag{26}$$

It's possible determine the ramp dimensions d and s by using more than one time lag, and then solutions, assuming $r<d$ and $r<s$:

$$\frac{S^3(br)}{S^3(r)} = \frac{-\frac{a^3 br}{(d+s)}\left[1-\frac{3}{2}\left(b\frac{r}{d}\right)+\frac{1}{2}\left(b\frac{r}{d}\right)^3\right]}{-\frac{a^3 r}{(d+s)}\left[1-\frac{3}{2}\left(\frac{r}{d}\right)+\frac{1}{2}\left(\frac{r}{d}\right)^3\right]} \equiv R \tag{27}$$

If $v \equiv \frac{r}{d}$ and $v^3 - \frac{12-3R}{16-R}v + \frac{4-2R}{16-R} = 0$

which yields a cubic solution for v and therefore d. The ramp amplitude can be obtained from

$$a^3 + \left[10S^2(r) - \frac{S^5(r)}{S^3(r)}\right]\frac{P_3}{P_5}a + 10S^3(r)\frac{P_3}{P_5} = 0 \tag{28}$$

Where: $P_2 = 1-\frac{1}{3}v^2$; $P_3 = 1-\frac{3}{2}v+\frac{1}{2}v^3$; $P_5 = 1-\frac{5}{2}v+\frac{10}{3}v^2-\frac{5}{2}v^3+\frac{2}{3}v^5$

And s is given by

$$s = \frac{-a^3 r}{S^3(r)}P_3 - d \tag{29}$$

Chen et al. (1997) noted that the ramps observed by Gao et al. (1989) and Paw U et al. (1992) had non-instantaneous terminations. They developed a set of equations to obtain the ramp dimensions, with the assumption that the spacing between ramps (s) was zero.

Surface Renewal (SR) analysis, as used in plant canopies, was originally conceived of as a simple "transilient" theory (Stull, 1984) in a pseudo-Lagrangian sense (Paw U et al. 1995). In SR only two heights are considered, some height above a plant canopy, and a height representing the entire plant canopy. The vertical motions are not followed. In figure 3 we may consider an air parcel, which instantaneously moves down and then it travels horizontally.

Parcel instantly drops from position 1 to position 2 in the canopy (upper cartoon). Canopy is a source of scalar, time goes on but the scalar does not increase because there is a speed of diffusion of the scalar (thermal inertia) (position 2 to 3 in the lower chart). Scalar starts to increase in the parcel (position 3, 4) to a peak (positions 5, 6 in the lower chart). After horizontally advecting some distance, the parcel instantly rises from position 5 to 6 in the upper cartoon. Simultaneously, from position 7, a new parcel instantly replaces the old parcel's position in the canopy, shown as position 8 (offset horizontally in the upper cartoon for clarity). This results in an instantly falling of the scalar value (lower graph), terminating the previous gradual scalar increase and forming the ramp pattern.

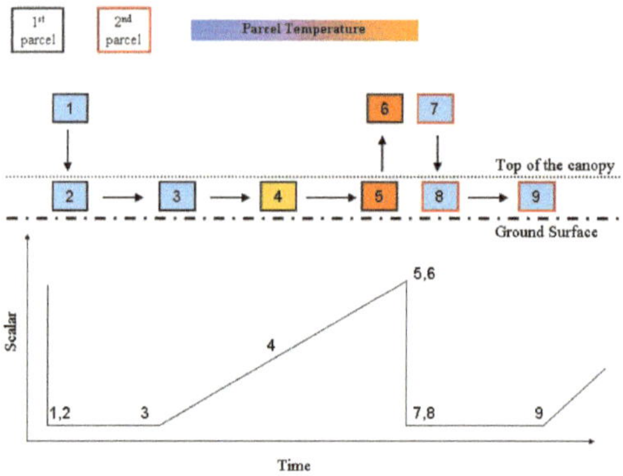

Fig. 3. Example of Surface Renewal process

3.2.2 Calibration

Under unstable conditions, (Fig. 3), a ramp will be determined by the warming caused by the plant canopy elements. In Paw U et al. (1995), the sensible heat flux density within a coherent structure, was derived as:

$$H' = \rho C_p \frac{dT}{dt}\left(\frac{V}{A}\right) \tag{30}$$

Where $\left(V/A\right)$ is the ratio of the volume over the horizontal area of the parcel (for a parcel in the canopy, this would be the canopy height). Practically for temperature recorded at the canopy top, the surface renewal equation is expressed as:

$$H = \alpha \rho C_p \frac{dT}{dt} h_c \tag{31}$$

Where T is measured at the canopy heigh, h_c is the canopy heigh and α is a calibration factor embodying temperature variation in the canopy, initially estimated at 0.5 to account for linear change in temperature with height. The time derivative represents the heating during the time period of the ramp, and during the spacing between ramps, when no heating occurs. Any advection or other processes not properly described in surface renewal analysis would also be included in α (Paw U et al. 1995). When the ramp dimensions are determined by structure functions, then the following equation for H is used to define the sensible heat flux density during a ramp:

$$H = \alpha \rho C_p \frac{dT}{dt} h_c \approx \alpha \rho C_p \frac{\partial T}{\partial t} h_c = \alpha \rho C_p \frac{a}{d} h_c \tag{32}$$

When the equation is weighted by the fraction of time during the ramp $\frac{d}{d+s}$, then the sensible heat flux density may be written as:

$$H = \alpha \rho C_p \frac{a}{d} h_c \left[\frac{d}{d+s} \right] = \alpha \rho C_p \frac{a}{d+s} h_c \tag{33}$$

Uncalibrated SR sensible heat flux density provides an estimate of H that is uncorrected for the mean parcel size, the unequal heating of air from the ground to the thermocouples, and micro-scale advection effects. When sensible heat flux density (H) from a sonic anemometer is plotted versus uncalibrated SR H', a good correlation is typically observed, the slope of the regression line through the origin is the α calibration factor.

It's possible to process the high frequency temperature data to obtain half hour means of the 2nd, 3rd, and 5th order moments of the time lag temperature differences (Snyder et al., 2000). The obtained moments are, then, used to determine the ramp amplitude (a) and inverse ramp frequency ($d+s$) using Van Atta (1977) structure function methodology. Generally two 76.2 μm diameter thermocouples are used to perform temperature measurements.

The second thermocouple is just a repetition, to take into account possible breaking of one of the two sensors. The thermocouples are mounted at the same height, and temperature was recorded at a frequency of 4 Hz. It seems that it is a sufficient high frequency for sampling over most crops. Typically, time lags of r=0.25 and r=0.50 are used for most canopies. The procedure to compute time lag temperature differences and the statistical moments is described below. If T_4, T_3, T_2 and T_1 are the previous readings, taken at 1.0, 0.75, 0.50 and 0.25 seconds and T_0 is the current temperature, then, before a new datum collection, T_4 is overwritten by T_3, T_3 by T_2, T_2 by T_1, T_1 by the previous T_0 and a new value of T_0 is recorded. This procedure is repeated for both thermocouples. Then, using time lags r=0.25 and r=0.50 the differences (T_1-T_0) and (T_2-T_0) are temporarily stored, then the second, third and fifth moments of these differences are calculated and temporarily stored. At the end of the half hourly period, the means of the 2nd, 3rd and 5th moments are calculated and stored in the output table.

3.3 Eddy covariance theory

The basis of the eddy flux method is to erect a notional control volume over a representative patch of surface, in order to measure the exchange across all the aerial faces of this volume, as well as by recording any accumulation within it, and then to infer the surface exchange by difference. We rely on turbulent mixing to act as a physical averaging operator, so that measurements at some height h capture exchange from a representative surface patch. If we assume that, when averaged over sufficiently long time, the flow field is effectively one-dimensional, we can write:

$$\frac{\overline{\partial c}}{\partial t} + \frac{\partial \overline{wc}}{\partial z} = \overline{S}\delta(z) \tag{34}$$

where $c(t)$ is a generic scalar, $w(t)$ is the vertical component of the velocity vector, the overbar denotes a filtering, detrending or averaging operation, $\overline{S}$ is the surface exchange, z the vertical or surface normal coordinate and $\delta(z)$ is the Dirac delta function. When the flow field is stationary, for example when there is no accumulation of the scalar c in the control volume, then the first term of the left side of eq. 34 is zero and thus:

$$\frac{\partial \overline{wc}}{\partial z} = \overline{S}\delta(z) \tag{35}$$

By integrating equation (35) from $z = 0$ to the sensor height h (the top of the control volume) we have:

$$\overline{wc}(h) = \overline{S} \quad (36)$$

The left hand side of equation (36) is the total covariance of $w(t)$ and $c(t)$ under a chosen averaging operator. One final step is required to replace $\overline{wc}(h)$ by the measured eddy flux.

$$\overline{wc} = \overline{\overline{wc}} + \overline{w'c'} \quad (37)$$

In horizontal homogeneous flows with z-axis normal to the surface, $\overline{w} \to 0$ and the integrated mass balance becomes a statement of the equality of the eddy flux at height h and of the surface exchange. So:

$$\overline{w'c'}(h) = \overline{S} \quad (38)$$

The presented theory explains how the eddy flux is equals to the surface exchange only under a set of quite restrictive conditions. Horizontal homogeneity is necessary if we ignore horizontal flux divergences; stationary is required to ignore storage term, and both coordinate rotation and an averaging operator (that obeys to Reynolds averaging rules) are required to replace the total covariance $\overline{wc}$ by the eddy covariance $\overline{w'c'}$.

4. Materials and methods

Micrometeorological experiments were conducted over orange orchards in the Catania Plain (Eastern Sicily, Italy). Energy balance stations were installed to measure the energy exchange fluxes in the soil-plant-atmosphere continuum. Surface Renewal was the main method for the sensible heat flux estimates. Eddy covariance stations were, also, used to compute the alpha calibration factor for surface renewal and to provide the closure of the energy balance equation.

4.1 Sites description

The experiment was carried out over a 120 ha (37° 16′ 41″N 14° 53′ 01″ E) orange orchard located in Lentini (Eastern Sicily, Italy) from September 2009 to September 2010 (Fig. 4). The orchard architecture consisted of mature trees, 3.7 m tall, with a mean leaf area index (LAI) of 4.25 m^2 m^{-2}, PAR light interception of 100% within rows and of 50% between rows. Orange orchards were surface drip irrigated with daily frequency during May-October period. The irrigation systems included on-line labyrinth drippers, in a number of four per plant, spaced at 0.80 m, with discharge rate of 4 l/h at a pressure of 100 kPa.

There was 4 meter of distance between trunks within rows and 5.5 m between rows. The field provides an opportunity for micrometeorological studies because of the flat, homogeneous and wide site. The site is located within the agricultural context of the Catania Plain (Eastern Sicily) where clear skies, high summer temperature, light wind, no rainfall during summer and regional advection were the typical weather conditions. Regardless of the wind direction, the fetch was large because the trees were similar for the adjacent plots.

Fig. 4. Experimental orange orchard at Lentini

The eddy covariance (EC) technique (Aubinet et al. 2000) was used to simultaneously measure the mass and energy exchange flux densities over the orchard field. It encompassed a 3-dimensional sonic anemometer (CSAT3) for measuring the components of wind and a fast-responding open-path gas analyzer LI-7500 (LI-COR, Lincoln, NE, USA) to measure carbon dioxide and latent heat flux. The EC equipment was mounted at 8 meter above the soil surface.

The net radiation (R_n) over the crop (at 8 meter from soil surface) was measured by using two net radiometer (CNR 1 Kippen&Zonen). Net radiation measurements were representative of the average mixed conditions characterizing the heterogeneous context under study. At the plot, soil heat flux (G) was measured using a network of three soil heat flux plates (HFP01, Campbell Scientific Ltd), which were placed horizontally 0.05 meter below soil surface. Three different measurements of G were selected: in the trunk row (shaded area), at 1/3 distance to the adjacent row, and at 2/3 of the distance to the adjacent row. The soil heat flux was measured as the mean output of three soil heat flux plates. The gradual build up of plant matter changed the thermal properties of the upper layers. Consequently, heat storage (ΔS) was quantified in the upper layer by measuring the time rate of change in temperature. The net storage of energy (ΔS) in the soil column was determined from the temperature profile taken above each soil heat flux plate. Three probes (TCAV) were placed in the soil to sample soil temperature. The sensors were placed 0.01-0.04 m (z) below the surface; the volumetric heat capacity of the soil C_v was estimated from the volumetric fractions of minerals (V_m), organic matter (V_0) and volumetric water content (θ). Therefore, G at the surface is estimated by measuring G′ at the depth of 0.05 m and the change in temperature with time of the soil layer above the heat flux plates to determine ΔS.

$$G = G' + \Delta S = G' + C_v \left(\frac{T_f - T_i}{t_f - t_i} \right) \cdot d_g \tag{39}$$

where G' is the heat flux density measured by the plate, ΔS is the heat storage, Tf is the final temperature at time tf, Ti is the initial temperature at time ti (the measurement time interval was of 30 min), dg is the depth (m) of the heat flux plates, and Cv is the volumetric heat capacity ($J \cdot m^{-3} \cdot K^{-1}$), which depends on the bulk density (ρb) of the soil and the volumetric water content (θ).

The SR method to estimate H is based on high frequency temperature measurements. When plotted, the temperature traces show ramp like characteristics, which are used to estimate heat fluxes using a conservation of energy equation. Fine-wire thermocouples (76.2 μm dia.)

were, thus, used to measure high frequency (10 Hz) temperature fluctuations. For the experimental site, two thermocouples were mounted 0.5 m above the canopy height and SR estimates of H were computed using a structure function (Van Atta, 1977) and time lags of 0.25 and 0.50 seconds for each thermocouple to determine the mean ramp-like temperature trace characteristics. A 3-D sonic anemometer (Windmaster Pro, Gill Instruments Ltd) was set up at 0.5 meter above the canopy top. The three wind components and air temperature were recorded at 10 Hz. Wind components were rotated to force the mean vertical wind speed to zero and to align the horizontal wind speed to the mean streamwise direction.

Volumetric water content was measured hourly from 0.3 to 0.6 meter below soil surface by using the time domain reflectometry theory (TDR) (CS 616, Campbell Scientific, Logan UT, USA). The site was also equipped with an automatic weather station to measure the values of ancillary meteorological features (i.e. solar radiation, precipitation, air temperature, relative humidity, pressure, wind speed and direction).

Air temperature and wind speed profiles were realized within the orchard in order to apply the aerodynamic analysis. Wind speed and air temperature sensors were installed at 4.5, 6.0, 8.0 and 13.0 meter above the soil surface; data were recorded at 10 minute intervals. To monitor canopy temperature and detect stress conditions onset, five infrared thermometers (IRTS-P, Apogee) were installed within the orchard.

Data gap-filling due to systems stop was performed. In order to take into account of the data gaps, a parametrization was used as outlined in Aubinet et al. (2000) when meteorological data was available. In the case of unavailable data, the missing flux was replaced by interpolation.

The second experimental site is a citrus orchard located in the north-east part of the Catania Plain near Motta Sant'Anastasia (37°29'36"N - 14°55'12"E). The study area has a surface of almost 2.5 ha (200 x 125 m), all around there are citrus orchards divided from little farm road. The experimental field is a fifteen year old orange orchard, the average height of the trees is 4.0 m; plant are spaced 5m x 5m with a density of 400 trees ha^{-1}, and 70% of ground cover (C_g) (Fig. 5).

Fig. 5. Experimental orange orchard at Motta S. Anastasia

The orchard is irrigated by micro-sprinklers system with a flow rate of 140 l/h, working at a pressure of 172 kPa, with a coverage angle of 360°. The irrigation schedule is not regular, and it is decided by the farmer. Usually the irrigation timing is 3-4 hours for each irrigation, which means around 20 mm, every 2 weeks. An energy balance station was installed in the field to monitor the exchange energy fluxes (net radiation, soil heat flux density, sensible heat flux density and latent heat flux density) in the soil-plant-atmosphere continuum.

Fig. 6. Fine wire thermocouples

The net radiation (R_N, W m^{-2}) was measured using a net radiometer (CNR1, Kipp&Zonen) placed at 1.5 m height above the canopy top. The soil heat flux (G, W m^{-2}) was measured as the mean output of three soil heat flux plates (model HFP-01, Hukseflux Thermal Sensors). They were buried 0.05 m below the surface. The heat storage (ΔS) was quantified in the upper layer by measuring the time rate of change in temperature (Fuchs and Tanner, 1967). Three probes (TCAV) were placed in the soil to sample soil temperature. The sensors were placed 0.01-0.04 m (z) below the surface. In particular the first measurement point was placed near the trunk of the tree (shaded), the second was palced at 1/4 of the distance to the next raw (middle), and the last at 1/2 of the distance to the next raw.

The sensible heat flux density (H) was measured using surface renewal technique. For this pourpose two Campbell's FW3 fine wire type E thermocouple were installed 0.5 m above the top of the canopy (Fig. 6).

The volumetric water content of the soil under study was monitored using 10 Campbell's TDR model CS616 for computing the soil thermal conductivity. Campbell Scientific IRR-P infrared thermometers were installed one meter above the canopy, looking downward to measure the canopy temperature. An infrared temperature sensor is a non-contact probe, that measures the surface temperature of an object by sensing the infrared radiation emitted by the target. The IRR-P is widely used to measure the surface canopy temperature. With contact sensors it's difficult to avoid influencing temperature, maintain thermal contact, and provide a spatial average. By mounting the IRR-P at an appropriate distance it is possible measure the temperature of just one leaf, a canopy or any surface of interest. The canopy temperature is useful for the computation of the Crop Water Stress Index (CWSI), which uses the difference between the canopy temperature and the air temperature to quantify the stress condition of the plant.

The latent heat flux density (LE) was finally computed as residual of the energy balance equation neglecting the small Misc. term. To convert the latent heat flux (LE) into actual water mass flux (ET_a), LE was divided by the latent heat of vaporization (L) equal to 2.45 [MJ kg^{-1}]. Estimates of crop coefficient K_c, obtained as ration between ET_a (assuming ET_a=ET_c due to well watered conditions) and reference ET (ET_0), were compared with K_c data from FAO papers.

Sensible heat flux data from SR technique (H_{SR}) were calibrated with independent measurements of H_{EC} by a 3-D sonic anemometer (Gill Wind-Master) located in the same area of study. The calibration data subset was used to derive the α value by simple linear regression forced through the origin. For this regression, H_{EC} was used as the dependent variable and H_{SR} as the independent one. In this way, the regression slope was the α value looked for, used to correct H from the uncalibrated SR analysis.

4.2 The simplified aerodynamic method

By adopting the micrometeorological aerodynamic method, the sensible heat flux H is determined by the flux-gradient relationship. From the applicative point of view, the main advantage of the aerodynamic method consists in avoiding humidity measurements. Nevertheless, its accuracy depends on the number of measurement levels of wind speed (u) and temperature profile (T).

A simplified version of the method has been proposed by Itier (1981) and Riu (1982). In this version, the measurement of Δu and ΔT is only necessary on two levels. The method is based on the flux-gradient relationship and the Monin-Oboukhov similarity theory. Calculation of H is dependent on atmospheric stability which is estimated by means of the Richardson number, R_i:

$$R_i = \frac{g}{T}\frac{\partial T / \partial z}{(\partial u / \partial z)^2} \tag{40}$$

where g (m s^{-2}) is the acceleration due to gravity. On the basis of R_i values, H calculation is made by four equations:

i. Moderate instability condition ($-0.3 \le R_i \le 0$, day situation):

$$H = K\Delta T\Delta u\left(1 - K_R\frac{\Delta T}{\Delta u^2}\right)^{3/4} \tag{41}$$

where K e K_R are coefficients depending on the position of the sensors:

ii. Very unstable condition (Ri<-0.3, day situation):

$$H = \alpha\Delta T^{3/2} \tag{42}$$

iii. Moderate stability condition (0<Ri≤0.15, night situation):

$$H = K\Delta T\Delta u\left(1 - \frac{\Delta T\Delta z}{6\Delta u^2}\right)^2 \tag{43}$$

iv. Very stable condition (Ri>0.15, end of clear nights):

$$H = K\frac{\Delta T\Delta u}{10} \tag{44}$$

In the application of the simplified aerodynamic method (SAM) two elements must be underlined: (i) measurements levels z_1 and z_2 must be chosen so that z_2 is high enough to keep $(z_1 z_2)^{1/2}$ outside the roughness layer; (ii) (z_2-z_1) must be large enough to measure wind and temperature differences with sufficient accuracy. In the proposed application of the method, for the case study of Lentini, z_1 was posed at 4.5 meter and z_2 assumed values of 6, 8 and 13 meters, respectively, from the soil surface layer.

4.3 Performance indicators

Linear regression analysis, L_{RA} (slope, intercept, determination coefficient R^2) and the root mean square error, RMSE, were used to compare the sensible heat flux estimates using SR (H_{SR}) and SAM (H_{SAM}) analyses against the eddy covariance (EC) method (H_{EC}). The coefficient $D=\Sigma y/\Sigma x$ which is the sum of the flux estimates (Σy) over the sum of fluxes taken as reference (Σx), where H_{EC} is the reference data, was also determined as an evaluation parameter. The bias is (D-1) time the mean value determined from the observations. Parameter α of eq. 33 is a factor that corrects the unequal amount of the scalar (T, air temperature) from the measurement height to the ground. If the scalar trace is measured at one height only, α must be determined by comparison of SR results with independently measured exchange rates. In the study, sensible heat flux data from SR technique (H_{SR}) were calibrated with independent measurements of H_{EC} by two 3-D sonic anemometers located, respectivley, at 4 and 8 meters. The calibration data subset was used to derive the α value by simple linear regression forced through the origin. For this regression, H_{EC} was used as the dependent variable and H_{SR} as the independent one. In this way, the regression slope was the α value looked for, used to correct H from the uncalibrated SR analysis.

5. Results and discussion

5.1 The case study of lentini

The values of sensible heat flux (H) estimated from SR and SAM were compared versus the H measured using the EC system deployed at z = 8 m, (i.e., it is the H desired to estimate LE). Figure 7 shows the energy balance closure at 8 meter. The H values estimated from SR and SAM (with z_2=13 meter) were close to H from EC method. Figures 8 and 9 show that the hourly H_{SR} and H_{SAM} versus H_{EC} were similar for a wide range of H. When the simplified aerodynamic method (SAM) was applied with z_2 at 8 or 6 meter, H_{SAM} resulted quite different from H_{EC}. Most likely (z_2-z_1) wasn't large enough to measure wind and temperature differences with sufficient accuracy. In fact, when the atmospheric surface boundary layer is moderate or quite unstable, similar studies indicate that the roughness sub-layer depth mostly varies from 1 to 2 times the canopy height (Castellvì and Snyder 2009). Fig. 10 shows the frequency of z = 8 m to be above the roughness sublayer versus time (GMT).

In general, it is shown that for hours with negative (Rn-G), the measurement height mainly remained within the inertial sub-layer. For positive (Rn-G) the roughness sublayer depth was oscillating around the upper level for about the 50% of the samples from sunrise until about two hours after noon, and during late afternoon the upper level tended to remain in the inertial sublayer.

Furthermore, some spurious H estimates were obtained from SAM during moderate and very stable atmospheric conditions, thus reducing the available data set for the comparison. In particular, the number of samples gathered under unstable atmospheric conditions was higher than that from stable conditions.

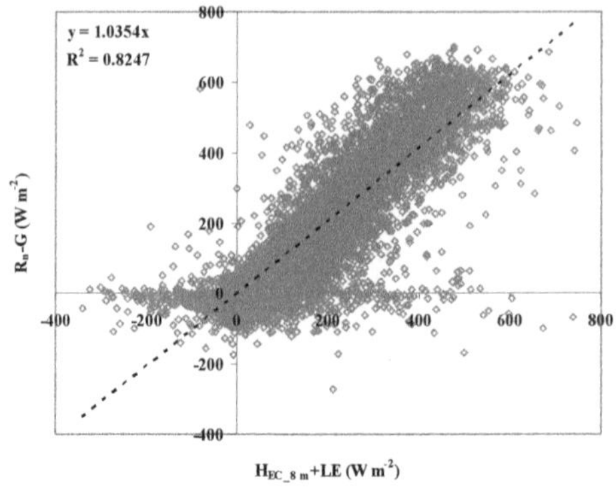

Fig. 7. Energy balance closure at 8 meter

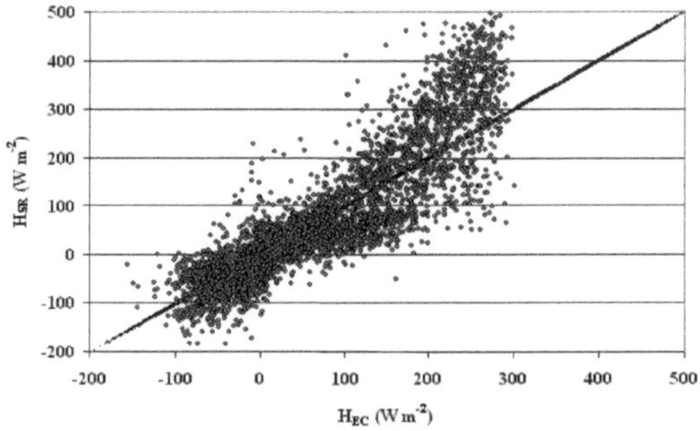

Fig. 8. H_{SR} versus H_{EC} at 8 meter

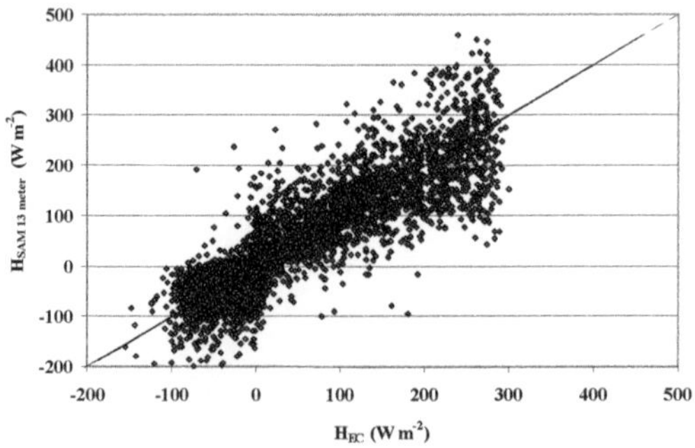

Fig. 9. H_{SAM} at 13 meter versus H_{EC} at 8 meter

Table 1 shows the results of the linear regression analysis (slope, intercept and coefficient of determination, R^2), root square mean error, RMSE, and D obtained for the analyzed data set. The largest RMSE values were obtained using the simplified aerodynamic method applied at 8 and 6 meters.

Fig. 10. Percentage of cases when z=8 m falls above z*

In a review of experiments where the measurements were taken in the roughness sublayer, the RMSE values found to adjust surface renewal method where lower than that in Table 1 (Paw U et al. 1995; 2005). However these experiments are not direclty comparable because H was measured using a one dimensional sonic anemometer.

Slight underestimations by the proposed alternative methods (SR and SAM) were observed. A 3.8% underestimation was found by SR for the whole data set including both stable and unstable cases; as a result that for small fluxes the ramp amplitude was often not in accord with the sign of H_{EC}. RMSE values from SR and SAM were not high. SR and SAM linear regression analyses showed slopes of, respectively, 1.12 and 0.95 very close to unit, intercepts near zero and high coefficients of determination.

For optimized drip irrigation, hourly LE estimates are desired. The best LE estimates from eq.2 are attained when all the terms are averaged hourly. A better closure may be achieved when turbulent fluxes are determined using longer block averages than half-hourly because lower frequencies are captured as well (Finningan et al. 2003). However, SR is based on the analysis of organized motion near the canopy-atmosphere interface. The continuous ramp patter exhibited in the scalar trace is a canopy-scale coherent structure, which is not associated with large circulations.

	H_{SR} versus H_{EC}	H_{SAM} (z_2= 13 m) versus H_{EC}	H_{SAM} (z_2= 8 m) versus H_{EC}	H_{SAM} (z_2= 6 m) versus H_{EC}
A	1.12	0.95	1.20	1.87
b (W m^{-2})	-5.80	-2.33	2.96	-4.58
R^2	0.80	0.77	0.63	0.75
RMSE (W m^{-2})	50.19	44.52	58.92	117.11
D	0.96	0.88	1.11	1.72

Table 1. Linear regression analysis, RMSE, D and, comparing the hourly H_{SR} and H_{SAM} (with different z_2) versus H_{EC} for the whole data set.

During the irrigation season (May-October period), latent heat flux density averaged 8.5, 7.0, and 7.6 MJ m^{-2} d^{-1} for the EC, SR and SAM, respectively. The mean (about 0.90) of daytime evaporative fraction (EF) during summer, which characterizes the partition of the energy budget at the daily time scale, varied little (0.06) based on average cloudiness. The temporal variability of the partitioning, expressed in terms of EF daily standard deviation, reached a

maximum of 14%. The experiment showed that the evaporative fraction computed from flux measurements at 4 hours past sunrise tends to increase very slowly, thus to assume that the underestimation in daytime average would be not significant. Actual crop ET (ET_a) was computed by dividing LE by the latent heat of vaporization: L = 2.45 MJ m^{-2} mm^{-1}. Generally, crop coefficients are determined by calculating the ratio $K_{co}=ET_c/ET_o$, where ET_c is the evapotranspiration of a well-watered crop. Since these orchards are well managed, it is assumed that there was little or no transpiration reducing water stress and $ET_a \approx ET_c$.
Hourly variations of crop coefficient (K_{co}) values, during the monitoring period, were determined using ET_c and reference evapotranspiration (ET_o) for a short canopy (Allen et al. 1998). Weather data used to calculate ET_o came from the SIAS station (the agro-meteorological service of the Sicilian region) which is located 4 km far from the site. Hourly ET_o values were summed over 24-hour periods to obtain daily ET_o.
During May-October period, average daily values of ET_c were of 3.5, 2.9 and 3.1 respectively from EC, SR and SAM, with corresponding values of crop coefficient of 0.54, 0.51, and 0.62. H_{EC_4m} and H_{EC_8m} were rather well correlated (R^2=0.93) considering the different amount of small eddies (much higher close to the canopy top) that cannot be properly sampled by sonic anemometer. Figure 11 shows H_{EC_4m} versus H_{EC_8m} for all the campaign. In general H_{EC_4m} underestimated H_{EC_8m} of 18%. For unstable cases, the α values for unequal heating was of 0.66 at 4 meter and of 0.68 at 8 meter.

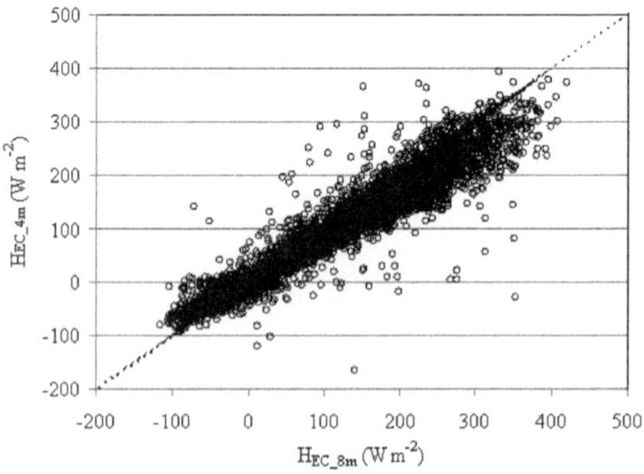

Fig. 11. Sensible heat flux measurements using the EC method at z=4 m, H_{EC_4m} versus at z=8 m, H_{EC_8m} for all the data.

5.2 The case study of Motta S. Anastasia

Table 2 shows the mean values for surface fluxes corresponding to four different datasets during the monitoring periods. Two datasets were formed from samples gathered under unstable and stable surface-layer atmospheric conditions. In particular, stable-case data set corresponded to night time periods with negative R_N-G (available energy) and small H and LE. The unstable data set is referred to diurnal periods where R_N-G is positive, with the exception of the periods close to sunrise, sunset and late afternoon that may be included in the stable-case data set. The dry and humid period data sets correspond, for the area under study, to May-September and October-April, respectively.

Year	atmospheric condition		H_{SR} (W m-2)	LE (W m-2)	R_N-G (W m-2)
2005	unstable	Dry period	77.4	255.8	331.6
		Humid period	27.9	116.1	143.8
	stable	Dry period	-11.9	-32.6	-45.2
		Humid period	-18.7	-23.7	-42.0
2006	unstable	Dry period	34.9	283.5	320.3
		Humid period	35.2	140.0	176.9
	stable	Dry period	-4.6	-29.4	-35.9
		Humid period	-10.5	-28.5	-39.6
2007	unstable	Dry period	34.6	262.9	299.6
		Humid period	19.4	157.0	177.8
	stable	Dry period	-4.4	-30.2	-36.5
		humid period	-6.4	-28.6	-36.0
2008	unstable	Dry period	77.5	146.0	223.6
		humid period	42.6	108.5	151.0
	stable	Dry period	-4.9	-29.7	-36.9
		humid period	-8.8	-24.7	-33.9

Table 2. Mean values of energy fluxes during unstable and stable conditions

As shown in Table 2, the energy partitioning of the available net surface energy was remarkably distinct for these different datasets. During both dry and humid periods well-formed ramp traces of air temperature were observed, which explains the good performance of LE values. Under unstable atmospheric conditions, when a cool air parcel moves instantaneously downward and travels horizontally, a positive amplitude ramp (a) in the temperature trace is observed (positive H). For this reason, in the unstable-case when RN-G is positive, most of the available net surface energy contributes to positive latent heat flux. It reached a maximum of around 650 W m-2 during late July of the monitoring periods. The average hourly ratio H/LE during dry periods was around 0.56 and reached the peak between 12.00 a.m. and 2.00 p.m. Under stable atmospheric conditions, the temperature, generally, risen as the warm air swept into the canopy an heat transferred from the air to the plant canopy elements, causing a slow temperature drop ($a<0$ and $H<0$). In the study, the latent heat flux, the energy transfer due to evaporation or condensation, was largely dependent on leaf area index (LAI) and PAR light interception (LI) by vegetation, with peaks when LAI and PAR LI were higher.

The hourly patterns of energy fluxes in the orchard is presented for the same days-period in Figure 12. In all the years the energy balance calculations were fairly similar. On the average, the evaporative fraction EF (%), the ration between LE and the available surface energy (R_N-G) was of about 80%, during the dry period and under unstable-case atmospheric conditions.

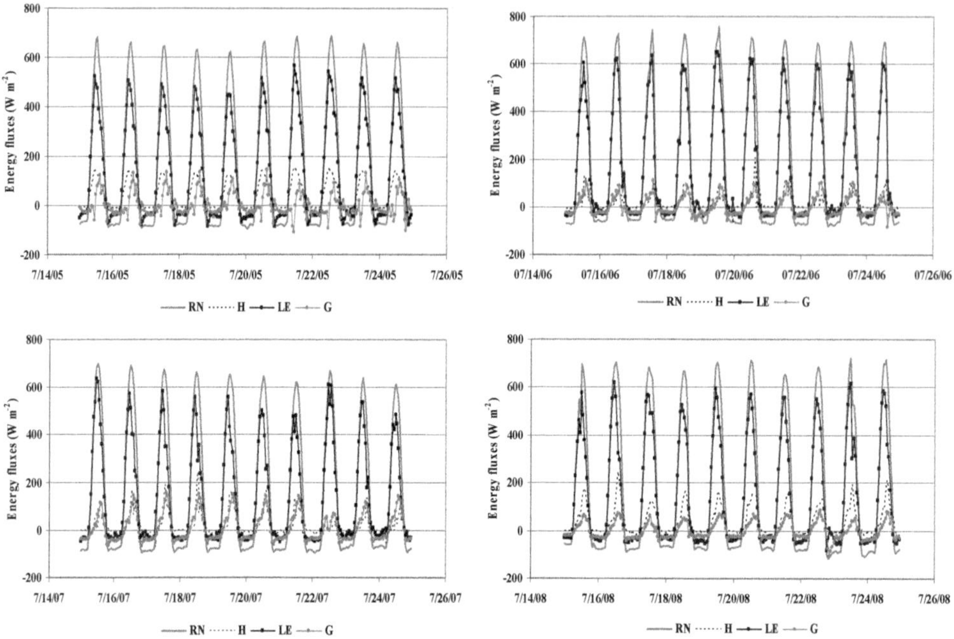

Fig. 12. Hourly patters of energy fluxes in the orchard for 10-days periods during the monitoring years

The regression of half hourly averages of sensible heat flux density (H_{EC}) from the sonic anemometer versus uncalibrated Surface Renewal measurements of H was carried out during the months of July of the years 2005 and 2006. The slope of the regression of H_{EC} versus uncalibrated H_{SR} was of about 0.24 with a determination coefficient R^2=0.85. This calibration for α was used during 2007 and 2008.

During summer, LE values had an average of about 11.0 MJ m^{-2} d^{-1} and a maximum of 17.4 MJ m^{-2} d^{-1}, with a variation of about 0.5. Study revealed that net radiation (R_N) was highest at 12.00 noon, with the daily mean value of about 13.2 MJ m^{-2} d^{-1}, maximum around 18.0 MJ m^{-2} d^{-1} and variation coefficient of 0.55. The course of soil heat flux (G) was affected by the development of crop canopy or leaf area index. On a daily basis, G density is generally close to zero. Sensible heat flux (H) had mean and maximum values of about 2.2 MJ m^{-2} d^{-1} and 4.7 MJ m^{-2} d^{-1}, respectively; daily H values resulted fairly scattered around the mean, with a variation of about 0.8.

In general, mean reference evapotranspiration ET_0 during dry periods tends to be between 5-6 mm d^{-1} in each year; ET_0 varied between a maximum of about 11 mm d^{-1} to a minimum of about 1.5 mm d^{-1}. The mean ET_a was about 4.5 mm d^{-1} (variation less than 10%) during the dry periods of the monitoring years. Before the experiment in 2005, the orchard was topped and pruned, which probably explains the slightly lower ET_a during that season. During humid periods, mean ET_0 and ET_a were of about 2.3 and 1.8 mm d^{-1}, respectively. Crop coefficient (K_c) values were higher than typical reported (Allen et al., 1998). The K_c values generally increased during the monitored dry periods with averages of 0.76, 0.89, 0.86 and 0.73 going from 2005 to 2008. In 2005, the increase in summer was due to the canopy growth following topping and pruning. In 2008, the experimental orchard was less vigorous than in the previous years due to a pathogenic diseases caused by the *Closterovirus Citrus Tristeza Virus* (CTV). It determined significant economical income reductions for the grower. The *virus* mainly affects the plant root system and its water supplying capacity; others evident effects concern the reduction of vegetation indicators such as LAI, WDVI, NDVI and PAR light interception by the monitored orchards.

The rows at the experimental orchard are east-west oriented rather than north-south and this might partially explain the higher crop coefficient values. Another possible reason for the higher K_c values observed in this experiment is the use of micro-sprayer irrigation rather than surface irrigation and higher application frequency in this orchard than reported in old experiments (Allen et al., 1998). In fact, these publications do not site the original source of information, and it is likely that the values came from research on flood or furrow irrigated citrus orchards prior to 1977, so those data may be too low for dense canopy, vigorous, sprayer-irrigated orchards. In the experimental orchard, mean K_c values varied from about 0.83 to 0.74 from June to September of the monitored years. In the widely used FAO papers (27 and 56), the estimated crop coefficient for clean-cultivated, mature (> 70% cover) citrus is given as K_c=0.65 for June through August. Nearly linear relationships between K_c value and LAI and PAR light interception were observed in the study. The peak K_c value occurred at LAI of 1.8 m^2 m^{-2} and PAR light interception of 80%.

6. Conclusions

This chapter gives the micrometeorological basics, which were applied in the presented experimental researches. The energy balance equation and the principal micrometerological methods to estimate its flux components were presented. The reliability of alternative micrometerological techniques, such as Surface Renewal for estimating sensible heat fluxes was proposed and discussed. Two different study cases were presented both realized within a Mediterranean environment, characterized by semi-arid climatic conditions and water availability restrictions for agricultural purposes. On the basis of the first experimental results, carried out within the orange orchard located in Lentini (Eastern Sicily), two alternative methods: Surface Renewal (SR) and a simplified aerodynamic method (SAM) for estimating sensible heat fluxes were proposed and compared with the more rigorous eddy covariance technique at the same site. The SAM analysis was accomplished by coupling different equations based on simple gradient flux expressions to account for atmospheric changes depending on stability or unstability conditions. In general H_{SR} and H_{SAM} were similar to H_{EC} regardless of the atmospheric stability conditions, demonstrating the potential of SR and SAM analyses as methods applicable to estimate sensible heat flux. The SAM was more sensitive to moderate and very stable atmospheric conditions, thus resulting in some spurious data. SR technique appears in advantage with respect to SAM and EC because it may operate close to the canopy, thus minimizing fetch requirements, which make it a useful micrometeorological method where fetch requirements limit the application of other techniques. The combination of the SR procedure and the simplified surface energy

balance equation appears to be an affordable alternative to be considered for estimating water use in agriculture.
According to the second experiment, data were collected over orange orchards and analyzed for different ranges available energy, sensible and latent heat fluxes during the monitoring period 2005-2008 in Motta S. Anastasia (Eastern Sicily). During unstable atmospheric conditions, the surface energy balance analysis revealed hourly sensible heat fluxes similar to soil heat flux data, with a mean partitioning of available energy into latent heat flux of about 80%. The crop coefficient (Kc) shows apparent correspondence with the development of canopy cover, LAI and PAR Light interception. The established relationships can be useful for development algorithm of crop simulation model for predicting LAI or PAR LI. Crop coefficient values estimated from the field experiment during 2005-2008 monitoring periods were quite different (means between 0.73 and 0.89) from the crop coefficients published by FAO (0.65) for citrus with ground cover by vegetation of about 70%. Differences between observed Kc and old Kc values from FAO papers are most likely due to differences in plant density, cultural practices, irrigation methods (micro irrigation techniques versus surface irrigation) and frequencies and methods to estimate references ET rates.
To conclude, measurements of high frequency temperature data for the case study showed that the structure function approach used in Surface Renewal analysis provide a good performance in terms of both H and LE. Fairly high correlation were found between calibrated Surface Renewal H_{SR} and Eddy Covariance H_{EC} (from sonic anemometer). By working at 4 Hz, Surface Renewal uses a smaller amount of data than Eddy Covariance (10 Hz) and allows an easier and cheap computation of sensible heat fluxes.

7. References

Allen, R. G.; Pereira, L. S., Raes, D. & Smith M. (1998). Crop evapotranspiration: Guidelines for computing crop water requirements. Irr. & Drain. Paper 56, FAO Rome

Aubinet MA; Grelle A. & Ibron. A (2000). Estimates of the annual net carbon and water exchange of forests: the EUROFLUX methodology. *Adv. Ecol. Res.,* Vol.30, pp. 113-175

Castellvi F.,; Snyder R.L. & Baldocchi D.D. (2008). Surface energy-balance closure over rangeland grass using the eddy covariance method and surface renewal analysis. *Agric. For. Meteorol.,* Vol.148 No.6-7, pp. 1147-1160

Castellvì F. & Snyder R.L. (2009). On the performance of surface renewal analysis to estimate sensible heat flux over two growing rice fields under the influence of regional advection. *J. Hydrol.,* Vol.375, pp. 546-553

Chen, W.,; Novak, M. D., Blanck, A. & Lee, X. (1997). Coherent eddies and temperature structure functions fro three contrasting surfaces - Part I: ramp model with finite microfront time. *Bondary-Layer Meteorol.* Vol. 66, No 1-2, pp. 65-80

Consoli, S.,; O'Connell, N.V. & Snyder, R.L. (2006). Estimation of evapotranspiration of different orange sized orchard canopies using energy balance. *J. Irr. and Drain. Engineer.* ASCE. Vol. 32, No 1, pp. 2-8

Desjardins, R.L. (1972). CO2 measurements by eddy correlation methods. *Bulletin of the American Meteorological Society*, Vol. 53, pp. 10-40.

Finnigan, J.J.,; Clements, R., Malhi, Y., Leuning, R. & Cleugh, H. (2003). A revaluation of long-term flux measurement techniques. Part I: averaging and coordinate rotation, *Boundary-Layer Meteorology*, Vol.107, pp. 1-48

Gao, W.,; Shaw, R.H. & Paw U, K.T. (1989). Observation of organized structure in turbulent flow within and above a forest canopy. *Bounday-Layer Meteorology*, Vol. 47, pp. 349-377

Itier, B. (1981). Une mèthode simple pour la mésure du l'evapotranspiration réelle à l'échelle de la parcelle, *Agronomie*, Vol.1, pp. 869-876

Paw U, K. T. & Brunet, Y. (1991). A surface renewal measure of sensible heat flux density. *In preprints, 20th Conference on Agricultural and Forest Meteorology*, pp. 52-53, Salt Lake City, Utah, September 10-13

Paw U, K.T.,; Brunet, Y., Collineau, S., Shaw, R.H., Mitani, T., Qui, J. & Hipps, L. (1992). On coherent structures in turbulence within and above agricultural plant canopies. *Agricultural and Forest Meteorology*, Vol. 61, pp. 55-68

Paw U, K. T.,; Qui, J., Su, H. B., Watanabe, T. & Brunet, Y. (1995). Surface renewal analysis: a new method to obtain scalar fluxes without velocity data. *Agric. For. Meteorol.* Vol. 74, pp. 119-137

Paw U, K.T., Baldocchi, D.D., Meyers T.P. & Wilson, K. (2000). Correction of eddy-covariance measurements incorporating both the advective effects and density fluxes. *Boundary-Layer Meteorology*, Vol. 97, pp. 487-511

Pruitt, W.O. (1963). Application of several energy balance and aerodynamic evaporation equations under a wide range of stability. *Final Report to USAEPG on contract no. DA-36-039-SC80334*. University of California, Davis (Ed.), pp 107-124

Riou, C. (1982). Une expression analytique du flux de chaleur sensible en conditions superadiabatiques à partir de mesures du vent et de la témperature à deux niveux. *J. Rech. Atmos.*, Vol. 16, pp. 15-22

Snyder, R.L.,; Spano, D. & Paw U, K.T. (1996) Surface Renewal analysis for sensible and latent heat flux density. *Boundary-Layer Meteorol.*, Vol. 77, pp. 249-266

Snyder, R.L.,; Bali, K., Ventura, F. & Gomes-MacPherson H. (2000). Estimating evapotranspiration from bare soil or nearly bare soil. *J. Irrig. & Drain. Eng.*, Vol.126 No(6), pp. 399-403

Spano D.,; Snyder, R.L., Duce, P. & Paw U, K.T. (1997). Surface renewal analysis for sensible heat flux density using structure functions. *Agric. For. Meteorol.* Vol. 86, pp. 259 - 271

Stull, R.B. (1984). Transilient turbulence theory. 1. The concept of eddy-mixing across finite distances. *Journal of Atmospheric Science*, Vol.41, pp. 3351-3367

Stull, R.B. (1988). *An Introduction to Boundary Layer Meteorology*, Kluwer Academic Publishers.

Swinbank, W.C. (1955). *Eddy transport in the lower atmosphere*. Technical paper No.2, Division of Meteorological Physics, Commonwealth Scientific and Industrial Research Organization, Melbourne, Australia. 30pp

Tillman, J. E. (1972). The Indirect Determination of Stability, Heat and Momentum Fluxes in the Atmospheric Boundary Layer from Simple Scalar Variables during Dry Unstable Conditions. *Journal of Applied Meteorology*, Vol.11, pp. 783-793.

Van Atta, C.W. (1977). Effect of coherent structures on structure functions of temperature in the atmospheric boundary layer. *Arch. of Mech.*, Vol.29, pp. 161-171

4

(Evapo)Transpiration Measurements Over Vegetated Surfaces as a Key Tool to Assess the Potential Damages of Air Gaseous Pollutant for Plants

Giacomo Gerosa[1], Angelo Finco[1,2], Simone Mereu[3], Antonio Ballarin Denti[1] and Riccardo Marzuoli[1]
[1]*Dep. of Matematics and Physics, Università Cattolica del Sacro Cuore BrescialI*
[2]*Ecometrics s.r.l., Spin-off company Università Cattolica del Sacro Cuore*
[3]*Department of Economics and Woody Plant Systems, University of Sassari*
Italy

1. Introduction

The term evapotranspiration (ET) is used to describe the contemporary evaporation of water from surfaces and transpiration of water trough stomata.

Evaporation (E) consists in the change of state of water from liquid to vapour that occurs when water molecules momentarily acquire high speed near the surface of water as the result of collisions with other molecules. This kind of process can be enhanced by environmental factors, such as direct solar radiation and temperature, which provide the required energy.

Leaf transpiration can be thought of as a necessary cost associated with the opening of stomata to allow the diffusion of carbon dioxide inside the leaf, for photosynthesis and plant growth. However, these low-resistance apertures also provide a favourable diffusional pathway for atmospheric gas pollutants, such as ozone, that can reach the substomatal cavity and the mesophyll inside the leaf, (causing negative effects at different biological and physiological levels).

Stomata opening is regulated by environmental factors such as light, air temperature and humidity, wind speed and water availability in the soil (Jarvis, 1976; Stewart, 1988).

The interpretation of the ET and the understanding of the potential influencing factors are important topics for ecophysiology, agriculture and agro-meteorology since the past century (Penman, 1948).

ET plays a key role in the water cycle, affecting the water balance from local up to regional scale and causing feedback between vegetation and climate (Wilske et al. 2010). Penman (1948), combining the energy balance with the mass transfer method, derived an equation to calculate the evaporation from an open water surface using meteorological data such as radiation, temperature, humidity and wind speed. Today the Penman-Monteith equation is considered the most reliable method to calculate evapotranspiration.

The FAO (Food and Agriculture Organization of the United Nations) Penman-Monteith equation states that evapotranspiration can be obtained as it follows:

$$\lambda E = \frac{(R_N - G)s + \frac{c_p \rho [e_s(T_a) - e]}{r_a}}{s + \gamma \left(\frac{r_a + r_s}{r_a} \right)} \tag{1}$$

where R_N is the net radiation, G is the soil heat flux, $(e_s - e_a)$ represents the vapour pressure deficit of the air, ρ is the mean air density at constant pressure, c_p is the specific heat of the air, s represents the slope of the saturation vapour pressure temperature relationship, γ is the psychrometric constant, and r_s and r_a are the surface and aerodynamic resistances.
The Penman-Monteith approach as formulated above includes all parameters that govern energy exchange and corresponding latent heat flux (evapotranspiration) from ecosystems.
Slightly different formulations and simplification of this equation have been proposed and used for different purposes.
In this chapter it will be showed how it is possible, starting from transpiration measurements at leaf level, at plant level or at ecosystem level, to estimate how much ozone is taken up by the vegetation through the stomata. Ozone is in fact widely recognized to be an air pollutant which may cause significant damages to the vegetation. At leaf level ozone can damage mesophyll cells, affecting the photosynthetic activity and causing the appearance of chlorosis and necrosis. This lead the plant to spend energy for the detoxification, causing, in turn, a decrease of the available energy for the plant growth. As a consequence, stem and root dimension, stem and root ratio, crown architecture and productivity can be influenced by ozone uptake (Musselman et al. 2006). A key passage for estimating the ozone uptake at each scale (leaf, plant or ecosystem level) is the determination of the stomatal conductance.

2. Measuring techniques of stomatal conductance

Evapotranspiration from an ecosystem is influenced by physiological and morphological properties of the canopy (Baldocchi & Vogel, 1996), and soil properties that govern the water vapor exchange at the soil-atmosphere interface (Baldocchi et al., 2000). Additionally, meteorological conditions within and above the canopy, such as wind speed, air temperature and humidity, are further influencing factors that must be taken into account. Different methodologies are available to obtain indirect measurements of stomatal conductance, all of them are linked to transpiration measurements with the same factors influencing both of them.

2.1 Leaf level measurements

Different instruments, such as porometers and gas exchange devices, allow to measure stomatal conductance at leaf level taking into account the environmental parameters (temperature, relative humidity, solar radiation) inside the measuring cuvette.
Three main types of porometers are available: steady-state porometers, dynamic porometers and null balance porometers.
Steady-state porometers (e.g. Leaf porometer, Decagon Devices, Pullman, USA) are based on the measurement of leaf conductance in series with two elements whose conductance is

known, using a sensor head with a fixed diffusion path. Since the conductance of the sensor elements are known, assuming that the leaf tissues are at a saturation point and that leaf temperature is equal to the air temperature of the nearer sensor, the leaf conductance can be calculated applying the Ohm's law and by measuring air temperature and relative humidity in two points.

The principle of operation of a null balance porometer (e.g. LI-1600, LI-COR, Lincoln, USA) is to manually maintain the air relative humidity at a set-point inside the cuvette. The transpiration from a leaf is obtained by measuring the flow rate of dry air which is necessary to keep constant the relative humidity inside the cuvette where the leaf is placed. Usually the environmental relative humidity of the air is assumed as a null point. The stomatal conductance is hence calculated from the measurements of relative humidity, leaf and air temperature and flow rate of dry air.

Dynamic porometers (e.g. AP4, Delta-T Devices LTD, Cambridge, UK) measure water stomatal conductance indirectly. In fact stomatal conductance is derived from the measurement of the time which is needed to the leaf to release inside the cuvette enough water vapour so that the relative humidity inside the cuvette increase by a fixed amount (2.3%). As a matter of fact, the measurement is constituted by a cycle of measurements to warrant reproducibility. The stomatal conductance is finally obtained by comparing the measurement with a calibration plate whose conductance is known (Monteith et al., 1988; Bragg et al., 2004).

Adding other environmental measurements like soil water content it is possible to apply a Jarvis-type model (1976) to estimate stomatal conductance, further developed by Stewart (1988).

The Jarvis-Stewart model hypothesize that the stomatal conductance of vegetation can be modelled as the product between the maximum stomatal conductance and a series of limiting functions whose values are between 0 and 1. These functions depend on environmental parameters such as air temperature, VPD, soil water content (or soil water potential) and light (PAR or global radiation).

In details the model foresee stomatal conductance g_S as:

$$g_s = g_{\max} \cdot [f_T \cdot f_{light} \cdot f_{VPD} \cdot f_{SWC}] \tag{2}$$

where g_{max} is the highest stomatal conductance of the studied vegetation, this is a value determined by the genetic characteristics of the species, it is often measured in laboratory experiment letting the studied species grow in optimal conditions; f_T, f_{light}, f_{VPD} and f_{SWC} are the limiting functions of temperature, PAR, VPD and soil water content respectively.

The first step in order to obtain these functions is to perform measurements of stomatal conductance at leaf level, by means of porometers or gas exchange analyzers, covering a large scale of different conditions for each environmental factor; then the relative stomatal conductance (g_s/g_{max}) is plotted versus the corresponding values of each variables. Finally the function is obtained by a boundary-line analysis of the plot. Leaf-level measurements should be performed in largely different conditions (T, PAR, VPD and SWC) in order to define the most representative model.

The Jarvis-Stewart approach is widely used especially in the UN/ECE (United Nations-Economic Commission for Europe) scientific community in order to calculate the ozone stomatal dose absorbed by plants. Recently the ICP Vegetation (The International

Cooperative Programme on Effects of Air Pollution on Natural Vegetation and Crops) proposed some modifications of the original Jarvis-Stewart model introducing new standard limiting functions (ICP Vegetation, 2004)
First of all, leaf stomatal conductance is modelled as it follows:

$$g_s = g_{\max} \cdot [\min(f_{O_3}, f_{PHEN})] \cdot f_{light} \cdot \max[f_{\min} \cdot (f_T \cdot f_{VPD} \cdot f_{SWC})] \quad (3)$$

where g_{max} is the maximum stomatal conductance of the studied specie and f_{O3}, f_{PHEN}, f_T, f_{light}, f_{VPD} and f_{SWC} are the limiting functions depending on ozone concentration, phenology, temperature, PAR, VPD and soil water content respectively. Function f_{min} represents a threshold value for the functions of the temperature, VDP and SWP. The part of the Eq. 3 related to f_{PHEN} and f_{O3} is a most limiting factor approach. Either senescence due to normal ageing is limiting or the premature senescence induced by ozone is limiting. Early in the growing season and at low ozone exposure f_{PHEN} is always limiting and f_{O3} then does not come into operation (ICP Modelling and mapping, 2004). Furthermore, it is worth to be noticed that the limiting function f_{O3} has been calculated only for potato and wheat.

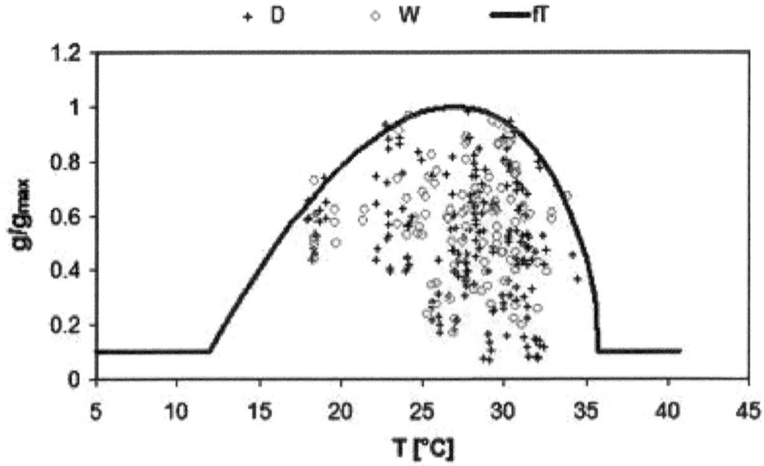

Fig. 1. Limiting function f_T for the stomatal conductance model of poplar. g/gmax is the relative stomatal conductance to water vapour while T is the temperature. D are measurements on non-watered plants while W are measurements on watered plants (modified from Gerosa et al., 2008).

As an example, the limiting function for the temperature adopted by Gerosa et al. 2008 in an Open-Top Chambers experiment is shown in Figure 1. The f_T function is described by:

$$f_T = \frac{(T - T_{\min})}{(T_{opt} - T_{\min})} \cdot \left[\frac{(T_{\max} - T)}{(T_{\max} - T_{opt})} \right]^b \quad (4)$$

where T represents the temperature of the air near the leaf; T_{opt} (27 °C), T_{max} (36 °C), T_{min} (11 °C) are 3 species-specific parameters, respectively representing optimal, maximum and minimum temperatures and b (adimensional) is defined as:

$$b = \frac{(T_{\max} - T_{opt})}{(T_{opt} - T_{\min})} \tag{5}$$

If T is greater than T_{max} or smaller than T_{min}, the function f_T is given the minimum value of 0.1.

In a similar way it is possible to obtain the other limiting functions and hence to estimate the stomatal conductance to water and the stomatal conductance to ozone by multiplying the former by 1.68, which represents the ratio between the diffusivity of water in air and the diffusivity of ozone in air.

The ozone stomatal dose is finally obtained as the integral of F_{ST} over the measuring period

$$D_{ab} = \int_{t_a}^{t_b} F_{ST}(t)dt$$

Further details about the intermediate passages and results can be found in Gerosa et al. (2008).

Leaf level measurements of stomatal conductance are a useful tool for studying physiological behaviour of vegetation both in natural and controlled environment. On the other hand, the importance of studying vegetation behaviour at ecosystem level is fundamental for many research fields such as climatology, hydrology and air quality. Hence, upscaling forest ecological processes and their link with canopy structure has puzzled researchers from different fields (Ewers *et al.*, 2007; De Pury & Farquhar, 1997; Cescatti & Zorer, 2003; Martens *et al.*, 2000; Law *et al.*, 2001). The development of a reliable upscaling method would permit the evaluation of important ecosystem functions such as: gross primary productivity (GPP), evapotranspiration, stomatal uptake of air pollutants, biogenic volatile organic compounds (BVOCs) emissions.

Upscaling conceptually consists of the integration of the physiological characteristics (for instance stomatal conductance or transpiration) of all the elements in which the canopy is divided. (Van de Zande et al., 2009). These physiological characteristics are controlled by the environmental factors surrounding that element and light is usually the most varying factor in space and time. The concept of upscaling is fundamental for the bottom-up approach and two main problems can be identified:

- the description of the light environment inside the canopy
- how physiological characteristics scale with that light environment.

Knowing that the light environment inside a canopy is the resultant of the interaction of the canopy structure with the incoming light, a full description of the light environment requires high detailed knowledge of the canopy structure. Canopy structure can be defined as the temporal and spatial organization of the above ground vegetation components including their position, extent, quantity, type, orientation, shape, and connectivity (Parker, 1995).

One of the main assumptions about light penetration is that the canopy is often considered as a turbid medium in order to simplify the light-vegetation interaction. In this way it is possible to apply the Lambert-Beer's law (Monsi and Saeki, 1953) to describe radiation attenuation in the canopy. In a canopy, the average (temporal and spatial) distribution of light shows a profile which decline with the height. This behaviour is well modelled by the Lambert-Beer's law. On the contrary instantaneous profiles are not well modelled.

Furthermore, the leaf angle is fundamental in determining the absorbed irradiance in canopies: top-canopy leaves could absorb less light from direct irradiation than leaves inside the canopy (De Pury and Farquhar, 1997).

Different upscaling approaches have been used in literature: the big-leaf model, two leaves model and multilayer model.

The widely used big-leaf approach considers the whole canopy equivalent to a single big-leaf. This approach allows a reliable upscaling, but only after parameterization over reference data. In fact one of the main disadvantages of this approach is that the model for different levels of irradiance through the canopy is solved by applying semi-empirical corrections for the canopy photosynthetic capacity. Furthermore these corrections may depend on environmental conditions or crown architecture, thus reducing the application of the model in different conditions. Despite this limitations, big-leaf models have been shown to well describe canopy photosynthesis (Amthor et al. 1994; Lloyd *et al.*1995) and thanks to their limited number of input, big-leaf models were employed for large scale modelling.

Norman (1993) and Raupach (1995) pointed out that big-leaf models do not apply a real scaling up procedure: the variables involved at leaf level have different relationships among them at canopy level and hence canopy dynamics may be misinterpreted in terms of the leaf level behaviour.

Interesting results were obtained by De Pury and Farquhar (1997): they proposed to split the big-leaf into sunlit and shade fractions, which are subject to changing during the day. Canopy irradiance absorption and photosynthetic capacity change during the day because of the variation of sunlit and shade leaves fractions. The significant improvement of this model is the simultaneous use of five different features that were already used separately in models. These features are: distinction between sun and shade leaves; change of photosynthetic capacity through the canopy; use of photosynthetic biochemical modelling; a temporal dependant partitioning between sun and shade leaves; a single layer model with relatively simple calculations. But probably, the most important feature of this model is its scaling approach that allows a direct scaling of parameters from leaf to canopy scale.

Before the introduction of the sun/shade model by De Pury and Farquhar (1997), multi-layer models were already used (e.g. Norman, 1982). In these models the canopy is split into multiple layers and each layer is usually separated into sunlit and shaded fractions, and the sunlit fraction is divided into different leaf angle classes. Absorbed irradiance is calculated for the shaded fraction and for each leaf class of the sunlit fractions separately. Vertical profile of leaf nitrogen and of photosynthetic capacity are well modelled by this approach but the number of required input is very large. De Pury and Farquhar (1997) compared first a big-leaf model and a multi-layer model with measurements run at Wagga Wagga, Australia (de Pury, 1995). Despite the previous mentioned fine tuning of the big-leaf model better results were obtained from the multi-layer model. The most significant result by De Pury and Farquhar (1997) was the good agreement between their sun/shade model and the multi-layer models and measurements.

2.2 Plant level measurements

Sap is the fluid flowing through the xylem of plants, and it is constituted by water and nutrients brought up from the roots through the plant and finally to the leaves from where most of it is evaporated. The flow of sap is, hence, more or less equivalent to the whole plant transpiration. The water transport through the xylem is not realized by energy spent by tracheary elements, which are dead, but by a passive mechanism due to the difference in water potential between the leaves and the soil.

The most important cause of xylem sap flow is the evaporation of water from the surfaces of mesophyll cells to the atmosphere. Liquid water reaches the transpiring surface through nanopores in the mesophyll cell walls and, when transpiration occurs, water forms menisci in each of the nanopores with a curvature depending mostly on the transpiration rate. Transpirational pull results from the evaporation of water from the surfaces of cells in the leaves. This evaporation causes the surface of the water to recess into the pores of the cell wall. By capillary action, the water forms concave menisci inside the pores. The high surface tension of water pulls the concavity outwards, generating enough force to lift water as high as a hundred meters from ground level to the highest branches of trees.

The resulting surface tension causes a negative pressure (or tension) in the xylem that pulls the water from the roots and soil. Sap flow hence occurs when the pressure in the leaves is lower than that of the soil. (Tyree, 1997).

In some circumstances (dry soil and fog for example), leaf water potential can be less negative than that of roots, and in this cases e reverse flow can occur (Burgess et al, 2004). If the water potential of the root cells is more negative than that of the soil, usually due to high concentrations of solute, water can still move by osmosis into the root from the soil. This causes a positive pressure that forces sap up the xylem towards the leaves. This phenomena, guttation, however is not usually observed in woody species.

Transpirational pull requires that the vessels transporting the water are very small in diameter, otherwise cavitation would break the water column. When the water pressure within the xylem reaches an extreme tension due to low water input from the roots (if, for example, the soil is dry and transpiration is high), gaseous species, dissolved in the xylem abandon the liquid, or are sucked in from near by tissues, forming an air bubble - an embolism forms and the vessel is no more functional. Although the mechanism has not been cleared yet, embolism can be cured by the plant possibly by an active mechanism of companion cells.

Methods for sap flow measurements, based on thermodynamic principles, are widely used in plant physiology and hydrology studies. They have many advantages compared to other techniques and they are easy to apply for long-term automated monitoring of water movement in plants. Detailed reviews of the most frequently applied methods (heat pulse velocity, HPV; tissue heat balance, THB; stem heat balance, SHB; heat dissipation, HD) have been presented in many papers (Pickard & Puccia, 1972; Pickard, 1973; Swanson & Whitehead, 1981; Swanson, 1983, 1994; Jones et al., 1988; Valancogne & Nasr, 1989; Campbell 1991; Groot & King, 1992; Barret et al., 1995; Grime et al., 1995; Edwards et al., 1996; Smith & Allen, 1996, Braun, 1997; Kostner et al., 1998; Wullschleger et al., 1998; Steppe et al., 2010). The theoretical background for heat flow in sapwood has been developed by Marshall (1958) , 26 years after convected heat pulses were first applied as indicators of sap velocity in trees (Huber, 1932). Nevertheless, empirical works within that period (Baumgartner, 1934; Dixon, 1936; Kuniya, 1950) were important for understanding the complex physical processes of the heat transfer in living wood.

Essentially two big groups can be distinguished among thermal methods: those applying pulse heating and those based on continuous heating.

Measurements of sap flow, based on heat transfer methods, display two different arrangements of differential thermocouples around the heater: a symmetrical one, with the ends of the thermocouple placed at equal distances below and above the heater along the axial direction (e.g. Daum 1967; Vieweg & Ziegler, 1960; Sakuratani, 1981; Nadezhdina et al., 1988) and an asymmetrical one with the upper end of the thermocouples placed at the same axial height of the heater and the lower (reference) end of the thermocouple placed at a certain distance below the heater (e.g. Ittner 1968; Cermak et al.1973; Granier 1985).

The symmetrical arrangement of thermocouples was first applied as a modification of the heat pulse method, termed the heat ratio method, HRM. The background theory for HRM was suggested by Marshall (1958). Marshall derived an equation for the heat pulse velocity, V_h, as the ratio of the increase in temperature after releasing a heat pulse at points equidistant downstream and upstream from the heater:

$$V_h = \frac{k_{st} \times \ln(T_1/T_2)}{\rho_{st} \times cp_{st} \times X} \tag{6}$$

or

$$V_h = \frac{D_{st}\,\mathbf{ln}(T_1/T_2)}{X} \tag{7}$$

where, k_{st}, ρ_{st}, cp_{st} and D_{st} are thermal conductivity, density, specific heat capacity and thermal diffusivity of green stem wood; T_1 and T_2 are increases in temperatures at equidistant points X cm above and below from the heater. This arrangement of thermocouples allows the HRM method to measure low and reverse flow rates.

Vieweg and Ziegler (1960) were the first who applied continuous heating for sap flow measurements. They were also the first who spoke of a deformation of the heat field ("cloud" in their terminology) caused by the moving of sap. Using a series of thermometers, placed in the axial direction above and below the heater, they showed that when the sap velocity was zero, the heat field around the heater was symmetrical, while when the sap was moving, the heat field becomes asymmetrical. No equations were given by the authors approving the method quantitatively, but experimentally it was found that the system is very sensitive and that it can measure sap velocities down to 3 cm h^{-1} in large larch and birch trees. Several decades later the infrared technology introduced by Anfodillo *et al.* (1992, 1993) and applied in further sap flow studies (Granier *et al.*, 1994; Nadezhdina et al., 2004) allowed to "see" *in situ* the deformation of heat field in live trees.

Vieweg and Ziegler (1960) suggested to take the extent of asymmetry of the heat field (and thus of the temperature differences in the xylem) as a measure of sap velocity. Ten years later, Saddler and Pitman (1970), revised the method of Vieweg and Ziegler (1960). Their novel contribution was to develop an analytical solution. The authors modified the equation of heat conduction in a woody stem solved by Marshall (1958) for the case of an instantaneous line source of heat released into an infinite block of wood substance. The solution of the authors was derived for steady-state conditions in an infinite uniform cylinder with uniform continuous heat supply within a cross-section. Their theoretical considerations showed that the logarithm of the ratio of the temperature difference (T-T_a) in the "downstream" position, T_+ to that in the "upstream" position, T_- gives a measure of the sap flux, u, (named by them as sap speed index, SSI) by the equation:

$$u = \frac{k_{st} \times \ln(T_+ / T_-)}{\rho_{st} \times cp_{st} \times X} \quad (8)$$

where ρ_t and cp_{st} are density and specific heat of sap, respectively, k_{st} is the thermal conductivity of the stem and X is the axial distance between the heater and the thermocouples. In addition, the authors also mentioned that the values ρ_{st} and c_{st} could be taken as those for water without appreciable error because the composition of sap would be sufficiently close to pure water.

When comparing both approaches, suggested by Marshall (1958) for heat pulse (HRM) and by Saddler and Pitman (1970) for steady-state heat flow, it could be observed that the derived Eq. 6 and Eq.8 are very similar. The main difference relies on the measured temperature variables around the heater source. Such similarity is explained by the use of a "compensation" approach for heat movement through conduction by introducing equidistant points for temperature measurements from a heat source in both cases.

The majority of methods that apply symmetrical arrangements of thermocouples around a heater, have very high sensitivity and good relation with the true sap flow under low flow rates (Vieweg and Ziegler, 1960; Saddler and Pitman, 1970; Nadezhdina, 1999). However, the situation is different under high flow rates. The output from a symmetrical sensor gives a diurnal variation with two peaks, one in the morning and one in the afternoon, and a drop around midday (e.g., Ittner, 1968 in spruce, Nadezhdina, 1988, 1999 in apple).

The temperature difference between the heater (or the needles, spaced in the middle of heated elements) and the reference needles, which measure the sapwood ambient temperature below the heater, is used for the sap flow calculation in both tissue heat balance, THB, and heat dissipation, HD, methods. This temperature difference is termed as asymmetrical, dT_{as}.

The THB method was developed for sap flow measurements in large trees in the late 1970th (Čermák et al., 1973; Kučera et al., 1977) and it is commercially available and rather widely used in field experiments.

For trees with stem diameter over 70 mm Granier (1985) described the continuously heated probe technique based on empirical calibration (HD technique). Thanks to its simplicity and robustness, this technique became one of the most used for sap flow measurements in the field.

However, it is problematic to measure extremely low flow rates using these methods. There are also problems to distinguish accurately between zero flow and fictitious flow (or a similar variable corresponding to the heat losses from the segment), which must be subtracted from the recorded data in order to get the true sap flow (Burgess et al., 2000). Do & Rocheteau (2002a, 2002b) underlined, that for the HD-method, zero flow must be assumed during certain nights and further external information (e.g., the natural defoliation) should be used to test the assumption. Both the THB and the HD methods cannot measure bi-directional flows, although possibility to use the modified thermal dissipation technique for bi-directional measurements in roots was reported more recently by Brooks *et al.* (2002).

In spite of the fact that numerous works improving the theory and instrumentation for thermodynamic methods have been done since the first application of heat as an indicator for sap movement in trees, there is still a lot to do in order to decrease limitations for the methods and instrument application and to increase their accuracy. Some of the problems cannot be avoided by nature (e.g. circumferential and radial variation of flow in trees), others we have to cope with (size of trees, wound effect, sensitivity under low rates, bi-lateral flow records, high time resolution, probe spacing, natural temperature gradients, variable water content, etc.).

In 1991, while working with the "sap flow index", Nadezhdina focused on a new method based on the heat field deformation (HFD) (Nadezhdina, 1989; 1992; 1999). This method is capable to measure over a wide range of flow rates without loosing the advantages of other methods based on a continuous heating, i.e. high time resolution. Since 1996, the method was further developed (Nadezhdina & Čermák, 2000; Nadezhdina *et al.*, 1998) and it was intensively tested in different species (up to 50) and different environmental conditions. The fundamental idea was to combine the advantages of the earlier methods based on the measurements of asymmetrical and symmetrical temperature gradients, dT_{sym} and dT_{as}, around a heat source and, hence, to avoid the related limitations of the methods, when applying each of them separately. Data of temperature gradients, measured simultaneously by differential thermocouples at different distances around a heater (Nadezhdina, 1998) or calculated from the absolute temperature grid around the heater (Nadezhdina et al., 1998), were used to characterize the shape of the heat field around it. Then measured temperature gradients were compared against SFI (or dT_{sym}) and sap velocity, measured by HPV method, and theoretical heat field shapes corresponding to different flow rates were reconstructed (Nadezhdina, 1998). This way, also the non-linear relationship between SFI and sap velocity as well as the sensitivity of SFI for measurements of very low rates, including zero and reverse flow, were explained.

It is assumed that the deformation of the heat field is caused by the real sap flow rate (and not by the heat velocity), in fact, heat dissipation is due to the flux and not to the velocity of the cooling medium.

The sap flow density (Q) is then scaled by sapwood / leaf area ratio (LA/SA) resulting equal to transpiration per unit leaf area, G_s can finally be derived from sap flow measurements, based on a simplification of the Penman-Monteith equation (Whitehead and Jarvis,1981; Pataki *et al.*, 1998; Martinez-Vilalta *et al.*, 2003):

$$Gs = \frac{\gamma \times \lambda \times Q_l}{\rho \times cp \times VPD} \tag{9}$$

where γ is the psychrometric constant (kPa K^{-1}), λ is the latent heat of vaporization of water (J kg^{-1}), ρ is the density of air (kg m^{-3}), cp is the specific heat of air at constant pressure (J kg^{-1} K^{-1}) and VPD=[esat(T)-e(T)] is the vapour pressure deficit (kPa) of the air. The simplification can be considered valid if G_s is predominant over the leaf boundary layer conductance, g_b (Whitehead and Jarvis 1981) i.e. when the canopy is strongly coupled with the atmosphere.

The ozone stomatal dose can be obtained as the integral of the product between Gs and the ozone concentration at canopy level (C_C):

$$D_{ab} = \int_{t_a}^{t_b} G_{ST}(t) C_C(t) dt \tag{10}$$

2.3 Ecosystem level measurements

Eddy covariance (EC) technique is the most direct technique, among the ones which have been developed through the last decades. in order to measure and calculate fluxes of a scalar over a surface. This technique was applied for the first time by Swinbank (1951) and only recently was largely used and it is considered as a reference method because of the absence of assumptions for its applicability.

This technique is based on the simultaneous measurement of the fluctuations both of the vertical component of the wind (w) and of the considered scalar whose the fluxes are measured, i.e. specific humidity for evapotranspiration and latent heat fluxes, air temperature for sensible heat fluxes, ozone concentration for ozone fluxes (Fowler & Duyzer, 1989). Vertical fluxes are obtained as the average of the product between the fluctuations of w and the fluctuations of the studied variable from their average values, i.e. their covariance.

EC technique requires fast-response instrumentation in order to measure the vertical component of the wind (ultrasonic anemometer) and the studied scalars (hygrometer and fast response analyzers) (Hicks et al., 1989). These instruments must have a response time less than 0.1 s in order to be able to measure all the spectrum of the vortexes which drive the flux and hence avoid the so called flux-loss and the subsequent empirical corrections (Moncrieff et al., 1997; Garratt, 1975). EC was first adopted only for short field campaigns (1/2weeks) but since some years it has been widely employed in permanent monitoring field stations.

At ecosystem level, stomatal conductance, and hence ozone stomatal fluxes, the dose taken up by the ecosystem, cannot be measured directly, but only calculated by means of the so called SVAT (Soil-Vegetation-Atmosphere-Transfer models) models. Even if there are differences in the parameterizations adopted by these models, they are all based on an electric analogy in the flux description (Monteith and Unsworth, 1990; Garratt, 1994). Ozone flux is equivalent to a current intensity, ozone concentration difference is equivalent to a voltage difference and the resistance to the ozone diffusion through the atmosphere are parameterized by means of micrometeorological variables. This analogy was first used by Chadwick and Chamberlan in 1953.

One of the models that uses this analogy is the big-leaf model. It must be noticed that the big-leaf model described in this section is used with different purposes: in the section about leaf level measurements big-leaf model is used in a prognostic way, that is from leaf level measurement the model foresees the whole canopy behaviour (a bottom-up approach), while hereafter the big-leaf model is used in a diagnostic way, that is stomatal conductance is inferred from measurements of quantities referred to the whole canopy (a top-down approach). In this case, the big-leaf model states that, concerning vertical exchange of energy and matter, the whole canopy of an ecosystem can be assumed to be equivalent to a big-leaf at the displacement height $z=d+z_{0X}$ above the soil, where d is the zero plane height ($\approx 2/3$ of canopy height) and z_{0X} is the roughness length for the X (momentum, heat, concentration of a gas) scalar.

The displacement height can change if fluxes of momentum or sensible heat or latent heat or matter are considered: in fact $d+z_0$ is the height where momentum become null inside the canopy, which usually differs from the height from where there is the source for sensible and latent heat or from the height where ozone concentration is null ($d+z_{0O3}$). Because of the similarity of the diffusion processes in Atmospheric Surface Layer (Monin and Obukhov, 1948) $d+z_{0O3}$ is usually set equal to $d+z_{0H}$ and simply indicated as $d+z_{0'}$. Furthermore, above the surface there is a thin sub-laminar layer. where the transport is no longer due to turbulence.

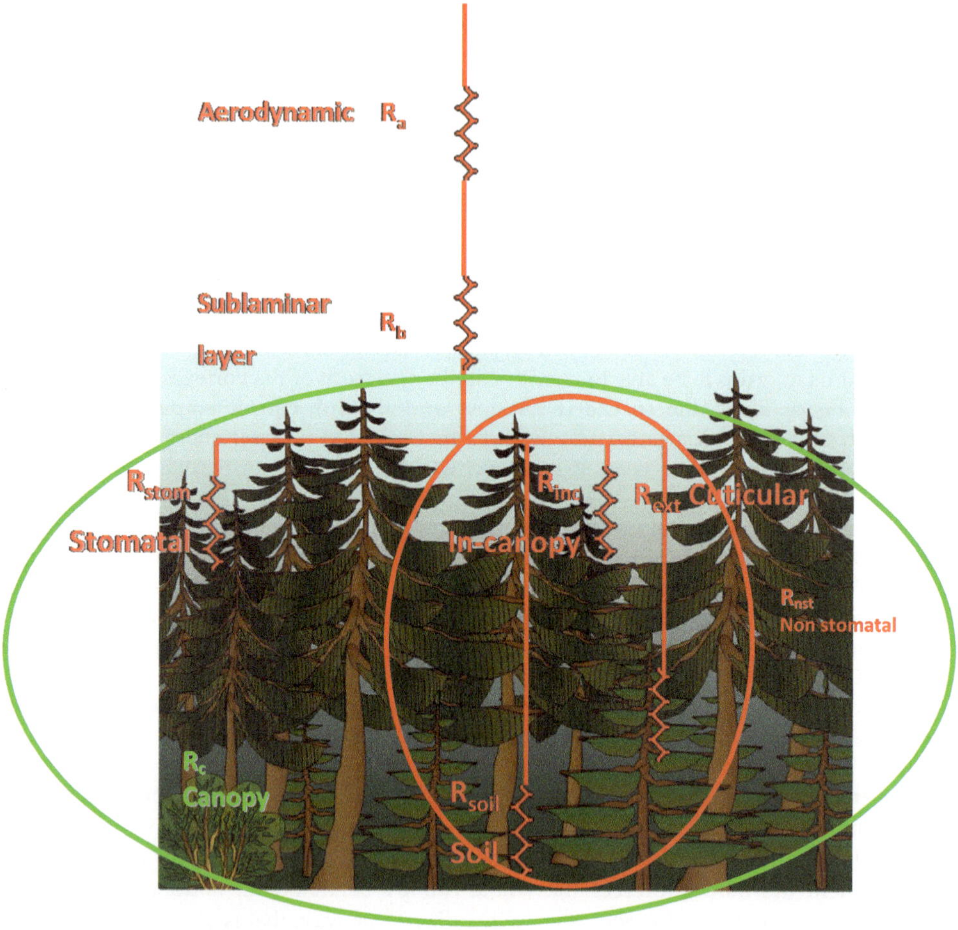

Fig. 2. Resistive scheme adopted: the aerodynamic resistance R_a, the sublaminar layer resistance R_b and the canopy resistance R_c. R_c is decomposed into two resistances (included in the green oval line): the stomatal resistance R_{stom} and the non stomatal resistance R_{nst}. The in-canopy resistance R_{inc}, the soil resistance R_{soil} and the cuticular resistance R_{ext} are some of the possible pathways in which the non stomatal deposition can be decomposed. In the big-leaf model described here these three resistances are not considered.

The total resistance to the ozone diffusion is equal to three series resistances:

$$R_{tot}(z) = R_a(d + z_0, z) + R_b + R_c \tag{11}$$

where R_a and R_b are atmospheric resistances and R_c is the integrated resistance of the exchange surface.

$R_a(d+z_0, z)$ is the aerodynamic resistance that ozone (but also any other scalar quantity) faces during the turbulent transport from the height z to the height $d+z_0$ (momentum sink). This resistance mirrors wind behaviour and thermodynamic characteristics of the atmosphere such as the stability. Dyer (1974) proposed for R_a the following formulation:

$$R_a = \int_{d+z_0}^{z_m} \frac{\Phi_H(\zeta)}{ku^* z} dz \tag{12}$$

where: $\Phi_H(\zeta) = [1+5\zeta]$ if $\zeta = (z-d)/L > 0$

$$\Phi_H(\zeta) = [1-16\zeta]^{-1/2} \text{ if } \zeta \le 0$$

k is the von Karman constant, u* is the friction velocity and is the Monin-Obukhov length. R_b is the resistance faced by ozone through the sub-laminar layer and it depends on ozone molecular diffusivity in the air and it is specific for the considered gas. Hicks et al. (1987) suggested for R_b:

$$R_b = \frac{2}{k\,u^*}(Sc/\Pr)^{2/3} \tag{13}$$

where k is the von Karman constant, u^* is the friction velocity, Sc is the Schmidt number for the ozone (≅1.07) and Pr is the Prandtl number for the ozone (≅ 0.72).
R_c represents the surface or canopy resistance to ozone deposition and it includes all the processes linked to the influence of the system plant-soil on vertical exchange of the gas (e.g. stomatal uptake, deposition on non transpiring surfaces).
R_c strongly depends on the physiological characteristic of the vegetation (e.g. stomatal conductance), on the phenological characteristics of the vegetation (e.g. crown architecture, LAI) and on the underlying type of soil. R_c values are one or even two orders of magnitude greater than R_a and R_b, highlighting the great importance of the role played by vegetation in the deposition process. The relatively less importance of the two other resistances explains why even using different parameterizations of R_a and R_b, some authors obtains similar results. The equivalent total resistance can be obtained as the ratio between the total ozone fluxes and the ozone concentrations at the measuring point. This latter term is as a matter of fact the difference between the ozone concentration at the measuring point and the ozone concentration in the substomatal cavity, but this is usually assumed null as proposed by Laisk (1989). R_c can be finally obtained as the residual between R_{tot} and R_a plus R_b.
Rc is hence considered equivalent to two parallel resistances:

$$R_c^{-1} = R_{ST}^{-1} + R_{NS}^{-1} \tag{14}$$

where R_{ST} is the stomatal resistance and R_{NS} which includes all the pathways of destruction of ozone such as deposition on non transpiring surfaces, on the soil or by means of chemical reactions. All these pathways are collectively named non-stomatal resistance (Cieslik, 2004). R_{ST} could be deduced from the Penman-Monteith equation (Monteith, 1981), where the latent heat flux (that is proportional to the evapotranspiration). The Penman-Monteith approach attributes the whole evaporative process to the stomatal activity and thus it is effective only if E from the soil is negligible. This condition is satisfied if the soil surface is dry or the canopy of the studied ecosystem is completely close (Biftu & Gan, 2000). In any case Choudhury e Monteith (1988) showed that evaporation from the soil is always less than 5% if the ecosystem LAI is greater than 1. R_S is obtained as:

$$\lambda E = \frac{(R_n - G)\cdot\frac{\partial e_s(T)}{\partial T} + \frac{\rho c_p[e_s(T)-e]}{R_a + R_b}}{\frac{\partial e_s(T)}{\partial T} + \gamma\left(1 + \frac{R_S}{R_a + R_b}\right)} \tag{15}$$

where R_n is the net radiation flux, G is the soil heat flux (both are expressed in W m^{-2}); c_p is the air specific heat capacity (J K^{-1} Kg^{-1}) and c_p is air density (kg m^{-3}); e and $e_s(T)$ are respectively water vapour pressure and water vapour pressure at saturation (both Pa); γ is the psychrometric constant (equal to 67 Pa K^{-1}). The terms λE, R_n, G, T of this equation are usually measured in eddy-covariance field campaigns and R_a and R_b are obtained from Eq. 12 and Eq. 13 using other micrometeorological data (e.g. u*). Saturation water vapour pressure can be obtained by Murray (1967) equation:

$$e_s(T) = 0{,}611\cdot\exp[17{,}269\cdot(T-273)/(T-36)] \tag{16}$$

where T is the air temperature in K. Water vapour pressure can be obtained applying this equation:

$$e = \frac{q\cdot P}{(q - 0.622\cdot q + 0.622)} \tag{17}$$

where q is the specific humidity and P is the atmospheric pressure.
Since all the other terms of the Eq. 15 are known it is possible to solve this equation in order to obtain R_S, the stomatal resistance (the reciprocal of the stomatal conductance) of water. Since in the sub-stomatal cavity the transport of the molecules is only diffusional, the stomatal conductance of the ozone can be calculated taking into account the following equation:

$$\frac{R_{ST}}{R_S} = \frac{D_{H_2O}}{D_{O_3}} = 1.65 \tag{18}$$

where R_{ST} is the stomatal resistance of ozone in the air, R_S is the stomatal resistance of water in the air, D_{H2O} is the diffusive coefficient of water in the air and D_{O3} is the diffusive coefficient of ozone in the air.
Using the Ohm's laws the stomatal fluxes is given by:

$$F_{ST} = \frac{C_c}{R_{ST}} = \frac{R_c}{(R_a + R_b + R_c)R_{ST}}C_m \tag{19}$$

where C_C is the ozone concentration at canopy level.
The stomatal dose is simply given by the integral of F_{ST} over the measuring period.

$$D_{ab} = \int_{t_a}^{t_b} F_{ST}(t)dt \tag{20}$$

3. Applications

Some significant applications at the different scales (leaf, plant and ecosystem level) and their main results will be showed here. A more detailed study on the Mediterranean vegetation will be presented. In fact, Mediterranean vegetation is often exposed to co-occurring prolonged drought and high ozone concentrations, the effects of drought on stomatal conductance and hence on ozone uptake will be showed.

Karlsson et al. (2000, 2004) developed a model to estimate the stomatal conductance and ozone flux of Norway spruce saplings in an open-top chamber experiment. A Jarvisian model was parameterized against needle conductance measurements, obtained by means of a portable porometer. These measurements were made on 4–6-year-old spruce saplings, grown in open-top chambers, in July–September during three different seasons. The spruce saplings were either maintained well watered or subject to a 7–8 week drought period in July–September each year. The modeled stomatal conductance showed a good agreement with the measured stomatal conductance for the well-watered as well as the drought stress-treated saplings, however significantly improvements, applying different VPD functions, were observed. The cumulated ozone uptake showed the occurrence of long-term 'memory-effects' from the drought stress treatments on the stomatal conductance. In this paper it was highlighted that these memory-effects should to be considered when simulation models for stomatal conductance are applied to long-lived forest trees under a multiple stress situation. In details, it was observed that the reduction of the total biomass was to 1% per 10 mmol m^{-2} of ozone uptake, on a projected needle area basis.

Plejel et al. (2000) reviewed six years of open top chambers experiments on wheat obtaining a specific parameterization of a Jarvisian model. The stomatal model was sensitive to light, temperature and VPD. The relative importance of these climatic factors on stomatal conductance varied considerably between years. VPD was a very important factor limiting daytime ozone uptake in warm and dry summers, while temperature was most important in cool and wet summers. The very strong potential modifying effect by air humidity on ozone effects was described already by McLaughlin & Taylor (1981). Plejel et al. found out that, although ozone concentrations were lower than in south and central Europe, the potential ozone uptake from a given concentration was higher. In fact the long photo-period and the mean values of temperature and humidity lead to relatively great values of stomatal conductance, allowing the vegetation to take up a lot of ozone through the stomatal pathway.

In a two years open-top chambers experiment Gerosa et al. (2009a) realized a reliable parameterization of a stomatal conductance model for four forest species (*Populus nigra L., F. excelsior L., Fagus sylvatica L., Quercus robur L.*) , under the typical meteo-climatic conditions of the Southern Alpine region. The experiment was focused on the determination of a critical level of O_3 uptake for leaf visible injury onset, in different ozone and water availability conditions. It was noticed that seedlings began to suffer the effect of water deficiency as soon as the soil moisture fell just below the field capacity, due both to the soil type and to the incomplete development of their root system. In fact, the correlation between soil water content and stomatal conductance is not linear, because its reduction rate increases as the water shortage persists. This experiment confirmed the influence of an important environmental factor such as soil water availability on plant response to ozone stress.

Sapflow measurements in tree trunks represent a useful link between the leaf and forest canopy level integrating over the entire tree crown and separating tree water flux from other evaporating components in ecosystems (Kostner et al, 2008). However in the same paper the authors highlight some limitations of the sapflow approach: stomatal conductance derived trunk sapflow may not include higher short term conductances of individual branches in the crown; furthermore, since the foliage transpiration, which starts at sunrise and usually ends at sunset, might be not synchronized with sapflow at breast height, because of the trunk storage capacitance, it was underlined the importance of a proper synchronization between them in order to realize a correct ozone risk assessment. The ozone flux data showed in this paper were in good agreement with the ones presented by Wieser et al. (2000) from gas exchange measurements, proving the goodness of the sapflow approach. It is worth to notice that Kostner et al. (2000) showed the feasibility of the sapflow approach also for measuring NOx, NH_3 and CO_2 fluxes.

During the exceptionally hot 2003 and the following year, Gerosa et al. (2009b) measured ozone fluxes over a coastal *Q.Ilex* forest. From August to October, in 2003 stomatal conductance at canopy level (Gs) was half or less than the following year in the same months. This low values in 2003 can be partially explained, only for August values, with higher temperature and lower water availability. In fact, even after strong rainfalls in September Gs remained low. The authors suggested as a possible explanation that, after a prolonged drought, damages modify the whole plant hydraulic conductivity. As a consequence ozone uptake was nearly double in 2004.

In 2007 a field campaign measuring ozone and energy fluxes over a Mediterranean Maquis ecosystem was run by Gerosa et al. (2009c). Here, additional and different analyses of the data from that field campaign are showed. The field campaign lasted nearly three months from the beginning of May until the end of July. This period corresponds to two different seasonal stages: late spring (until the half of June) and early summer (from the half of June until the end of July). The measuring site was located inside the Castelporziano estate (N 41° 40' 49.3'', E 12° 23' 30.6''), 20 km far from the center of Rome (Italy). The studied ecosystem is composed by two different succession stages of the maquis (low and medium maquis). Ninety percent of the ground is covered by 6 main species: *Quercus ilex, Arbutus unedo, Rosmarinus officinalis, Cistus spl, Phyllirea latifolia, Erica multiflora.* Most of the vegetation height was between 90 cm and 150 cm, with the exception of some higher *Quercus ilex* and *Arbutus unedo* individuals. The eddy covariance technique was used to measure ozone and energy fluxes and a big-leaf approach (Wesely and Hicks, 2000; Gerosa et al., 2003; 2004; 2005) was used to estimate the ozone dose which was taken up by the ecosystem.

The eddy covariance instrumentation (an ultrasonic anemometer (USA-1, Metek, Elmshorn, Germany), a CO_2/H_2O open path fast sensor (LI-7500, LI-COR, Lincoln, Neb., USA) and a fast ozone analyser (COFA, Ecometrics, Italy) were mounted on scaffold and the sampling or measuring point of these instruments was at 3.8 m height. An additional O_3 analyzer (SIR S-5014, DASIBI e, Spain) was used as a reference; the sampling point of the latter instrument was at the same height of the former ones. Additional probes were used to measure net radiation (NR lite, Kipp&Zonen, Holland), photosyntetically active radiation (190SA, LI-COR, Lincoln, Neb., USA), soil water content (C616, Campbell Scientific, Shepshed, United Kingdom), soil heat flux (HFP01SC, Hukseflux, Delft, Holland), leaf temperature (Pt100, DeltaT, United Kingdom), leaf wetness (237, Campbell Scientific, United Kingdom). A temperature and humidity profile (0.1 m, 1 m, 3.8 m) was equipped with three probes (50Y, Campbell Scientific, Shepshed, United Kingdom).

Eddy covariance data were recorded on a computer while for all the other measurements a CR10x datalogger (Campbell Scientific, Shepshed, United Kingdom) was used. Further details about other additional measurements (NO and NO_2 fluxes, which were measured using the aerodynamic gradient technique) and the data selection and gap-filling can be found in Gerosa et al. (2009c).

The measuring site was characterized by a Mediterranean climate with a strong and prolonged drought, very few rainfall events. A wind breeze regime with wind coming from the inland by night and from the sea during the daylight was observed (Fares et al., 2009). The average temperature in the late spring was 24.0 °C, while in the second part of the field campaign was 28.8 °C. Rainfall was low during the field campaign: 14.5 mm in the first period and nearly null (0.2 mm in the second period) (Fig. 3). Due to the sea breeze regime and the closeness to the sea relative humidity was nearly always above 60% during the day and close to the saturation during the night (Mereu et al., 2009). For this reason, dew formation on leaves happened almost all the nights and, in order to avoid that dew evaporation might be considered as ecosystem transpiration the approach of Gerosa et al. (2009c), considering only data with dry canopy for the resistance analysis and the estimate of the stomatal ozone fluxes, is applied here too.

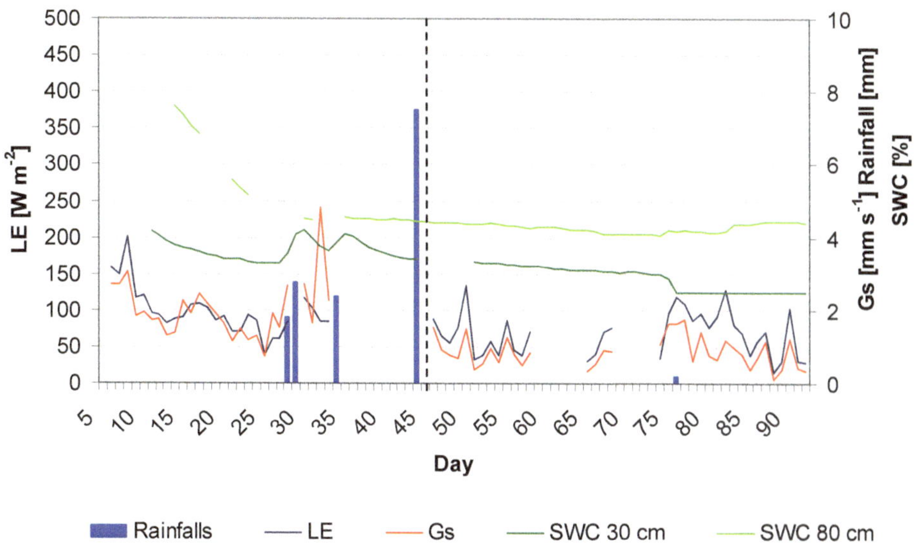

Fig. 3. Daily averages of latent heat flux, stomatal conductance, rain and soil water content (v/v) during the whole measuring period. Soil water was measured at two different levels: 30 cm, 80 cm. Vertical dashed line is the separation between the two periods.

The average ozone concentration on the whole period was 35.9 ppb, lower in the first period (32.6 ppb) and higher in the second (38.6 ppb) (Gerosa et al., 2009c). Ozone concentration showed the typical bell-shaped daily course: from very low concentrations (less than 10 ppb) ozone concentration increased until early in the afternoon reaching its maximum. After that it decreased until sunset, reaching low values for all the night. This behaviour was observed nearly all the days with the exception of few nights when ozone concentration was above 30 ppb (Figure 4).

The total ozone fluxes showed a great daily variability, doubling or halving the daily maximum in one or two days (Figure 4). This great variability is typical of measurements at ecosystem level where turbulence regulates gas exchange between atmosphere and the ecosystem. The average daily course of the total ozone fluxes course showed two slightly different behaviours in the first and in the second period. In the first one they increased after 9.00 AM and then decreased until midday, after that they increased again until 3.00 PM and then they decreased irregularly until 8.00 PM; the range of the variations was on average between 15 nmol m^{-2} s^{-1} and 20 nmol m^{-2} s^{-1} (Figure 5). In the second part of the field campaign the total ozone fluxes were on average almost constant around 12.5 nmol m^{-2} s^{-1} until 3.00 pm and then decreased until sunset at about 10 nmol m^{-2} s^{-1}.

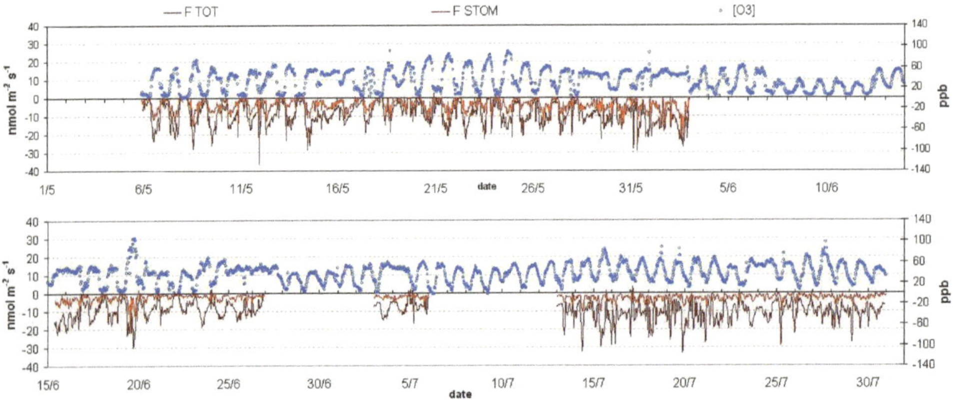

Fig. 4. Ozone concentration and ozone fluxes. The red dark line is the total ozone flux to the ecosystem and the red line is the stomatal flux (left axis), i.e. the amount of ozone taken up by vegetation through stomata. Blue circles represent ozone concentrations (right axis). Modified from Gerosa et al. (2009c).

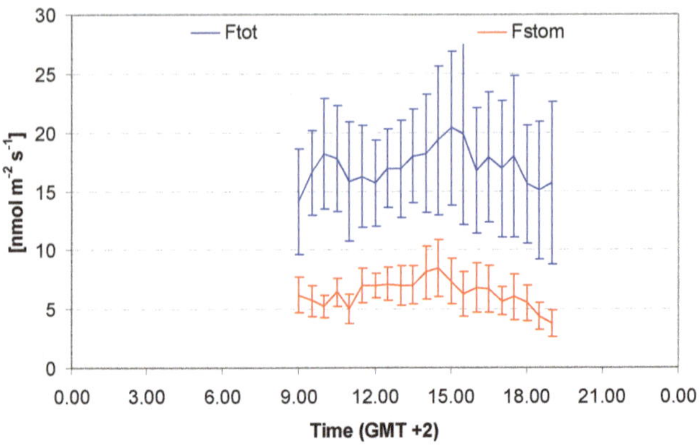

Fig. 5. Mean daily course of the absolute values of total and stomatal ozone fluxes in the first period F_{tot} (blue line) and F_{stom} (red line) are the total and the stomatal ozone fluxes when the canopy were completely dry, i.e. excluding the periods where dew was found on the leaves. Vertical bars are the standard deviations.

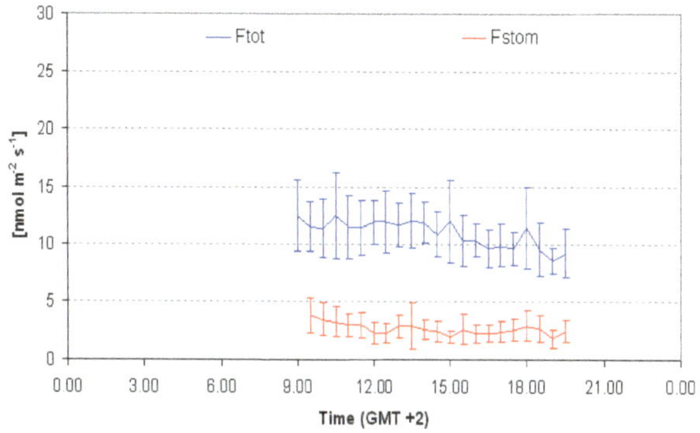

Fig. 6. Mean daily course of the absolute values of total and stomatal ozone fluxes in the second period F_{tot} (blue line) and F_{stom} (red line) are the total and the stomatal ozone fluxes when the canopy were completely dry, i.e. excluding the periods where dew was found on the leaves. Vertical bars are the standard deviations.

The stomatal ozone fluxes showed a daily course similar to the total fluxes of ozone (Figure 5, Figure 6). In the first period the stomatal ozone fluxes were on average around 5 nmol m^{-2} s^{-1} early in the morning, increasing until 8 nmol m^{-2} s^{-1} around 2.00 PM and then decreasing until less than 5 nmol m^{-2} s^{-1} at the sunset. In the second period the stomatal ozone fluxes reached their maximum in the morning (less than 4 nmol m^{-2} s^{-1}) and slowly declined for all the day until sunset when they were around 2 nmol m^{-2} s^{-1}. Another significant difference about stomatal fluxes between the two periods was the stomatal fraction of the ozone fluxes: as reported by Gerosa et al. (2009c) the stomatal fraction was between 40% and 50% in the first period and between 20% and 30% in the second period. These substantial differences can find an explanation looking at the values of the resistances and conductances calculated by means of the big-leaf model.

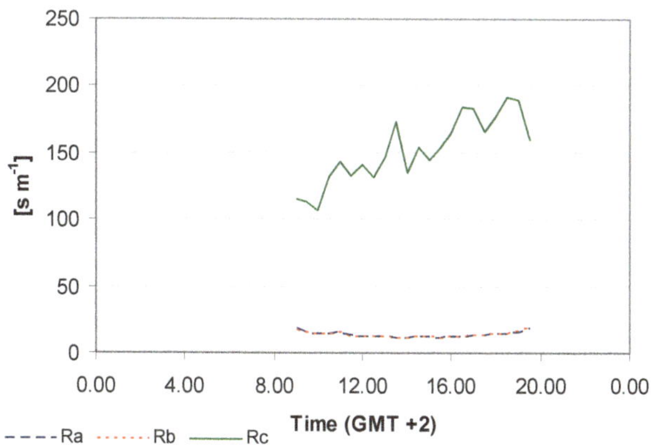

Fig. 7. Mean daily course of the aerodynamic resistance R_a (dashed blue line), of the sublaminar layer resistance R_b (dashed red line) and of the canopy resistance R_c (green line) in the first period.

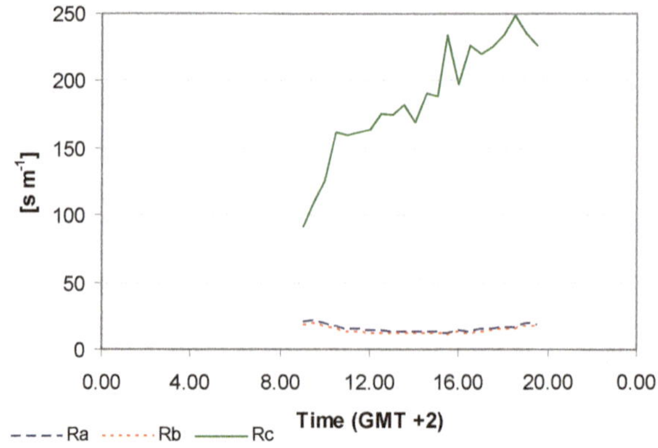

Fig. 8. Mean daily course of the aerodynamic resistance R_a (dashed blue line), of the sublaminar layer resistance R_b (dashed red line) and of the canopy resistance R_c (green line) in the second period.

In both periods R_a and R_b values were significant lower than R_c (Figure 7, Figure 8), showing the major role played by the canopy resistance, and hence by the stomatal and non stomatal pathways, for the ozone deposition. A slightly greater R_c was observed in the second period from late in the morning until the end of the day (Figure 7, Figure 8). Considering now the stomatal and the non stomatal conductance in both periods (Figure 9, Figure 10), a substantial reduction of the stomatal conductance was observed. Its average values in the second period were more than 50% less than in the first period. On the contrary non stomatal conductance showed similar values in the central hours of the day in both period, but was much higher in the morning and in late afternoon in the first half of the measurement campaign.

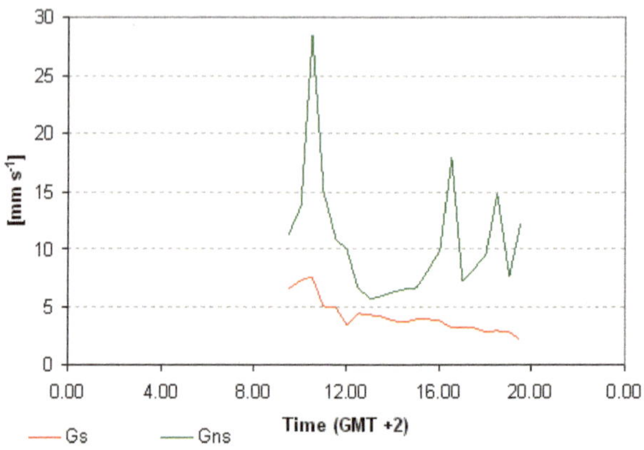

Fig. 9. Mean daily course of the stomatal conductance G_s (red line) and of the non stomatal conductance G_{ns} (green line) in the first period.

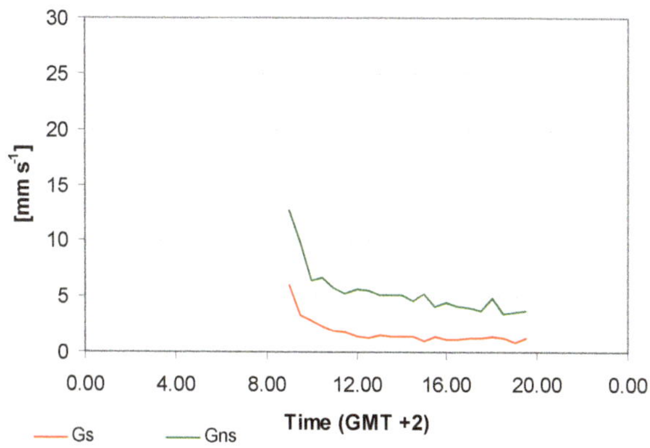

Fig. 10. Mean daily course of the stomatal conductance G_s (red line) and of the non stomatal conductance G_{ns} (green line) in the second period

It is further worth to notice the good agreement between the daily averages of stomatal conductance, soil water content (at two depth) and latent heat fluxes, which is proportional to (evapo)transpiration. Soil water shortage affected stomatal conductance, which in turn affected transpiration: in fact, in the first half of the field campaign soil water content decrease at both depth was in good agreement with the decrease of both stomatal conductance and latent heat fluxes, while in the second half they were all nearly constant (Figure 3).

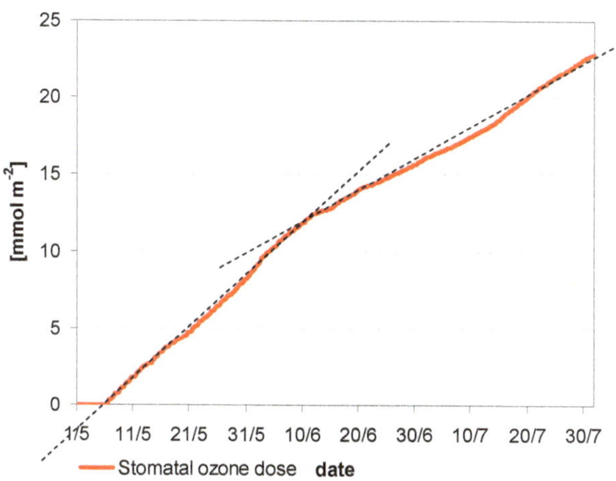

Fig. 11. Evolution of ozone stomatal dose during the measuring period.

The stomatal fluxes, after gap-filling procedure, were all summed up to calculate the ozone dose taken up by the ecosystem, the ozone dose resulted 22.8 mmol m^{-2}. Considering the accumulation of the ozone dose during the field campaign (Figure 11) two different growth rates of the cumulated dose were observed, mirroring the different behaviours of the stomatal conductance and thus of the fluxes in the two periods of the field campaign.

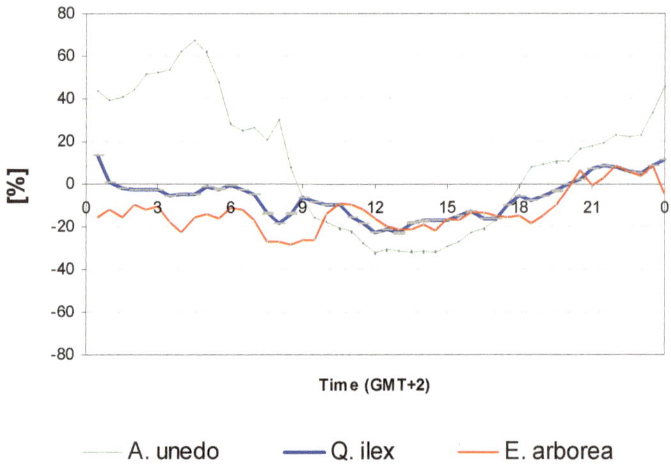

Fig. 12. Percentage difference between the two periods for the transpiration of the three species. Negative values correspond to a decrease in the second period and positive values correspond to an increase.

During this field campaign, Mereu et al. (2009) performed sap flow measurement over three species (*A. unedo, Q. ilex, E. arborea*). Unfortunately these three species represented only 38% of the total cover and hence it was not possible to test any up-scaling procedure, however some interesting features were observed. First of all it was possible to realize that the stomatal conductance peak, observed early in the morning by means of eddy covariance measurements and the following analyses, was probably due to the evaporation of dew from leaves (Gerosa et al., 2009c). In fact none of the three studied species showed any stomatal conductance peak early in the morning and the other three main species are known to not show any behaviour like this (Gerosa et al., 2009c). For this reason only data when canopy was dry where used to estimate canopy stomatal conductance and fluxes. Another important feature that was highlighted by sap flow measurements was the differences among the studied species. For instance, in both periods *Q.ilex* transpiration (which is directly proportional to stomatal conductance) increased earlier than the two other species, but, in the central hours of the day, it was from 10% to 25% less than *E. arborea* transpiration and from 40% to 50% less than *A. unedo*. Furthermore, if we consider the percentage variation of the transpiration of each species some interesting features can be observed. All the three species showed a sensible decrease in the daylight hours (from 15% to 30%) while after the sunset for all the three species an increase of the transpiration (more remarkable for *A. unedo*) was observed during the second period. From noon to the dawn the behaviour of the three species was different in the second period: *A. unedo* increased the night time transpiration of 40% on average, *Q.ilex* showed a slight reduction (about 5%) and *E. arborea* a more marked reduction (15-20%).

As already observed by Gerosa et al. (2009c), this analysis remarks the importance of considering plant physiology when studying multi-species ecosystem, in fact the different behaviours of these three species led to different ozone uptake by them and thus to different level of ozone risk for each species.

4. References

Amthor, J.S., Goulden, M.L., Munger, J.W. & Wofsy, S.C., 1994. Testing a mechanistic model of forest-canopy mass and energy-exchange using Eddy-Correlation - carbon-dioxide and ozone uptake by a mixed Oak Maple stand. *Australian Journal of Plant Physiology* 21, 623-651

Anfodillo, T., Sabatti, M., Sigalotti, G.B. & Valentini, R., 1992. An application of infrared thermal image to monitor water transport in plants. In: Carlomagno, G.M., Corso, C. (Eds), *Advanced Infrared Technology and Applications*. Firenze, pp. 427-437.

Anfodillo, T., Sigalotti, G.B., Tomasi, M., Semenzato, P. & Valentini, R., 1993. Application of thermal imaging in the study of sap flow in woody species. *Plant Cell Environment* 16, 997-1001.

Baldocchi, D.D. & Vogel,C.A.,1996. Energy and CO2 flux densities above and below a temperate broadleaved forest and a boreal pine forest. *Tree Physiology*. 16,5–16.

Baldocchi, D.D., Law, B.E. & Anthoni, P.M., 2000. On measuring and modeling energy fluxes above the floor of a homogeneous and heterogeneous conifer forest. *Agriculture and Forest Meteorology*. 102, 187–206.

Barret, D.J., Hatton, T.J., Ash, J.E. & Ball, M.C., 1995. Evaluation of the heat-pulse velocity technique for measurement of sap flow in rainforest and eucalyptus forest species of south-eastern Australia. *Plant Cell and Environment*. 18, 463-469.

Baumgartner, A., 1934. Thermoelektrische Untersuchungen uber die Geschwindigkeit des Transpirationsstromes. Z.Bot., 28, 81-136.

Biftu G.F. & Gan T.Y., 2000. Assessment of evapotranspiration models applied to a watershed of Canadian Prairies with mixed land-uses. *Hydrological Processes* 14, 1305-1325

Bragg, T., Webb, N., Spencer, R., Wood, J., Nicholl & C., Potter, E., 2004. AP4 Porometer Manual, Version 3. Delta-T Devices Ltd., Cambridge, UK.

Braun, P., 1997. Sap flow measurement in fruit trees - advantages and shortfalls of currently used systems. *Acta Horticulturae*. 449, 267-272.

Brooks, J.R., Meinzer, F.C., Coulombe, R. & Gregg, J., 2002. Hydraulic redistribution of soil water during summer drought in two contrasting Pacific Northwest coniferous forests. *Tree Phisiology*. 22, 1107-1117.

Burgess, S.S.O., Adams, M.A. & Bleby, T.M., 2000. Measurement of sap flow in roots of woody plants: a commentary. *Tree Physiology* 20, 909-913.

Burgess S.S.O. & Dawson T.E.. 2004 The contribution of fog to the water relations of Sequoia sempervirens (D. Don): foliar uptake and prevention of dehydration. *Plant, Cell and Environment*. 27, 1023–1034.

Campbell, G.S., 1991. An overview of methods for measuring sap flow in plants. Collected summaries of papers at the 83rd ann. Meet. of the American Society of Agronomy, Division A-3: *Agroclimatology and agronomic modeling*. Denver, Colorado, pp. 2-3.

Čermák, J., Deml, M. & Penka, M., 1973. A new method of sap flow rate determination in trees. *Biologia Plantarum* 15, 171-178.

Cescatti, A. & Zorer, R., 2003. Structural acclimation and radiation regime of Silver Fir (Abies alba Mill.) shoots along a light gradient. *Plant Cell and Environment* 26, 429-442

Chamberlain A.C. & Chadwick R.C., 1953. Deposition of airborne radioiodine vapour, *Nucleonics* 11, 22-25

Choudhury, B.J. & Monteith, J.L. 1988. A four-layer model for heat budget of homogeneous land surfaces. *Quarterly Journal of the Royal Meteorological Society* 114, 373-398

Cieslik, S., 2004. Ozone uptake at various surface types: a comparison between dose and exposure. *Atmospheric Environment* 38, 2409–2420.

Daum, C.R., 1967. A method for determining water transport in trees. *Ecology* 48, 425-431.

De Pury, D.G.G., 1995. Sealing photo.synthesis and water use from leaves to paddoeks. PhD thesis. The Australian National University, Canberra.

De Pury, D.G.G. & Farquhar, G.D., 1997. Simple scaling of photosynthesis from leaves to canopies without the errors of big-leaf models. *Plant Cell and Environment* 20: 537-557

Dixon, H.H., 1936. The convection of heat and materials in the stem of a tree. Scient. Proc.R.Dubl.Soc.(N.S.), 21, 477-488.

Do, F. & Rocheteau, A., 2002a. Influence of natural temperature gradients on measurements of xylem sap flow with thermal dissipation probes. 1. Field observations and possible remedies. *Tree Physiology* . 22, 641-648.

Do, F. & Rocheteau, A., 2002b. Influence of natural temperature gradients on measurements of xylem sap flow with thermal dissipation probes. 2. Advantages and calibration of a noncontinuous heating system. *Tree Physiology* . 22, 649-654.

Dyer, A.J., 1974. A review of flux-profile relationships, *Boundary-Layer Meteorology* 7, 363-372

Edwards, W.R.N., Becker, P. & Čermák, J., 1996. A unified nomenclature for sap flow measurements. Tree Physiology . 17, 65-67.

Ewers, B.E., Oren, R., Kim, H.S., Bohrer, G. & Lai, C.T., 2007. Effects of hydraulic architecture and spatial variation in light on mean stomatal conductance of tree branches and crowns. *Plant Cell and Environment* 30: 483-496

S. Fares, S., Mereu, S., Scarascia Mugnozza, G., Vitale, M., Manes, F., Frattoni, M., Ciccioli, P., Gerosa, G. & Loreto, F.. 2009. The ACCENT-VOCBAS field campaign on biosphere-atmosphere interactions in a Mediterranean ecosystem of Castelporziano (Rome): site characteristics, climatic and meteorological conditions, and eco-physiology of vegetation. *Biogeosciences*, 6, 1043-1058, 2009

Fowler, D.& Duyzer, J.H. 1989. Micrometeorological techniques for the measurement of trace gas exchange. In Andreae M.O. & Schimel D.S. (eds), "Exchange of Trace Gas between Terrestrial Ecosystems and the Atmosphere", John Wiley & Sons Ltd., 189-207

Garratt, J.R., 1975. Limitations of the eddy-correlation technique for the determination of turbulent fluxes near the surface. *Boundary-Layer Meteorology* 8, 255-259

Garratt, J.R., 1994. The atmospheric boundary layer. Cambridge University Press, UK

Gerosa, G., Cieslik, S & Ballarin-Denti, A. 2003. Micrometeorological determination of time-integrated stomatal ozone fluxes over wheat: a case study in Northern Italy, *Atmospheric Environment*, 37, 777–788,.

Gerosa, G., Marzuoli, R., Cieslik, S. & Ballarin-Denti, A. 2004. Stomatal ozone fluxes over a barley field in Italy."Effective exposure" as a possible link between exposure-and flux-based approaches, Atmos. Environ., 38, 2421–2432,.

Gerosa, G., Vitale, M., Finco, A., Manes, F., Ballarin Denti A. & Cieslik, S. 2005. Ozone uptake by an evergreen Mediterranean forest (Quercus ilex). I. Micrometeorological flux measurements and flux partitioning. *Atmospheric Environment*. 39, 3255–3266.

Gerosa, G., Marzuoli, R., Desotgiu, R., Bussotti, F., Ballarin-Denti, A., 2008. Visible leaf injury in young trees of Fagus sylvatica L. and Quercus robur L. in relation to ozone uptake and ozone exposure. An Open-Top Chambers experiment in South Alpine environmental conditions. *Environmental Pollution* 152, 274–284.

Gerosa, G., Marzuoli, R., Desotgiu, R., Bussotti, F., and Ballarin-Denti, A, 2009a Validation of the stomatal flux approach for the assessment of ozone effects on young forest trees, A summary report of the TOP (Transboundary Ozone Pollution) experiment at Curno, Italy. Environmental. *Pollution* 157, 1497–1505.

Gerosa, G., Finco, A., Mereu, S., Vitale, M., Manes, F. & Ballarin Denti, A.. 2009b. Comparison of seasonal variations of ozone exposure and fluxes in a Mediterranean Holm oak forest between the exceptionally dry 2003 and the following year, *Environmental Pollution*, 157, 1737–1744.

Gerosa, G., Finco, A., Mereu, S., Marzuoli, R. & Ballarin Denti, A.. 2009c. Interactions among vegetation and ozone, water and nitrogen fluxes in a coastal Mediterranean maquis eco system. *Biogeosciences*, 6, 1783-1798

Granier, A., 1985. A new method to measure the raw sap flux in the trunk of trees, *Annals of Forest Science* 42, 193-200.

Granier, A., Anfodillo, T., Sabatti, M., Cochard, H., Dreyer, E., Tomasi, M., Valentini, R. & Breda, N., 1994. Axial and radial water flow in the trunks of oak trees: a quantitative and qualitative analysis. *Tree Physiology* . 14, 1383-1396.

Grime, V.L., Morinson, J.I.L. & Simmonds, L.P., 1995. Sap flow measurements from stem heat balance: a comparison of constant with variable power methods. *Agriculture and Forest Meteorology* 74, 27-40.

Groot, A. & King, K.M., 1992. Measurement of sap flow by the heat balance method: numerical analysis and application to coniferous seedlings. *Agriculture and Forest Meteorology* 59, 289-308.

Hicks B.B., Baldocchi D.D., Meyers T.P., Hosker R.P. & Matt D.R., 1987. A Preliminary multiple resistance routine for deriving dry deposition velocities from measured quantities, *Water, Air & Soil Pollution* 36, 311-330

Hicks B.B., Matt D.R. & Mc Millen R.T., 1989. A micrometeorological investigation of surface exchange of O3, SO2 and NO2: a case study. *Boundary-Layer Meteorology* 47, 321-336

Huber, B., 1932. Beobachtung und Messung pflanzlicher Saftstrome. Ber.dt.bot. Ges., 50, 89-109. ICP Modelling and Mapping, 2004. Manual on methodologies and criteria for modelling and mapping critical loads and levels and air pollution effects, risks and trends.

Ittner, E., 1968. Der Tagesgang der Geschwindigkeit des Transpirationsstromes im Stamme einer 75-jahrigen Fichte. Oecol. Plant. 3, 177-183.

Jarvis P.G., 1976. The interpretation of leaf water potential and stomatal conductance found in canopies in the field. *Phil. Trans. R. Soc. London*, Ser. B, 273, 593-610.

Jones, H.G., Hamer, P.J.C. & Higgs, K.H., 1988. Evaluation of various heat-pulse methods for estimation of sap flow in orchard trees: comparison with micrometeorological estimates of evaporation. *Trees* 9, 250-260.

Karlsson, P.E., Pleijel, H., Pihl Karlsson, G., Medin, E.L. & Skarby, L., 2000. Simulations of stomatal conductance and ozone uptake to Norway spruce saplings in open-top chambers. *Environmental Pollution* 109, 443–451.

Karlsson, P.E., Medin, E.L., Ottosson, S., Sellde´n, G., Wallin, G., Pleijel, H. & Ska¨rby, L., 2004. A cumulative ozone uptake e response relationship for the growth of Norway spruce saplings. *Environmental Pollution* 128, 405-417.

Kostner, B., Granier, A. & Čermák, J., 1998. Sap flow measurements in forest stands-methods and uncertainities. *Annals of Forest Science*. 55, 13-27.

Köstner, B., Matyssek, R., Heilmeier, H., Clausnitzer, F., Nunn, A.J., & Wieser, G., 2008.Sap flow measurements as a basis for assessing trace-gas exchange of trees, *Flora* 203. 14–33.

Kučera, J., Čermák, J. & Penka, M., 1977. Improved thermal method of continual recording the transpiration flow rate dynamics. *Biologia Plantarum* 19, 413-420.

Kuniya, Y., 1950. Thermoelectric studies on sap streaming in plants. Sci.Rep.Tohoko Univ.(Ser.4), 18, 527-432.

Laisk A., Kull O. & Moldau H., 1989. Ozone concentration in the leaf intercellular air spaces is close to zero. *Plant Physiology* 90, 1163-1167

Law, B.E., Cescatti, A. & Baldocchi, D.D., 2001. Leaf area distribution and radiative transfer in open-canopy forests: implications for mass and energy exchange. *Tree Physiology* 21: 777-787

Lloyd, J., Grace, J., Miranda, A.C., Meir, P., Wong, S.C., Miranda, B.S., Wright, I.R., Gash, J.H.C. & Mcintyre, J., 1995. A simple calibrated model of amazon rain-forest productivity based on leaf biochemical-properties. *Plant Cell and Environment* 18, 1129-1145

Marshall, D.C., 1958. Measurements of sap flow in conifers by heat transport. *Plant Physiology*. 33, 385-396.

Martens, S.N., Breshears, D.D. & Meyer, C.W., 2000. Spatial distributions of understory light along the grassland/forest continuum: effects of cover, height, and spatial pattern of tree canopies. *Ecological Modelling* 126: 79-93

Martinez-Vilalta, J., Mangiròn, M., Ogaya, R., Sauret, M., Serrano, L., Penuelas, J. & Pinol, J. 2003. Sap flow of three co-occurring Mediterranean woody species under varying atmospheric and soil water conditions. *Tree Physiology* 23 747-758

McLaughlin, S.B., & Taylor, G.E., 1981. Relative humidity: important modifier of pollutant uptake in plants. *Science* 211 167–169

Mereu, S., Gerosa, G., Finco, A., Fusaro, L., Muys, B. & Manes F.. 2009. Improved sapflow methodology reveals considerable night-time ozone uptake by Mediterranean species. *Biogeosciences*, 6, 3151-3162

Moncrieff J., Valentini R., Greco S., Seufert G. & Ciccioli P., 1997. Trace gas exchange over terrestrial ecosystems: methods and perspectives in micrometeorology. *Journal of Experimental Botany* 48, 1133-1142

Monin A.S. & Obukhov A.M., 1954. Basic laws of turbulent mixing in the atmosphere near the ground. Translated in Aerophysics of Air Pollution (Fay J.A. & Hoult D.P. Ed., 1969), AIAA, New York, pp 90-119). Akademija Nauk CCCP, Leningrad, Trudy Geofizicheskowo Instituta 151 (N.24), 163-187

Monsi, M. & Saeki, T., 1953. Uber den Lichtfaktor in den Pflanzengesellschaften ubd seine Bedeutung fur die Stoffproduktion. *Japanese Journal of Botany* 14: 22-52

Monteith, J.L., 1981, Evaporation and surface temperature. Quarterly Journal of the Royal Meteorological Society. 107, 1-27

Monteith, J.L., Campbell, G.S., Potter, E.A., 1988. Theory and performance of a dynamic diffusion porometer. *Agricultural and Forest Meteorology*. 44, 27-38.

Monteith J.L. & Unsworth M., 1990. Principles of Environmental Physics. 2nd Edition. Arnold, London.

Murray F.W., 1967. On the computation of saturation vapour pressure. *Journal of Applied Meteorology* 6, 203-204

Musselman, R.C., Lefohn, A.S., Massman, W.J. & Heath, R.J.. 2006. A critical review and analysis of the use of exposure- and flux-based ozone indices for predicting vegetation effects. *Atmospheric Environment*, 40, 1869–1888

Nadezhdina, N., 1988. Apple tree water relations and their optimization under conditions of southern Ukraine (in Russian). Ph.D. Thesis. Kiev.

Nadezhdina, N.E., 1989. A physiological algorithm of woody plant irrigation control under air drought (in Russian). *Fiziologiya rastenii* 36, 972-979.

Nadezhdina, N.E., 1992. Apple tree water relations studied by means of the relative rate of water flow in the trunk xylem. *Biologia Plantarum*. 34, 431-437.

Nadezhdina, N., 1998. Temperature gradients around a linear heater in stems due to mowing sap. In: Čermák, J., Nadezhdina, N. (Eds), Measuring Sap Flow in Intact Plants. Proc. 4th. Int. Workshop, Zidlochovice, Czech Republic, IUFRO Publications. Brno, Czech Republic, Publishing House of Mendel University, pp. 65-71.

Nadezhdina, N., Čermák, J. & Nadezhdin, V., 1998. Heat field deformation method for sap flow measurements. In: Čermák, J., Nadezhdina, N. (Eds), Measuring Sap Flow in Intact Plants. Proc. 4th. Int. Workshop, Zidlochovice, Czech Republic, IUFRO Publications. Brno, Czech Republic, Publishing House of Mendel University, pp. 72-92.

Nadezhdina, N., 1999. Sap flow index as an indicator of plant water status. *Tree Physiology* . 19, 885-891.

Nadezhdina, N. & Čermák, J., 2000. The technique and instrumentation for estimation the sap flow rate in plants (in Czech). Patent No.286438 (PV-1587-98).

Nadezhdina, N., Tributsch, H. & Čermák, J., 2004. Infra-red images of heat field around a linear heater and sap flow in stems of lime trees under natural and experimental conditions. *Annals of Forest Science*. 61, 203-214.

Norman J.M., 1982. Simulation of microclimates. In *Biometeorology in Integrated Pest Management* (Eds J.L. Hatfield and I.J.Thomson, pp65-99, Academic Press, New York.

Norman J.M., 1993. Scaling processes between leaf and canopy levels. In *Sealing Physiologieal Proeesses: Leaf to Globe* (eds J.R. Ehleringer & C. B. Field), pp. 41-76. Academic Press, Inc.,San Diego.I. J. Thomason), pp, 65-99. Academic Press, New York.

Parker, G.G., 1995. Structure and microclimate in forest canopies. In: Lowman, M., Nadkarni, N. (Eds.), Forest Canopies. Academic Press, California, USA, pp. 73-106.

Pataki, D.E., Oren, R., Katul, G. & Sigmon, J. 1998. Canopy conductance of *Pinus taeda*, *Liquidambar styraciflua* and *Quercus phellos* under varying atmospheric and soil water conditions. *Tree Physiology* 18 307-315

Penman, H.L. (1948). "Natural evaporation from open water, bare soil, and grass". *Proc. Roy. Soc.* (London, U.K.) A193 (1032): 120–145.

Pickard, W.F., 1973. A heat pulse method of measuring water flux in woody plant stems. *Mathematical Biosciences*. 16, 247-262.

Pickard, W.F. & Puccia, Ch.J., 1972. A theory of the steady-state heat step method of measuring water flux in woody plant stems. *Mathematical Biosciences*. 14, 1-15.

Pleijel, H., Danielsson, H., Phil Karlsson, G., Gelang, J., Karlsson, P.E. & Sellden, G. 2000. An ozone flux-response relationship for wheat. *Environmental Pollution* 109, 453-462.

Raupach, M.R., 1995. Vegetation-atmosphere interaction and surface conductance at leaf, canopy and regional scales. *Agrieultural and Forest Meteorology* 73, 151-179.

Saddler, H.D.W. & Pitman, M.G., 1970. An apparatus for the measurement of sap flow in unexcised leafy shoots. *Journal of Experimental Botany*. 21, 1048-1059.

Sakuratani, T., 1981. A heat balance method for measuring water flux in the stem of intact plants. *Journal of Agricultural Meteorology* (Japan), 37, 9-17.

Smith, D.M. & Allen, S.J., 1996. Measurement of sap flow in plant stems. *Journal of Experimental Botany*. 47, 1833-1844.

Steppe, K., De Pauw, D.J.W., Doody T.M. & Teskey, R.O.. 2010. A comparison of sap flux density using thermal dissipation, heat pulse velocity and heat field deformation methods.). *Agricultural and Forest Meteorology*. 150:1046–1056.

Stewart, J.B. 1988. Modelling surface conductance of pine forest. *Agricultural and Forest Meteorology*. 43, 19–35.

Swanson, R.H., 1983. Numerical and experimental analysis of implanted-probe heat-pulse theory. Ph.D. Thesis, University of Alberta, Edmonton, Canada.

Swanson, R.H., 1994. Significant historical development in thermal methods for measuring sap flow in trees. *Agriculture and Forest Meteorology* 72, 113-132.

Swanson, R.H. & Whitfield, D.W.A., 1981. A numerical analysis of heat-pulse velocity theory and practice. *Journal of Experimental Botany*. 32, 221-239.

Swinbank W.C., 1951. The measurement of vertical transfert of heat and water vapor by eddies in the lower atmosphere. *Journal of Meteorology* 8, 135-145.

Tyree, M.T. 1997. The Cohesion-Tension Theory of sap ascent: current controversies. *Journal of Experimental Botany* 48 1753-1765.

Van de Zande, D., Mereu, S., Nadezhdina, N., Cermak, J., Muys, B., Coppin, P., Manes, F., 2009. 3D upscaling of transpiration from leaf to tree using ground-based LiDAR: Application on a Mediterranean Holm oak (*Quercus ilex* L.) tree. *Agricultural and Forest Meteorology* 149, 1573-1583

Valancogne, C. & Nasr, Z., 1989. Measuring sap flow in the stem of small trees by a heat balance method. *Horticultural Science* 24, 383-385.

Vieweg, G.H., Ziegler, H. & 1960. Thermoelektrische Registrierung der Geschwindigkeit des Transpirationsstromes. Ber.Deut.Bot.Ges. 73, 221-226.

Wullschleger, S.D., Meinzer, F.C. & Vertessy, R.A., 1998. A review of whole-plant water use studies in trees. *Tree Physiology*.18, 499-512.

Whitehead, D. & Jarvis, P.G. 1981. Coniferous forests and plantations. Water Deficits and Plant Growth. Ed. T.T. Kozlowski. Academic Press, New York, pp. 49-152

Wieser, G., Häsler, R., Götz, B., Koch, W. & Havranek, W.M., 2000. Role of climate, crown position, tree age and altitude in calculated ozone flux into needles of *Picea abies* and *Pinus cembra*: a synthesis. *Environmental Pollution* 109 415–422.

Wilske, B., Kwon, H., Wei, L., S Chen, S., Lu, N., Lin, G., J Xie, J., Guan, W., Pendall, E., Ewers, B.E. & Chen, J. 2010. Evapotranspiration (ET) and regulating mechanisms in two semiarid Artemisia-dominated shrub steppes at opposite sides of the globe. *Journal of Arid Environments*. 74(11):1461-1470.

5

Evapotranspiration Partitioning Techniques for Improved Water Use Efficiency

Adel Zeggaf Tahiri
Manzanar-Mauritania. B-151, F-Nord, Tevragh Zeina, Nouakchott, Mauritania

1. Introduction

Improvement of agricultural water use efficiency is a key issue to alleviate the pressure of the ever-expanding world population on water resources (Zeggaf and Filali, 2010). In areas such as the Southern Mediterranean, where the agricultural sector consumes more than 80 % of renewable water resources in most countries, as little as 10 % increase in water use efficiency by the agricultural sector would provide 40 % more water for domestic and industrial use (Lacirignola et al., 2003). In other words, the challenge for the 21st century will be undoubtedly to produce more food with less water to cope with the increasing water demand by the water sector usages (agriculture, industry, domestic).

An efficient irrigation scheduling at crop field level minimizes water losses by soil evaporation and maximizes water uptake by crop transpiration. To be achieved, monitoring separately crop transpiration and soil evaporation, and quantifying their inter-relations is paramount. However, most of the scientific and technical literature concerned with crop water requirements and irrigation scheduling foster soil water consumption at field level as a whole "evapotranspiration" (ET), which is the water amount used at field level for plant transpiration and soil evaporation.

2. Separate measurements of transpiration and soil evaporation

Evaporation is the process by which liquid water is converted to water vapor (vaporization) and then removed from the evaporating surface (vapor removal). Water evaporates from a variety of surfaces, such as lakes, rivers, pavements, soils and wet vegetation. Where the evaporating surface is soil, the degree of shading of the canopy and the amount of water available at the evaporating surface are other factors that affect soil evaporation process.

Transpiration consists of the vaporization of liquid water contained in plant tissues and the vapor removal to the atmosphere. Transpiration, like direct evaporation, depends on the energy supply, vapor pressure gradient and wind. Hence, radiation, air temperature, air humidity and wind terms should be considered when assessing transpiration. The soil water content and the ability of the soil to conduct water to the roots also determine the transpiration rate, as do water logging and soil water salinity. The transpiration rate is also influenced by crop characteristics, environmental aspects and cultivation practices.

In the following, a number of measuring methods of transpiration and soil evaporation are listed.

2.1 Transpiration

Transpiration has been measured on surfaces varying in area from part of a leaf, entire fields to forests, and the methods used have varied equally widely. Originally, most measurements were made on individual plants, but in agriculture and forestry interest has turned toward study of the water balance of large stands of plants (Kramer, 1983). In the following, some methods for measuring transpiration are listed.

2.2 Gravimetric method

From the time of Hales (1727) to the present, investigators had grown plants in containers and measured transpiration by weighing the containers at appropriate intervals. It was necessary to grow the plants in waterproof containers and to cover the soil to prevent loss by soil evaporation (Kramer, 1983). Soil moisture had to be replenished frequently so that water supply did not become limiting, a common defect of many early experiments (Raber, 1937). This technique was routinely used for measuring trees transpiration (Fritschen and Gay, 1979). The containers in which plants were growing should be protected from direct sun to prevent overheating, and the ideal arrangement was to have them set with the tops flush with the surrounding soil in the habitat where they would normally grow.

2.3 Cut-shoot method

Measurements of transpiration were made on detached leaves weighted at intervals of a minute or two on a sensitive balance. Such measurements could proceed for only a few minutes after cutting the leaf because transpiration tends to decline with decreasing leaf water content (Kramer, 1983). Sometimes, there was a transient increase in transpiration shortly after detaching a leaf or branch, the Ivanov effect, probably resulting from release of tension in the xylem. This method was used for measuring transpiration of trees (Roberts, 1977). However, the tree-cutting procedure could affect the entry of water to the conducting tissues. Also, large differences could be caused by detaching the plant organ, and by measuring transpiration in an environment different from that of its location on the plant. Hence, extrapolation of results could not be attempted. In spite of its inherent errors, the cut-shoot method was used to measure differences in transpiration among species (Hygen, 1953; Kaul and Kramer, 1965).

2.4 Measurement of water vapor loss

Measurement of transpiration could be made by monitoring the change in humidity of an air stream passed through a container enclosing the plant material. The containers were usually made of plastic and vary from tiny cuvettes holding one leaf or part of a leaf (Slavik, 1974) to those holding a branch (Kaufmann, 1981). This method eliminated errors caused by detaching leaves or branches, but imposed a somewhat artificial environment on the leaf or plant enclosed in the container. Grieve and Went (1965) described the use of cuvettes containing a humidity sensor to enclose a single leaf for short-term measurements. This method has been developed into equipment that can make a measurement of transpiration and stomatal resistance in less than one minute. Several porometers are described by Jarvis and Mansfield (1981) and Kaufmann (1981).

2.5 Canopy-chamber method

Canopy-chamber method remains an appropriate approach for plot-sized experimental agriculture (Steduto et al., 2002). Two major canopy-chamber systems can be identified for field applications:

Steady-state open-systems include the open-top chambers, most widely used for long-term studies of field-grown plants mainly exposed to elevated CO_2 or atmospheric polluting gases (Leadley and Drake, 1993). These chambers have the advantage of continuously monitoring the plant response throughout the season, but with the drawback of altering the microclimate of the crop. Moreover, they require flow measurements and, most of the times, climate control (Steduto et al., 2002).

The canopy-chambers operating as transient-state closed-systems, instead, do not need any flow measurement or climate conditioning and are mainly used for ambient-level CO_2 and water vapor gas-exchange measurements. These chambers are placed over the crop for a very short time (about a couple of minutes) and then removed for a subsequent measurement, allowing enough replicates and minimal disturbance of the plant environment. Nevertheless, during the time of measurement, the natural gradients of temperature, CO_2 and water vapor are reduced due to forced ventilation (Held et al., 1990), and the orientation pattern of leaves at the chamber borders can be modified during the placement (Reicosky et al., 1990).

2.6 Sap flow method

A method that has shown promise is the steady-state heat balance method developed by Sakuratani (1981, 1984). Use of this method does not alter any of the environmental or physiological factors affecting the transpiration process and Sakuratani (1981) reported an accuracy of ±10 %. This result was supported by Baker and Van Bavel (1987). The method works in the following way. A steady, known amount of heat is applied to a small segment of the stem from a thin flexible heater that encircles the stem and is itself encircled by foam insulation. In the steady state, this heat input to the segment must be balanced by four heat fluxes out of the segment: conduction up the stem, conduction down the stem, conduction outward through the foam sheath and convection in the moving transpiration stream. Subtraction of the conductive fluxes from the known heat input yields the heat transported by the moving sap flow (Baker and Nieber, 1989). The method is direct, requires no calibration or knowledge of the cross-sectional area of the xylem vessels. However, some authors reported that high sap flow rates may cause some systematic errors in estimating the heat balance components (Baker and Nieber, 1989). Also, Ishida et al. (1991) suggested that the gauge accuracy may be influenced by stem vascular anatomy, with potentially greater accuracy in dicotyledons than in monocotyledons.

2.7 Soil evaporation

Most soil evaporation (E_s) takes place in two stages: the constant and the falling rate stages (Philip, 1957):

In the constant rate stage (stage 1), the soil is sufficiently wet for the water to be transported to the surface at a rate at least equal to the evaporation potential. In this stage, evaporation is determined by atmospheric demand and soil conditions, rather than the conductive properties of the soil profile. The transition to the second stage of drying occurs when cumulative soil evaporation reaches a soil-specific threshold (Ritchie, 1972).

In the falling rate stage (stage 2), the surface soil water content has decreased below a threshold value, so that E_s depends on the flux of water through the upper layer of soil to the evaporating site near the surface (Ritchie, 1972). The cumulative soil evaporation was found to be proportional to the square root of time (Philip, 1957). The proportionality between second stage soil evaporation and the square root of time has been supported by several laboratory and field studies (Hillel, 1980). In the following, some methods for measuring soil evaporation are listed.

2.8 Micro-lysimeter

The theory and inherent assumptions of micro-lysimetry have been examined by Boast and Robertson (1982) and Walker (1983). To ensure accurate measurement of soil evaporation, the soil core within a micro-lysimeter must have a moisture content profile similar to that of the surrounding soil. Once the core has been cut from the plot, the two profiles begin to diverge as extraction of water by roots and vertical fluxes are prevented by the walls and the base of the micro-lysimeter. Daily soil evaporation from the micro-lysimeters can be calculated from weight loss and rainfall. Micro-lysimeters have been used to measure soil evaporation from both bare soil and soil beneath sparse canopy. Some authors reported that the micro-lysimeters can measure soil evaporation accurately during dry periods, but are unreliable on days when rainfall is present (Allen, 1990). Unless soluble by changes in design, this problem seriously limits the usefulness of micro-lysimeters for evaluating the contribution of soil evaporation to the seasonal water use by crops in rainfed dryland agriculture (Allen, 1990).

2.9 Energy balance method

Ben-Asher et al. (1983), building on work by Fox (1968), developed an energy balance method (EBM) for measuring soil evaporation. This method used average daily wind speed and the difference between midday maximum soil surface temperatures of a reference dry soil and a drying soil to estimate daily soil evaporation from the drying soil. Soil surface temperature can be measured by infrared thermometry, and then soil evaporation can be calculated by the energy balance method described by Ben-Asher et al. (1983). The Ben-Asher method calculates evaporation using the difference between dry and drying soil surface temperatures.

3. Combined measurements of transpiration and soil evaporation

Evaporation and transpiration occur simultaneously and there is no easy way of distinguishing between the two processes (Allen et al., 1998). Apart from the water availability in the topsoil, soil evaporation from a cropped soil is mainly determined by the fraction of the solar radiation reaching the soil. This fraction decreases over the growing period as the crop develops and the canopy shades more and more of the ground area. When the crop is small, water is predominately lost by soil evaporation, but once the crop is well developed and completely covers the soil, transpiration becomes the main process.

Latent heat fluxes from the canopy and the soil are complex processes governed by energy exchange between the soil, canopy, and the aerial environment. Investigating evaporation and energy exchanges in the crop field requires energy balance of the soil and the canopy to be examined separately (Ham et al., 1991).

Several methods are used to measure or determine evapotranspiration (*ET*) components simultaneously. Ham et al. (1991) showed that energy balances of canopy and soil in a cotton field could be determined by combining sap flow with BREB measurements of transpiration (*T*) and evapotranspiration respectively. Also, *ET* in a drip-irrigated vineyard was determined from separate measurements of *T* by sap flow gauges and soil evaporation

(E_s) by micro-lysimeters at various positions (Yunusa et al., 2004). However, when different methods are used to determine ET components, the consistency among these methods deserves attention (Jara et al., 1998). Ham et al. (1990) reported a comparison of E_s calculated by difference between ET measured by BREB method and T measured by heat balance stem flow measurements, with that measured by micro-lysimeters. A rapid decline in the precision of calculated E_s at low soil evaporation levels was reported, which indicated that the utility of the approach might be limited when E_s is less than 20 % of ET. This restriction might hinder quantification of surface energy balance relationships and transport processes during certain soil and canopy conditions (Ham et al., 1990). The expensiveness of the measurement equipments involved in similar experiments and the different scales at which these measurements are performed limit the large scale adoption of these techniques by research scientists. Cheaper and precise methods for studying energy balance exchange in the crop field are needed.

The BREB method is considered to be fairly robust for measuring ET (Steduto and Hsiao, 1998b), and has compared favorably with other methods, e.g., soil water balance (Malek and Bingham, 1993), aerodynamic method (Malek, 1993), eddy covariance method (Dugas et al., 1991), and weighing lysimeter (Prueger et al., 1997; Tanner, et al., 1960). The validity of this method has been established over various vegetation stands (Ham et al., 1991; Heilman et al., 1994), and natural vegetation (Kalthoff et al., 2006). Ashktorab et al. (1989) used a micro-Bowen ratio system for energy balance determination close to bare soil, and reported E_s readings within 10 % of the weighing lysimeter measurements. Accordingly, it was suggested that the BREB method should be considered an excellent candidate for the determination of the soil component of ET from row crops (Ashktorab et al., 1989).

4. Double Layer Bowen Ratio Energy Balance system (DOLBOREB): A case study

4.1 Energy budget at crop field level

a. Energy balance of maize field

The energy balance of maize field can be expressed as:

$$R_n = \lambda E + H + G \tag{1}$$

where

R_n: net radiation above canopy, λE: latent heat flux, H: sensible heat flux, and G: soil heat flux, all units of W m^{-2}.

In Eq. (1), the convention used for the signs of the energy fluxes is R_n positive downward and G is positive when it is conducted downward from the surface. λE and H are positive upward, with a direction opposite to that of the temperature and vapor pressure gradients. Over an averaging period, assuming equality of the eddy transfer coefficients for sensible heat and water vapor (Verma et al., 1978), and measuring the temperature and vapor pressure gradients between two levels within the adjusted surface layer, the Bowen ratio (β) is calculated by:

$$\beta = \gamma(\delta T / \delta\zeta) / (\delta\varepsilon / \delta\zeta) = \gamma \Delta T / \Delta\varepsilon \tag{2}$$

Where

ΔT and Δe: temperature and vapor pressure differences between two measurement levels (z), respectively, $\gamma = c_p\, p \,/\, \varepsilon\, \lambda$: psychrometric constant, c_p: specific heat of air at constant pressure (1.01 kJ kg^{-1} ^{o}C^{-1}), p: atmospheric pressure (kPa), ε: ratio between molecular weights of water vapor and air (0.622), and λ: latent heat of vaporization (kJ kg^{-1}).
The partition of energy between λE and H is determined by the BREB method (Tanner et al., 1960; Kustas et al., 1996, Perez et al., 1999) by means of β as:

$$\beta = H \,/\, \lambda E \tag{3}$$

The Bowen ratio (Eq. 3) is used with the energy balance (Eq. 1) to yield the following expressions for λE and H:

$$\lambda E = (R_n - G) \,/\, (1 + \beta) \tag{4}$$

$$H = (R_n - G)\, \beta \,/\, (1 + \beta) \tag{5}$$

The energy balance of the maize field is measured by a BREB unit. Air temperature and vapor pressure gradients are determined from two dry and wet bulb ventilated psychrometers. The distance between the two psychrometers is 1 m, and the lowest psychrometer is positioned at 0.2 m above the canopy. Net radiation at 1 m above the canopy, is measured by a net radiometer. Soil heat flux is calculated as an average value of two or more heat flux plates measurements at 2 cm below soil surface. Wind speed is measured, 1 m above the canopy, by a wind speed sensor. All data are measured every minute by a datalogger and multiplexer and averaged over 10 minutes's time interval.

b. Energy balance over soil surface

The energy balance over soil surface can be expressed as:

$$R_{ns} = \lambda E_s + H_s + G \tag{6}$$

where
R_{ns}: R_n to soil surface, λE_s: soil latent heat flux, and H_s: sensible heat flux from soil, all units of W m^{-2}.
Similar to maize field, the Bowen ratio at soil surface level (β_s) was calculated by:

$$\beta_s = H_s \,/\, \lambda E_s \tag{7}$$

where
λE_s and H_s: determined from Eq. 6 and 7 as by Eq. 4 and 5, respectively.
The energy balance over soil surface is measured at the same location as that of the maize field. Following a similar set-up made by Ashktorab et al. (1989) over bare soil, air temperature and vapor pressure gradients within the rows are determined from two dry and wet bulb ventilated psychrometers. The distance between the two psychrometers is 0.1 m, and the lowest psychrometer is positioned 0.05 m above soil surface.

c. Energy balance of canopy

The energy balance of canopy can be expressed as:

$$R_{nc} = \lambda E_c + H_c \tag{8}$$

where

R_{nc}: R_n intercepted by canopy, and λE_c and H_c: fluxes of latent and sensible heat from canopy, respectively.

Applying the principle of continuity and the definition of R_n, it can be shown that R_{nc} is the difference between R_n above and below the canopy (Ham et al., 1991).

$$R_{nc} = R_n - R_{ns} \tag{9}$$

where

R_n and R_{ns} are measured by net radiometers.

Canopy latent heat flux is calculated by Eq. 10, while H_c is calculated as a residual from Eq. 8.

$$\lambda E_c = \lambda E - \lambda E_s \tag{10}$$

4.2 DOLBOREB set-up

A DOLBOREB system consists of four dry and wet bulb ventilated psychrometers mounted on moveable arms, two net radiometers, and two or more soil heat flux plates (figure 1). The first two dry and wet bulb ventilated psychrometers are used to determine air temperature and vapor pressure gradients above the crop. The distance between the two psychrometers is 1 m, and the lowest psychrometer should be positioned at 0.2 m above the canopy. The other two dry and wet bulb ventialetd psychrometers are used to determine air temperature and vapor pressure gradients above the soil surface. Following a similar set-up made by Ashktorab et al. (1989) over bare soil, air temperature and vapor pressure gradients within the rows are determined from two dry and wet bulb ventilated psychrometers. The distance between the two psychrometers is 0.1 m, and the lowest psychrometer is positioned 0.05 m above soil surface.

Net radiation at 1 m above the canopy and over soil surface is measured by net radiometers. Soil heat flux is calculated as an average value of two or more heat flux plates measurements at 2 cm below soil surface. Wind speed is measured, 1 m above the canopy, by an anemometer. All data are measured every minute by dataloggers and multiplexers and averaged over 10 minutes's time interval.

4.3 DOLBOREB system outputs

Use of the DOLBOREB system enables measuments of all energy components at crop field level. It also permits to explore energy exchange at crop field level (Zeggaf et al., 2008). In the following, diurnal trend of energy balances of maize field, soil and canopy by the DOLOBOREB system for a sample day will be presented and discussed (Fig. 2). This day was selected because of clear sky and variable wind speed. Maximum air temperature and R_n were 32°C and 645 W m^{-2}, respectively. Wind speed at 1 m above the canopy ranged from 1.1 m s^{-1} at early morning to 3.9 m s^{-1} around noon.

During most part of the day, λE was less than net radiation at maize field level, except at early morning and late afternoon (Fig. 2A). λE exceeding R_n suggested that there might be some brief periods when advection of sensible heat supported evapotranspiration. Similar observations were reported for vineyard (Yunusa et al., 2004), and for cotton (Ham et al., 1991). During daytime, most of R_n was used to drive λE (Fig. 2A). Only 8.5 % of available energy (R_n - G) was used to generate H. Similar results have been reported for cotton

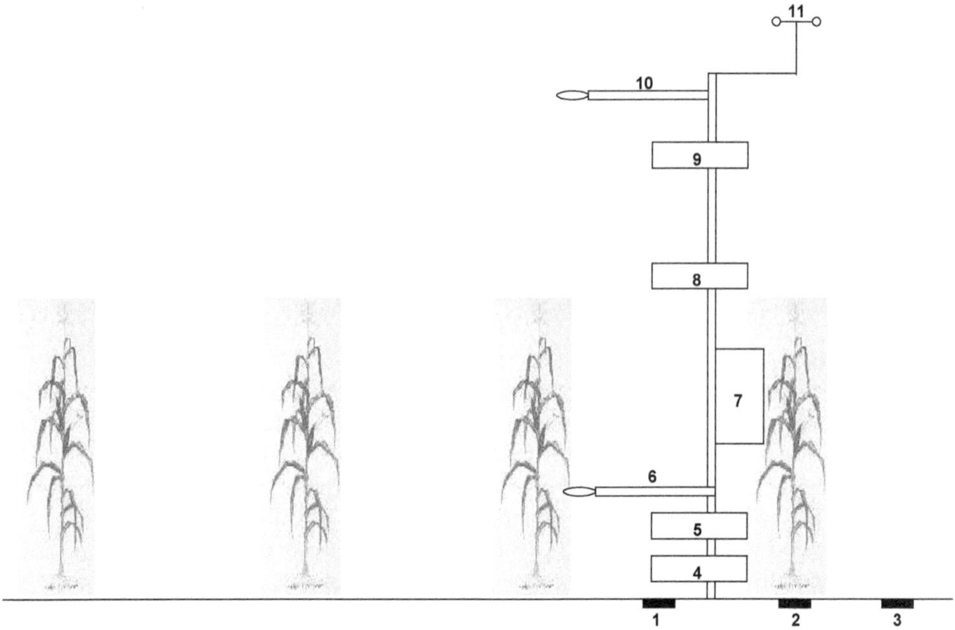

Fig. 1. DOLBOREB system set-up. 1,2,3 : Soil heat flux plates. 4 : Lower ventilated psychrometer over soil surface. 5: Upper ventilated psychrometer over soil surface.6: Net radiometer over soil surface. 7: Datalogger. 8: Lower ventilated psychrometer over crop canopy. 9: Upper ventilated psychrometer over crop canopy. 10: Net radiometer over crop canopy. 11: Anemometer.

(Ritchie, 1971; Ham et al., 1991) and for maize (Steduto and Hsiao, 1998a). During daytime, β ranged from -0.3 to 0.1. Accordingly, Steduto and Hsiao (1998a) reported positive β values less than 0.25 under incomplete maize canopy (L = 0.58). Soil heat flux was less than 10 % of R_n, this value is commonly found in the literature (Yunusa et al., 2004).

During daytime, λE_s was less than R_{ns}, and H_s remained positive (Fig. 2B), indicating convective transport of heat away from soil surface. R_{ns} was used almost equally to drive λE_s (52 %) and to generate H_s (48 %), with β_s ranging from 0.20 to 0.96 as shown in Fig. 2. Other authors reported λE_s accounting for 29 to 47 % of R_{ns} for vineyard (Heilman et al., 1994).

At canopy level, λE_c exceeded R_{nc} during most of the daytime period. Negative H_c values were obtained as shown in Fig. 2C indicating the canopy was absorbing convective heat from soil surface, which provided supplement energy for λE_c. Redistribution of available energy by sensible heat transfer occurs when there is (i) abundant supply of soil water (ii) absence of significant physiological restraint of water vapor flux through stomata, and (iii) high evaporative demand (Oke, 1987). In similar conditions, this redistribution of energy has been reported to supply up to one third of the energy needed for transpiration (Hicks, 1973; Heilman et al., 1996; Sene, 1996).

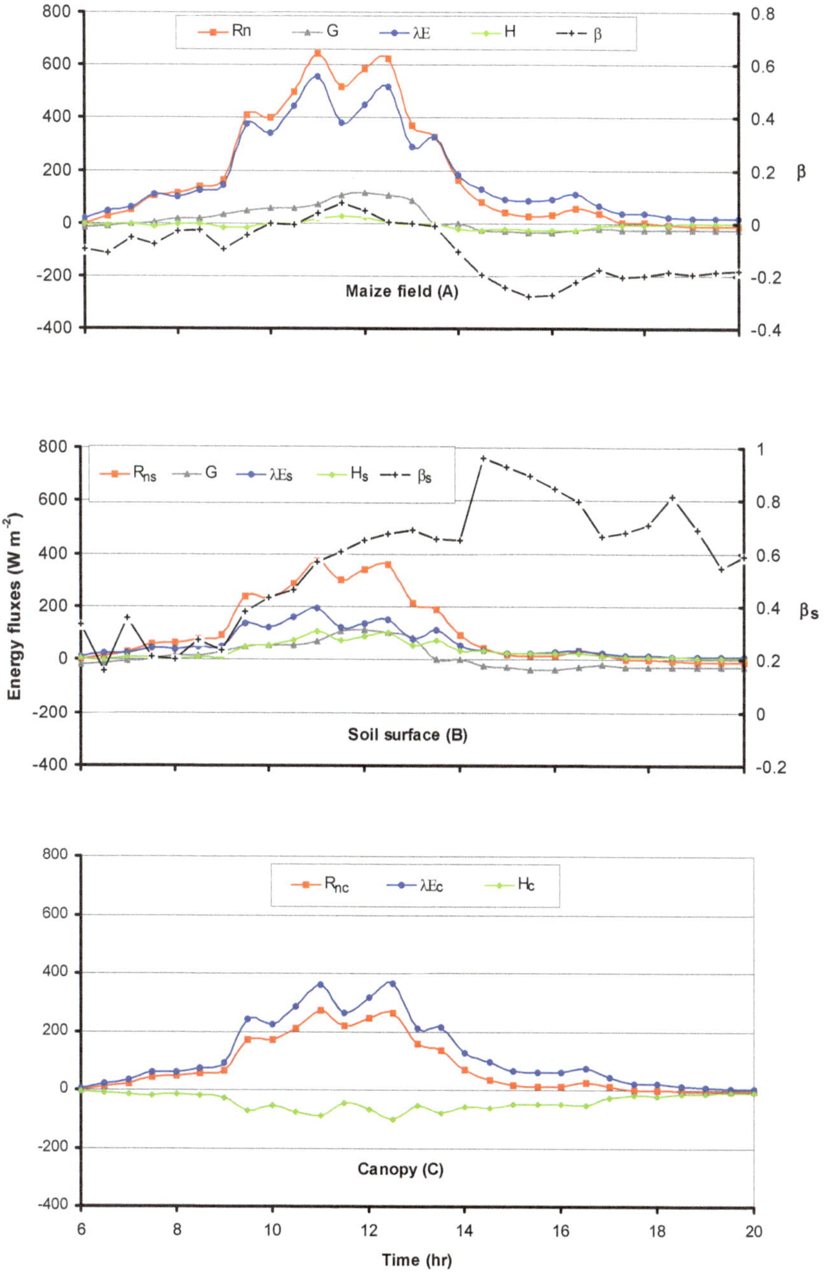

Fig. 2. Diurnal trend of energy balances of maize field, soil surface and canopy by the DOLOBOREB system for a sample day.

For 12 days, results of energy balance measurements from the DOLBOREB system compared positively to other direct methods measurements (Zeggaf et al., 2008). Latent heat flux over maize field from The DOLBOREB system showed a coefficient of determination of 0.72 with that measured with weighing lysimeter. Also, a coefficient of determination of 0.77 was obtained between DOLBOREB system and sap flow measurements of latent heat flux from maize transpiration. This shows that the DOLBOREB system can be used with reasonable accuracy in place of other expensive and separate measurements of evapotranspiration components to measure all energy balance components at crop field level.

5. Use of the DOLBOREB system for water use efficiency at crop field level

A comparison between energy balance patterns at maize field level between 2 periods: a wet period where maize field was irrigated daily and a dry period where irrigation was halted, was studied (Zeggaf et al., 2007).

5.1 Diurnal pattern of energy fluxes from maize field for the wet and dry periods

Diurnal patterns of energy fluxes from maize field for the wet and dry periods by the DOLBOREB system are shown in Fig. 3. Soil heat flux ranged from 7 to 15 % of R_n for both periods, which is close to the common value of 10 % reported by Yunusa et al. (2004). In fact some authors reported very small G for dense maize canopy (L = 5.3), but this component was larger for incomplete canopy (L = 0.58) because of the exposed and dry soil (Steduto and Hsiao, 1998b). As reported by Steduto and Hsiao (1998b), latent heat flux from maize field (λE) was closely coupled to R_n, giving rise to a nearly perfect coincidence between R_n and λE in their rise and fall shown in Fig. 3, when changing clouds effected rapid fluctuation of radiation. This result is expected as the net radiation is the main source of energy for evapotranspiration.

For the wet period, λE was at its full rate as shown in Fig. 3-wet. Sensible heat flux was very small and could be accounted as negligeable. During this period, λE was always smaller or equal to net radiation indicating no major advective conditions prevailed. Similar results have been reported for other crops as cotton and vineyard (Ham et al., 1991; Yunusa et al., 2004).

For the dry period, H slightly increased relatively to the wet period but was still low as shown in Fig. 3-dry. Sensible heat was almost positive during daytime and ranged from 5 and 8 % of R_n. The Bowen ratio (β) for the dry period was slightly greater than that for the wet period. Similar result was reported by Steduto and Hsiao (1998) for maize for the dry soil water regime. However, even when irrigation was halted, λE still represented a large part of the available energy (around 93 %). This result suggested that water stress was not evident on evapotranspiration for a period of six days after irrigation was halted.

Linear regression lines between available energy (R_n - G) and λE from maize field for the wet and dry periods produced high values of r^2 as shown in the following Eqs. 11 and 12. Similar results were obtained by Ham et al. (1991) who reported that within row advection increased λE_c, and that the difference in total λE from the wet and dry soil was not significant. They concluded that management practices aimed at reducing soil evaporation might increase canopy transpiration and not reduce total evapotranspiration. As reported by Steduto and Hsiao (1998) who wrote about the pivotal role of radiation in latent heat flux, our data confirmed the strong dependence of evapotranspiration on the amount of available energy during both periods. However, this dependence was much higher for the wet than for the dry period. Also, greater data scatter was observed during the dry period, especially when energy fluxes were low.

For the wet period:

$$\lambda E = 0.97\left(R_n - G\right),\ \text{with}\ r^2 = 0.99 \tag{11}$$

For the dry period:

$$\lambda E = 0.95\left(R_n - G\right),\ \text{with}\ r^2 = 0.97 \tag{12}$$

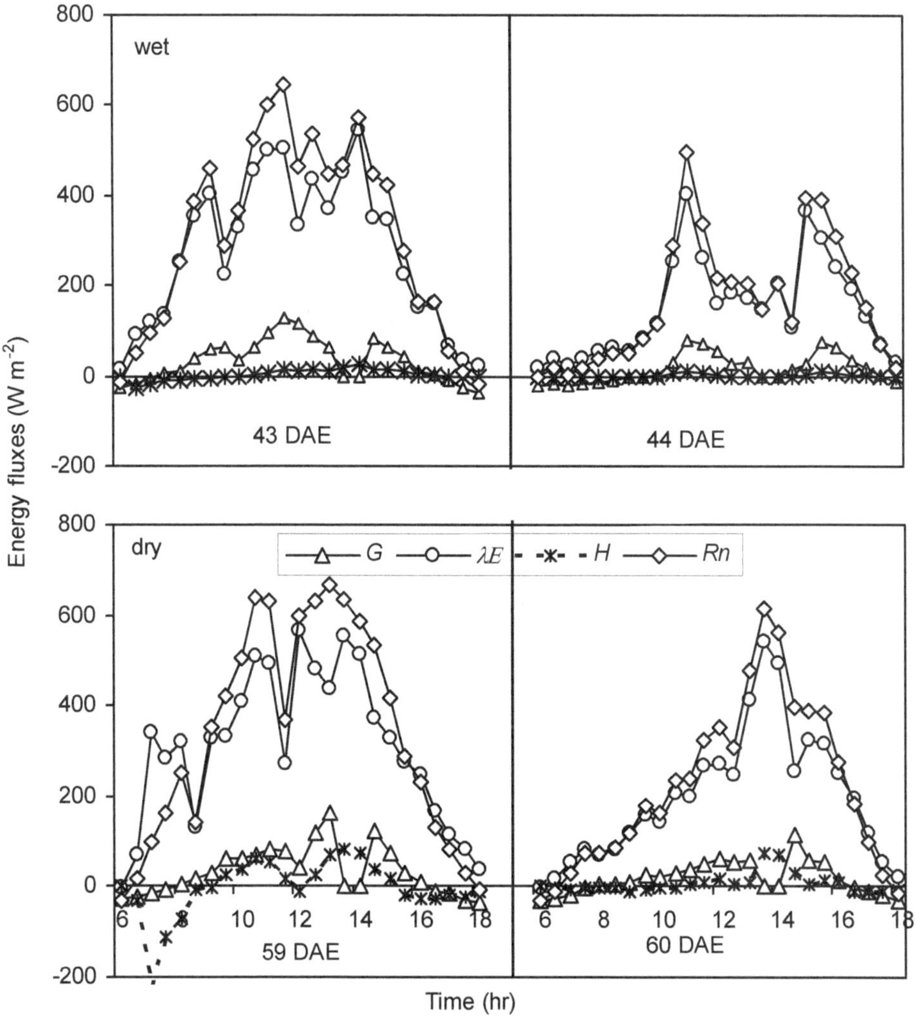

Fig. 3. Diurnal patterns of energy fluxes from maize field for the wet and dry periods

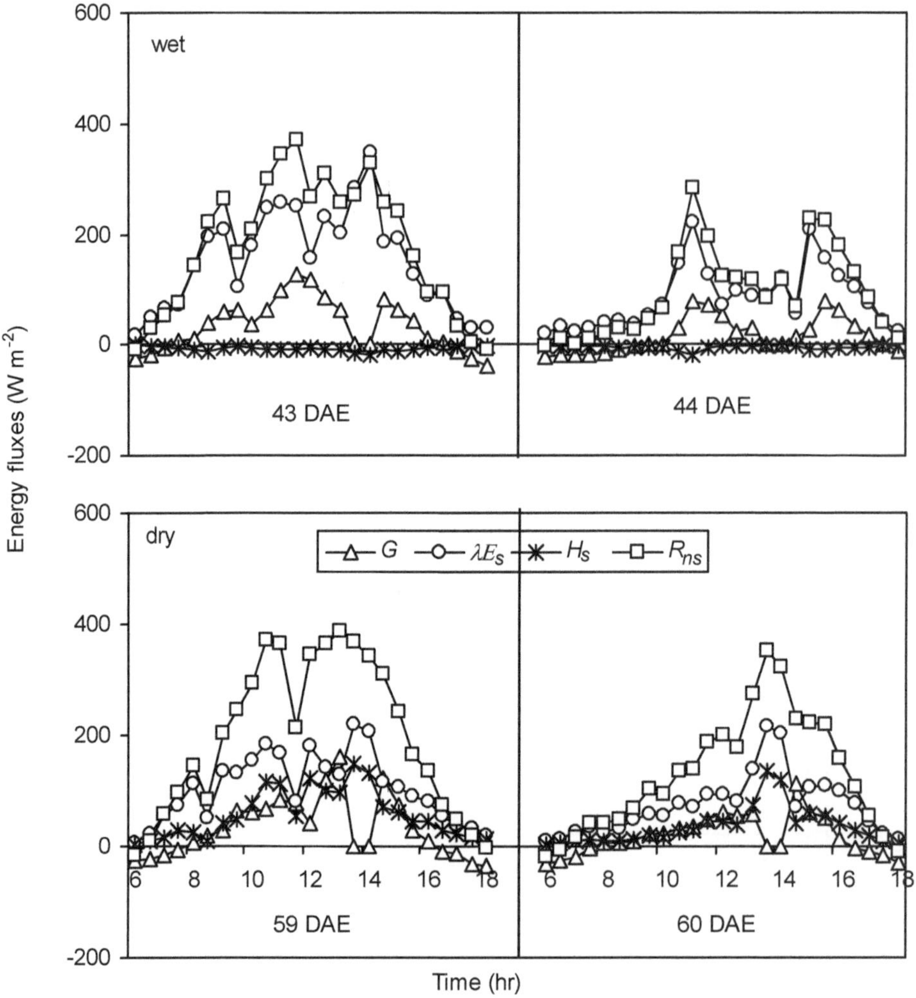

Fig. 4. Diurnal patterns of energy fluxes from soil surface for the wet and dry periods

5.2 Diurnal pattern of energy fluxes from soil surface for the wet and dry periods

Diurnal patterns of energy fluxes from soil for the wet and dry periods by the DOLBOREB system are shown in Fig. 4. There were large differences in energy flux patterns between the wet and dry periods. For the wet period, almost all available energy was directed to generate latent heat flux, while soil sensible heat flux (H_s) remained negligible during daytime. At morning, soil sensible heat flux was low and negative indicating that soil surface temperature was low, creating an energy sink at soil surface. The ratio of H_s to net radiation to soil (R_{ns}) was less than 5 % and therefore was negligible. Similar conditions were reported for cotton by Ham et al. (1991) after irrigation. They concluded that a wet soil appears to reduce λE_c by acting as a sink for advective energy, while reducing the radiation load on the canopy. For the dry period, R_{ns} was almost equally divided into outgoing latent and sensible heat fluxes. This suggested that soil was not evaporating at its potential rate. During this period, a shortage of soil water content at the soil upper layer reduced soil evaporation and much energy was directed to warm the soil rather than to evaporate soil water. Similar results were reported by Ham et al. (1991) on cotton, who reported that soil evaporation proved to be the primary form of latent heat flux when soil was wet, even when the L was between two and three, and that soil evaporation was markedly reduced by dry surface conditions.

Linear regression between available energy (R_{ns} - G) and latent heat flux from soil for the wet and dry periods showed a reduction of λE_s for the dry period of about 35 % of available energy to soil surface. Also, more scattered data were observed for the dry period, indicating lower dependence of latent heat flux from soil on R_{ns}.

For the wet period:

$$\lambda E_s = 1.07\left(R_{ns} - G\right), \text{ with } r^2 = 0.99 \tag{13}$$

For the dry period:

$$\lambda E_s = 0.65\left(R_n - G\right), \text{ with } r^2 = 0.94 \tag{14}$$

5.3 Diurnal pattern of energy fluxes from canopy for the wet and dry periods

Diurnal patterns of energy fluxes from canopy for the wet and dry periods by the DOLBOREB system are shown in Fig. 5. There were large differences in energy flux patterns from canopy between the wet and dry periods.

For the wet period, canopy latent heat flux (H_c) was low and most of the available energy for canopy was directed to generate λE_c, mainly because of sparse canopy. During this period no major energy exchanges occurred between soil and canopy.

Negative values of H_c, and positive values of H and H_s, indicated that the canopy was absorbing sensible heat that was generated at soil surface during the dry period. The within-row advection occurred during most of the day. However, Heilman et al. (1994) for vineyard reported similar observations occurred mainly in the afternoon where canopy temperature was as much as 5°C lower than air temperature. Also, Ham et al. (1991) reported for cotton that a wet soil appears to reduce λE_c by acting as a sink for advective energy, while also reducing the radiation load on the canopy. Extensive literature concerning radiation balance studies of row crops indicated soil and canopy can influence λE_c and λE_s (Tanner, 1960; Fuchs, 1972). However, inadequate measurements techniques have limited research to a specific set of conditions or the examination of a singular process (Ham et al., 1991).

The linear regression lines between λE_c and R_{nc} were obtained with high values of r^2 as shown in the following Eqs. 15 and 16.

For the wet period:

$$\lambda E_c = 0.87\, R_{nc}, \text{ with } r^2 = 0.99 \tag{15}$$

For the dry period:

$$\lambda E_c = 1.26\, R_{nc}, \text{ with } r^2 = 0.93 \tag{16}$$

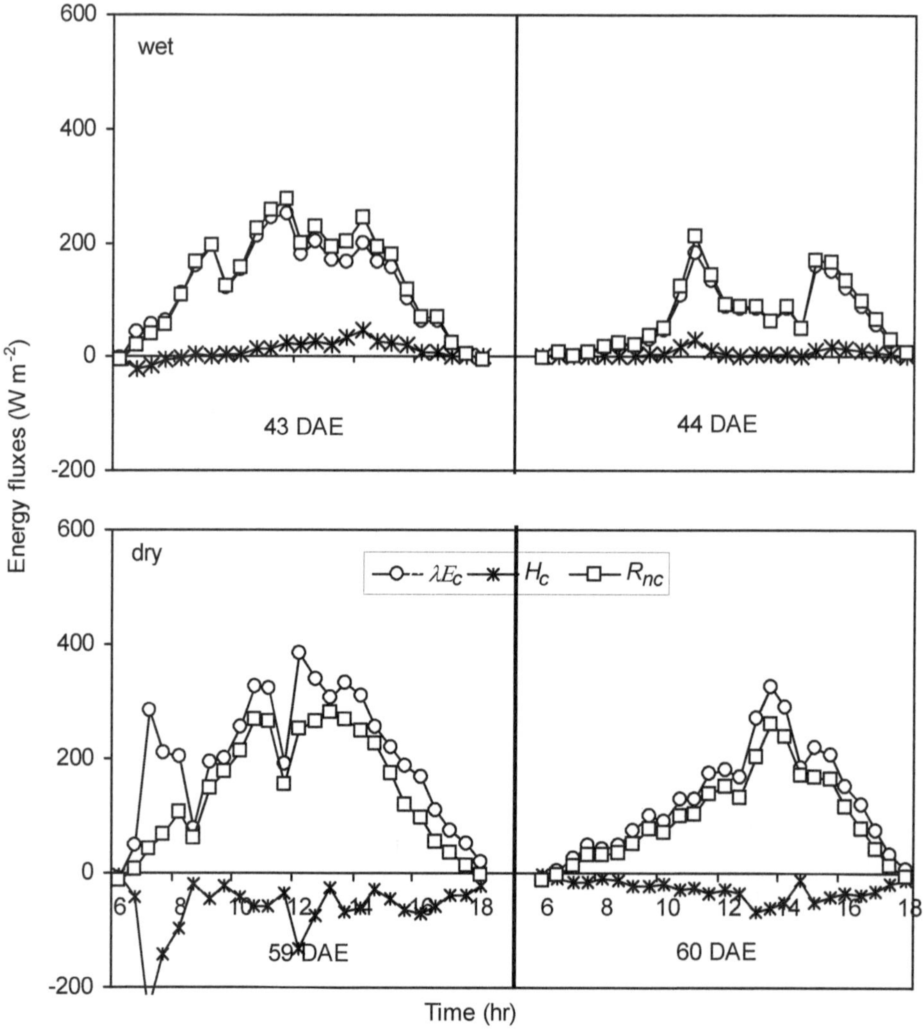

Fig. 5. Diurnal patterns of energy fluxes from canopy for the wet and dry periods

5.4 Summary of energy balance differences between wet and dry periods

Figure 6 shows a summary of the typical patterns of energy balances over maize field, soil surface and maize canopy by the DOLBOREB system during wet and dry periods. No major differences were observed, at maize field level, for energy balance patterns between wet and dry period. Evapotranspiration at maize field level remained high during both periods, and ratio of sensible heat, and soil heat fluxes to daily net radiation too. In fact, maize field energy balance measurements alone provide virtually no information on how energy balances of soil surface and canopy are partitioned. This shows clearly the limitations of considering crop field evapotranspiration as a whole, especially when addressing such important issues as would be water use efficiency improvement. A number of factors have contributed to this situation. The high cost of the equipment involved in such experiments and the inherent errors associated with the use of different measurement devices and measurement scales tremendously hinted the large-scale adoption of such techniques either by research scientists and/or by irrigation practitioners (Zeggaf et al., 2008). The DOLBOREB system indicated that soil had a major impact on the energy balance over maize canopy level. Also, the experiment shows that a frequent irrigation regime, as during the wet period, is not necessarily a synonym of maximum plant transpiration. In accordance with these results, Ham et al., (1991) concluded that a wet soil appears to reduce λE_c by acting as a sink for advective energy, while reducing the radiation load on the canopy.

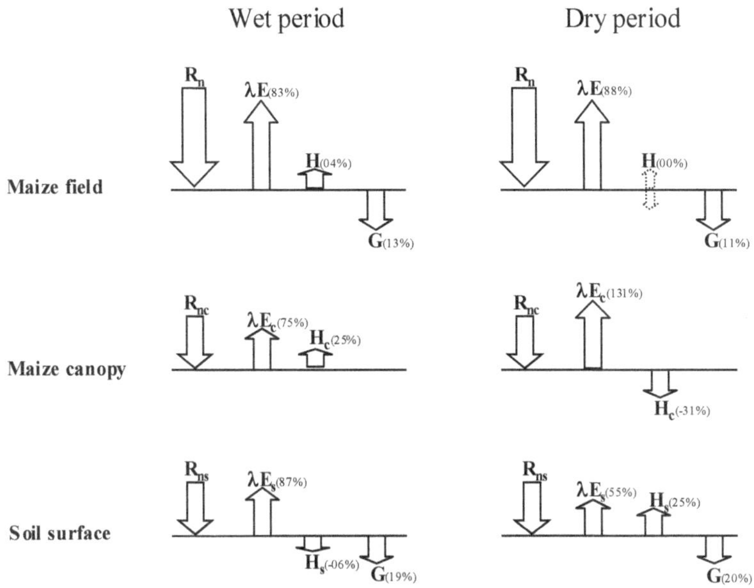

Fig. 6. Typical patterns of energy balances over maize field, soil surface and maize canopy by the DOLBOREB system during wet and dry periods.

The DOLBOREB system proved effective in depicting energy exchange phenomena between maize canopy and soil surface. These energy exchanges, mainly soil surface advection, boosted maize canopy transpiration during the dry period. Sensible heat flux generated from soil surface was absorbed by maize canopy, which increased dramatically latent heat flux from maize canopy. Similar results were obtained by Ham et al. (1991) who reported that within row advection increased λE_c. This finding corroborates the concept of "more crop per drop" which summarizes the objective of water use efficiency studies. In fact, proper irrigation management practices aimed at improving water use efficiency at crop field level might increase canopy transpiration, reduce soil evaporation, but not reduce total evapotranspiration. We suppose that this objective could be achieved through adoption of appropriate irrigation scheduling aiming to take advantage of soil surface advection or supplemental irrigation in arid and semiarid areas. Finally, the use of the DOLOBOREB system for irrigation scheduling could improve water use efficiency at crop field level and save up to 56 % of irrigation water without compromising crop production.

6. Conclusions

Future studies for other crops and under different climatic conditions are needed to improve our knowledge of water relations at crop field level. Examining the effect of factors such canopy size, crop type, and plant water stress...etc. on soil surface and canopy energy balances is of considerable importance. Energy flux data generated by the DOLBOREB system would be useful for building evapotranspiration, and crop growth models.

7. References

Allen, R.G., Pereira, L.S., Raes, D., Martin, M., 1998. Crop evapotranspiration. Guidelinesfor Computing Crop Water Requirements, Food and Agriculture Organization of the United Nations, FAO Irrigation and Drainage Paper 56, Rome, 300p.

Allen, S.J., 1990. Measurement and estimation of evaporation from soil under sparse barley crops in northern Syria. Agric. For. Meteorol. 49, 291-309.

Ashktorab, H., W.O. Pruitt, K.T. Paw U, and W.V. George. 1989. Energy balance determinations close to the soil surface using a micro-Bowen ratio system. Agric. For. Meteorol. 46:259-274.

Baker, J.M., Nieber, J.L., 1989. An analysis of the steady-state heat balance method for measuring sap flow in plants. Agric. For. Meteorol. 48, 93-109.

Baker, J.M., Van Bavel, C.H.M., 1987. Measurement of mass flow of water in the stems of herbaceous plants. Plant Cell. Environ. 10, 777-782.

Ben-Asher, J., Matthias, A.D., Warrick, A.W., 1983. Assessment of evaporation from bare soil by infrared thermometry. Soil Sci. Soc. Am. J. 47, 185-191.

Boast, C.W., Robertson, T.M., 1982. A "micro-lysimeter" method for determining evaporation from bare soil: description and laboratory evaluation. Soil Sci. Soc. Am.J. 46, 689-696.

Dugas, W.A., J.L. Fritshen, L.W. Gay, A.A. Held, A.D. Matthias, D.C. Reicosky, P. Steduto, and J.L. Steiner. 1991. Bowen ratio, eddy correlation and portable chamber measurements of sensible and latent heat flux over irrigated spring wheat. Agric. For. Meteorol. 56:1-20.

Fox, M.J., 1968. A technique to determine evaporation from dry stream beds. J. Appl. Meteorol. 7, 697-701.

Fritschen, L.J., Gay, L.W., 1979. Environmental instrumentation. New York: Springer. Grieve, B.J., Went, F.W., 1965. An electric hygrometer apparatus for measuring water vapour loss from plants in the field. Arid Zone Res. 25, 247-256. Hales, S. (1727). "Vegetable Staticks" W. & J. Inneys and T. Woodward. London.

Ham, J.M., J.L. Heilman, and R.J. Lascano. 1990. Determination of soil water evaporation and transpiration from energy balance and stem flow measurements. Agric. For. Meteorol. 59:287-301.

Ham, J.M., J.L. Heilman, and R.J. Lascano. 1991. Soil and canopy energy balances of a row crop at partial cover. Agron. J. 83:744-753.

Heilman, J.L., C.L. Brittin, and C.M.U. Neale. 1989. Fetch requirements of Bowen ratio measurements of latent and sensible heat fluxes. Agric. For. Meteorol. 44:261-273.

Heilman, J.L., K.J. McInnes, and R.W. Gesch, R.J. Lascano, and M.J. Savage. 1996. Effects of trellising on the energy balance of a vineyard. Agric. For. Meteorol. 81:79-93.

Heilman, J.L., McInnes, K.J., Savage, M.J., Gesch, R.W., Lascano, R.J., 1994. Soil and canopy energy balances in a west Texas vineyard. Agric. For. Meteorol. 71, 99-114.

Held, A.A., Steduto, P., Oegaz, F., Matista, A., Hsiao, T.C., 1990. Bowen-ratio/energy-balance technique for estimating crop net CO_2 assimilation, and comparison with a canopy-chamber. Theor. Appl. Climatol. 42, 203-213.

Hicks, B.B. 1973. Eddy fluxes over a vineyard. Agric. Meteorol. 12:203-215. Hillel, D., 1980. Application of soil physics. Academic Press. New York.

Hygen, G., 1953. Studies in plant transpiration. II. Physiol. Plant. 6, 106-133. Ishida, T., Campbell, G.S., Calissendorff, C., 1991. Improved heat balance method for determining sap flow rate. Agric. For. Meteorol. 56, 35-48.

Jara, J., C.O. Stockle, and J. Kjelgaard. 1998. Measurement of evapotranspiration and its components in a corn (Zea Mays L.) field. Agric. For. Meteorol. 92:131-145.

Jarvis, P.G., Mansfield, T.A., 1981. Stomatal Physiology. Cambridge Univ. Press. London and New York.

Kalthoff, N., M. Fiebig-Wittmaack, C. Meißner, M. Kohler, M. Uriarte, I. Bischoff-Gauß, and E. Gonzales. 2006. The energy balance, evapo-transpiration and nocturnal dew deposition of an arid valley in the Andes. J. Arid Environ. 65:420-443.

Kaufmann, M.R., 1981. Automatic determination of conductance, transpiration, and environmental conditions in forest trees. For. Sci. 27, 817-827.

Kaul, O.N., Kramer, P.J., 1965. Comparative drought resistance of two woody species. Indian For. 91, 462-469.

Kramer, P.J., 1983. Water Relations of Plants. Academic Press, Inc. (London) LTD. 489 pp. Hales, S. (1727). "Vegetable Staticks" W. & J. Inneys and T. Woodward. London.

Kustas, W.P., D.I. Stannard,and K.J. Allwine. 1996. Variability in surface energy flux partitioning during Washita'92: resulting effects on Penman-Monteith and Priestley-Taylor parameters. Agric. For. Meteorol. 82:171-193.

Lacirignola, C., Hamdy, A., Todorovic, M., 2003. Regional action program on "water resources management": an overview of actions towards better water use in Mediterranean agriculture. In A. Hamdy (eds.) Regional Action Program (RAP): Water resources management and water saving in irrigated agriculture (WASIA PROJECT). Bari: CIHEAM-IAMB, 1-18p.

Leadley, P.W., Drake, B.G., 1993. Open-top chambers for exposing plant canopies to elevated CO_2 concentration and for measuring net gas-exchange. Vegetatio. 104/105, 3-15.

Malek, E., 1993. Comparison of the Bowen ratio-energy balance and stability-corrected aerodynamic methods for measurement of evapotranspiration. Theor. Appl. Climatol. 48, 167-178.

Malek, E., and G.E. Bingham. 1993. Comparison of the Bowen ratio-energy balance and the water balance methods for the measurement of evapotranspiration. J. Hydrol. 146:209-220.

Oke, T.R. 1987. Boundary Layer Climates, second ed. Rouledge, London. 435 p.

Perez, P.J., F. Castellvi, M. Ibanez, and J.I. Rosell. 1999. Assessment of reliability of Bowen ratio method for partitioning fluxes. Agric. For. Meteorol. 97:141-150.

Philip, J. R., 1957. Evaporation and moisture and heat fields in the soil, J. Meteorol. 14, 354-366.

Prueger, J.H., J.L. Hatfield, J.K. Aase, and J.L. Pikul Jr. 1997. Bowen-ratio comparisons with lysimetric evapotranspiration. Agron. J. 89:730-736.

Raber, O., 1937. Water utilization by trees, with special reference to the economic forest species of the north temperate zone. Misc. Publ.-U.S., Dep. Agric. 257.

Reicosky, D.C., Wagner, S.W., Devine, O.J., 1990. Methods of calculating carbon dioxide exchange rates for maize and soybean using a portable field chamber. Photosynthetica. 24, 22-38.

Ritchie, J.T. 1971. Dryland evaporation flux in a subhumid climate: I. Micrometeorological influences. Agron. J. 63:51-55.

Ritchie, J.T., 1972. Model for predicting evaporation from a row crop with incomplete cover. Water Resour. Res. 8, 1204-1213.

Roberts, J.M., 1977. The use of tree-cutting techniques in the study of the water relations of Pinus sylvestris L. I: the techniques and survey of the results. J. Exp. Bot. 28, 751- 767.

Sakuratani, T., 1981. A heat balance method for measuring water flux in the stem of intact plants. J. Agric. Meteorol. (Japan). 37, 9-17.

Sakuratani, T., 1984. Improvement of the probe for measuring water flow rate in intact plants with the stem heat balance method. J. Agric. Meteorol. (Japan). 40, 273-277.

Sene, K.J. 1996. Meteorological estimates for the water balance of a sparse vine crop growing in a semiarid condition. J. Hydrol. 179:259-280.

Slavik, B., 1974. Methods of Studying Plant Water Relations. Springer-Verlag, Berlin and New york.

Steduto, P., Cetinkoku, O., Albrizio, R., Kanber, R., 2002. Automated closed-system canopy-chamber for continuous field-crop monitoring of CO_2 and H_2O fluxes. Agric. For. Meteorol. 111, 171-186.

Steduto, P., Hsiao, T.C., 1998a. Maize canopies under two soil water regimes. I. Diurnal patterns of energy balance, carbon dioxide flux, and canopy conductance. Agric. For. Meteorol. 89, 169-184.

Steduto, P., Hsiao, T.C., 1998b. Maize canopies under two soil water regimes. IV. Validity of Bowen ratio-energy balance technique for measuring water vapor and carbon dioxide fluxes at 5-min intervals. Agric. For. Meteorol. 89:215-228.

Tanner, C.B., 1960. Energy balance approach to evapotranspiration from crops. Soil Sci. Soc. Am. Proc. 24, 1-9.

Tanner, C.B., Peterson, A.E., Love, J.R., 1960. Radiant energy exchange in a corn field. Agron. J. 52, 373-379.

Verma, S.B., Rosenberg, N.J., Blad, B.L., 1978. Turbulent exchange coefficients for sensible heat and water vapor under advective conditions. J. Appl. Meteorol. 14, 330-338.

Walker, G.K., 1983. Measurement of evaporation from soil beneath crop canopies. Can. J. Soil Sci. 63, 137-141.

Yunusa, I.A.M., R.R. Walker, and P. Lu. 2004. Evapotranspiration components from energy balance, sap flow and microlysimetry techniques for an irrigated vineyard in inland Australia. Agric. For. Meteorol. 127:93-107.

Zeggaf, T.A., Filali, M. Y., 2010. Problématique et perspectives de L'efficience d'utilisation de l'eau agricole au Maghreb. In l'Etat des Ressources en Eau au Maghreb en 2009. UNESCO publication. ISBN-978-9954-8068-3-0. 203-217.

Zeggaf, T.A., H. Anyoji. S. Takeuchi. T. Yano. 2007. Partitioning energy fluxes between canopy and soil surface under sparse maize during wet and dry periods. In 'Water saving in Mediterranean agriculture and future research needs'. Options Méditerranéennes Série B n° 56. Vol I, 201-211.

Zeggaf, T.A., S. Takeuchi, H. Dehghanisanij, H. Anyoji, T. Yano. 2008. A Bowen ratio technique for partitioning energy fluxes between maize transpiration and soil surface evaporation. Agronomy Journal. 100, 988-996.

6

Daily Crop Evapotranspiration, Crop Coefficient and Energy Balance Components of a Surface-Irrigated Maize Field

José O. Payero[1] and Suat Irmak[2]
[1]*The University of Queensland, Queensland Alliance for Agriculture and Food Innovation (QAAFI), 203 Tor St, Toowoomba, QLD 4350,*
[2]*University of Nebraska-Lincoln, 241 L.W. Chase Hall, Lincoln, NE 68583-0726,*
[1]*Australia*
[2]*USA*

1. Introduction

Irrigation water shortage is affecting many areas of the world. This is the case in Nebraska and in other areas of the USA High Plains. In Nebraska, field maize (*Zea mays L.*) is the primary irrigated row crop and is one of the most important drivers of the state's economy. The most typical cropping rotations in Nebraska are continuous maize and maize following soybeans. With the development of ethanol and bio-fuel industries, continuous maize farming practices are increasing. Over-pumping of groundwater for irrigation, coupled with recent drought conditions, and increasing continuous maize production are causing serious concerns about the future availability and sustainability of water resources. Increasing fuel prices, which increases pumping costs should provide an economic incentive for farmers to match irrigation applications to crop water requirements. Accurate determination of actual crop evapotranspiration (ETc) can improve utilization of water resources through well-designed irrigation management programs. Reliable estimates of ETc are also vital for developing criteria for in-season irrigation management, water allocations, long-term estimates of water supply, water demand and use, design and management of water management infrastructures, and the effect of changes in land use and management on water balances (Irmak and Irmak, 2008).

Water shortage is creating a need for growers to produce crops using less water. An option is to use more efficient irrigation systems, which has already occurred to a large extend over the last four decades in the USA Midwestern states by converting from surface irrigation to overhead sprinkler systems. For example, water and labour shortages have already driven growers in Nebraska to switch from surface irrigation to more efficient overhead sprinkler irrigation systems (mainly center pivots). Currently, of the 3.5 million ha of irrigated land, 75% is irrigated using center pivots and 25% still uses surface irrigation, mainly furrow irrigation. Adoption of more efficient drip systems, like subsurface drip irrigation, to irrigate maize, however, has been limited due mainly to their high initial cost and maintenance requirements. Therefore, additional gains in irrigation efficiency will have to come from improved management of existing irrigation methods and by using sound irrigation scheduling techniques. A potentially useful irrigation scheduling technique can be implemented by developing and adopting computer tools to estimate daily soil water status. These tools normally use daily weather data as input to estimate crop evapotranspiration (ETc) and irrigation requirements from a daily soil water balance. For these tools to be accurate, they need to be validated with direct measurements of daily ETc from local crops and weather conditions.

Direct measurement of ETc can be performed using techniques such as weighing lysimeters, eddy covariance, Bowen ratio energy balance, surface renewal, and soil water balance. While each method has its own advantages and drawbacks, they have been successfully used to quantify ETc under different conditions. However, their practical and widespread application is limited, mainly due to their high cost and the need for skilled personnel to install and maintain equipment and to interpret the resulting data. Because of the lack of direct ETc measurements, ETc is usually estimated from weather data using a two-step approach (Doorenbos and Pruitt, 1975; 1977, FAO-56, 1998). In this approach, ETc is estimated by adjusting the calculated reference evapotranspiration (ETref), either grass- or alfalfa-reference ET (ETo or ETr, respectively), with a crop coefficient (Kc) value for the specific crop and growth stage (i.e., ETc = Kc x ETref). The Kc value takes into account the morphological and physiological characteristics of the crop and, to a limited degree, the effect of management practices (Wright, 1982; FAO-56, 1998).

Carefully developed Kc values, along with ETref values, can provide robust and accurate ETc estimates. This method relies heavily on published Kc values for different crops. While there are significant assumptions imbedded in the Kc values, which may not apply to many locations/conditions, the simplicity of this method has been one of the main reasons for its widespread adoption. However, accuracy for local conditions could be compromised by uncertainties in generalized Kc values. Kc values for the same crop can show significant variation between locations due to differences in crop variety, soil properties, irrigation method and frequency, climate, and perhaps more importantly, crop management practices. Consequently, the Kc values reported in the literature can vary significantly from actual values if crop growing conditions differ from those where they were experimentally derived. Thus, while it is a difficult and challenging task, locally-derived Kc values are preferable to produce more robust and accurate ETc estimates. The objectives of this study were to: (1) compare grass-reference and alfalfa-reference ET estimates under the conditions of West-Central Nebraska, (2) quantify the daily magnitude and seasonal trends of energy balance components and ETc for maize and, (3) derive daily maize Kc values under local conditions and compare these values with those reported in FAO-56.

2. Methods

2.1 Site description and crop management

Surface energy balance components were measured during 2001 at the University of Nebraska-Lincoln West Central Research and Extension Center, at North Platte, Nebraska (41.1° N, 100.8° W, 861 m above sea level). Measurements were made on a surface-irrigated maize field that was under a ridge-tilled maize-soybean rotation system (Fig. 1). The soil at this site is classified as Cozad silt loam (*Fluventic Haplustolls*), but profile sampling has shown that the soil in this field is predominantly loam to a depth of 1.2 m.

Maize was planted at 0.76-m row spacing on 8 May 2001 [Day of year (DOY) = 128], within the normal planting window for maize in the region. The planting density was 7.6 seeds per m^2 with a north-south row direction. The maize hybrid planted was DeKalb DKC63-22 (YG), which had a relative maturity of 113 days. The maize field was irrigated with gated pipes, every other furrow, using groundwater pumped from a deep well. The field was dedicated to commercial maize production; therefore, the crop was fully-irrigated to maximize yield by preventing water stress during the entire season. Only three irrigations were needed to supplement timely rainfall events, which were applied on 8 July, 8 August, and 25 August. Water application depths were not recorded, but neutron probe soil water measurements after the first irrigation event (data not shown) indicated that the applied irrigation water in each application was enough to refill the soil profile.

2.2 Energy balance measurements

Energy balance variables were measured using an eddy covariance system (Campbell Scientific, Inc., Logan, Utah) installed on a tower at the center of the experimental field (Fig. 1). The system had about 150 m of fetch in all directions. The field was surrounded by soybean to the north and east and by native grasses to the west and south. Predominant winds were in the north-west to south-east direction. The measurements were made during most of the year, from DOY 11 (11 January) to DOY 353 (19 December). There was a gap in data collection from DOY 38-85 (7 February–26 March) and several small gaps due to system malfunction. All system components, except the soil moisture, soil heat flux, and soil temperature sensors, were kept approximately 1 m above the crop canopy throughout the measurement period. Table 1 provides the list of instrumentation type and variables measured.

Data were sampled, processed and recorded with a CR23X datalogger (Campbell Scientific, Inc., Logan, Utah). The datalogger was housed in an environmental enclosure and had a SM16M data storage module. A 12-Volt (75-Amp) marine deep cycle battery, charged by a solar panel supplied power to the system. The sonic anemometer (CSAT3), fine wire thermocouple (FW05), and a krypton hygrometer (KH20) sensors were sampled at a frequency of 10 Hz (10 times per second). All other sensors were sampled once per minute. The 30-min averages were stored, and daily averages were calculated during post-processing.

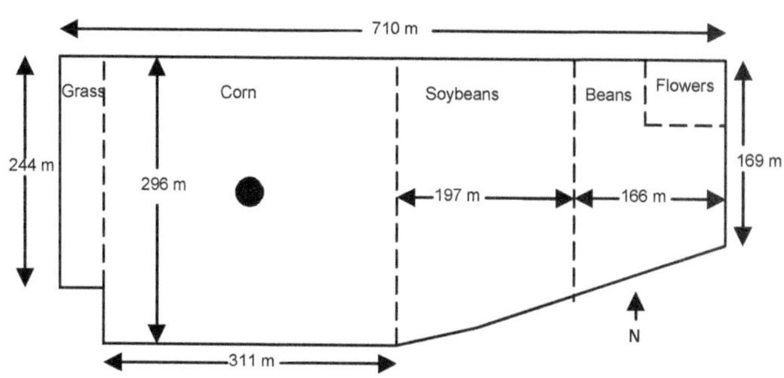

Fig. 1. Layout of the experimental field at North Platte, Nebraska, during the 2001 study. The black dot shows the location of the eddy covariance system used to measure the energy balance variables.

Sensor Model	Sensor Type	Manufacturer	Variable Measured
LI200X	Silicon pyranometer	Licor, Inc. Lincoln, Nebraska	Solar radiation
Q7.1	Net radiometer	Radiation and Energy Balance Systems, Inc., Bellevue, Washington	Net Radiation
CS500	Temperature/RH sensor	Campbell Scientific, Inc. Logan, Utah	Air temperature/Relative Humidity
CSAT3	3-D sonic anemometer	Campbell Scientific, Inc. Logan, Utah	3-D wind speed/temperature
FW05	Fine-wire thermocouple	Campbell Scientific, Inc. Logan, Utah	Air temperature
KH20	Krypton hygrometer	Campbell Scientific, Inc. Logan, Utah	Absolute air humidity
CS615	Water content reflectometer	Campbell Scientific, Inc. Logan, Utah	Volumetric soil water content
HFT3	Soil heat flux plate	Radiation and Energy Balance Systems, Inc., Bellevue, Washington	Soil heat flux
TCAV	Soil thermocouple	Campbell Scientific, Inc. Logan, Utah	Soil temperature
03001	3-cup anemometer/wind vane	RM Young Company Traverse City, Michigan	Horizontal wind speed and direction

Table 1. Sensors used and variables measured at North Platte, Nebraska, during 2001.

The 30-min averages included all of the energy balance components (in units of W m^{-2}), such as incoming shortwave solar radiation (Rs), net radiation (Rn), latent heat flux (LE), sensible heat flux (H), and soil heat flux (G). G was calculated from the output of the two soil heat flux plates, the soil temperature sensor (which included four thermocouples), the soil water content reflectometer, and soil physical properties (Hanks and Ashcroft, 1980) as:

$$G = SHF + S \tag{1}$$

$$S = (T_i - T_{i-1}) \times D \times C_s/t \tag{2}$$

$$Cs = BD \times (C_{sd} + W \times C_w) \tag{3}$$

where, SHF = flux measured by the soil heat flux plates (W m^{-2}), S = change in heat stored above the soil heat flux plates (W m^{-2}), T_i = soil temperature during current time interval (°C), T_{i-1} = soil temperature during previous time interval (°C), D = depth to soil heat flux plates (m), C_s = heat capacity of soil (J m^{-3} °C^{-1}), t = time interval (s), BD = soil bulk density (kg m^{-3}), C_{sd} = specific heat of mineral soil (840 J kg^{-1} °C^{-1}), W = soil water content on a mass basis (kg kg^{-1}), and C_w = specific heat of water (4,190 J kg^{-1} °C^{-1}). The two soil heat flux plates were installed 1 m apart at a depth of 0.08 m from the soil surface. The two soil thermocouples were installed next to each soil heat flux plate at depths of 0.02 and 0.06 m.

The soil water content reflectometer was installed horizontally midway the two soil heat flux plates at a depth of 0.025 m.

The experimental site was visited at least weekly to download data, provide regular sensor maintenance, and measure crop canopy height. Crop canopy height was measured 17 times throughout the season. The 30-min energy fluxes for each day were plotted and visually inspected to detect inconsistent data. The LE values calculated from the KH20 sensor were rejected since they were unreasonably low, and daily LE values were instead calculated as a residual from the one dimensional energy balance equation (LE = Rn- G - H). As part of data validation, data were also excluded if the daily LE or Rn were greater than Rs, or if the daily average Rn was negative.

2.3 Weather data

In addition to the eddy covariance measurements, weather data for calculating daily reference ET were measured with an electronic weather station located at the research station, which was part of the High Plains Regional Climate Center (HPRCC) network. The data were downloaded from the HPRCC (http://www.hprcc.unl.edu/home.html), including daily maximum (Tmax) and minimum (Tmin) air temperature, relative humidity (RH), wind speed (measured at 3-m height), solar radiation, rainfall, and alfalfa-reference ET (ETr-Neb). The HPRCC calculated ETr-Neb using an equation developed by Kincaid and Heermann (1974) by modifying the Penman (1948) equation with an empirical wind function for Mitchell, Nebraska. Kincaid and Heermann (1974), however, suggested that the coefficients used in the new equation were nearly the same as those reported by Jensen (1969) for Twin Falls, Idaho (Irmak et al., 2008).

2.4 Reference ET, ETc and Kc

Daily grass- and alfalfa-reference ET were calculated using the standardized FAO-56 Penman-Monteith method (FAO-56, 1998) as:

$$\text{ETref} = \frac{0.408\Delta(R_n - G) + \gamma \dfrac{C_n}{T+273} U_2\,(e_s - e_a)}{[\Delta + \gamma\,(1 + C_d U_2)\,]} \tag{4}$$

where, ETref = reference evapotranspiration, either grass-reference (ETo) or alfalfa-reference (ETr) evapotranspiration (mm d^{-1}), Δ = slope of the saturation vapour pressure versus air temperature curve (kPa °C^{-1}), Rn = net radiation at the crop surface (MJ m^{-2} d^{-1}), G = heat flux at the soil surface (MJ m^{-2} d^{-1}), T = mean daily air temperature at 1.5 to 2.5 m height (°C), U_2 = mean daily wind speed at 2 m height (m s^{-1}), e_s = saturation vapor pressure (kPa), e_a = actual vapor pressure (kPa), e_s - e_a = vapour pressure deficit (kPa), γ = psychrometric constant (kPa °C^{-1}), C_n = numerator constant (°C mm s^3 Mg^{-1} d^{-1}), C_d = denominator constant (s m^{-1}), 0.408 = coefficient having units of m^2 mm MJ^{-1}. Daily Rn, e_s, and e_a were calculated using the equations given by FAO-56 (1998) and ASCE-EWRI (2005) using measured RH, Tmax, and Tmin, and constant albedo (α = 0.23). Values for the Stefan-Boltzmann constant (σ = 4.901 × 10^{-9} MJ K^{-4} m^{-2} d^{-1}) (for calculating net outgoing long-wave radiation (R_{nl})), specific heat at constant temperature (c_p = 1.013 × 10^{-3} MJ kg^{-1} °C^{-1}), and latent heat of vaporization (λ = 2.45 MJ kg^{-1}) followed FAO-56 and ASCE-EWRI (2005). The psychrometric constant (γ) was computed as a function of atmospheric pressure (P), λ, c_p, and the ratio of molecular weight of water vapour to dry air (ε = 0.622). P was calculated as a function of station elevation (z), and daily G = 0 MJ m^{-2} d^{-1} was assumed. Wind speed measured at a height of 3 m was converted to the standard height of 2 m using equation 47 in FAO-56 (1998). Parameters for calculating daily ETo and ETr are shown in Table 2.

Parameter	Grass-reference	Alfalfa-reference
C_n	900 °C mm s^3 Mg^{-1} d^{-1}	1600 °C mm s^3 Mg^{-1} d^{-1}
C_d	0.34 s m^{-1}	0.38 s m^{-1}
Canopy Height	0.12 m	0.50 m
Surface resistance (r_s)	70 s m^{-1}	70 s m^{-1}

Table 2. Parameters used in the FAO-56 Penman-Monteith method for daily calculations of grass-reference and alfalfa-reference evapotranspiration (ETo and ETr).

In addition to the measured LE values, daily maize ET (ETc) for a well-watered maize crop was estimated using the reference ET and dual crop coefficient (Kc) approach [i.e., ETc = ETo x (dual Kc)] described in FAO-56. The basal Kc (K_{cb}) curve developed from values given in Table 11 and 17 of FAO-56, using the length of growth stages in Table 3 was used in the calculations. From the calculated ETo and measured ETc (LE) values, daily Kc values for maize were also derived (Kc = ETc/ETo), both as a function of days from emergence (DFE) and as a function of cumulative growing degree days from crop emergence (C-GDD). Measured Rn values over the maize canopy were also compared with those estimated using the FAO-56 Rn calculation method for a reference grass surface.

Growth Stage (GS)	Definition	Length (Days)	K_{cb}
Initial	Planting to 10% ground cover	30	0.15 (K_{cb} ini)
Crop Development	10% ground cover to effective full cover	40	-
Mid-Season	Effective full cover to start of maturity	50	1.15 (K_{cb} mid)
Late Season	Start of maturity to harvest or full senescence	50	0.15 (K_{cb} end)

Table 3. Lengths of crop growth stages and basal crop coefficient (K_{cb}) values for field maize given in FAO-56 (1998). Growth stage lengths are for Kimberly, Idaho.

3. Results and discussion

3.1 Weather conditions

Table 4 shows the monthly averages of weather variables during 2001 at North Platte, Nebraska. Freezing temperatures occurred from January to March and during November and December. July had the highest monthly temperatures (Tmax, Tmin, and T) and ETr, although the highest average solar radiation (Rs) occurred in June. Annual rainfall during 2001 was 564 mm, which was 33% (141 mm) above the long-term average (normal) of 423 mm for North Platte for the period of 1982-2006 (Fig. 2). Above-normal rainfall occurred mainly in April and August, with rainfalls of more than twice the long-term average (Fig. 3). Rainfall in April refilled the soil profile prior to planting in early May. Rainfall was about normal in May and July, just above normal in September, and about half of normal in June and October. Rainfall in October had no effect on crop development, since by that time the crop had already reached physiological maturity. There were 30 rainfall events during the growing season, with several events occurring between 18 and 32 days after crop emergence when the crop was small and the soil surface was still exposed to solar energy (Fig. 4).

Month	ETr (mm month^{-1})	Tmax (°C)	Tmin (°C)	T (°C)	U_2 (m s^{-1})	Rs (MJ m^{-2} d^{-1})	RH (%)
Jan	43.8	6.0	-7.8	-0.9	2.1	7.6	67.3
Feb	35.9	1.8	-9.4	-3.8	2.8	9.0	77.1
Mar	78.9	10.3	-2.8	3.8	2.5	13.3	71.2
Apr	147.8	18.0	3.1	10.6	3.1	18.6	62.2
May	170.8	22.0	8.8	15.4	2.6	20.4	61.9
Jun	227.3	28.8	12.6	20.7	2.8	24.9	61.0
Jul	233.4	33.7	18.0	25.9	2.7	24.0	67.7
Aug	200.9	30.7	14.9	22.8	2.3	21.5	65.5
Sep	142.6	25.2	9.8	17.5	2.4	16.1	68.6
Oct	112.6	19.5	1.4	10.4	2.1	12.5	57.2
Nov	62.7	14.1	-3.2	5.4	2.2	8.0	66.8
Dec	47.0	7.0	-8.4	-0.7	1.9	7.1	59.5
Avg	125.3	18.1	3.1	10.6	2.5	15.2	65.5
Min	35.9	1.8	-9.4	-3.8	1.9	7.1	57.2
Max	233.4	33.7	18.0	25.9	3.1	24.9	77.1

ETr = alfafa-reference evapotranspiration, Tmax = maximum air temperature, Tmin = minimum air temperature, T = average air temperature, U_2 = wind speed at 2 m height, Rs = solar radiation, RH = relative humidity.

Table 4. Average weather variables for each month during 2001 at North Platte, Nebraska

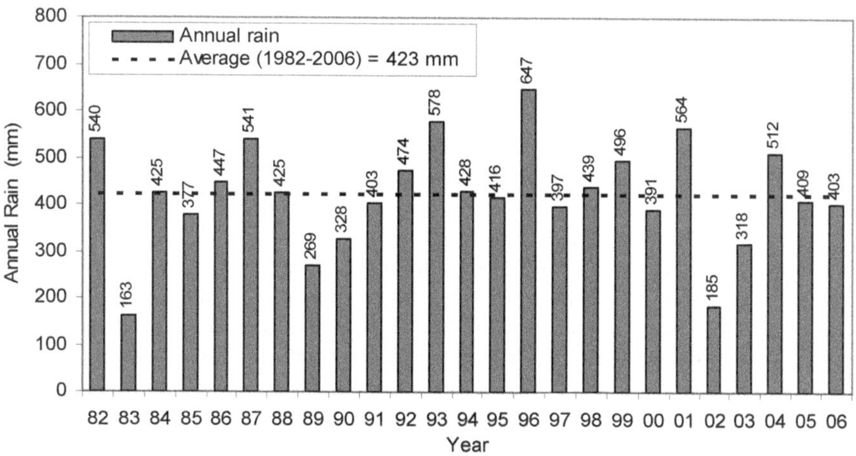

Fig. 2. Annual rainfall and long-term average (1982-2006) at North Platte, Nebraska

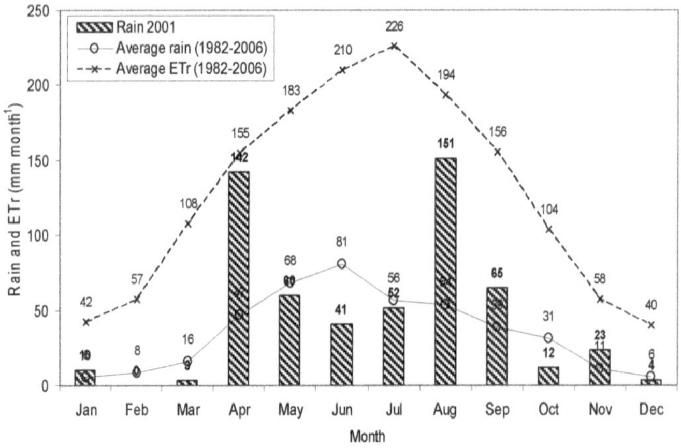

Fig. 3. Monthly rainfall at North Platte for 2001, and the long-term averages (1982-2006) for rainfall and alfalfa-reference evapotranspiration (ETr)

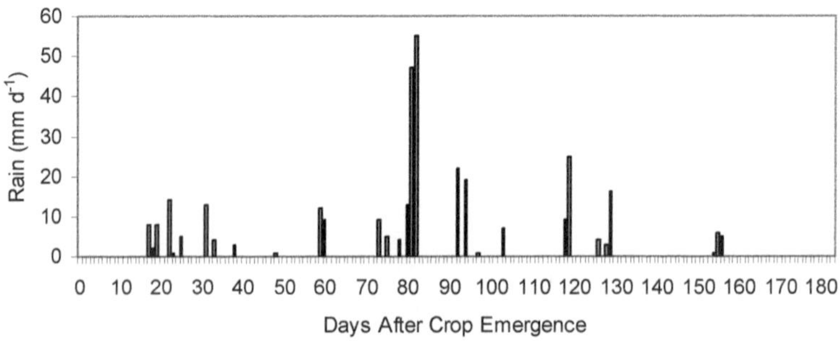

Fig. 4. Daily rainfall during the 2001 growing season at North Platte, Nebraska

3.2 Crop development

The crop emerged on 12 May (DOY 132), started tasseling on 13 July and about 90% of plants had tasseled by 16 July. The crop was at the silking stage by 19 July, matured at the end of September (about 25 September), and was harvested on 10 November. Figure 5 shows the progression in crop canopy height as a function of day of the year, day from emergence (DFE), and cumulative growing degree days from emergence (C-GDD). The crop grew slowly early in the season, but the growth rate increased as the season progressed until the crop reached maximum canopy height (2.7 m) at about 68 DFE (DOY 200, July 19, C-GDD = 723 °C-days). The crop had three periods with distinct growth rates (Fig. 6). The slopes of the three regression lines in Fig. 6 indicate that for DFE = 0 to 25 days, the crop grew at a rate of 0.008 m d^{-1}. The growth rate increased to 0.030 m d^{-1} during the period of DFE = 25 to 54 days. The highest growth rate was 0.114 m d^{-1} during DFE = 54 to 68 days.

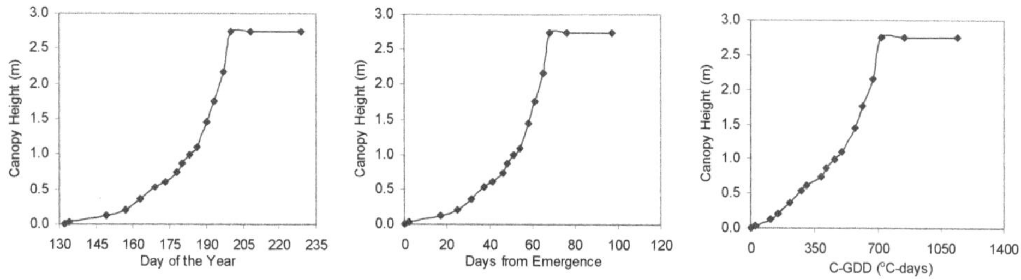

Fig. 5. Maize canopy height as a function of day of the year, days from emergence, and cumulative growing degree days from emergence (C-GDD) measured at North Platte, Nebraska, during the 2001 growing season. Daily growing degree days were calculated using a lower limit (base temperature) of 10 °C (50 °F), with no upper limit imposed.

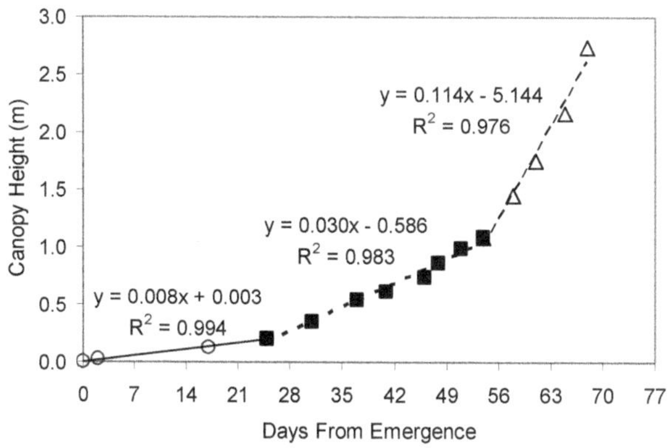

Fig. 6. Maize canopy height as a function of days from emergence measured in 2001 at North Platte, Nebraska. The lines are regression lines fitted for three different growing periods.

3.3 Reference ET

Calculated daily grass-reference ET for the whole year (ETo) (Fig. 7) varied from less than 0.5 mm d^{-1} during the winter to a maximum of about 10 mm d^{-1} during the summer. There were considerable day-to-day variations in ETo due to variations in climatic factors. Although linearly-related, there were considerable differences between the calculated grass-reference (ETo) and alfalfa-reference ET (ETr and ETr_Neb) values. Figure 8 shows good correlation of ETr and ETr_Neb[1] with ETo (R^2 of 0.97 and 0.98, respectively). The slopes of the lines, however, indicate that on average, ETr and ETr_Neb were 29% and 37% higher than ETo, respectively.

---

[1] ETr_Neb is alfalfa-reference evapotranspiration as reported by the High Plains Regional Climate Center (HPRCC). The HPRCC calculates ETr_Neb using a modified Penman method with an empirical wind function. The ETr values, on the other hand, were calculated using Equation 4.

Our results are in agreement with those observed by Irmak and Irmak (2008) who found an average ratio of standardized ASCE-PM ETr to ETr_Neb of 1.29 for the conditions of south central Nebraska. Figure 9 also shows that the magnitude of the ETr/ETo ratio varied throughout the year from 1.03 to 1.68, reaching a minimum value in the middle of the summer and a maximum during the winter. Understanding the magnitude of the differences between the ETo and ETr are important when selecting crop coefficients for a local region. Practitioners need to make sure that Kc values are matched with the appropriate reference surface for which they were developed. Under the conditions of this study, failure to do so could potentially result in errors in ETc estimation as high as 37%.

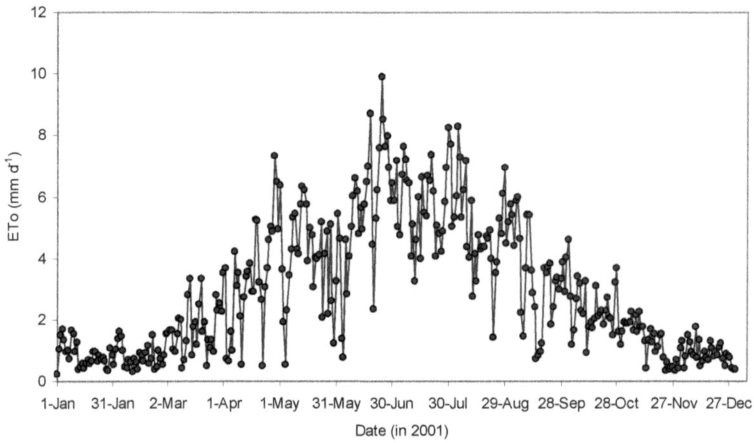

Fig. 7. Daily Grass-reference evapotranspiration (ETo) for 2001 at North Platte, Nebraska (n = 365).

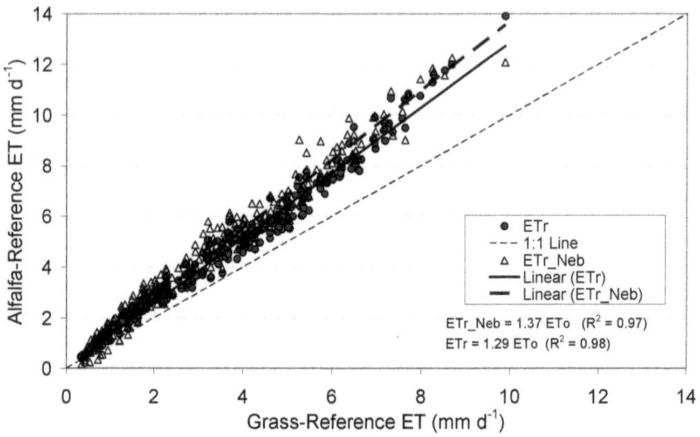

Fig. 8. Relationship between daily grass-reference and alfalfa-reference evapotranspiration values calculated for 2001 at North Platte, Nebraska. ETo and ETr are the grass- and alfalfa-reference evapotranspiration calculated using the FAO-56 procedure, and ETr_Neb are the alfalfa-reference evapotranspiration values obtained from the High Plains Regional Climate Center (n = 278).

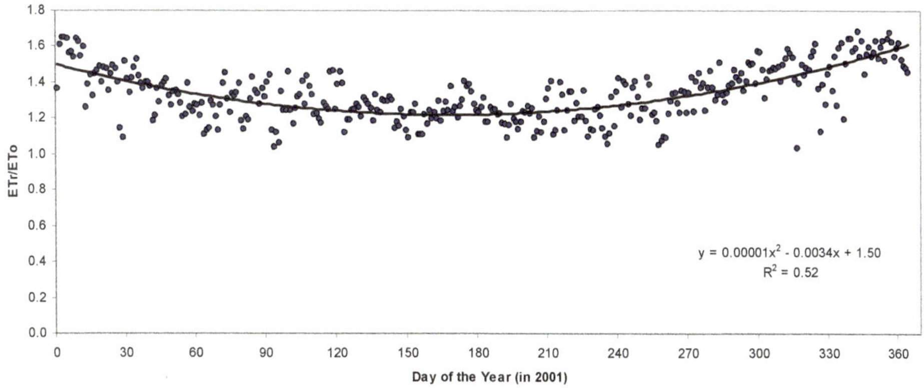

Fig. 9. Distribution of the ETr/ETo ratio as a function of day of the year during 2001 at North Platte, Nebraska. ETr and ETo are the estimated alfalfa-reference and grass-reference evapotranspiration, respectively, calculated with the standardized FAO-56 Penman-Monteith method (n = 365).

3.4 Energy fluxes

Figure 10 shows the trend in 30-min average energy fluxes for three cloud-free days during the maize growing season. In addition to clear-sky conditions, these days were presented because they represent a range of canopy cover conditions. Measurements on DOY 120 were taken prior to planting (with a bare soil surface), DOY 161 represents a maize canopy height of about 0.30 m, and DOY 216 represents a maize crop at full cover. The contribution of H and G with respect to the other energy fluxes tended to decrease as canopy cover increased. Also, worth noticing is that positive LE values were recorded in this environment during the night, which would result from advection of heat. During DOY 216, LE exceeded Rn during the afternoon hours (i.e., LE/Rn > 1.0), which is another indication of advection of heat.

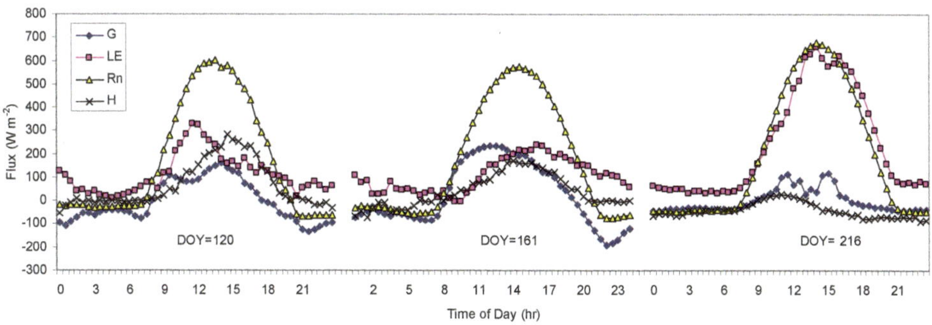

Fig. 10. Values of energy balance components measured during three clear-sky days on a surface-irrigated maize field at North Platte, Nebraska, during 2001. LE = latent heat flux (ETc), H = sensible heat flux, G = soil heat flux, Rn = net radiation. Each data point represents the 30-min average flux. DOY =day of the year. Maize was planted on DOY 128.

Daily values of energy fluxes (Fig. 11) show that all variables had considerable day-to-day fluctuation due to daily changes in weather and surface soil water conditions. Measurement of Rs during cloud-free days conformed to the theoretical bell-shaped clear-sky solar radiation (Rso) "envelope". As expected, significant reduction in daily Rs with respect to Rso resulted from the presence of clouds. The maximum energy input from solar radiation measured during this study was equivalent to a water loss of 12.5 mm d-1. Advection can also supply additional energy to the environment for evapotranspiration. However, its magnitude would be expected to be minor compared with that supplied directly by Rs. Therefore, the maximum measured value of 12.5 mm d-1 of Rs would place an approximate upper limit on the ET that can effectively occur at this site, considering that a portion of Rs would be reflected and would not be entirely available for LE.

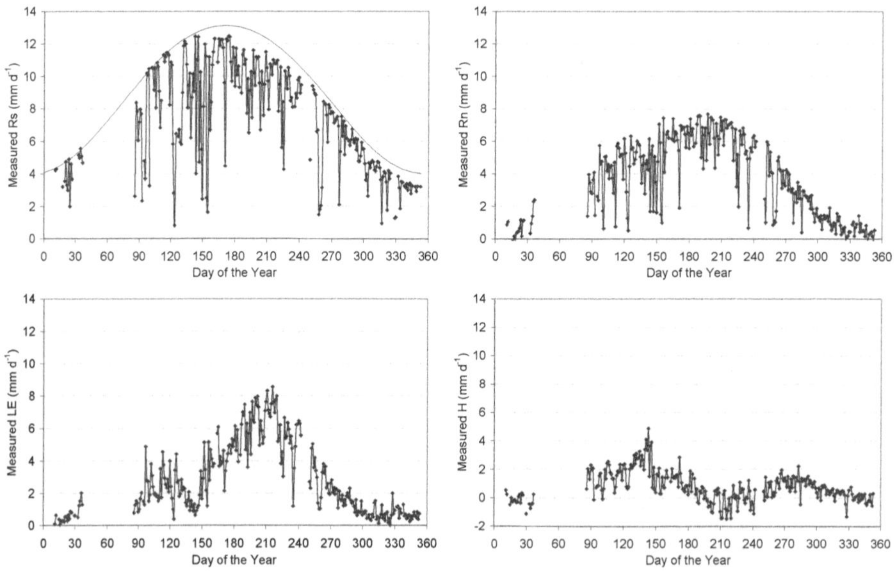

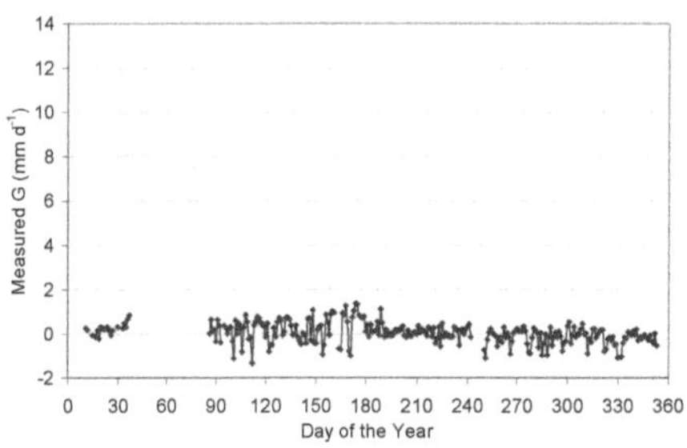

Month	G/Rs	LE/Rs	Rn/Rs	H/Rs	G/Rn	LE/Rn	H/Rn
Jan	0.05	0.13	0.17	-0.01	0.45	0.94	-0.39
Feb	0.10	0.26	0.29	-0.07	0.44	1.06	-0.50
Mar	0.02	0.20	0.46	0.25	0.03	0.42	0.55
Apr	0.02	0.30	0.49	0.17	0.04	0.60	0.35
May	-0.02	0.26	0.55	0.31	-0.04	0.46	0.58
Jun	0.01	0.44	0.57	0.12	0.02	0.77	0.21
Jul	0.01	0.65	0.66	0.00	0.02	0.97	0.01
Aug	0.01	0.63	0.63	-0.01	0.01	1.02	-0.02
Sep	-0.04	0.45	0.54	0.13	-0.08	0.82	0.25
Oct	-0.03	0.24	0.39	0.17	-0.08	0.63	0.44
Nov	-0.10	0.24	0.25	0.11	-0.45	0.99	0.46
Dec	-0.04	0.21	0.18	0.01	-0.33	1.30	0.03
Total	**0.00**	**0.33**	**0.43**	**0.10**	**0.00**	**0.83**	**0.16**

Fig. 11. Daily energy balance components measured on a surface-irrigated maize field at North Platte, Nebraska, during 2001. Rs = solar radiation, LE = latent heat flux (ETc), H = sensible heat flux, G = soil heat flux, Rn = Net radiation. The table shows the average monthly values of each variable as a fraction of Rs and Rn calculated from the daily measurements. The fractions for February and March are based on very few measurements. The solid line in the Rs graph is the calculated clear-sky solar radiation (Rso) or Rs envelope (n = 258).

Net radiation (Rn) measured over the maize canopy followed a similar trend as Rs during the year, but, as expected, it was always lower than Rs. The maximum measured Rn was equivalent to a water depth of 7.7 mm d^{-1}. Figure 12 shows that the Rn measured over the maize canopy was well-correlated to the values calculated for grass using the FAO-56 procedure (R^2 =0.92). Although there were some differences in measured daily Rn over the maize canopy and Rn calculated for grass, on average, the relationship followed the 1:1 line almost perfectly. Similar results between non-stressed maize and grass surface in south central Nebraska were reported by Irmak et al. (2010). Irmak et al. (2010) stated that the Rn values measured over maize canopy vs. grass canopy would be expected to be similar due to the fact that both plant species have similar biophysical attributes and that the albedo values for grass and non-stressed maize canopy were similar enough to only cause minor differences in Rn.

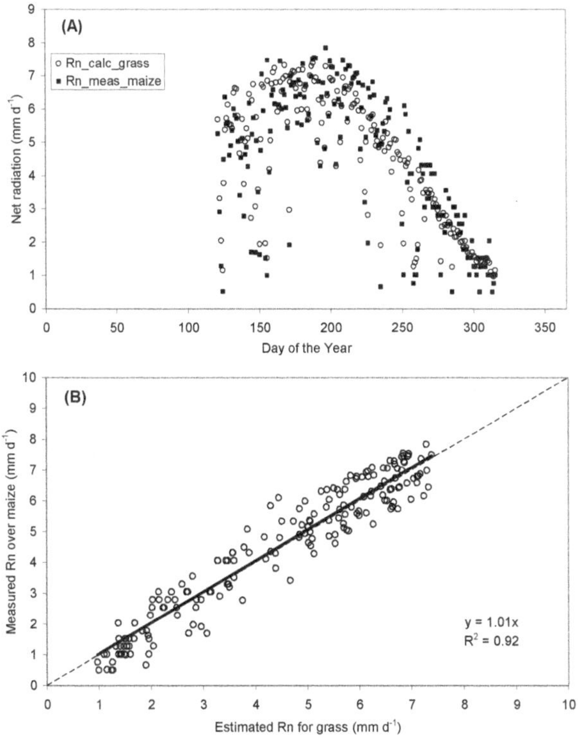

Fig. 12. (A) Comparison of daily net radiation (Rn) measured over maize (Rn_meas_ maize) with Rn calculated for grass (Rn_calc_grass) using the procedure described in FAO-56 (1998), and (B) regression analysis of measured and estimated Rn data in (A) (— Linear regression line, ----1:1 line, n = 185).

The measured daily LE values (the same as ETc when converted to a water equivalent, such as mm d^{-1}) followed a similar trend throughout the year as the energy available from Rs and Rn (Figure 11). However, it is important to note that the LE curve peaked much later than the Rs and Rn curve. Rs peaked at about DOY 170, Rn at about DOY 190, and LE at about

DOY 210. The delay on the peak of LE with respect to that of Rs and Rn is, in part, a reflection of the marked impact of the properties of the crop and soil surface on LE.

The maximum LE measured during the study was 8.5 mm d^{-1}, which was 0.8 mm d^{-1} higher than the maximum Rn value. LE values of 5 mm d^{-1} or higher were observed in April (DOY 191-120) from bare soil. These relatively high LE values were due to the presence of a wet soil surface resulting from spring rain. Relatively high values of ET early in the season in a maize field were also measured by Howell et al. (1998). Li et al. (2003), using lysimeter ET measurements in a semi-arid region, found that maize approached a peak ET rate of 48.45 mm per week (6.92 mm d^{-1}) at the 12th week after sowing (94 days). These values are much lower than the values obtained in this study, and the fact that the crop took much longer to reach the peak ETc in the Li et al. (2003) study, and points to a cooler environment, compared with the conditions of this study. However, Howell et al. (1998) reported daily ET values (measured with weighing lysimeters) for maize at Bushland, Texas, of close to 14 mm d^{-1} (see Fig. 4 of Howell et al., 1998), which were much higher than the maximum ET values measured in this study.

The magnitude of sensible heat flux (H) was comparable to that of LE early in the growing season, but tended to decrease as the crop cover increased. Usually, an opposite trend between LE and H was observed. When LE was low, H was high, and when LE was high during the mid and late season, H was at its lower values. H fluctuated from approximately -2 mm d^{-1} on DOY 210 to as high as 5 mm d^{-1} on DOY 140. H usually had negative values during the winter and during the growing season when the crop was at full canopy cover. Although G can be a significant component of the energy balance of a crop canopy when measured in short time steps, such as hourly, the daily average is usually close to zero. This is because positive values obtained in the middle of the day will cancel out negative values obtained during the rest of the day, and especially at night (Payero et al., 2001; Payero et al., 2005). Figure 11 shows that daily G averages usually fluctuated between small negative and positive values (about $-1 \text{ mm d}^{-1} \leq G \leq +1 \text{ mm d}^{-1}$), but the average was 0.0 mm d^{-1}.

Figure 11 also shows the ratios of the different energy-balance fluxes with respect to Rs and Rn for each month. For each of the variables, the magnitude of the ratios varied considerably from month to month. Overall, the G/Rs and G/Rn ratios were very close to zero. For the entire year, about 43% of Rs was available to the surface as Rn. A higher percentage of Rs was available as Rn during the summer months, and a lower percentage was available during the winter moths due to higher albedo. Furthermore, 33% of Rs was partitioned into LE and 10% into H. That means that, on average, about 57% of daily Rs was reflected and/or diffused. Also, on average for the entire year, 83% of the available daily Rn energy was used for LE and 16% for H.

Figure 13 shows the relationships between the daily values of Rs, Rn, and LE. Although the relationship had some scatter, there was a good linear relationship between daily Rs and the daily Rn values measured over the maize canopy ($R^2 = 0.87$). The relationship was offset from the origin reflecting the fact that Rn can be negative while Rs cannot. The R^2 value indicated that 87% of the variability in Rn could be explained by variability in Rs, while the remaining 13% of the variability is due to other factors. The main factors include the changes in the characteristics of the surface, and changes in the angle of incidence of the incoming Rs. Both of these factors suffer from daily and seasonal variations, which would affect albedo. Although there was a reasonable linear relationship between Rn and LE ($R^2 = 0.76$), the relationship was better explained by a curvilinear function that could be fitted to a second degree polynomial ($R^2 =0.80$). These results suggest that Rn explained about 80% of the variability in LE.

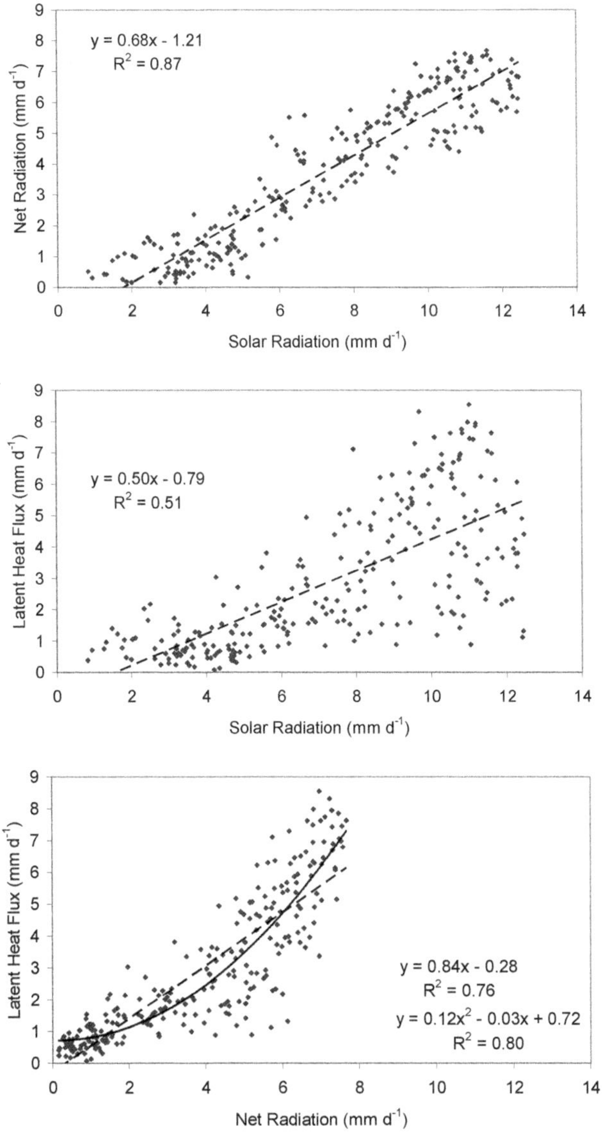

Fig. 13. Relationships between daily values of energy balance components measured in a surface-irrigated maize field at North Platte, Nebraska, during 2001 (— polynomial, ---- linear, n = 258).

In Figure 13, there were a considerable number of days when LE did not respond to increasing Rn, especially between the Rn range of 4 to 6 mm d^{-1} and the LE range of 1 to 2 mm d^{-1}. Similarly, there were days when an increase in LE was observed for this same range of Rn. This is because although Rn is one of the main drivers of LE in most cases, in dormant (non-growing) periods, as well as in conditions with extremely high atmospheric demand, vapour pressure deficit, wind speed, and air temperature can play crucial roles in the magnitude of LE. Thus, in these conditions, the variations in LE are more often due to variations in these other factors than to variations in Rn. Figure 13 also shows that the relationship between Rs and LE was relatively poor (R^2 = 0.51). The correlation between LE and Rs gets weaker with increases in both variables (i.e., when Rs > 6 mm d^{-1} and when LE > 2 mm d^{-1}). This is because, while Rs is the main source of available energy for LE, the amount of energy actually utilized (Rn-G) for LE is dictated or controlled by many factors that are dynamic in space and time. These factors include, but are not limited to, soil water status, soil and crop management practices, climatic and microclimatic factors.

3.5 Crop coefficient curve

Measured daily crop coefficients are presented in Figure 14 as a function of day from emergence (DFE) and cumulative growing degree days from emergence (C-GDD), along with the basal Kc curve for maize given in FAO-56 for Kimberly, Idaho. Although the time scale in FAO-56 was only given in days from planting, in this study, the C-GDD corresponding to the days from planting were calculated, and Kc curves were presented in both DFE and C-GDD "time" scales. The basal Kc curve for Kimberly, Idaho, was selected because Kimberly has a similar climate to North Platte and it is located at a similar latitude (North Platte = 41.1° N, Kimberly = 42.4° N).

There was significant variability in the daily Kc values, which made it challenging to fit an average Kc curve to the observed data. For instance, for the mid-season development stage, the daily Kc values ranged from 0.93 to 1.44. Also, early in the season, from emergence to about 40 days from emergence, frequent rain events wetted the soil surface, which resulted in high evaporation rates and high Kc values. This made it difficult to obtain a good estimate of the basal Kc early in the season without independent measurements of the evaporation and/or transpiration components of ET. Because of the daily variability in Kc, no attempt was made to fit a function to the data. However, Figure 14 shows that the basal Kc values given in FAO-56 for Kimberly fitted the general trend in Kc obtained in this study reasonably well, except for the periods when the soil surface was wet, as would be expected. However, in the mid-season stage, the Kimberly curve corresponded more to the average Kc values than to the basal Kc values, but early and late in the season, the Kimberly curve followed the apparent basal values reasonably well, especially when using the C-GDD "time" scale.

Parkes et al. (2005) using lysimeter and soil water measurements, found that the Penman-Monteith model provided the best model efficiency when used with a peak Kc of 1.25 for maize in China. Also, Li et al. (2003) found Kc values for maize to be used with the FAO-56 Penman-Monteith model of 0.5, 1.02, 1.26, and 0.68 during the initial, crop development, mid-season, and late-season stages, respectively. The 1.25 or 1.26 values for the mid-season are within the range of values obtained in this study. Jensen et al. (1990) suggested that grass-reference Kc values for a full-cover crop should not exceed 1.25 if this value is to represent large expanses of vegetation. Also, the "spikes" in the Kc curve following wetting events exceeded the maximum 1.20 value suggested by FAO-56 and Jensen et al. (1990).

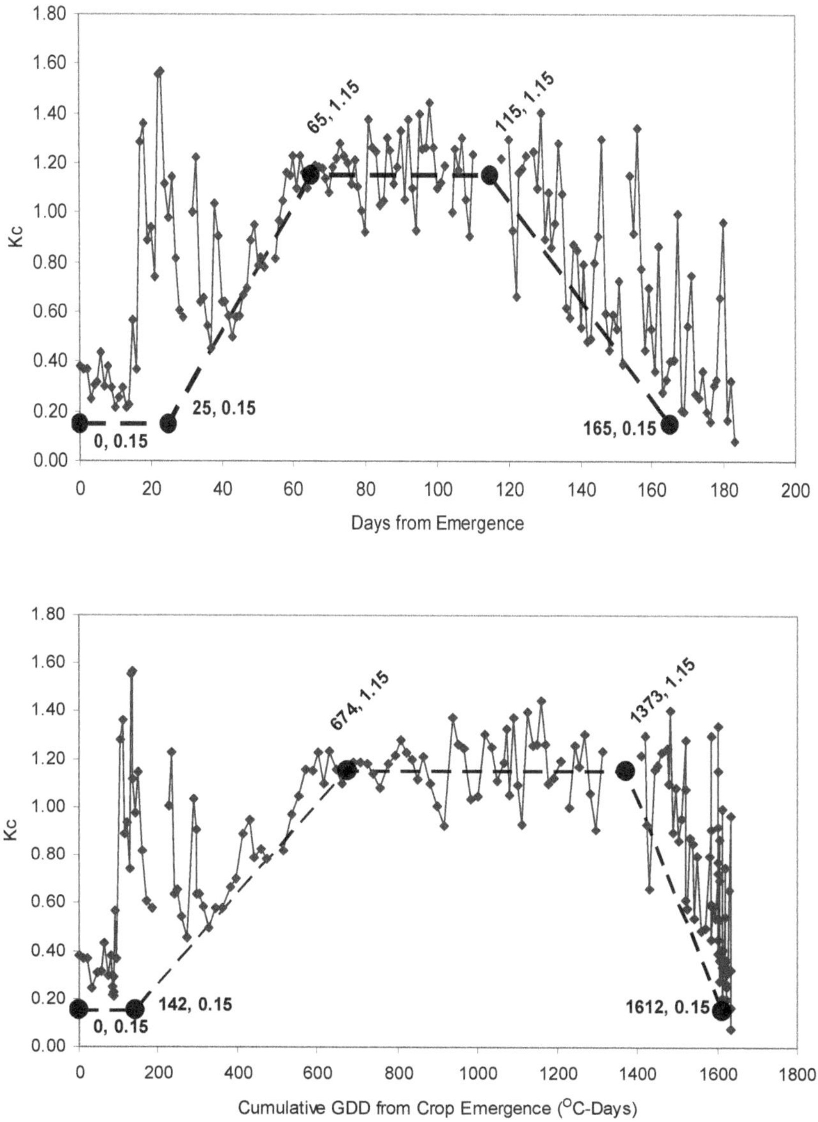

Fig. 14. Crop coefficient (Kc) curves for field maize measured at North Platte, Nebraska, during 2001, as a function of days from emergence and cumulative growing degree days (GDD) from crop emergence. The Kc values were calculated based on a grass reference (ETo), which was calculated with the FAO-56 method. The broken line represents the Kc function proposed in FAO-56 with growth stage lengths for Kimberly, Idaho. The number pairs on the graphs are the coordinates for each of the black dots for the Kimberly curves from FAO-56 (n =182).

3.6 Measured and estimated maize ETc

A comparison of the daily measured and calculated maize ETc values is presented in Figs. 15 and 16. The calculated daily ETc values correlated reasonably well with measured values throughout the season. Regression analysis showed a good linear relationship (R^2 = 0.86) between measured and estimated values, and the slope of the line (1.06) showed just a 6% departure from the 1:1 line. The departure seemed to be created by an overestimation of the calculated ETc values at high ETc rates. For instance, although measured values were limited to a maximum of 8.5 mm d^{-1}, estimated values were as high as 10.7 mm d^{-1}. Differences in the estimated and measured ETc could be due to potential errors in the water balance procedure used to estimate ETc and to potential measurement errors.

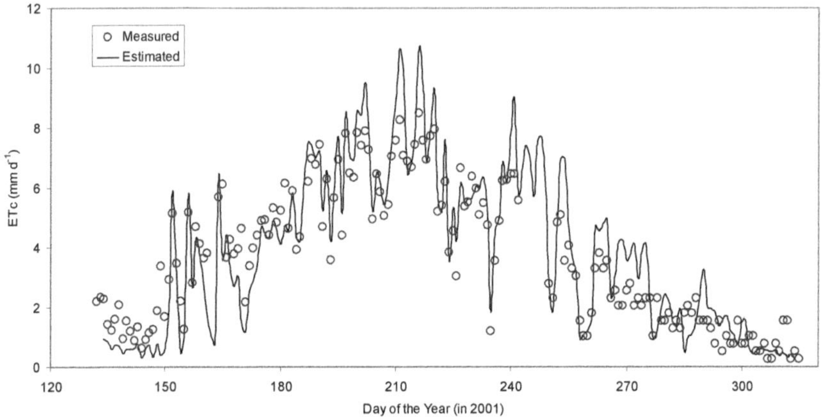

Fig. 15. Comparison of measured and estimated daily maize evapotranspiration (ETc) at North Platte, Nebraska, during 2001. ETc was estimated using the FAO-56 dual crop coefficient procedure (n = 182).

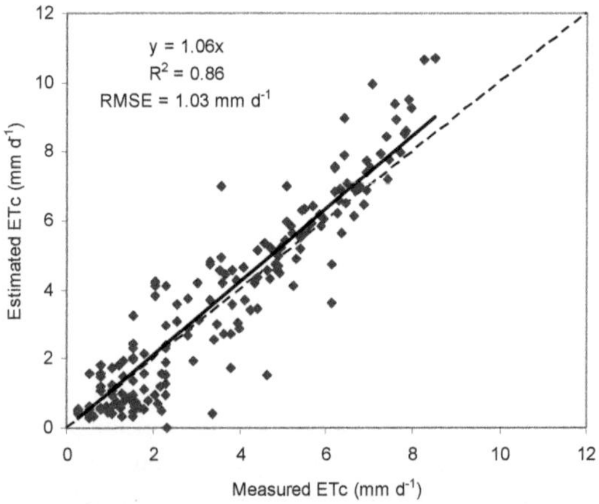

Fig. 16. Relationship between measured and estimated daily maize evapotranspiration (ETc) during 2001 at North Platte, Nebraska. The solid and dashed lines are the regression and 1:1 lines, respectively. RMSE = root mean squared error (n = 182).

4. Conclusions

Daily surface energy balance components, including crop evapotranspiration, for field maize were measured from a surface-irrigated field at North Platte, Nebraska, during 2001. Using these measurements and weather variables from a nearby weather station, other variables such as daily ETo, ETr, ETr/ETo, Kc, and maize ETc were calculated. We found that daily ETo in 2001 varied from less than 0.5 mm d^{-1} during winter to a maximum of about 10 mm d^{-1} during summer. ETr and ETr_Neb were linearly related with ETo, but ETr and ETr_Neb were 29% and 37% higher than ETo. The ETr/ETo ratio varied during the year from 1.03 to 1.68, with lowest values in summer and highest in winter.

All measured energy balance components had considerable day-to-day variability due to rapid changes/fluctuations in weather conditions, which is typical to west-central Nebraska. Although Kc values had considerable daily variations, the values given in FAO-56 for maize, using the lengths of growth stages from Kimberly, Idaho, fitted the measured data reasonably well, except early in the season when the soil surface was wet. The fit improved when we used cumulative growing degree days from crop emergence as the "time" scale.

Daily maize ETc calculated with FAO-56 [ETc = ETo x (dual Kc)] also fitted measured daily ETc values reasonably well (R^2 = 0.86, slope = 1.06). Locally-developed Kc curves can be vital for more accurate quantification of ETc for an effective irrigation management. Locally-derived Kc curves can minimize the errors associated with using curves developed for other climates and for other soils and crop management conditions.

5. Acknowledgements

The authors would like to acknowledge Don Davison for providing technical support for this project. Names of commercial products are solely provided as information to the reader and do not imply an endorsement or recommendation by the authors or their organizations.

6. References

FAO-56. (1998). *Crop evapotranspiration - guidelines for computing crop water requirements. (Irrigation and Drainage Paper No. 56).* Food and Agriculture Organization of the United Nations (FAO), Rome, Italy.

ASCE-EWRI. (2005). *The ASCE Standardized Reference Evapotranspiration Equation.* Environmental and Water Resources Institute (EWRI) of the Am. Soc. of Civil. Engrs., ASCE, Standardization of Reference Evapotranspiration Task Committee Final Report. 213 pp. ASCE, Reston, VA.

Doorenbos, J. & Pruitt, W.O. (1975). *Guidelines for prediction of crop water requirements.* FAO Irrig. and Drain. Paper No. 24, Food and Agriculture Organization of the United Nations (FAO), Rome, Italy. 179 pp.

Doorenbos, J. & Pruitt, W.O. (1977). *Guidelines for prediction of crop water requirements.* FAO Irrig. and Drain. Paper No. 24 (revised), Rome, Italy. 144 pp.

Hanks, R.J. & Ashcroft, G.L. (1980). *Applied Soil Physics.* New York: Springer-Verlag.

Howell, T.A. ; Tolk, J.A. ; Schneider, A.D. & Evett, S.R. (1998). Evapotranspiration, yield, and water use efficiency of corn hybrids differing in maturity. *Agronomy Journal*, 90, 3-9.

Irmak, S. ; Mutiibwa, D. & Payero, J.O. (2010). Net radiation dynamics: Performance of 20 daily net radiation models as related to model structure and intricacy in two climates. *Transactions of the ASABE* 53(4):1059-1076.

Irmak, A. ; Irmak, S. & Martin, D.L. (2008). Reference and crop evapotranspiration in south central Nebraska: I. Comparison and analysis of grass and alfalfa-reference evapotranspiration. *Journal of Irrigation and Drainage Engineering* 134(6) : 690-699.

Irmak, A. & Irmak, S. (2008). Reference and crop evapotranspiration in south central Nebraska: II. Measurement and estimation of actual evapotranspiration. *Journal of Irrigation and Drainage Engineering* 134(6) : 700-715.

Jensen, M.E. (1969). Scheduling irrigations using computers. *Journal of Soil and Water Conservation*, 24, 193-195.

Jensen, M.E. ; Burman, R.D. & Allen, R.G. (1990). *Evapotranspiration and Irrigation Water Requirements*. ASCE Manuals and Reports on Engineering Practices No. 70. ASCE, New York, NY, 360 p.

Kincaid, D.C. & Heermann, D.F. (1974). Scheduling irrigations using a programmable calculator. USDA-ARS Report No. ARS-NC-12. 10 pp.

Li, Y.L. ; Cui, J.Y. ; Zhang, T.H. & Zhao, H.L. (2003). Measurement of evapotranspiration of irrigated spring wheat and maize in a semi-arid region of north China. *Agricultural Water Management*, 61, 1-12.

Parkes, M. ; Jian, W. & Knowles, R. (2005). Peak crop coefficient values for Shaanxi, North-west China. *Agricultural Water Management*, 73, 149-168.

Payero, J.O. ; Neale, C.M.U. & Wright, J.L. (2001). Estimating diurnal variation of soil heat flux for alfalfa and grass. ASAE Paper No. 017007. St. Joseph, Mich.: ASAE.

Payero, J.O. ; Neale, C.M.U. & Wright, J.L. (2005). Estimating soil heat flux for alfalfa and clipped tall fescue grass. *Applied Engineering in Agriculture*, 21, 401-409.

Penman, H.L. (1948). Natural evaporation from open water, bare soil and grass. *Proceedings of the Royal Society of London*, A193:120-146.

Wright, J.L. (1982). New evapotranspiration crop coefficients. Journal of the Irrigation and Drainage Division, ASCE, 108(IR2), 57-74.

7

Evapotranspiration and Transpiration Measurements in Crops and Weed Species by the Bowen Ratio and Sapflow Methods Under the Rainless Region Conditions

J. Pivec, V. Brant and K. Hamouzová
University of Life Sciences Prague, The Czech Republic

1. Introduction

The aim of this chapter is to provide a systematic description of the measurements of total evapotranspiration and transpiration of selected agricultural crops and weeds, results of investigation, and perspectives of these methods for agricultural usage. This study provides a contribution towards increased knowledge on the consumptive water use of arable crops and weeds within the temperate climatic zone under specified weather conditions and actual crop structure given by biometric observation. The water consumption of plants represents a significant part of the landscape water balance (Merta et al. 2001). An important factor influencing the water balance of the plant stands on agricultural soil and thereby in the countryside is the species composition of phytocoenosis. Within the framework of phytocoenosis, the cultivated plants and weeds take share in influencing the water balance (Pivec & Brant 2009). Competition between plants to capture the resources essential to plant growth (i.e. light, water and nutrients) is one of the key processes determining the performance of natural, semi-natural and agricultural ecosystems (Kropff & van Laar, 1993). The issue of evapotranspiration and transpiration demands of field crops is a subject of intensive study especially in arid and semi-arid areas. In terms of eliminating the negative impact of agriculture on the environment and in terms of increasing the efficiency of the production systems, its monitoring is important for the temperate climate as well.

2. Used methods of actual evapotranspiration and transpiration measurements

The sensible heat flux (H) and the latent heat flux (λE) were measured by Liu & Foken (2001) using the eddy covariance method (EC) and the Bowen ratio/Energy balance method (BREB). The results indicate that H (BREB) is about 30 ± 20 W m^{-2} higher than H (EC) and λE (BREB) is about 180 ± 40 W m^{-2} higher than λE (EC) during the daytime. Liu & Foken (2001) proposed a modified Bowen ratio method (MBREB) to determine sensible and latent heat fluxes without using the surface energy balance equation. Their findings are to the contrary to the findings of Brotzge & Crawford (2003), who comment that the EC system favours latent heat flux and the BREB system favours sensible heat flux. Perez et al. (1999) show that,

if advection is considered negligible, the BREB method is able to determine correctly the surface flux partitioning or the flux values when certain conditions, consistent with the flux-gradient relationship, are fulfilled. San José et al. (2003) postulates, that different architecture of the canopy had a minor effect on the flux densities of net radiation as well as the partitioning of available energy into sensible and latent heat. His results indicate that the phenological trend of the daily λE was controlled by the leaf area index (LAI) development. When LAI reached its maximum value at the flowering and pod-filling stages, λE was controlled mainly by the available energy and temperature. The BREB method (eq. 1) was used to measure latent heat fluxes above the *Zea mays* canopy as well as between the soil surface and the canopy by Zeggaf et al. (2008). Then, the latent heat flux from *Z. mays* transpiration was calculated by the difference between that of the *Z. mays* field and soil surfaces. In method 2, a weighing lysimeter and sap flow gauges were used to measure latent heat fluxes from the maize field and *Z. mays* transpiration, respectively. Then, latent heat flux from the soil surface was calculated by the difference between that of the *Z. mays* total evapotranspiration and *Z. mays* transpiration. The coefficient of determination between latent heat fluxes by the two methods was 0.72 from the *Z. mays* field and 0.77 from the *Z. mays* transpiration. However, results indicated a low correlation between the latent heat fluxes from the soil surface by the two methods (coefficient of determination = 0.36).
Sap flow measurements (the heat balance method) may be used in determining plants' water demands. A survey of the literature has shown that information about the moisture requirements of herbal species, particularly their determination under natural conditions, is relatively much less abundant. Kjelgaard et al. (1997) and Jara et al. (1998) reported that sap flow measurements at the same plant were practicable for one week in dependence on weather conditions and stem thickening. For both the gas exchange and sap flow methods, scaling up from leaf to plant and to canopy is difficult to carry out because measurements with this method reflect only the reactions of single plants (e.g. Köstner et al. 1996). Data on weeds' water consumption represent a basic parameter for determining the ecological and economic functions of agriculture.

2.1 BREB measurement method

The BREB method is based on the precondition of the coefficients of the apparent and latent heat being equal (1), when it is possible to determine the ratio of the sensible and latent heat by measuring the gradients of the air temperature and humidity above the evaporating surface (Woodward & Sheehy 1983):

$$\beta = \frac{H}{\lambda ET} = \gamma \frac{dt}{de} \tag{1}$$

in which H is the flow of sensible heat, λ is the specific heat of the water vapour, ET is evapotranspiration, γ is a psychrometric constant 0.66 hPa °C^{-1}, dt/de is a temperature/ humidity gradient of air at two levels above the evaporating surface. Fig. 1 documents the instruments settings.

2.2 Sap flow method

The use of the heat balance method is based on the relation (2) between the entering heat amount and the increase in temperature within a defined space (Kučera et al. 1977, Tatarinov et al. 2005):

$$P = Q \cdot dT \cdot c_w + dT \cdot z \qquad (2)$$

Fig. 1. BREB installation over winter rape-seed field: radiation balance gauge on left, pair of temperature/relative humidity sensor in the middle (upper with global radiance gauge), anemo-indicator on right. Locality: Červený Újezd 2006, photo by authors.

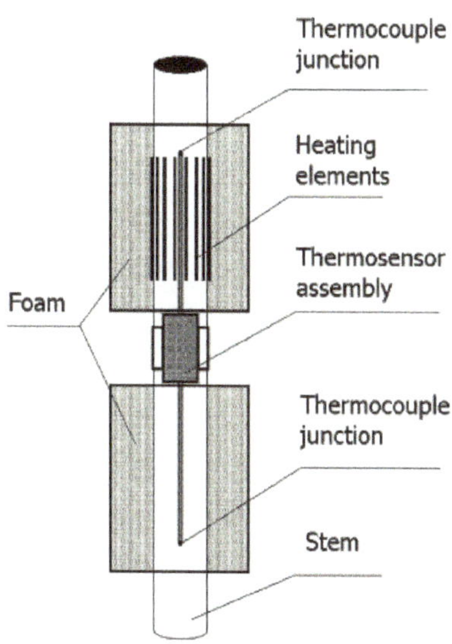

Fig. 2. Diagram of the EMS "baby sensor" for shoots or small stems (adapted from Čermák et al. 2004).

in which P is the heat energy input (W), Q is the sap flow (kg s^{-1}), dT is the temperature difference within the measured space (K), c_w is specific heat of water (J $kg^{-1}K^{-1}$) and z is a coefficient of the heat losses in the measured space (W K^{-1}). During our experiments that had taken place from 2005 to 2010, the sap flow values were evaluated in selected cultivated and weed plants under field conditions. The Q values were measured by a 12-channel T4.2 flow meter for the stems of 6 to 20 mm diameter, made by the EMS Brno (CZ) firm (see the diagram in Fig. 2). The values obtained during the measurements were recorded at 10 minute intervals during the entire period of individual measurements. The measurements point was always located at the base of the plant or stalk of the plant.

3. Evapotranspiration of field crops

Knowledge of the arable crops evapotranspiration is the **basis** of the understanding of the influence of agriculture on the environment and the basis for the elimination of the agricultural activities negative influences on the landscape water balance. The exchange of water vapour and CO_2 between the crops and surrounding air can also be perceived as an important factor for the photosynthetic assimilation and, consequently, for the biomass production. From the practical point of view, the knowledge of the evapotranspiration demands can be used for the water balance optimization through the finally structured crop and growing phases duration and growth access periods (San José et al. 2003).

3.1 Actual values of evapotranspiration

The crop transpiration depends on the management, such as a supply of nutrients (Shepherd et al. 1987), seeding days (Connor et al. 1992) and the plant species or cultivars (Eastham & Gregory 2000). Additionally, the energetic fluxes and the water use efficiency (e.g. Corbeels et al. 1998; Asseng et al. 2000) as well as the dissipation of the energy within the landscape (Ripl 1995) are evaluated. The energy balance components are strongly affected by the leaf area index and plant height during all developmental stages of the canopy, especially the sensible heat flux. Table 1 demonstrates the average values of the ET_c, ET_0 and Bowen ratio (β) on the Budihostice site for selected field crops from 2007 to 2010. The site is situated at an altitude of 220 m a.s.l. and the soil type is Haplic chernozem. Potential evapotranspiration slightly exceeds the precipitation totals (P/ET_0), and in the normal period (1961-1990) this ratio ranged from 0.7 to 0.8 (Pivec et al. 2006). ET_0 values were determined by an algorithm used by FAO (Allen et al. 1998).

3.2 Reference evapotranspiration and its relationship to the actual evapotranspiration

Also important for the estimation and verification of the crop coefficients values is the actual evapotranspiration assessment (Inman-Bamber & McGlinchey 2003; Hanson & May 2006; Kato & Kamichika 2006). Crop coefficients are classified as single coefficients or dual coefficients (Allen et al. 1998). Single coefficients include both, evaporation from the soil and plant transpiration. Dual crop coefficients consist of basal crop coefficients and coefficients that describe evaporation from the soil. The basal coefficients reflect the conditions of a dry soil surface and sufficient soil water to maintain maximum plant transpiration (Allen et al. 1998; Hanson & May 2006). No limitations are placed on crop growth or evapotranspiration from soil water and salinity stress, crop density, pests and diseases, weed infestation or low fertility. ET_c is determined by the crop coefficient (K_c) approach whereby the effect of the various weather conditions is incorporated into the reference crop evapotranspiration (ET_0)

(Allen et al. 1998). Values of Kc determined for most agricultural crops will typically vary in relation to the changes in vegetative growth until effective full cover is attained (Hunsaker et al. 2003). After full cover, the K_c will tend to decline, the extent of which is primarily dependent on the particular growth characteristics of the crop (Jensen et al. 1990).

Year	Crop		Period of DOY																		
			81-90	91-100	101-110	111-120	121-130	131-140	141-150	151-160	161-170	171-180	181-190	191-200	201-210	211-220	221-230	231-240	241-250	251-260	261-270
2007	*Hordeum vulgare* 1	β				0.9	0.8	0.7	1.1	1.0	1.8	1.9	1.7	1.8							
		ET_0				4.4	3.4	4.1	4.1	4.3	4.9	4.2	4.0	4.7							
		ET_c				3.9	3.5	4.0	2.7	3.0	2.1	2.2	2.2	2.3							
2008	*Beta vulgaris* var. *altissima* 2	β										0.7	0.7	0.5	0.6	0.9	0.5	0.9	0.9	1.0	1.0
		ET_0										4.7	4.7	3.1	4.0	4.7	3.1	3.6	3.3	2.6	1.5
		ET_c										4.8	4.8	3.4	4.2	3.4	3.0	3.5	3.0	2.2	1.6
	Medicago sativa 3	β														1.1	0.9	0.6	0.8	1.1	1.2
		ET_0														4.9	3.3	3.9	3.4	2.7	1.5
		ET_c														3.7	2.5	3.8	3.2	1.9	1.4
2009	*Medicago sativa* 4	β	0.8	1.1	0.8	0.7	1.0	1.0	0.7	0.7	0.6	0.7	0.6	0.5	0.8	1.2	1.1	1.7	2.3	1.3	1.5
		ET_0	1.4	3.3	3.6	4.1	3.4	3.3	3.5	3.3	3.9	2.7	4.0	3.5	4.2	4.5	4.2	4.2	3.4	2.7	2.7
		ET_c	1.8	3.1	3.9	4.2	3.2	2.9	3.5	3.6	4.3	2.5	3.8	3.6	4.0	2.8	2.7	2.0	1.6	2.0	1.6
	Zea mays 5	β				1.7	1.6	1.1	1.0	1.6	1.1	0.8	0.5	1.5	1.6	1.7	1.5	1.8			
		ET_0				4.7	3.6	3.6	3.5	3.1	3.8	2.8	4.0	3.5	4.0	4.5	4.1	4.1			
		ET_c				4.9	2.8	3.2	3.1	2.5	2.9	2.3	3.3	2.0	2.3	2.5	2.4	2.1			
	Triticum aestivum 6	β	0.7	1.2	1.0	0.7	0.8	0.9	0.8	1.2	1.0	0.9	1.4	1.5	1.7	1.7	1.7				
		ET_0	2.4	2.9	3.3	4.1	3.6	3.2	3.4	2.7	3.4	2.3	2.3	2.2	2.4	2.5	2.1				
		ET_c	1.5	3.3	3.5	4.1	3.4	3.3	3.4	3.3	4.0	2.7	4.2	3.7	4.3	4.4	4.1				
2010	*Zea mays* 7	β								1.0	1.3	1.3	0.9	0.6	0.5	0.6	0.5	0.6			
		ET_0								3.3	3.8	4.7	5.2	4.9	3.5	3.1	3.0	3.5			
		ET_c								2.5	2.3	2.8	3.6	3.9	3.1	2.6	2.6	3.0			
	Sorghum bicolor 8	β									1.0	1.5	1.0	0.6	0.7	0.8	1.1	1.2			
		ET_0									3.8	4.4	4.9	4.8	3.5	3.0	2.7	3.4			
		ET_c									2.7	2.2	3.0	3.7	2.7	2.1	1.7	2.2			
	Triticum aestivum 9	β	1.0	1.1	0.9	0.9	0.8	0.7	0.8	0.7	0.8	0.9	1.0	1.4	1.3	0.9	0.9	1.0			
		ET_0	2.3	2.3	2.5	3.8	1.9	1.8	2.5	3.1	3.5	4.6	4.8	4.9	3.4	3.0	3.1	3.4			
		ET_c	2.5	1.9	2.7	3.7	2.0	2.0	2.6	3.0	3.1	3.6	3.2	2.6	2.1	2.0	2.1	1.9			

Table 1. Average daily values of the Bowen ratio (β), reference evapotranspiration (ET_0, mm period^{-1}) and actual evapotranspiration (ET_c, mm period^{-1}) for selected periods at selected stands in the years of 2007 to 2010. For the determination of the ET_C values the BREB method was used. DOY means day of the year. 1 – Spring barley, harvest 194 DOY, 2 - harvest 291 DOY, 3 - cutting 218 DOY, 4 - cutting 134, 171 and 208 DOY, 5 - harvest 243 DOY, 247 DOY stubble ploughing, 6 - Winter wheat, harvest 205 DOY, 7 - harvest 245 DOY, 8 - harvest 245 DOY, 9 - Winter wheat, harvest 226 DOY, 231 DOY stubble ploughing.

Year	Crop		Period of DOY															
			81-90	91-100	101-110	111-120	121-130	131-140	141-150	151-160	161-170	171-180	181-190	191-200	201-210	211-220	221-230	231-240
2007	*Hordeum*	*BBCH*				23	33	43	61	71	76	85	89	92				
	vulgare	K_c				0.90	1.03	0.97	0.66	0.69	0.43	0.52	0.53	0.48				
2009	*Medicago*	*a)*						cutting I.				cutting II.			cutting III.			
	sativa	K_c	1.27	0.96	1.09	1.02	0.94	0.86	1.03	1.11	1.12	0.93	0.96	1.03	0.95	0.63	0.65	0.49
	Zea	*BBCH*				13	14	16	30	31	33	35	39	57	65	73	79	83
	mays	K_c				1.04	0.78	0.88	0.89	0.81	0.75	0.85	0.84	0.58	0.57	0.56	0.58	0.51
	Triticum	*BBCH*	22	23	25	28	31	33	55	61	69	73	83	87	91			
	aestivum	K_c	0.64	1.10	1.07	1.01	0.95	1.00	0.98	1.22	1.17	1.17	1.80	1.67	1.84			
2010	*Zea*	*BBCH*								16	31	32	34	39	59	67	75	81
	mays	K_c								0.75	0.61	0.58	0.70	0.80	0.86	0.85	0.86	0.86
	Sorghum	*b)*									2		3		5		6	
	bicolor	K_c									0.71	0.49	0.61	0.77	0.76	0.68	0.64	0.64
	Triticum	*BBCH*	20	21	23	25	31	32	45	51	61	71	83	85	92	93		
	aestivum	K_c	1.06	0.83	1.10	0.96	1.08	1.09	1.07	0.96	0.90	0.79	0.67	0.54	0.61	0.67		

a) for *M. sativa* the cutting dates are introduced
b) for *S. bicolor* the following growth stages were estimated: 2 – five leaf stage, 3 – growing point differentiation, 5 – boot stage and 6 – half bloom
K_c values were determined on the basis of the relationship $ET_c = ET_0 \times K_c$.

Table 2. Average daily values of K_c for the period and growth stage BBCH (BBCH principal stage in a given period) for selected crops in the years of 2007, 2009 and 2010. K_c values were determined on the basis of the relationship $ET_c = ET_0 \times K_c$. ET_0 is the average daily reference evapotranspiration (mm period-1) and ET_c the average daily actual evapotranspiration (mm period-1). DOY means day of the year. BBCH stages were estimated by Meier (2001).

ET_0 is the average daily reference evapotranspiration (mm period-1) and ET_c the average daily actual evapotranspiration (mm period-1). Table 2 documents the K_c values for selected field crops in central Europe in relation to crop growth stages. The table clearly shows a decrease in the K_c values in case of cereal crops just at the onset of the grain maturation stage. The stands of *M. sativa* illustrated the value of K_c around 1 throughout the whole vegetation period. Cutting reduces water demands and hence the decrease of K_c values. Stands with *Z. mays* and *S. bicolor* have a typical lower K_c values provided in the main crop growth compared with cereals.

4. Transpiration of field crops

The current knowledge of the plant species moisture requirements has been obtained predominantly within the framework of the study of forest communities while the transpiration values are known in wood species (e.g. Čermák et al. 1992 and 1995; Schulze et al. 1985). Information on the moisture requirements of the herbal species by using the sap flow method, particularly their determination under natural conditions, are relatively, on the basis of literature survey, not so abundant. Much more frequent are data of moisture demands on field crops set out in laboratory conditions (e.g. Dugas 1990; Angadi et al. 2003).

4.1 Transpiration of crop-plants

Moisture requirements of crops in relation to different abiotic and biotic factors are intensively investigated. Table 3 summarizes the values of water flow through several crop plant species under laboratory or field conditions. Longer-term measurements allow the determination of moisture needs based on daily values of *Q*, particularly in relation to the growth phase. Table 3 shows the average daily values of *Q* for selected cultivated plants established under the field conditions around the world, while Table 4 includes the values measured by authors.

Species	Variety/Cultivar	*Q*	Conditions	Source
Brassica napus	Quantum	to 39	greenhouse	Angadi et al. 2003
	Arrow	0 - 27	field	
Glycine soja		0 - 95	plastic chamber	Cohen et al. 1993
Gossypium sp.		0 - 75	greenhouse	Dugas 1990
	Deltapine 77	0 - 95	field	Dugas et. al. 1994
Helianthus annuus		0 - 200	greenhouse	Kjelgaard et al. 1997
Solanum tuberosum	Atlantic	0 - 55	greenhouse	Gordon et al. 1997
	Monona	0 - 25		
		0 - 35	greenhouse	Kjelgaard et al. 1997
Triticum sp.		0 - 5	field	Senock et al. 1996
Zea mays		0 -175	greenhouse	Gavloski et al. 1992
		0 - 150	greenhouse	Kjelgaard et al. 1997

Table 3. Large range of sap flow rates (*Q*, g h^{-1}) by crop-plants.

4.2 Transpiration of weeds

Weedy plants are a permanent part of the plant-based agricultural soil communities. In terms of water demands determination of agrophytocoenosis is also important to determine the transpiration of wild plants. Knowledge of weed transpiration plays an important role in assessing the competition of weeds against cultivated plants. Table 5 demonstrates the values of transpiration flow of select weeds using sap flow method (Pivec & Brant 2009). Based on these results, it is possible to make a detailed comparison of water demands of weeds and cultivated plants. If, for example, we compare the transpiration requirements of

Plant species	Date of measurement	BBCH stage	*n*	*Q*	Q_{max}
Brassica napus	9.6. - 22.7.2005	71 - 88	6	0.044	0.121
	5.6. - 25.7.2006	75 - 97	6	0.092	0.187
	26.4. - 29.6.2007	64 - 86	24	0.030	0.079
	29.5. - 14.7.2008	71 - 87	17	0.085	0.203
Helianthus annuus	7.7. - 22.7.2009	53 - 59	3	0.337	0.731
Sorghum bicolor	14.8. - 31.8.2010	-	12	0.177	0.816
Zea mays	15.7. - 3.9.2008	63 - 75	11	0.080	0.201
	12.8. - 30.8.2009	75 - 83	11	0.081	0.244
	24.7. - 31.8.2010	63 - 81	10	0.178	0.885

Table 4. Averages of daily values of transpiration flow (*Q*, kg day^{-1}), their maxima (Q_{max}, kg day^{-1}) and BBCH stages for the evaluated plant species for the period under observation. *n* - number of measured plants.

the plants of *B. napus* and those of *Lactuca serriola*, which can become a weed in the stands of *B. napus*, we will find out that they are similar. We can then express an assumption that the occurrence of one plant of *L. serriola* per unit of area of the *B. napus* stand has the same effect on the transpiration requirements of the stand and competition relations for water, as the increase in the numbers of individuals of *B. napus* per given area unit by one plant. A more distinct effect on the transpiration requirements of the growth stand will be found if we evaluate the influence of the occurrence of *Artemisia vulgaris* plants in the stands of *B. napus*. If the daily average value of the transpiration flow in *A. vulgaris* reached 0.077 to 0.084 kg H_2O per single stalk, then with the average number of stalks, which can range from 3 to 7 in *B. napus*, the moisture requirements of this weed are considerably higher in comparison with a single plant of *B. napus* (Pivec & Brant 2009).

Plant species	Date of measurement	*n*	*Q*	Q_{max}	*Rg*	*P*	Notes
Amaranthus retroflexus	2.8. - 27.8.2006	2	0.018	0.080	14.1	99.0	*1*
Artemisia vulgaris	2.8. - 27.8.2006 19.7. - 17.8.2007	7* 7*	0.077 0.084	0.150 0.157	14.1 17.8	99.0 79.0	*1*
Cirsium arvense	2.8. - 8.8.2005	1*	0.016	0.025	14.8	20.6	2
Conyza canadensis	2.8. - 27.8.2006 19.7. - 17.8.2007	6 9	0.046 0.078	0.116 0.174	14.1 17.8	99.0 79.0	*1*
Lactuca serriola	2.8. - 27.8.2006 19.7. - 17.8.2007	9 8	0.068 0.025	0.153 0.093	14.1 17.8	99.0 79.0	*1*

Table 5. Averages of daily values of transpiration flow (*Q*, kg day^{-1}), their maxima (Q_{max}, kg day^{-1}) for the evaluated weed species and the average daily sums of global solar radiation (*Rg*, MJ m^{-2} day^{-1}) and daily totals of precipitation (*P*, mm) for the period under observation (modified by Pivec & Brant 2009). *n* - number of measured plants or stalks*, *1* - measured in solitary plants, 2 - measured in the stand of *Z. mays*.

4.3 Transpiration modelling

One way or determining the influence of different factors on the plant water consumption is the model estimation of the calculated value of sap flow (Q_{calc}). An actual value of Q depends strongly on the input of the solar radiation and vapour pressure deficit (e.g. Gordon et al. 1999; Pivec et al. 2009; Pivec et al. 2010). One of the possibilities of Q_{calc} determination is to use the algorithm (3) as shown below (Kučera, EMS Brno, pers. comm.; Pivec et al. 2010):

$$Q_{calc} = par1 \frac{Rg}{(Rg + par2)} \frac{VPD}{(VPD + par3)} \tag{3}$$

where Rg is global solar radiation (W m^{-2}) and VPD is vapour pressure deficit (hPa). The parameters (par) 1–3 for the Q_{calc} calculation were estimated for the entire measurement period.

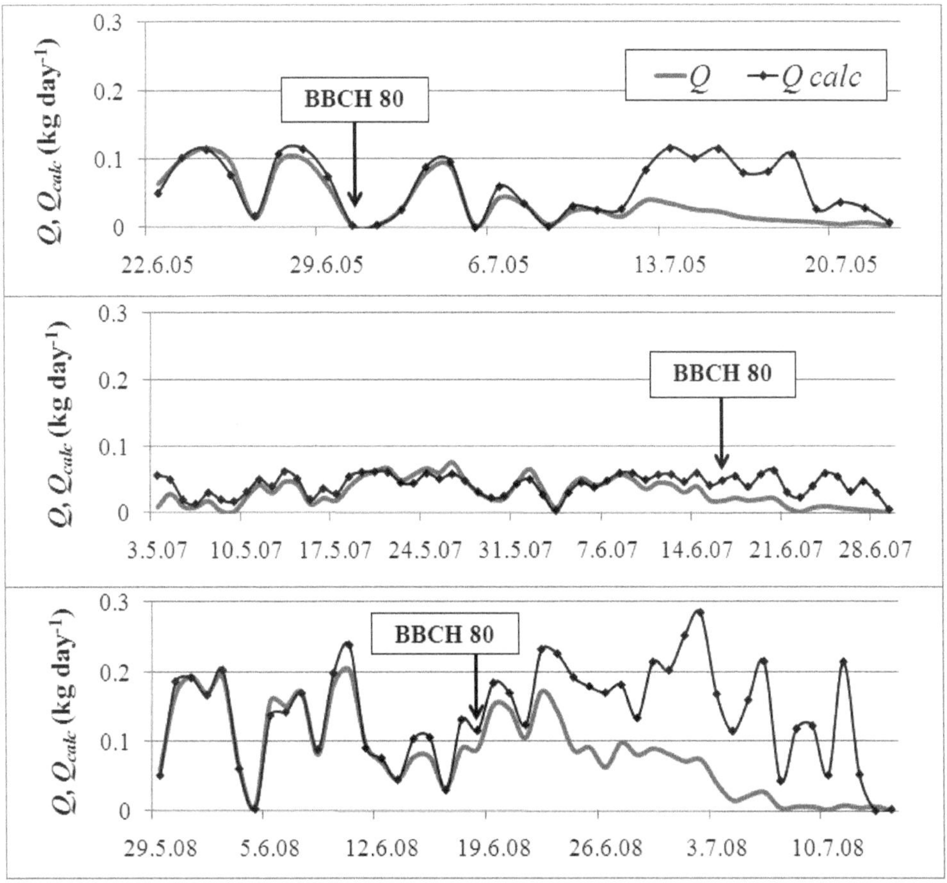

Fig. 3. Daily values of Q and Q_{calc} (kg day^{-1}) in the *B. napus* plant during the observed period in the years 2005, 2007 and 2008 (Pivec et al. 2010).

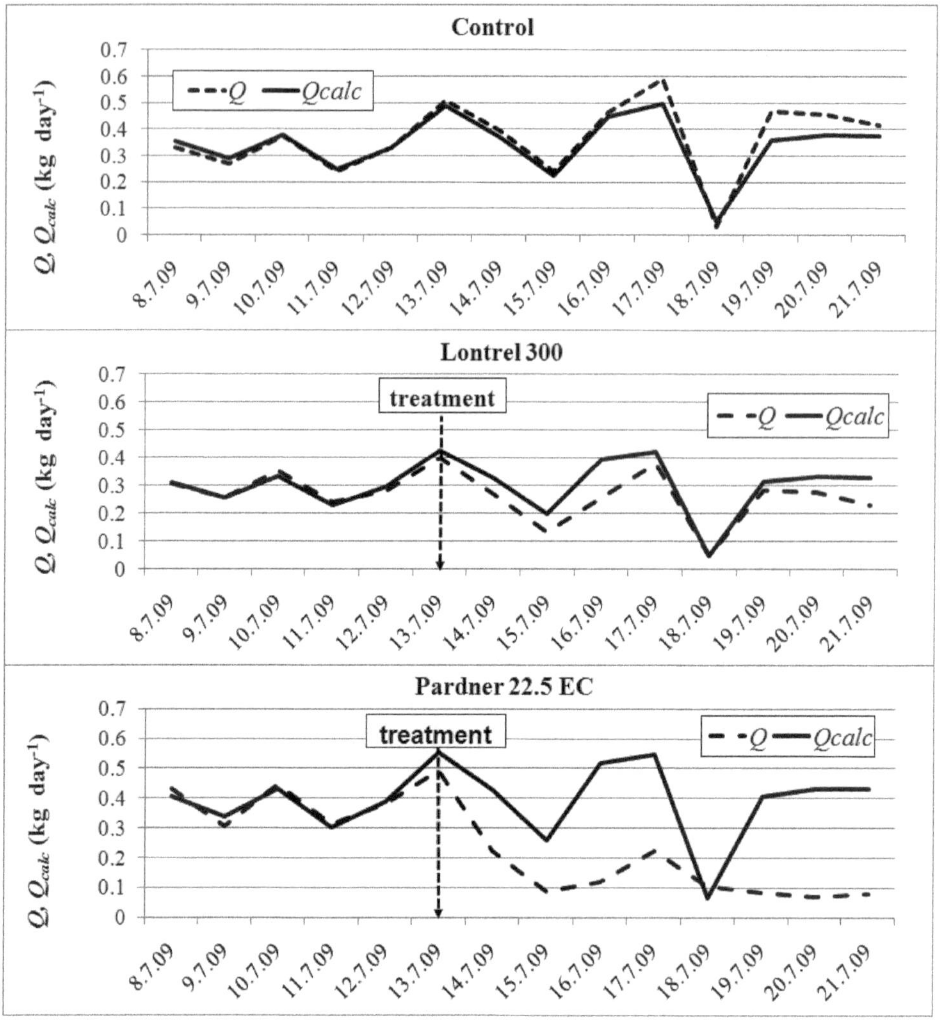

Fig. 4. Influence of herbicide treatment on average daily values of Q and Q_{calc} (kg day^{-1}) in the *Helianthus annuus* plant. Influence of herbicide treatment on water flow decline was proved by computing correlation coefficients comparing transpiration average daily values (Q) - in the period from 8.7. to 13.7.2009 with the calculated values of (Q_{calc}). Modified by Brant et al. (2010).

An example is the usage of Q_{calc} calculation for determining the moisture changes in plants, depending on the growth stage. It is obvious from the different course of daily averages of Q and Q_{calc} values (kg day^{-1}) during the years under observation (Fig. 3) that sap flow decreases from the beginning of the maturation stage of *B. napus* plants. Values of the parameters (*par*) 1-3, used for the *Qcalc* computation were, in 2005, 0.257554/5205.436/22.10980 (corresponding period for the pars' estimation 22-29 June 2005), in 2007, 0.019516/992.2398/4.741211 (corresponding period 29 May - 11 June 2007), and in 2008, 0.101538/778.5762/17.45747 (corresponding period 29 May–10 June 2008). In terms of regression analysis a closer dependence between Q_{calc} and Q was confirmed from the start of the measurements up to the BBCH 83 stage in 2005, and up to the BBCH 81 stage in 2007 and 2008 (Pivec et al. 2010).

Another possibility for using the calculation of Q_{calc} is, for example, assessment of the effect of herbicides on the change of water demands of the plant. Effect of herbicides was tested on the plants of *Helianthus annuus* (the modelled plant). Herbicide treatment was carried out on 13.7. 2009. Three plants were untreated, three plants were treated with the herbicide Pardner 22.5 EC (225 g a.i. *bromoxynil* l^{-1}, active ingredient inhibiting PSII) at 1.5 l ha^{-1} while the three remaining plants were treated with Lontrel 300 (300 g a.i. *clopyralid* l^{-1}, synthetic auxin) at 0.4 l ha^{-1}. The growth stage of *H. annuus* was BBCH 56 at the beginning of the experiment. Mean values of Q in untreated plants exceeded the values of Q_{calc} (Fig. 4). This can be explained by an unlimited growth of the control plants. Average daily Q values in the plants treated with herbicide Lontrel 300 was lower on sunny days (14.7.-21.7.) than Q_{cal} before the herbicide treatment. This illustrates that plants transpired less than before the herbicide treatment and their growth was reduced, perhaps even stopped. Strong herbicide effect on Q decrease was evident following an application of Pardner 22.5 EC (Fig. 4).

5. Relationship between transpiration and evapotranspiration

In terms of actual evapotranspiration it is necessary to remember the contribution of its components, transpiration and evaporation, to its total value. Under annual field crops, the soil surface remains bare during fallow, preparatory tillage, planting, germination, and seedling stages. Most water is lost during these periods by direct evaporation from soil (Jalota & Prihar 1998). During the growing season, characterized by the highest evapotranspirational demands of crops, however, a proportion of evaporation to the total value of evapotranspiration is fundamental. Lösch (2001) states that on the land covered by vegetation the share of water delivered from the soil into the atmosphere via plants represents 2/3 up to 3/4 of the total evapotranspiration. An important role in terms of the proportion of evaporation to total evapotranspiration is played by tillage, crop architecture (row crops or densely sown crops), mulching technologies etc. During a normal growing season, evaporation from the soil surface may reach up to 50% of evapotranspiration (Peters 1960). Russell & Peters (1959) and Pivec & Brant (2009) points out the high proportion of evaporation to evapotranspiration, approximately 50% in crops such as *Z. mays*. Crop residues, applied to the soil surface (mulching), prevent water loss by evaporation (Brussiere & Cellier 1994; Gill & Jalota 1996).

Figure 5 illustrates the daily totals of Q/ETC measured by the sap flow/BREB technique in *B. napus* and *Z. mays* plants. Q values of *Z. mays* achieved 35% of ETC values. The amount of water passing through the *Z. mays* plant stems on 1 m^2 of crop as measured by the sap flow, when compared with the evapotranspiration values measured by BREB technique, denotes a higher evaporation than we had expected. This suggests that the heat balance method of the sap flow rate measurement can be disputed in respect to *Z. mays* plants, which are monocotyledonous and in which, therefore, the water flow runs across the whole cross-section of the vascular bundles in the stem. On the other hand, *Z. mays* is a representative of C4 plants with a smaller water consumption and a higher water use efficiency than revealed by C3 plants. In any case, the study of *Z. mays* will require a much greater effort and more detailed observation since there are few literature references on this subject. From Figure 5 is it clear that *B. napus* transpiration rates decline with the advancing maturation stage (BBCH phase 84) according to the results of Pivec & Brant (2009). After the maturation stage, the crop transpiration still drops and the values of evapotranspiration are probably influenced by evaporation.

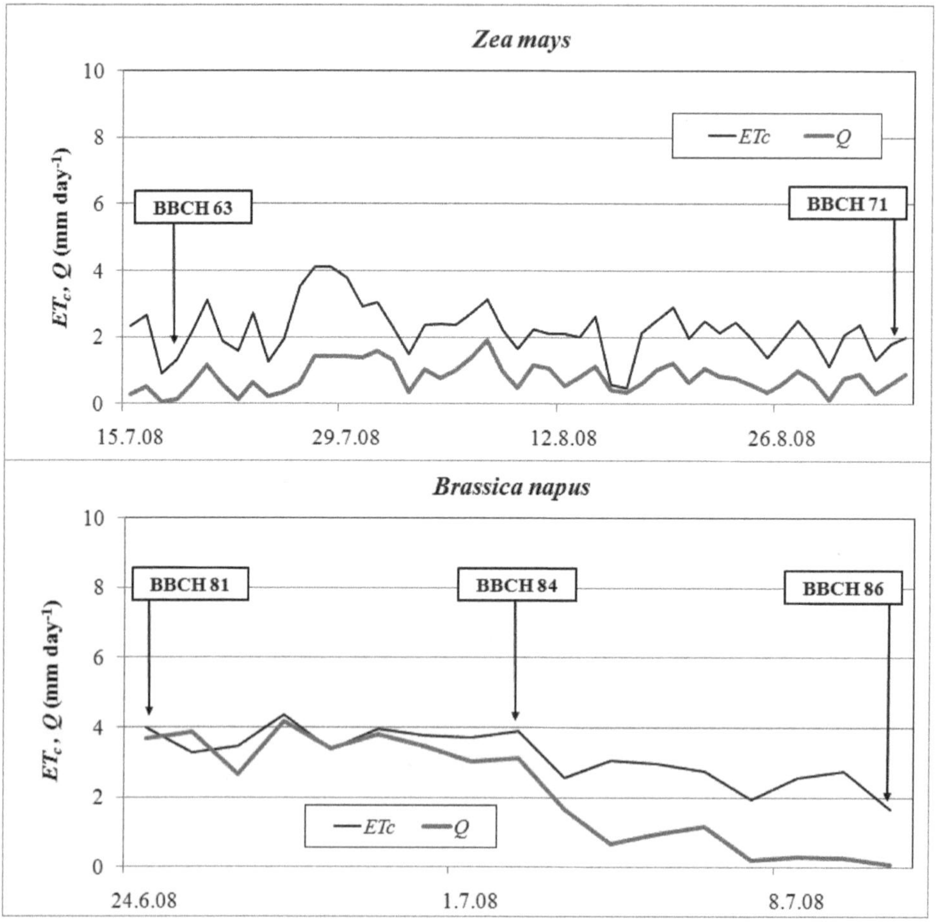

Fig. 5. Daily values of evapotranspiration (ET_c, mm) and sap flow (*Q*, mm) of *Zea mays* and *Brassica napus* plants. The average number of individuals in *Z. mays* was 96 000 ha^{-1} and in *B. napus* 42 individuals m^{-2} (Pivec & Brant 2009). Growth stages of plants are expressed by the BBCH growth scale.

6. Conclusions

This study presents the values of evapotranspiration and transpiration of field crops under the temperate climate conditions set out in the field. In practical terms, the usage of published results is important to determine the ratio between the actual and potential evapotranspiration of evaluated crops. The material can also be considered for determining the value of transpiration for selected field crops and weeds, which makes it possible to specify partially competitive relationships between plant species within agrophytocoenosis. The most crucial conclusion of this work is a comparison of actual evapotranspiration values measured by both the BREB method and the sap flow. Simultaneous use of these methods provides also the verification of the results obtained.

7. Acknowledgments

This contribution originated from work supported by the Research Project MSM 6046070901 of the Ministry of Education, Youth and Sports of the Czech Republic and the Project QH 82191 of the National Agency for Agriculture Research under the Ministry of Agriculture of the Czech Republic.

8. References

Allen, R.G.; Pereira, L.S., Raies, D. & Smith, M. (1998). *Crop evapotranspiration: guidelines for computing crop water requirements.* FAO Irrigation and Drainage Paper 56, ISBN 92-5-104219-5 United Nations, Rome, Italy

Angadi, S.V.; Cutforth, H.W. & Mcconkey, B.G. (2003). Determination of the water use and water use response of canola to solar radiation and temperature by using heat balance stem flow gauges. *Canadian Journal* of *Plant Science,* Vol. 83, No. 1, pp. 31-38, ISSN 1918-1833

Asseng, S. & Hsiao, T.C. (2000). Canopy CO_2 assimilation, energy balance, and water use efficiency of an alfalfa crop before and after cutting. *Field Crops Research,* Vol. 67, No. 3, pp. 191-206, ISSN 0378-4290

Brant, V.; Pivec, J., Hamouzová, K. & Satrapová, J. (2010). Herbicide efficacy evaluation using sap flow method. *15th EWRS Symposium 2010,* pp. 294, ISBN 978-963-9821-24-8, Kaposvár, Hungary, July 12-15, 2010

Brotzge, J.A. & Crawford, K.C. (2003). Examination of the surface energy budget: A comparison of eddy correlation and Bowen ratio measurement systems. *Journal of Hydrometeorology,* Vol. 4, No. 2, pp. 160-178, ISSN 1525-755X

Brussiere, F. & Cellier, P. (1994). Modification of the soil temperature and water content regimes by a crop residue mulch: experiment and modelling. *Agricultural and Forest Meteorology,* Vol. 68, No. 1-2, (March 1994), pp. 1-28, ISSN 0168-1923

Čermák, J.; Cienciala, E., Kučera, J. & Hällgren, J.E. (1992). Radial velocity profiles of water flow in trunks of Norway spruce and oak and the response of spruce to severing. *Tree Physiology,* Vol. 10, No. 4, (June 1992), pp. 376 – 380, ISSN 0829-318X

Čermák, J.; Cienciala, E., Kučera, J., Lindroth, A. & Bednářová, E. (1995). Individual variation of sap-flow rate in large pine and spruce trees and stand transpiration: a pilot study at the central NOPEX site. *Journal of Hydrology,* Vol. 168, No. 1, (June 1995), pp. 17-27, ISSN 0022-1694

Čermák J.; Kučera, J. & Nadezhdina, N. (2004). Sap flow measurements with some thermodynamic methods, flow integration within trees and scaling up from sample trees to entire forest stands. *Trees,* Vol. 18, No. 5, pp. 529–546, ISSN 0931-1890

Cohen, Y.; Takeuchi, S., Nozaka, J. & Yano, T. (1993). Accuracy of sap flow measurement using heat balance and heat pulse methods. *Agronomy Journal,* Vol. 85, No. 5, pp. 1080-1086, ISSN 0002-1962

Connor, D.J.; Theiveyanathan, S. & Rimmington, G.M. (1992). Development, growth, water-use and yield of a spring and a winter wheat in response to time of sowing. *Australian Journal of Agricultural Research, Vol.* 43, No. 3, pp. 493–516, ISSN 0004-9409

Corbeels, M.; Hofman, G. & Van Cleemput, O. (1998). Analysis of water use by wheat grown on a cracking clay soil in a semi-arid Mediterranean environment: weather and nitrogen effects. *Agricultural Water Management*, Vol. 38, No. 2, (December 1998), pp. 147–167, ISSN 0378-3774

Dugas, W.A. (1990). Comparative measurement of stem flow and transpiration in Cotton. *Theoretical and Applied Climatology*, Vol. 42, No. 4, pp. 215 - 221, ISSN 0177-798X

Dugas, W.A.; Heuer, M.L., Hunsaker, D., Kimball, B.A., Lewin, K.F., Nagy, J. & Johnson, M. (1994). Sap flow measurements of transpiration from cotton grown under ambient and enriched CO_2 concentrations. *Agricultural and Forest Meteorology*, Vol. 70, No. 1-4, pp. 231-245, ISSN 0168-1923

Eastham, J. & Gregory, P.J. (2000). The influence of crop management on the water balance of lupin and wheat crops on a layered soil in a Mediterranean climate. *Plant and Soil*, Vol. 221, No. 2, pp. 239–251, ISSN: 0032-079X

Gavloski, J.E.; Whitfield, G.H. & Ellis, C.R. (1992). Effect of restricted watering on sap flow and growth in corn (*Zea mays* L.). *Canadian Journal of Plant Science*, Vol. 72, No. 2, pp. 361-368, ISSN: 0008-4220

Gill, B.S. & Jalota, S.K. (1996). Evaporation from soil in relation to residue rate, mixing depth, soil texture and evaporativity. *Soil Technology*, Vol. 8, No. 4, (March 1996), pp. 293-301, ISSN: 0933-3630

Gordon, R.; Brown, D.M., Madani, A. & Dixon, M.A. (1999). An assessment of potato sap flow as affected by soil water status, solar radiation and vapour pressure deficit. *Canadian Journal of Soil Science*, Vol. 79, No. 2, pp. 245–253, ISSN 1918-1841

Gordon, R.; Dixon, M.A. & Brown, D.M. (1997). Verification of sap flow by heat balance method on three potato cultivars. *Potato Research*, Vol. 40, No. 3, pp. 267-276, ISSN 0014-3065

Hanson, B.R. & May, D.M. (2006). Crop coefficients for drip-irrigated processing tomato. *Agricultural Water Management*, Vol. 81, No. 3, (March 2006), pp. 381–399, ISSN 0378-3774

Hunsaker, D.J.; Pinter, P.J., Barnes, E.M. & Kimball, B.A. (2003). Estimating cotton evapotranspiration crop coefficients with a multispectral vegetation index. *Irrigation Science*, Vol. 22, No. 2, pp. 95–104, ISSN: 0342-7188

Inman-Bamber, N.G. & McGlinchey, M.G. (2003). Crop coefficients and water-use estimates for sugarcane based on long-term Bowen ratio energy balance measurements. Field Crops Research, Vol. 83, No. 2, pp. 125–138, ISSN: 0378-4290

Jalota, S.K. & Prihar, S.S. (1998). *Reducing soil water evaporation with tillage and straw mulching.* ISBN 0813828570, Iowa State University Press, Ames, Iowa, USA

Jara, J.; Stockle, C.O. & Kjelgaard, J. (1998). Measurement of evapo-transpiration and its components in a corn (*Zea mays* L.) field. *Agricultural and Forest Meteorology*, Vol. 92, No. 2, (July 1998), pp. 131–145, ISSN 0168-1923

Jensen, M.E.; Burman, R.D. & Allen, R.G. (1990). *Evapotranspiration and irrigation water requirements. ASCE – Manuals and Report on Engineering Practice, no. 70.* ISBN 0872627632, American Society of Civil Engineers, New York, USA

Kato, T. & Kamichika, M. (2006). Determination of a crop coefficient for evapotranspiration in a sparse sorghum field. *Irrigation and Drainage*, Vol. 55, No. 2, (April 2006), pp. 165–175, ISSN 1531-0353

Kjelgaard, J.F.; Stockle, C.O., Black, R.A. & Campbell, G.S. (1997). Measuring sap flow with the heat balance approach using constant and variable heat inputs. *Agricultural and Forest Meteorology*, Vol. 85, No. 3-4, pp. 239-250, ISSN 0168-1923

Köstner, B.; Biron, R., Siegwolf, R. & Granier, A. (1996). Estimates of water vapour flux and canopy conductance of Scots Pine at the tree level utilizing different xylem sap flow methods. *Theoretical and Applied Climatology*,Vol. *53*, No. 1-3, pp. 105–113, ISSN 0177-798X

Kropff, M.J. & Van Laar, H.H. (1993). *Modelling crop-weed interactions*. ISBN 0851987451,CAB International, Wallingford, UK

Kučera, J.; Čermák, J. & Penka, M. (1977). Improved thermal method of continual recording the transpiration flow rate dynamics. *Biologia Plantarum (Praha)*, Vol. 19, No. 6, pp. 413–420, ISSN 0006-3134

Liu, H.P. & Foken, T. (2001). A modified Bowen ratio method to determine sensible and latent heat fluxes. *Meteorologische Zeitschrift*, Vol. 10, No. 1, pp. 71-80, ISSN 0941-2948

Lösch, R. (2001). Wasserhaushalt de Pflanzen. Quelle und Meyer, Wiebelsheim, Germany, ISBN 3-494-02238-0

Meier U. (ed.) (2001). Growth stages of mono-and dicotyledonous plants. BBCH Monograph, 2. Edition, Federal Biological Research Centre for Agriculture and Forestry. (http://syntechresearch.hu/sites/default/files/

Merta M.; Sambale, C., Seidler, C. & Peschke G. (2001). Suitability of plant physiological methods to estimate the transpiration of agricultural crops. *Journal of Plant Nutrition and Soil Science*, Vol. 164, No. 1, pp. 43–48, ISSN 1522-2624

Perez, P.J.; Castellvi, F., Ibanez, M. & Rosell, J.I. (1999). Assessment of reliability of Bowen ratio method for partitioning fluxes. *Agricultural and Forest Meteorology*, Vol. 97, No. 3, pp. 141-150, ISSN 0168-1923

Peters, D.B. & Russell, M.B. (1959). Relative water losses by evaporation and transpiration in field corn. *Proceedings - Soil Science Society of America*, Vol. 23, No. 2, pp. 170-173, ISSN 0038-0776

Peters, D.B. (1960). Relative magnitude of evaporation and transpiration. *Agronomy Journal*, Vol. 52, No. 9, pp. 536-538, ISSN 0002-1962

Pivec, J.; Brant, V. & Moravec, D. (2006). Analysis of the potential evapotranspiration demands in the Czech Republic between 1961 - 1990. *Biologia*, Vol. 61, (Suppl.19), pp. 294–299, ISSN 0006-3088

Pivec, J. & Brant, V. (2009). The actual consumption of water by selected cultivated and weed species of plants and the actual values of evapotranspiration of the stands as determined under field conditions. *Soil and Water Research*, Vol. 4, (Special Issue *2)*, pp. 39–48, ISSN 1801-5395

Pivec, J.; Brant, V. & Bečka, D. (2009). The influence of weather conditions on the sap flow of *Brassica napus* L. during the fructification and maturation stages. *Ekológia (Bratislava)*, Vol. 28, No. 1, pp. 43–51, ISSN 1335-342X

Pivec, J.; Brant, V., Bečka, D. & Cihlář, P. (2010). Consumptive use of water in *Brassica napus* L. from flowering to ripening stage under the rainless region conditions. *Irrigation and Drainage*, DOI: 10.1002/ird.598publikaciok/bbch.pdf

Ripl, W. (1995). Management of water cycle and energy flow for ecosystem control - The Energy-Transport-Reaction (ETR) Model. *Ecological Modelling*, Vol. *78*, No. 1-2, pp. 61–76, ISSN 0304-3800

San José, J.J.; Bracho, R., Montes, R. & Nikonova, N. (2003). Comparative energy exchange from cowpeas (*Vigna unguiculata* (L.) Walp cvs. TC-9-6 and M-28-6-6) with differences in canopy architectures and growth durations at the Orinoco llanos. *Agricultural and Forest Meteorology*, Vol. 116, No. 3, pp. 197–219, ISSN 0168-1923

Schulze, E.D.; Čermák, J., Matyssek, R., Penka, M., Zimmemann, R. & Vašíček, F. (1985). Canopy transpiration and water fluxes in the xylem of the trunk of Larix and Picea trees - a comparison of xylem flow, porometer and cuvette measurements. *Oecologia*, Vol. 66, No. 4, pp. 475–483, ISSN 0029-8549

Senock, R.S.; Ham, J.M., Loughin, T.M., Kimball, B.A., Hunsaker, D.J., Pinter, P.J., Wall, G.W., Garcia, R.L. & LaMorte, R.L. (1996). Sap flow in wheat under free-air CO_2 enrichment. *Plant, Cell and Environment*, Vol. 19, No. 2, pp. 147-158, ISSN 0140-7791

Shepherd, K.D.; Cooper, P.J.M., Allan, A.Y., Drennan, D.S.H. & Keatinge, J.D.H. (1987). Growth, water use and yield of barley in Mediterranean-type environments. *Journal of Agricultural Science*, Vol. 108, No. 2, pp. 365–378, ISSN 0021-8596

Tatarinov, F.A.; Kučera, J. & Cienciala, E. (2005). The analysis of physical background of tree sap flow measurement based on thermal methods. *Measurement Science and Technology*, Vol. 16, No. 5, pp. 1157–1169, ISSN 0957-0233

Woodward, F.I. & Sheehy, J.E. (1983). *Principles and Measurements in Environmental Biology*. ISBN 0-408-10637-9, Butterworth & Co., Ltd., London, UK

Zeggaf, A.T.; Takeuchi, S., Dehghanisanij, H., Anyoji, H. & Yano, T. (2008). A Bowen ratio technique for partitioning energy fluxes between maize transpiration and soil surface evaporation. *Agronomy Journal*, Vol. 100, No. 4, pp. 988-996, ISSN 0002-1962

Part 2

Crop ET: Water Use, Water Quality and Management Aspects

8

Crop Evapotranspiration and Irrigation Scheduling in Blueberry

David R. Bryla
U.S. Department of Agriculture, Agricultural Research Service
USA

1. Introduction

There are currently 139,000 ha of blueberry worldwide, including 66,000 ha of highbush [comprises northern highbush (*Vaccinium corymbosum*), southern highbush (*Vaccinium* sp.), and rabbiteye (*V. virgatum* formerly *V. asheii*) cultivars] and 73,000 ha of lowbush blueberry (*V. angustifolium*) (Strik & Yarborough, 2005; USHBC, 2009). The majority of the fruit is produced in North and South America and Europe, although production is increasing in Asia and Africa. No matter where blueberry is grown, proper irrigation management is critical for producing high yields and good fruit quality. Even within a few days without rain or irrigation, water stress develops quickly in blueberry, reducing photosynthesis and leading to less growth and fruit production. Over irrigation, however, reduces blueberry root function, increases soil erosion and nutrient leaching, and enhances the probability of developing crown and root rot infection by soil pathogens such as *Phytophthora*. Developing accurate irrigation regimes requires knowledge of both the timing and amount of water needed to replenish any lost by crop transpiration and soil evaporation.

In this chapter, I discuss the importance of irrigation on growth and development in blueberry and examine its relationship to plant water relations. Identified are symptoms of water stress, the most critical stages of water limitations, and various techniques used to monitor plant water status throughout the growing season. I then discuss irrigation scheduling for blueberry, including procedures used to calculate crop evapotranspiration and estimate total irrigation requirements, and finally present recent data on the best methods to apply irrigation. Information is provided on the response of blueberry to not only different irrigation systems and configurations but also of when and where to apply the water. Throughout the chapter, irrigation methods and practices are related to other factors essential to consider when growing blueberries, including interactions with field establishment, planting bed management, nitrogen nutrition, and root disease.

2. Growth and development of blueberry in relation to irrigation

2.1 Seasonal patterns of root and shoot growth

The typical pattern of growth and development of highbush blueberry over an annual cycle are illustrated in Fig. 1. According to Abbott & Gough (1987), new root production begins in early spring when soils reach a temperature threshold of approximately 8 °C. This is then followed by leaf bud swell. Root growth peaks at two times during the growing season. The first peak occurs in late-spring and the second, the largest, occurs after harvest. Interestingly, both peaks occur when soil temperatures are at 14 to 18 °C, strongly suggesting that root growth is regulated, at least in part, by soil temperature in blueberry. Similar soil temperature optimums for root growth were found in other temperate fruit species such as apple (Nightingale, 1935; Rogers, 1939) and peach (Nightingale, 1935). Shoot growth, by comparison, appears less controlled by temperature and more controlled by availability of plant resources. Shoot growth first peaks after the initial peak in root growth but then declines when fruit maturation begins. During fruit maturation in mid-summer, fruit provide a highly competitive sink for carbohydrates and nutrients, considerably reducing the availability of resources to other parts of the plant. Because of the decline in vegetative growth during this period, fruit removal is often recommended during the first 2 years of orchard establishment in order to increase growth of new plantings and improve yields during following years (Strik & Buller, 2005). Aside from the beginning and end of the growing season, shoot and root growth are lowest just prior to fruit harvest (Fig. 1). Once harvest is complete, a second flush of new shoots and roots occur. Often, more than

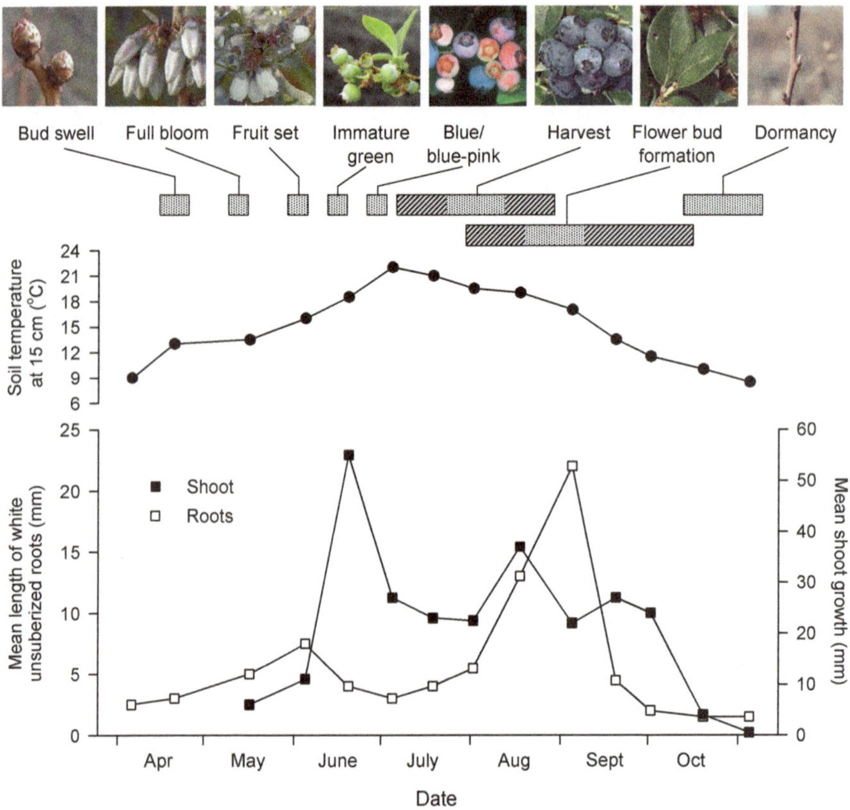

Fig. 1. Elongation of white unsuberized roots in relation to shoot growth, soil temperature, and stage of development of highbush blueberry plants. Adapted from Abbott and Gough (1987).

one flush of shoots can happen after harvest, although the number of flushes varies depending on cultivar and cultural practices. Flower bud induction overlaps with fruit harvest and coincides with the second peak in root growth. Shoot and most root growth finally ceases in late-autumn as the plant enters dormancy and does not resume until the following spring.

2.2 Flowering and fruit development

Bud break and bloom in blueberry occur in early spring when evaporative demand is usually low and leaf size is still small. Thus, aside from any water needed for fertilizer application, irrigation requirements prior to pollination and fruit set are minimal and often unnecessary, depending on spring precipitation. However, once the fruit are set and the canopy develops, sufficient rain or irrigation becomes critical. It is at this early stage of fruit development, often referred to as Stage I, that rapid cell division takes place in the fruit (Fig. 2). Cell division is very sensitive to water stress, and if diminished, will reduce the size of the berries at harvest. Following this stage, the berries enter Stage II, a period of slow growth for several weeks followed by a final stage of rapid cell expansion and fruit ripening

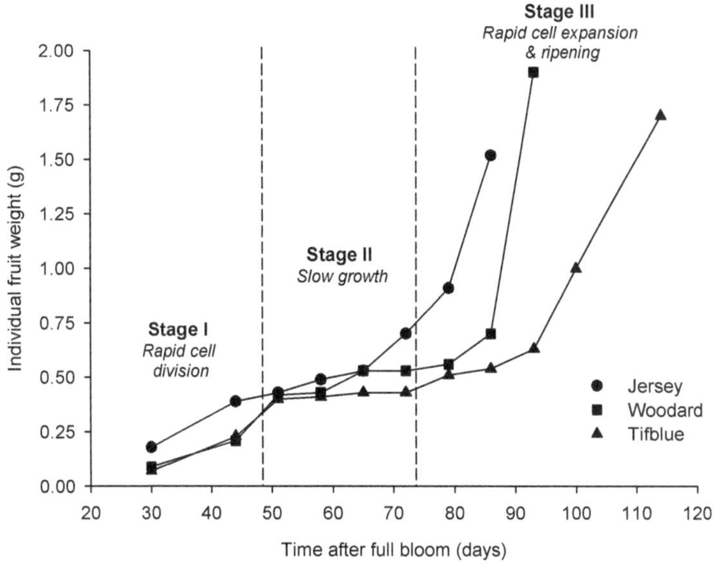

Fig. 2. Individual fruit weight of 'Jersey' highbush blueberry and 'Woodard' and Tifblue' rabitteye blueberries. Adapted from Tamada (2002).

(Fig. 2). Numerous studies on grape and tree fruit crops, which display similar double sigmoidal patterns of fruit growth as blueberry, indicate that effects of moderate water stress during this middle lag phase period has little effect on the size of the fruit at harvest. Theoretically, water stress at this stage should also have minimal effects on fruit size in blueberry. Abbott & Gough (1987), however, indicate that it is precisely at this stage of berry development that vegetative growth is at its maximum peak (Fig. 1). Blueberry is dependent on new wood for production of fruit the following year. Conceivably, any water stress occurring during this peak in shoot production could limit production of new canes for next year's crop.

Irrigation during Stage III is also critical and is perhaps the most sensitive period to water stress, as any water limitations at this point will reduce cell expansion and berry size and therefore have a large impact on yield. Mingeau et al. (2001) examined the effects of water deficits at various phenological stages in 'Bluecrop' blueberry and found that even moderate water stress (i.e., enough to reduce transpiration by 35%) during the final stage of fruit growth and ripening strongly influenced yield by reducing both mean fruit weight and fruit diameter. They also found that water stress after harvest reduced the number of flower buds. Flower bud induction occurs in mid- to late-summer in most cultivars and overlaps with late fruit development (Fig. 1). Thus, in addition to reducing yield of the current year's crop, water stress during the final stage of berry development will also reduce the number of flowers and fruit produced the following year.

Nutrient requirements also vary over the growing season but do not necessarily correlate with water demands. This difference is an important consideration when using irrigation to fertigate (Bryla et al., 2010). Unlike water, the largest demands for many nutrients, including nitrogen (N), typically occur early in the season during canopy development and at the beginning of fruit production (Throop & Hanson, 1997).

3. Plant water relations and response of blueberry to drought

3.1 Fundamentals of plant water potential

The growth, function, productivity, and water use of a plant are intimately related to its water status. Various parameters are used as indicators of plant water status, the most common of which is tissue or organ water potential. Values are typically expressed in units of pressure such as megapascals (MPa), bars, or atmospheres or in units of height or hydraulic head. In plants, the principle components affecting water potential is solute concentrations in cell water and turgor pressure caused by rigidity of the cell wall. For practical purposes, the water potential of free water is considered zero. Therefore any movement of water from wet soil to the plant requires a negative potential. Water potential measured at any point in the soil, plant, and atmosphere, referred to as the soil-plant-atmosphere continuum, is a measure of the tendency of water to move away from that point. Water tends to move from places where its potential is high (e.g., moist soil) to places where its potential is lower (e.g., ambient air with relative humidity less than 99%). The difference between leaf water potential and soil water potential (the latter near zero for moist soils) is an estimate of the driving force for water movement from soil to the foliage. Water readily moves from foliage to the atmosphere (via stomatal openings on the leaf surface; see below) due to relatively higher vapor pressure deficits in the atmosphere.

Plant water potential is often measured using a pressure chamber, sometimes referred to as a "pressure bomb" or a "plant water status console". To make a measurement, a severed part of a plant such as a leaf or branch is placed in an enclosed chamber with its freshly cut end protruding through a rubber seal. The air pressure in the chamber is then gradually increased until it just causes the exudation of xylem sap at the cut end (generally viewed with a magnifying glass). At this point, the resulting pressure of the sap is zero, so xylem pressure equals negative air pressure. If xylem osmotic potential can be ignored (which is often the case as it's usually near zero in most plants), xylem pressure is equal to xylem water potential, which can be the same as the water potential of the other tissues in the chamber (if water equilibration has been achieved) (for details, see Scholander et al., 1964).

Marked daily changes in the water potentials occur in the soil-plant-atmosphere continuum (Fig. 3). In most plants, leaf stomata close at night and as a result, transpiration essentially ceases, allowing root and leaf water to equilibrate with the soil water. The equilibration process may take hours to occur but generally happens before dawn. When the soil is wet and near field capacity, e.g., shortly after a rain or irrigation event, water potentials in the soil, root, and leaf approach zero at night. The stomata then open at dawn and transpiration begins, resulting in a decline in leaf water potential. Root and soil water potentials also decline shortly thereafter. If there is no additional rain or irrigation, leaf, root, and soil water potential becomes more and more negative. As the soil dries, the difference between root and soil water potential must become larger each day in order to sustain water movement from soil to the roots. In contrast, the difference between leaf and root water potential remains constant until the plant is no longer able to sustain a water potential gradient sufficient to absorb enough water to maintain leaf turgor, e.g., when leaf water potential reaches -1.5 MPa. The leaf thus wilts at this point but recovers at night. If drought persists, the leaf may wilt permanently and tissue damage will result.

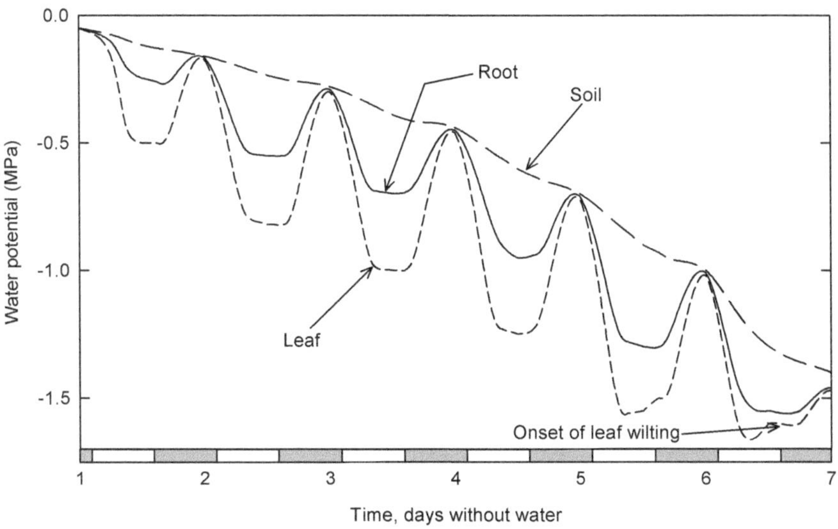

Fig. 3. Daily changes in soil, root, and leaf water potential following irrigation or a rain event. The shaded regions on the x-axis represent night and the white regions represent daytime. The figure is adapted from Slayter (1967).

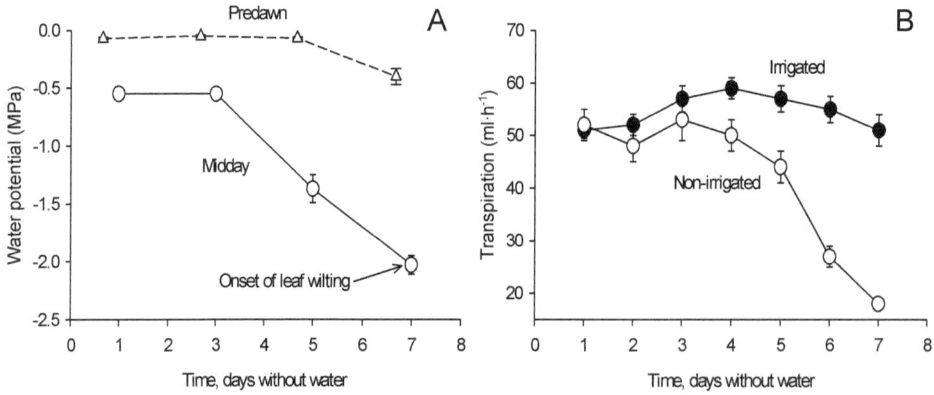

Fig. 4. A) Daily changes in predawn (0500 h) and midday (1400 h) leaf water potential and B) evapotranspiration of 3-year-old 'Elliott' blueberry plants grown in 23 L pots filled with sandy soil. Plants were either irrigated daily (B only) or exposed to drought for 7 days (A and B). Each symbol represents the mean of six plants and error bars represent one standard error.

3.2 Relationship between plant water potential and evapotranspiration in blueberry

An example of how water potential changes in a blueberry plant during the onset of drought is shown in Fig. 4. Although changes will differ somewhat among cultivars (Bryla and Strik, 2007), leaf water potential, measured either at predawn or at midday, declines as predicted when soil water is depleted over time. After 3 to 4 days without water, evapotranspiration also declines, demonstrating the proportional relationship between plant water potential and crop water use. This relationship is well illustrated by examining the response of stomatal conductance to changes in leaf water potential. Stomatal conductance is used to quantify gas diffusion processes, such as transpiration and CO_2 assimilation, between plants and the atmosphere. The usual pathway for CO_2 to enter a plant during photosynthesis is through controllable openings on the leaf surface known as stomata. Transpiration is an unavoidable consequence of water loss through these same openings. The openings are controlled by the presence of two guard cells surrounding a stomatal cavity inside the leaf. Stomatal conductance is most commonly measured using a diffusion porometer, which consists of a chamber for clamping onto the leaf surface and sensors to monitor changes in humidity inside the chamber (for details, see Pearcy et al., 1989). In most plants, including blueberry, stomata open during the day and close at night but will also close in response to water deficits during the day to help prevent excessive water loss during drought (Anderson et al., 1979). Under field conditions, stomatal conductance declined rapidly as leaf water potential approached values as high as -0.6 to -0.8 MPa, indicating highbush blueberry is quite sensitive to even moderate levels of water stress (Fig. 5). Davies & Johnson (1982) also determined that 'Bluegem' rabbiteye blueberry was sensitive to water potential changes but estimated that the critical water potential for total stomatal closure was at -2.2 MPa. By comparison, critical water potentials as low as -3.5 MPa have been reported in apple (Davies & Lasko, 1979).

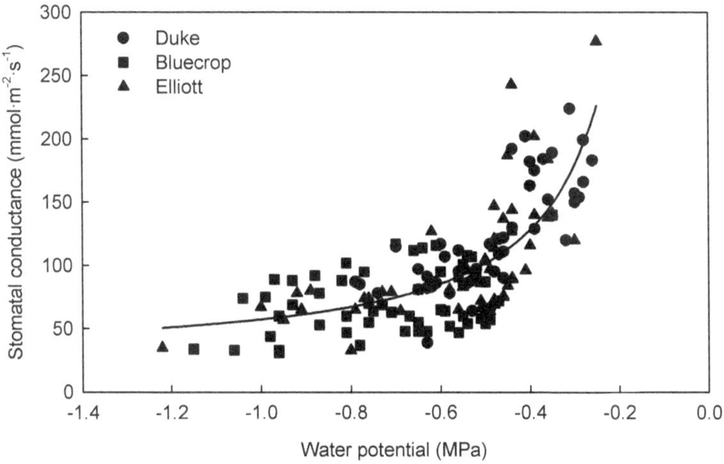

Fig. 5. Relationship between leaf stomatal conductance and midday (1400 h) leaf water potential in mature 'Duke', 'Bluecrop', and 'Elliott' blueberry plants grown under field conditions. Adapted from Bryla & Strik (2006).

Under field conditions, water stress often develops in blueberry within 3 to 7 days without rain or irrigation during summer, varying depending on plant age, cultural practices, phenological development, soil texture, and weather conditions (Hess et al., 1997). Stress symptoms include reduced shoot growth, increased root growth, lower water use, and less photosynthesis. Young, succulent shoots and leaves wilt readily under dry conditions, and if drought persists, leaf margins and tips may become necrotic and scorched. This scorching is similar in appearance to salt injury often associated with over-fertilization (Caruso & Ramsdell, 1995). Internode length is shortened by water deficits, as is the duration of shoot growth when these deficits occur early in the growing season (B. Strik, personal communication). Susceptibility to water deficits may increase after the initiation of fruit ripening. Berries of small fruit crops, including blueberry, however, have few stomata. The majority of the water lost by the plant occurs through the leaf surfaces with fruit playing a minor direct role in plant water losses. Resistance to water deficits may be enhanced by osmotic adjustment (e.g., Zhang & Archbold, 1993) or by increased root to shoot ratios (e.g., Renquist et al., 1982), leaf thickness and waxiness (Anderson et al., 1979), and cell wall elasticity (e.g., Savé et al., 1993).

In France, mature 'Bluecrop' blueberries exposed to drought closed their stomates and reduced transpiration gradually within 9 days after withholding irrigation (Améglio et al., 2000). Upon rewatering, recovery was slow,with stomatal conductance and transpiration returning to normal after 7 to 9 days. A vulnerability curve presented in the same study indicated that embolism in the xylem vessels was negligible when leaf water potential was -1.2 MPa or higher but increased rapidly at lower water potentials. To develop the curve, hydraulic conductance was measured at different applied pressures on 2 to 3 cm-long stem segments excised under water (Sperry et al., 1988). Percent loss of hydraulic conductance was 50% at -1.4 MPa and 100% at -2.1 MPa. However, *in situ* embolism measured during actual water stress was usually less than 30%. Apparently, rapid reduction in stomatal conductance reduced water loss and maintained water potential at the threshold of cavitation in 'Bluecrop', protecting it from total xylem cavitation and enhancing its ability to recover from drought.

Bryla & Strik (2007) examined the onset of water stress in three cultivars of 5-year-old highbush blueberry plants in Oregon, USA, including 'Duke', an early-season cultivar that ripens in late June to mid July, 'Bluecrop', a mid-season cultivar that ripens in mid July to early August, and 'Elliott', a late-season cultivar that ripens in early August to early September. Plants were exposed to water stress during each ripening period. During each period, stem water potential dropped only slightly within the first 3 to 4 days after irrigation was withheld but declined substantially, in many cases, after 5 to 7 days without irrigation (Fig. 6A-C). This later decline was associated with reduced rates of root water uptake, indicated by smaller changes in soil water content in each treatment. Within each cultivar, the most apparent decline in water potential occurred when fruit were in their final stages of ripening, just prior to harvest. The differences in water potential were attributed to seasonal variation in water use among the cultivars (Fig. 6D-F). 'Duke' acquired the most water, using 5 to 10 mm per day from mid-May to mid-August, while 'Elliott' acquired the least, using only 3 to 5 mm per day. Water use by 'Bluecrop' was intermediate. Water use was highest during fruit filling and ripening but declined markedly after harvest, especially in 'Duke', which ripened earliest. A sharp decline in water use was less apparent in 'Elliott', which had the latest and most extended fruit ripening period. Mingeau et al. (2001) reported that almost 55% of the total seasonal water requirements of 'Bluecrop' occurred in June and July during fruit ripening; once fruit were picked, plant water requirements decreased to nearly half. Higher rates of stomatal conductance and water use have been associated with increased photosynthetic activity during fruit ripening in lowbush blueberry (Hicklenton et al., 2000). Thus, as ripening periods differ among cultivars, water requirements at any given time of the year will also differ.

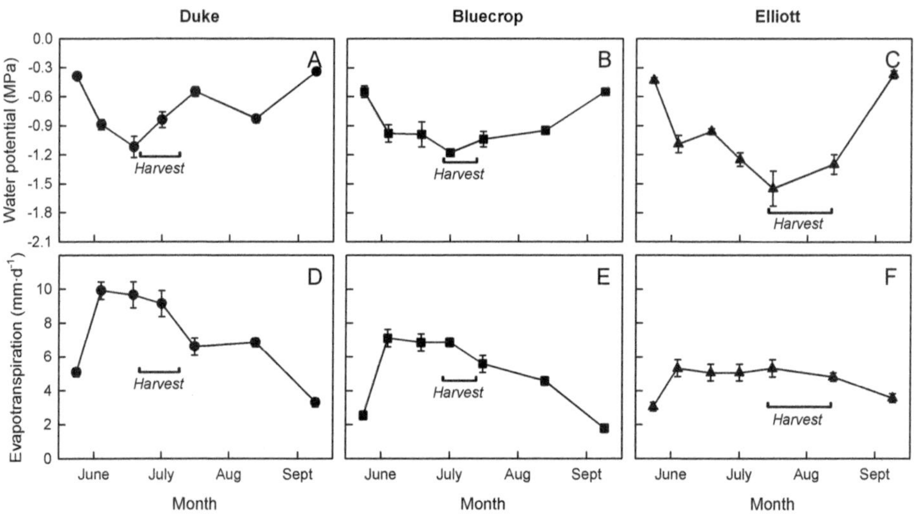

Fig. 6. Seasonal changes in (A-C) leaf water potential and (D-F) evapotranspiration in mature (A, D) 'Duke', (B, E) 'Bluecrop', and (C, F) 'Elliott' blueberry plants. Adapted from Bryla & Strik (2007).

4. Estimating evapotranspiration for irrigation scheduling in blueberry

Irrigation scheduling, a key element of proper water management, is the accurate forecasting of water application (amount and timing) for optimal crop production (yield and fruit quality). The goal is to apply the correct amount of water at the right time to minimize irrigation costs and maximize crop production and economic return. Many techniques and technologies can forecast the date and amount of irrigation water to apply. The appropriate technique or technology is a function of the irrigation water supply, technical abilities of the irrigator, irrigation system, crop value, crop response to irrigation, cost of implementing technology, and personal preference. This section illustrates tools and techniques available for improving irrigation scheduling in blueberry.

4.1 Procedures for calculating blueberry evapotranspiration

Irrigation is required of course whenever precipitation is inadequate to meet the water demands of the crop, which, depending on latitude and weather patterns, can occur anytime from March through October in the northern hemisphere and from September to May the in southern hemisphere. In Oregon, USA, average seasonal water requirements for blueberry range from 15 to 49 mm per week (Hess et al., 2000). The highest irrigation requirements typically occur in July, although actual peak irrigation demands vary considerably throughout the summer depending on weather, location, and stage of fruit development. Nearly all water taken up by a crop is lost by transpiration, a process that consists of the vaporization of liquid water contained in the plant to the atmosphere; only a tiny fraction is used within the plant. The water, together with some nutrients, is absorbed by the roots and transported through the plant. The water is vaporized within the leaves and transferred to the atmosphere through the leaf stomata. Water use by the crop is fairly complicated to estimate and will depend on numerous factors, including weather, plant age and cultivar, soil conditions, and cultural practices. Water is also lost from the soil surface by evaporation, particularly within the first few days after rain or irrigation. Crop transpiration and soil evaporation occur simultaneously and there is no easy way of distinguishing between the two processes. Therefore, crop water requirements are typically estimated as the combination of the two processes, collectively termed crop evapotranspiration (ET).

Weekly estimates of crop ET are often accessible on the internet from weather-based websites, e.g., AgriMet (Pacific Northwest Cooperative Agricultural Weather Network; http://www.usbr.gov/pn/agrimet/) and CIMIS (California Irrigation Management Information System; http://wwwcimis.water.ca.gov/cimis/welcome.jsp). These sites obtain data from a satellite-based network of automated agricultural weather stations located throughout a region of interest. Weather data are used to estimate ET of a reference surface such as grass (ET_o) or alfalfa (ET_r), which is then converted to crop ET using an appropriate crop coefficient (K_c) for blueberry (for details, see Allen et al., 1998). A crop coefficient represents the relative amount water used by a crop (e.g., blueberry) to that used by grass or alfalfa. Therefore, the value will change over the season as the crop canopy develops. Crop coefficients will also differ depending on whether crop ET is calculated using ET_o or ET_r. An example of crop coefficients used for calculating blueberry ET based on weather-based estimates of ET_r is shown in Fig. 7A. The coefficients increase as the canopy develops from bud break to the beginning of fruit ripening and then gradually declines until leaf senescence and dormancy. Blueberry reaches full effective canopy cover when the first blue fruit appear and it is at this stage that water use by blueberry is equal to alfalfa and $K_c = 1$.

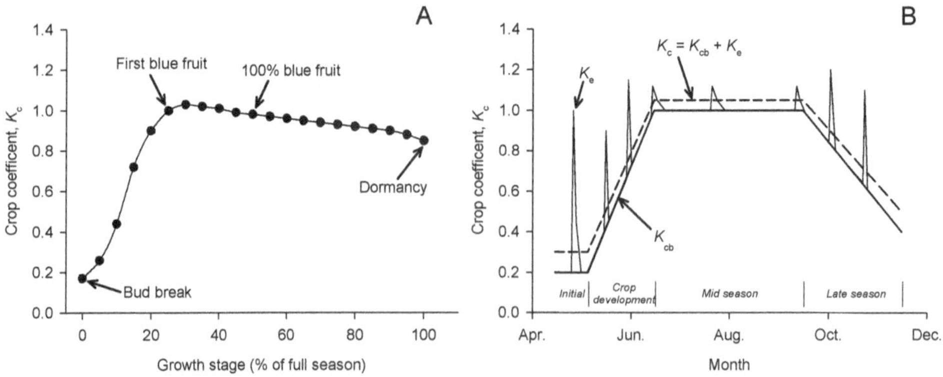

Fig. 7. Crop coefficient curves for highbush blueberry from (A) AgriMet and (B) FAO-56. Blueberry ET is calculated at various stages of crop development by multiplying K_c by ET_r or ET_o, respectively.

The FAO-56 guidelines for calculating crop ET recommends using a simplified segmented K_c curve approach whereby the growing season is divided into four distinct stages: initial, crop development, mid season, and late season (Allen et al., 1998; Fig. 7B). In perennial crops, the initial stage begins at bud break or the green-up date when new leaves are initiated and continues to about 10% ground cover. The K_c during the initial stage ($K_{c\ ini}$) is predominated by soil evaporation and therefore is large when the soil is wet from rain or irrigation and small when the soil is dry. The crop development stage runs from 10% ground cover to effective full cover. Again, full cover in blueberry occurs about when the fruit just begin to turn blue but may be prior to fruit ripening in later season cultivars. The mid-season stage runs from full cover to the beginning of leaf yellowing, i.e., the start of senescence. It is the longest stage during the growing season and is the period in which K_c reaches its maximum value ($K_{c\ mid}$). The late-season stage runs from leaf yellowing to complete leaf senescence ($K_{c\ end}$). The Kc values increase linearly from $K_{c\ ini}$ to $K_{c\ mid}$ and decrease from $K_{c\ mid}$ to $K_{c\ end}$; however, the slopes will vary depending on the length of each stage. Crop transpiration and soil evaporation may be combined into a single coefficient, K_c, (single crop coefficient approach) or separated into two coefficients: a basal crop coefficient (K_{cb}), which represents primarily the transpiration component of ET, and a soil evaporation component (K_e) (dual crop coefficient approach). In this later case, K_c is replaced by $K_{cb} + K_e$. The K_c values listed for berries (bushes) in FAO-56 are 0.30 ($K_{c\ ini}$), 1.05 ($K_{c\ mid}$), and 0.50 ($K_{c\ end}$); the K_{cb} values are 0.20 ($K_{cb\ ini}$), 1.00 ($K_{cb\ mid}$), and 0.40 ($K_{cb\ end}$). Blueberry ET in this case is calculated at each stage of development by multiplying K_c or $K_{cb}+K_e$ by ET_o. See Allen et al. (1998) for procedures on calculating K_e.

To adjust for smaller plant size in new plantings, Fereres et al. (1982) developed a correction factor, Fc, by correlating crop ET to canopy development using data from young almond trees. The relationship was modified by Holzapfel et al. (2004) to estimate ET for young blueberries using the following equation:

$$ET_{blueberry} = K_c ET_o F_c \tag{1}$$

where

$$F_c = K_1 S_h + K_2 \tag{2}$$

and

$$S_h = \frac{A_s}{HL} \tag{3}$$

S_h is percent shade ($10 \le S_h \le 70$), K_1 and K_2 are constants of the shadow factor adjusted for irrigation method (equal to 0.0118 and 0.25, respectively, for drip and 0.0127 and 0.1125, respectively, for microsprays), A_s is the area of the soil surface shaded by the crop canopy at 1200 h (m^2), H is the distance between rows (m), and L is the distance between plants within the row (m). The correction factor, F_c, is a function of cultural practices, the type of irrigation system used, planting density, and climatic conditions of the area. Once a planting has 70% cover or larger, it reaches an adult condition where crop ET is no longer a function of plant size (Bryla & Strik, 2007). However, because blueberry is a relatively short crop (<2.0 m tall and <1.5 m wide) with a fairly wide rows (3.0-3.6 m apart), F_c can also be used to adjust for lack of canopy cover between rows in mature plantings.

Normally, irrigation should be scheduled to replace any water lost by crop ET. Keep in mind, however, that these are ET estimates for mature, healthy, well-irrigated blueberry plants. Adjustments to these values are needed when plants are young or stressed (e.g., nutrient deficient). Under these circumstances, irrigators should reduce the amount of irrigation water applied but pay close attention to soil moisture conditions to avoid under- or over-irrigating their crop. There are numerous devices available for monitoring soil moisture, although some are more accurate and reliable than others. Many of these monitoring devices need to be calibrated to a particular site so that gathered data can be related to actual soil moisture conditions. Soil moisture monitors should be installed within the root zone (usually 0.15-0.30 m deep) of a representative plant and should not be located directly beneath an irrigation emitter.

4.2 Adjusting water applications for irrigation system efficiency

It is important to understand that a crop's irrigation requirements differ considerably from its water requirements. Crop water requirements indicate the total amount of water directly used by a crop but do not account for any extra water needed to compensate for non-beneficial water use or loss, e.g., run-off, deep percolation, evaporation, wind drift, ground cover, weeds, etc. Additionally, irrigation systems do not apply water with 100% uniformity (Burt et al., 2000). For accurate irrigation scheduling, these losses must be evaluated for each system. The most common systems used to irrigate blueberry are sprinklers and drip.

Average irrigation application efficiencies for well-maintained solid set sprinkler systems generally range from 65-75%, which largely depends on the quality of sprinkler overlap. Close spacing and newer sprinkler heads help improve sprinkler water application efficiency. Brand new drip systems, on the other hand, can generally be designed with 85-93% efficiency, except in cases with major elevation changes. Beware that neglected drip systems have been shown to have actual efficiencies closer to 60-80%. Primary causes for low efficiencies include flow variation due to poor system design, emitter plugging, and pressure differences within the field.

In northwest Oregon, USA, average irrigation requirements throughout the growing season are estimated to range from 4-16 mm of water per day with sprinklers and 1-4 mm of water per day by drip (Table 1). The highest irrigation requirements typically occur in July, although actual peak irrigation demands will vary throughout the summer depending on

	Average daily water requirements							
Month	Early-season cultivars (mm/day)	In-row spacing¶ 0.75 m	0.9 m	1.2 m	Late-season cultivars (mm/day)	In-row spacing¶ 0.75 m	0.9 m	1.2 m
		-------- (L/plant/day) ----				-------- (L/plant/day) ----		
				Sprinkler irrigation§				
May	6	15	18	24	4	9	11	14
June	12	27	33	43	10	22	27	36
July	16	37	45	59	16	38	45	60
August	11	27	32	42	12	28	33	44
September	9	21	25	33	9	21	25	34
Max. demand†	22	52	62	83	22	52	62	83
				Drip irrigation§				
May	2	4	4	6	1	2	3	4
June	3	7	8	11	3	6	7	9
July	4	9	11	15	4	10	12	16
August	3	7	8	11	3	7	8	11
September	2	5	6	8	2	5	6	8
Max. demand†	6	14	16	22	6	14	16	22
				According to AgriMet#				
ay	3	7	8	11	2	4	5	7
June	5	13	15	20	5	11	13	17
July	7	17	21	28	8	18	21	28
August	5	12	15	20	6	13	16	21
September	4	10	12	16	4	10	12	16
Max. demand†	10	24	29	39	10	24	29	39

Calculations are based on a 3.0-m wide between-row spacing.
§Values should be adjusted for precipitation before scheduling irrigation.
#Obtained from AgriMet website (http://www.usbr.gov/pn/agrimet/). These values must be adjusted for water application efficiency of the irrigation system in order to estimate irrigation water requirements.
†Occurs when conditions are hot (>35°C), dry, and windy in mid-July to early-August.

Table 1. Average daily water requirements for healthy, mature highbush blueberry plants in northwest Oregon. Note that values will differ at other locations, depending on latitude, elevation, and local weather conditions.

weather and stage of fruit development. A well-maintained drip system generally requires only about 25% of the water needed with sprinklers due to the higher efficiency associated with drip irrigation. It should be noted, however, that the actual crop water use by sprinkler and drip-irrigated blueberries is theoretically identical. Irrigation requirements will also vary of course with location but are easily adjusted when calculating crop ET.

4.3 Timing of water applications

The timing or frequency of water applications will depend on soil texture (e.g., sand versus clay), the irrigation system used (e.g., drip versus sprinkler), the rate at which the plant is using water, and the overall development of the plant's root system. Blueberry is a shallow-rooted plant compared to many perennial fruit crops. The roots of highbush blueberry are usually located in the top 0.5 m of soil and are often most concentrated near the soil surface (Fig. 8). Patten et al. (1988) found that 90% of the roots in rabbiteye blueberry, which tends to produce deeper roots than highbush cultivars, were less than 0.45 m deep even when plants were not mulched (drier soil surface) and were irrigated by drip (concentrated soil wetting pattern). Consequently, when water demands are high, blueberry plants quickly depletes the water from their root zone and require frequent applications of water in order to avoid water stress.

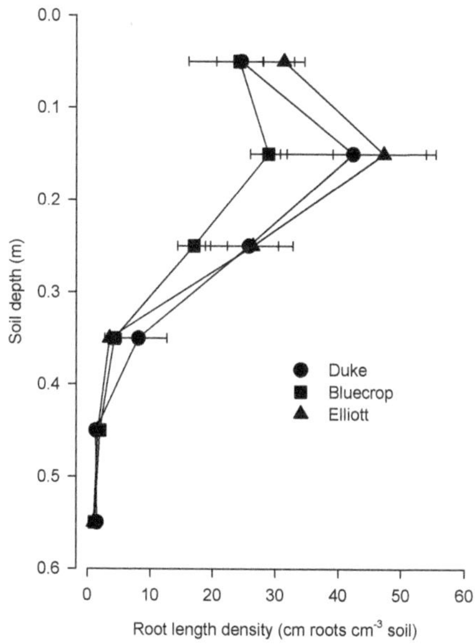

Fig. 8. Root length density of mature 'Duke', 'Bluecrop', and 'Elliott' blueberry plants. Roots were collected at 0.1-m depth increments. Adapted from Bryla & Strik (2007).

Frequent water applications are especially important when using drip, which tends to restrict soil wetting and thus produces a smaller root system. When done properly, frequent irrigations are beneficial and often increase growth and yield in many horticultural crops. For example, frequent irrigation by drip in peach increased fruit size and yield compared to other irrigation methods by maintaining higher tree water status between irrigations (Bryla et al., 2005). It may also be important to apply water to both sides of the plant. Abbott & Gough

(1986) found that when water was applied to only one side of the blueberry bush, growth and production was severely restricted on the other side. Irrigators, however, particularly those using drip, should be careful to avoid the temptation to over-irrigate. Over-irrigation depletes the root zone of much-needed oxygen, thus reducing both root growth and nutrient uptake and leading to a host of potential root disease problems. Davies & Flore (1986) observed, in both highbush and rabbiteye blueberry, that stomatal conductance declined within 5 days and photosynthesis declined within 9 days when plants were grown in flooded soil, and 18 days or more were required for each process to recover to pre-flood conditions.

High-frequency irrigation may be especially beneficial and perhaps even required when organic matter is incorporated into the planting bed. Organic matter often reduces water holding capacity of the soil and can lead to problems with hydrophobicity. Soil hydrophobicity is the lack of affinity of soil to water and is thought to be caused primarily by a coating of long-chained hydrophobic organic molecules, such as those released from decaying organic matter, on individual soil particles (DeBano, 2000). Hydrophobic soils often become very difficult to rewet once they dry out. White (2006) found that even with drip irrigation, sawdust incorporated into raised planting beds made it difficult to retain adequate moisture in the upper portions of the soil where many of the blueberry roots were located. To compensate, much longer and more frequent irrigation was required in beds with incorporated sawdust than without. Personal observations indicate that, even after 50 mm of rainfall, dry beds with incorporated sawdust tend to remain dry and do not become fully saturated until the following season. Krewer et al. (2002) found that water infiltrated much more readily through sandy soil amended with pine post and pole peelings (2-45 mm long pieces of bark and wood) than soil amended with milled pine bark, although plant growth was slightly less in the soil with the larger-sized product.

4.4 Other tools and techniques

Other potential methods available to schedule irrigation in perennial crops include soil-based and plant-based monitoring approaches. The soil-based approach relies on soil moisture monitoring devices set in a feedback mode to automatically open and close irrigation valves when soil moisture reaches a certain level of dryness. Unfortunately, these systems are not universal and require careful calibration from site to site to operate properly. Little research has been done with these soil-based systems on blueberry. The plant-based approach may be the most accurate method to schedule irrigations and avoid water stress during critical stages of growth but is probably also the most complex and labor intensive. This type of approach uses measures of plant growth or water status to determine exactly when irrigation is needed. Leaf water potential measurements are currently used as a successful tool for scheduling irrigation in fruit trees. Interpretation of water potentials for irrigation scheduling is complicated however by the fact that values are influenced by weather conditions. For example, leaf water potential tends to decrease with time over the season regardless of adequacy of irrigation due to increasing evaporative demand. To overcome this problem, a fully irrigated baseline (reference) value of stem water potentials must be calculated for any given value of midday air vapor pressure deficit (VPD). A baseline value is applicable to a wide variety of soil and irrigation conditions and has provided stem water potential guidelines for fully irrigated fruit trees grown in California using relative humidity and air temperature (Shackel et al., 1997). Once developed, data collected from weather stations can be used for baseline estimates in commercial fields throughout a region. Irrigation scheduling is accomplished by comparing actual water potentials to reference values; when actual values fall below reference values, irrigation is increased. Typically, irrigation is increased by 5-10% above the previous week's rate when mean weekly stem water potentials are lower than reference values, and decreased by 5-10% when actual and reference values are equal for two consecutive weeks. To ensure plants are not over or under irrigated, soil water content should also be monitored at least monthly. Soil water content measurements may also provide information to help determine initial irrigation rates based on root-zone changes in the soil water profile during the first few weeks of the growing season.

5. Irrigation systems and considerations for water application

Most commercial blueberry fields in the U.S. are irrigated by overhead sprinklers or drip (Strik & Yarborough, 2005). Water is typically applied one to two times per week as needed with sprinklers and every one to three days with drip. Sprinkler systems are relatively simple to install and maintain, and when designed properly, obtain reasonable uniformity of water application. Some major advantages of sprinklers are that they can be used to maintain a cover crop, protect the crop from frost damage during subfreezing temperatures, cool the crop during hot conditions, and wash dust off the crop before harvest. Drip systems are somewhat more expensive to install and more difficult to maintain than sprinklers but offer superior water control and distribution uniformity, lower energy costs, improved application of fertilizer and other chemicals, improved cultural practices, including the ability to irrigate during harvest, fewer weed and disease problems, and reduced food safety risks when using surface water to irrigate (Kruse et al., 1990). A few growers are also using microsprays on blueberry. Microspray irrigation offers advantages similar to drip but applies the water to the soil surface by a small spray. Although not commonly used in blueberry, Holzapfel et al. (2004) found in Chile that production was higher with microsprays than with drip. Because microsprays wet more soil volume than drip, plants tend to produce a larger root system, which may be a considerable advantage in a shallow, densely-rooted crop like blueberry (Patten et al., 1988).

5.1 A comparison of irrigation methods

Bryla et al. (2011) compared the water requirements for growing blueberry with sprinklers, drip, and microsprays to determine which method produces the most growth after planting. Two cultivars, 'Duke' and 'Elliott', were evaluated. By the end of the second growing season, drip irrigation produced the largest 'Elliott' plants among the irrigation methods with 42% less water than microsprays and 56% less water than sprinklers. The benefit of drip in 'Elliott' was likely a result of superior plant water status due to higher soil water content in the vicinity of the roots. Drip also maintained higher plant water potentials than microsprays in other perennial fruit crops, including peach and almond (Bryla et al., 2005; Edstrom & Schwankl, 2004). Drip irrigation, however, was not beneficial in 'Duke' (Bryla & Linderman, 2007). In this case, plants irrigated by drip were only half the size of those irrigated by sprinklers or microsprays. Root sampling revealed that 'Duke' was infected by *Phytophthora cinnamomi*, the causal organism often associated with root rot in blueberry, and the wetter soil conditions with drip were more favorable to the disease. Therefore, in terms of early plant growth and water use efficiency, drip irrigation was the best method out of the three to establish healthy blueberry plants. However, sprinklers and microsprays may be better alternatives for cultivars such as 'Duke' that are highly susceptible to root rot, especially at sites with heavy soils or a history of the disease.

Fruit were first harvested beginning the third year after planting in 'Elliott' and fourth year after planting in 'Duke' (Fig. 9). During the first 4 years of production, yields were similar in 'Duke' whether plants were irrigated by sprinklers or microsprays but lower when irrigated by drip due again to higher incidence of root rot. Root rot does not usually result in plant death in blueberry, although it will reduce growth and production even when plants are treated with fungicide, as we did each year beginning the third year after planting. In 'Elliott', yields were slightly higher with drip than with sprinklers and microsprays during the first year of production and still higher than sprinklers the second year. However, by the third year, yield was similar between drip and sprinklers but higher when plants were irrigated by microsprays. This latter result agrees with that of Holzapfel et al. (2004), who compared drip and microsprays in 'Bluecrop'. They positioned the microsprays under the canopy on each side of the plants whereas we hung the microsprays above the canopy between every other plant. Hanging the microsprays reduced the number of microsprays needed and reduced problems with plants interfering with the microsprays.

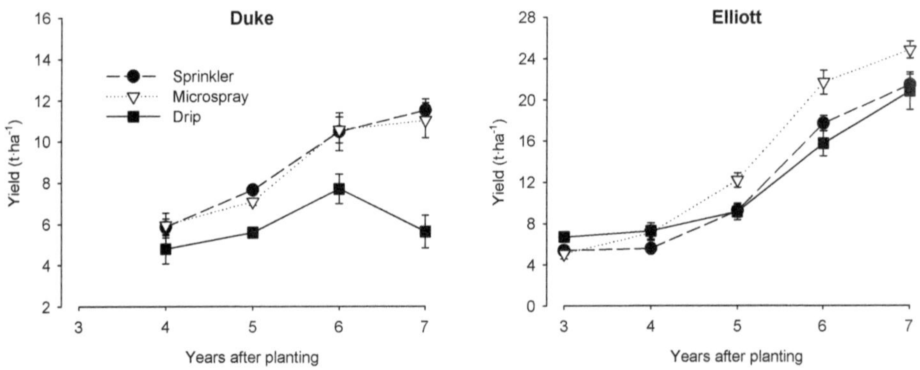

Fig. 9. Marketable fruit yield of 'Duke' and 'Elliott' blueberry irrigated by sprinklers, microsprays, or drip. Each symbol represents the mean of five plots with six plants each, and error bars represent one standard error.

5.2 Drip lateral placement

Most new plantings of blueberry are irrigated by drip. One or two laterals of drip tubing is used per row, and the tubing is usually either laid on the ground or hung on a trellis wire with one or two laterals of tubing used per row. Drip emitters are often spaced 0.3-0.6 m apart and range from 1-4 LPH, depending on the design of the system. The goal is to distribute water evenly around the plants; thus the optimum number and placement of emitters will vary depending on plant size, soil type, cultural practices, and weather conditions. Proper lateral placement improves growth and production and increases water use efficiency. It may also reduce problems with soil pathogens (Café-Filho & Duniway, 1996).

We examined the potential of using different drip configurations to reduce the incidence of root rot in 'Duke'. The configurations included two laterals of drip tubing placed on the soil surface on each side of the plants, two laterals buried 0.15 m deep on each side of the plants (approximately 0.3 m from the crown), and one lateral suspended 1.2 m above the plants.

After 2 years, plants irrigated by buried drip were larger and healthier than those irrigated by surface drip, particularly when the laterals were placed near the base of the plants. Signs of water stress, including marginal leaf necrosis, were evident in plants irrigated by surface drip, even after 3 years. In comparison, there was no evidence of water stress in plants irrigated by subsurface drip. The use of subsurface drip maintained lower soil water content near the plants, which reduced root rot and encouraged more lateral root development. Subsurface drip also eliminated water runoff and soil erosion observed with both the surface drip configurations.

6. Conclusions and future research needs

Blueberry is a shallow-rooted crop highly susceptible to water deficits. Within 3 to 4 days without irrigation, both plant water potential and transpiration steadily decline. The plants appear to be most affected by soil water limitation in the later stages of berry development, particularly during fruit ripening, as well as after harvest during fruit bud set. During fruiting, early-season cultivars with a compressed fruiting period have higher water requirements than later-season cultivars and therefore may be more readily exposed to water stress without rain or irrigation. Crop coefficients are available to estimate crop water requirements for irrigation scheduling in blueberry; however, the accuracy of these coefficients requires further testing using various cultivars and cultural practices (e.g., flat versus raised beds, different mulch materials, etc.).

The most common methods used to irrigate blueberry are sprinklers and drip. The amount of water applied by sprinklers or drip must be adjusted for application efficiency when irrigation is scheduled based on estimated water requirements. In general, estimated water requirements are less than the irrigation requirements when blueberry is irrigated by sprinklers but are higher than when irrigated by drip. The higher requirements with sprinklers are due to the relatively low application efficiency (approximately 50% water is applied between rows where there are no roots) while the lower requirements with drip are due to high application efficiency (water is applied directly to the roots) and the fact that canopy cover generally averages less than 50% even as plants approach full maturity, which thereby reduces the actual crop ET.

Drip irrigation improves growth and early production compared to sprinklers, provided the blueberry plants are healthy. Drip, however, may also increase incidence of root rot in susceptible cultivars and is not recommended at sites with heavy soils or a history of the disease. In healthy plants, yield differs little whether plants are irrigated by sprinklers or drip but may be higher when plants are irrigated by microsprays. Microsprays are not traditionally used in blueberry, but the method shows considerable promise and requires further study to determine its potential in commercial blueberry production systems. Work is still needed on the impacts of irrigation methods and scheduling strategies on fruit quality in blueberry, including such irrigation practices as cooling the fruit during hot weather.

7. References

Abbott, J.D. & Gough, R.E. (1986) Split-root water application to highbush blueberry plants. *HortScience*, Vol. 21, pp. 997-998, ISSN 0018-5345

Abbott, J.D. & Gough, R.E. (1987) Seasonal development of highbush blueberry roots under sawdust mulch. *Journal of the American Society for Horticultural Science*, Vol.112, pp. 60-62, ISSN 0003-1062

Allen, R.G., Pereira, L.S., Raes, D. & Smith, M. (1998) *Crop Evapotranspiration. Guidelines for Computing Crop Water Requirements*, Food and Agriculture Organization of the United Nations Irrigation and Drainage Paper 56, ISBN 92-5-104219-5, Rome, Italy

Améglio, T., Le Roux, X., Mingeau, M. & Perrier, C. (2000) Water relations of highbush blueberry under drought conditions. *Acta Horticulturae*, Vol.537, pp. 273-278, ISSN 0567-7572

Anderson, P.C., Buchanan, D.W. & Albrigo, L.G. (1979) Water relations and yields of three rabbiteye blueberry cultivars with and without drip irrigation. *Journal of the American Society for Horticultural Science*, Vol.104, pp. 731-736, ISSN 0003-1062

Bryla, D.R. & Linderman, R.G. (2007) Implications of irrigation method and amount of water application on *Phytophthora* and *Pythium* infection and severity of root rot in highbush blueberry. *HortScience*, Vol.42, pp. 1463-1467, ISSN0018-5345

Bryla, D.R. & Strik, B.C. (2006) Variation in plant and soil water relations among irrigated blueberry cultivars planted at two distinct in-row spacings. *Acta Horticulturae* Vol. 715, pp. 295-300, ISSN 0567-7572

Bryla, D.R. & Strik, B.C. (2007) Effects of cultivar and plant spacing on the seasonal water requirements of highbush blueberry. *Journal of the American Society for Horticultural Science*, Vol.132, pp. 270-277, ISSN 0003-1062

Bryla, D.R, Gartung, J.L. & Strik, B.C. (2011) Evaluation of irrigation methods for highbush blueberry—I. Growth and water requirements of young plants. *HortScience*, Vol.46, pp. 95-101, ISSN 0018-5345

Bryla, D.R., Dickson, E., Shenk, R., Johnson, R.S., Crisosto, C.H. & Trout, T.J. (2005) Influence of irrigation method and scheduling on patterns of soil and tree water status and its relation to yield and fruit quality in peach. *HortScience*, Vol.40, pp. 2118-2124, ISSN 0018-5345

Café-Filho, A.C. & Duniway, J.M. (1996) Effects of location of drip irrigation emitters and position of *Phytophthora capsici* infections in roots on Phytophthora root rot of pepper. Phytopathology 86:1364–1369

Caruso, F.L. & Ramsdell, D.C. (1995) *Compendium of Blueberry and Cranberry Diseases*. American Phytopathological Society Press, ISBN 978-0-89054-173-9, St. Paul., Minnesota, USA

Davies, F.S. & Flore, J.A. (1986) Gas exchange and flooding stress of highbush and rabbiteye blueberries. *Journal of the American Society for Horticultural Science*, Vol.111, pp. 565-571, ISSN 0003-1062

Davies, F.S. & Johnson, C.R. (1982) Water stress, growth, and critical water potentials of rabbiteye blueberry (*Vaccinium ashei* Reade). *Journal of the American Society for Horticultural Science*, Vol.107, pp. 6-8, ISSN 0003-1062

Davies, F.S. & Lakso, A.N. (1979) Water stress responses of apple trees. I. Effects of light and soil preconditioning treatments on tree physiology. *Journal of the American Society for Horticultural Science*, Vol.104, pp. 392-395, ISSN 0003-1062

DeBano, L.F. (2000) Water repellency in soils: A historical overview. *Journal of Hydrology*, Vol.231-232, pp. 4-32, ISSN 0022-1694

Edstrom, J. & Schwankl, L. (2004) Nickels Soil Lab Project, In: *Years of Discovery. A Compendium of Production and Environmental Research Projects, 1972-2003*, pp. 337-346, California Almond Board, Modesto, California, USA

Fereres, E., Martinich, D., Aldrich, T., Castel, J., Holzapfel, E. & Schulbach, H. (1982) Drip irrigation saves money in a young almond orchard. *California Agriculture*, Vol. 36, pp. 12-13, ISSN 0008-0845

Hess, M., Strik, B., Smesrud, J. & Selker, J. (2000) Blueberry, In: *Western Oregon Irrigation Guides*, pp. 11-12, Oregon State University Extension Service, EM 8713, Corvallis, Oregon, USA

Hicklenton, P.R., Reekie, J.Y., Gordon, R.J. & Percival D.C. (2000) Seasonal patterns of photosynthesis and stomatal conductance in lowbush blueberry plants managed in a two-year production cycle. *HortScience*, Vol.35, pp. 55-59, ISSN 0018-5345

Holzapfel, E.A., Hepp, R.F. & Mariño M.A. (2004) Effect of irrigation on fruit production in blueberry. *Agricultural Water Management*, Vol.67, pp. 173-184, ISSN 0378-3774

Krewer, G., Ruter, J., NeSmith, D.S., Clark, J., Otts, T., Scarborough, S. & Mullinix, B. (2002) Performance of low cost organic materials as blueberry substrates and soil amendments. *Acta Horticulturae*, Vol.574, pp. 273-279, ISSN 0567-7572

Kruse, E.G., Bucks, D.A. & von Bernuth R.D. (1990) Comparison of irrigation systems, In: *Irrigation of Agricultural Crops*, B.A. Stewart & D.R. Nielson (eds.), pp. 475–508, Agronomy Monograph No. 30, American Society of Agronomy, Inc. Publishers, ISBN 0-89118-102-4, Madison, Wisconsin, USA

Mingeau, M., Perrier, C. & Améglio, T. (2001) Evidence of drought-sensitive periods from flowering to maturity on highbush blueberry. *Scientia Horticulturae*, Vol.89, pp. 23-40, ISSN 0304-4238

Nightingale, G.T. (1935) Effects of temperature on growth, anatomy, and metabolism of apple and peach roots. *Botanical Gazette*, Vol.96, pp. 58-637, ISSN 0006-8071

Patten, K.D., Neuendorff, E.W., & Peters, S.C. (1988) Root distribution of 'Climax' rabbiteye blueberry as affected by mulch and irrigation geometry. *Journal of the American Society for Horticultural Science*, Vol.113, pp. 657-661, ISSN 0003-1062

Pearcy, R.W., Schulze, E.-D. & Zimmermann, R. (1989) Measurements of transpiration and leaf conductance, In: *Plant Physiological Ecology. Field Methods and Instrumentation*, R.W. Pearcy, J. Ehleringer, H.A. Mooney & P.W. Rundel (eds.), Chapman and Hall, ISBN 0-412-40730-2, New York, New York, USA

Renquist, A.R, Breen, B.J. & Martin, L.W. (1982) Influences of water status and temperature on leaf elongation in strawberry. *Scientia Horticulturae*, Vol.18, pp. 77-85, ISSN 0304-4238

Rogers, W.S. (1939) Root studies: VIII. Apple root growth in relation to rootstock, soil, seasonal, and climatic factors. *Journal of Pomology and Horticultural Science*, Vol.17, pp. 99-130, ISSN 0028-0836

Savé, R., Peñuelas, J., Marfà, O. & Serrano, L. (1993) Changes in leaf osmotic and elastic properties and canopy structure of strawberries under mild water stress. *HortScience*, Vol.28, pp. 925-927, ISSN 0018-5345

Scholander, P.F., Hammel, H.T., Hemmingsen, E.A. & Bradstreet, E.D. (1964) Hydrostatic pressure and osmotic potential in leaves of mangrove and some other plants. *Proceedings of the National Academy of Sciences, USA*, Vol.52, pp. 119-125, ISSN 0027-8424

Shackel, K.A., Ahmadi, H., Biasi, W., Buchner, R., Goldhamer, D., Gurusinghe, S., Hasey, J., Kester, D., Krueger, B., Lampinen, B.B., McGourty, G., Micke, W., Mitcham, E., Olsen, B., Pelletrau, K., Phillips, H., Ramos, D., Schwankl, L., Sibbert, S., Snyder, R., Southwick, S., Stevenson, M., Thorpe, M., Weinbaum, S. & Yeager, J. (1997) Plant water status as an index of irrigation need in deciduous fruit trees. *HortTechnology* Vol.7, pp. 23-29, ISSN 1063-0918

Slayter, R.O. (1967) *Plant-Water Relationships*. Academic Press, New York, New York, USA

Sperry, J.S., Donnelly, R.R. & Tyree, M.T. (1988) A method for measuring hydraulic conductivity and embolism in xylem. *Plant, Cell & Environment*, Vol.11, pp. 35-40, ISSN 0140-7791

Strik, B. & Buller, G. (2005) The impact of early cropping on subsequent growth and yield of highbush blueberry in the establishment years at two planting densities is cultivar dependent. *HortScience*, Vol.40, pp. 1998-2001, ISSN 0018-5345

Strik, B. & Yarborough, D. 2005. Blueberry production trends in north america, 1992 to 2003 and predictions for growth. *HortTechnology* Vol.15, pp. 391-398, ISSN 1063-0918

Strik, B., Brun, C., Ahmedullah, M., Antonelli, A., Askam, L., Barney, D., Bristow, P., Fisher, G., Hart, J., Havens, D., Ingham, R., Kaufman, D., Penhalgon, R., Pscheidt, J., Scheer, B., Shanks, C. & William, R. (1993) *Highbush Blueberry Production*. Oregon State University Extension Service Publication PNW 215

Tamada, T. (2002) Stages of rabbiteye and highbush blueberry fruit development and associated changes in mineral elements. *Acta Horticulturae*, Vol.574, pp. 129-137, ISSN 0567-7572

Throop, P.A. & Hanson, E.J. (1997) Effect of application date on absorption of ^{15}nitrogen by highbush blueberry. *Journal of the American Society for Horticultural Science*, Vol.122, pp. 422-426, ISSN 0003-1062

USHBC (2009) 2009 World blueberry acreage and production report. U.S. Highbush Blueberry Council.

White, L.D. (2006) The effect of pre-plant incorporation with sawdust, sawdust mulch, and nitrogen fertilizer rate on soil properties and nitrogen uptake and growth of 'Elliott' highbush blueberry. MS Thesis, Oregon State University, Corvallis, Oregon, USA

Zhang, B. & Archbold, D.D. (1993) Solute accumulation in leaves of a *Fragaria chiloensis* and a *F. virginiana* responds to water deficit stress. *Journal of the American Society for Horticultural Science*, Vol.118, pp. 280-285, ISSN 0003-1062

9

Evapotranspiration and Water Management for Crop Production

André Pereira and Luiz Pires
State University of Ponta Grossa
Brazil

1. Introduction

1.1 Evaporation (E)

Evaporation is the physical process through which liquid water is converted to water vapor. Water evaporation in the atmosphere is produced by oceans, lakes, rivers, soil, and wet vegetation (evaporation from dew and intercepted rainfall).

1.2 Transpiration (T)

Transpiration is the loss of water under the form of vapor by the plants, predominantly by means of leaves, although in woody plants a tiny loss might also occur through the lenticels of the bark of branches. On the leaves evaporation befalls from the cell walls into the direction of air intercellular spaces, coming up the diffusion process through the stomata to atmosphere. The stomata acts as a fundamental regulator of transpiration rates along with the adjacent air layer to the leaf. An alternative path to the stomata is the foliar cuticle, although under good water supply conditions the preferential via is the stomachic.

The maintenance of transpiration is achieved by the reposition of lost water, at the vapor phase, by the water from the transpiration current that takes place throughout the conductor system from the roots up to the leaves, as a function of a water potential gradient from the soil (ψ_{soil}) to the air (ψ_{air}) as shown in Figure 1. The atmosphere with its water potential highly negative performs as a drain for water vapor. The drier the air is (low relative humidity), the higher (more negative) the suction force of such a drain will be.

1.3 Evapotranspiration (ET)

Evapotranspiration is the simultaneous process of water transfer to the atmosphere both by soil water evaporation and plants transpiration. Depending on the vegetation conditions, size of the vegetated area, and soil water supply, different conceptions are to be defined, such as potential, actual, oasis, and crop evapotranspiration. Such particular terms are described as follows:

1.3.1 Potential or reference evapotranspiration (ETo)

Potential evapotranspiration is the amount of water taken up by a large surface vegetated by grass, with a height between 8 and 15 cm, at an active growth stage, covering completely the soil surface, and with no restriction of soil water supply. Conceptually, ETo is limited only by the vertical energy balance, i.e., by the conditions of local ambient. It can be estimated by empirical formulae developed and tested for several climatic conditions. Evapotranspiration under such conditions is referred to as reference when the goal is to determine the evapotranspiration of a crop under non standard conditions. Therefore, ETo is an indicative value of the atmospheric demand of a given site throughout a period of time. It is well known that a surface vegetated with grass, under the defined conditions for ETo, has a leaf area index (LAI) of 3 and a reflection coefficient (albedo) for solar radiation corresponding to 23%.

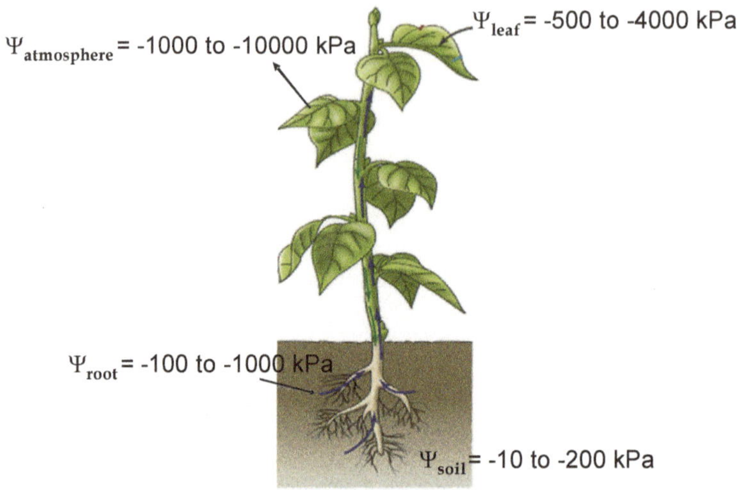

Fig. 1. Schematic representation of the water motion in the soil-plant-atmosphere system under optimal development conditions. Adapted by Reichardt (1985) and http://www.netxplica.com/manual.virtual.

1.3.2 Actual or real evapotranspiration (ETa)

Actual evapotranspiration is the amount of water actually utilized by an extensive surface vegetated with grass, at an active growth stage, covering completely the soil surface, however with or without water restriction conditions. Whenever there is not soil water restriction ETa = ETo. Thus, ETa ≤ ETo. At this point it is important to emphasize that, by definition, the concepts of ETa and ETo are applicable only to a surface vegetated with grass. Therefore, there is no rational in referring to potential evapotranspiration of a particular crop.

1.3.3 Oasis evapotranspiration (ETO)

Oasis evapotranspiration is the amount of water consumed by a small vegetated area (under irrigation) that is surrounded by an extensive dry area at which energy comes from advection (lateral transport of heat by the displacement of air mass), increasing the amount of available energy to evapotranspiration. Thus, by definition, ETO > ETo.

Figure 2 shows the border area necessary for minimizing the lateral transport of energy from the dry to the wet area (irrigated). At such an area, the ET that will take place is the oasis evapotranspiration. The size of this area depends on the climate of the region and height of vegetation. Tall vegetations by interacting more efficiently with the atmosphere require a larger border area than that for most grasses. The plants that are closer to the transition line (dry/irrigated) receive an extra amount of energy coming from the dry area, increasing the water consumption of the plants. Plants that are located further the transition spots are less influenced by dry areas and take up less water during the same period. In the case of irrigation, such a management practice should be adopted in such way as to take into account the variation of water loss along the irrigated area. For the central pivot system, the border area is circular. Therefore, the amount of water to be applied has to be calculated adequately in compliance with different demands along the pivot system.

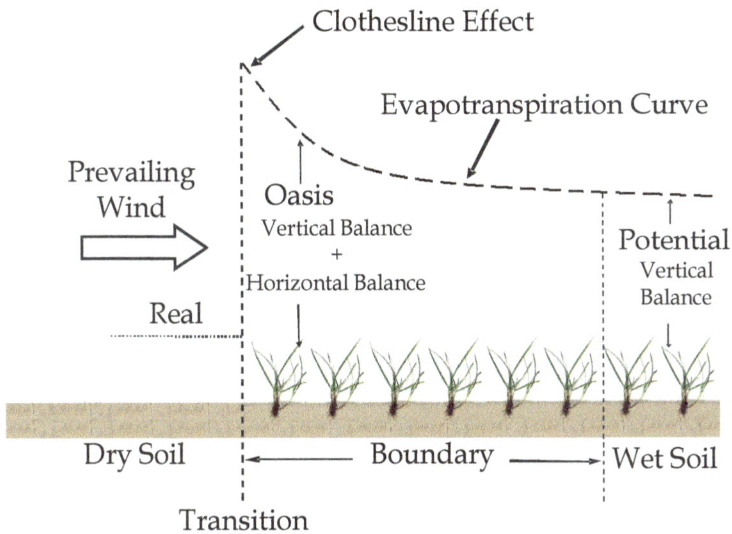

Fig. 2. Schematic representation of the ETO and ETo. Adapted by Camargo & A.R. Pereira (1990)

1.3.4 Crop evapotranspiration (ETc)

Crop evapotranspiration is the amount of water used by a crop at any growth stage, since the sowing / planting date up until the harvest, whenever there is no water restriction in the soil. This process is also called crop maximum evapotranspiration. ETc is a function of leaf area (transpiring surface), because the bigger the leaf area, the higher ETc will be for the same atmospheric demand. ETc might be obtained from ETo by means of the following expression: ETc = Kc * ETo. Kc is the crop coefficient and varies with the phenological stage of the crop, and also among species and varieties (cultivars), being a function of LAI. Figure 3 shows the effect of foliar area on water consumption of annual and perennial plants, as well as the variation of Kc throughout the growth/development of such hypothetical crops.

In annual crops, in so far as the plant grows the LAI increases until reaching a maximum value, decreasing then afterwards during the period of leaves senescence. Subperiod I depicts the crop establishment (sowing to germination); subperiod II characterizes the vegetative development (germination to flowering); subperiod III represents the reproductive period (flowering to grain filling); and subperiod IV is the harvest.

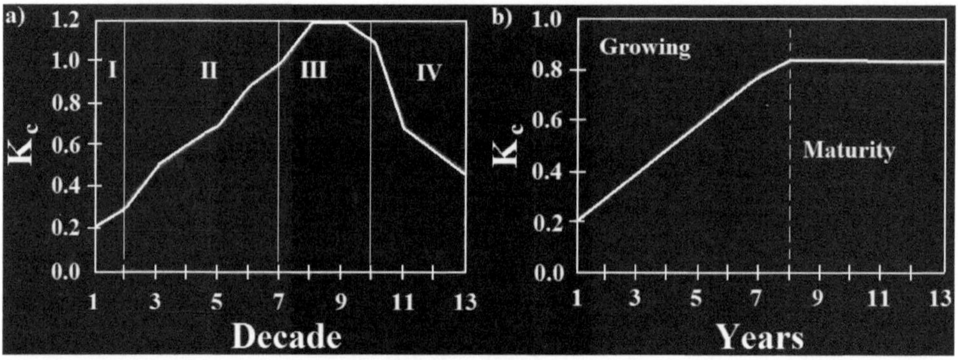

Fig. 3. Relationship between the phenological subperiods and Kc for annual crops, and between age and Kc for perennial crops.

In perennial crops, as a result of a continual growth of the plants, the value of Kc is crescent throughout the years that precede maturity, and from this moment on it turns out to be practically constant with a little seasonal variation, as a function of LAI. One example is the rubber tree, which loses its leaves in fall and also coffee tree that due to harvest and hibernal resting have a reduction in LAI.

When crop evapotranspiration does not occur under the ideal conditions aforementioned, i.e., under water stress condition, ET is denominated crop real evapotranspiration (A.R. Pereira et al., 2002).

2. Evapotranspiration determining factors

Weather parameters, crop characteristics, management and environmental aspects are factors affecting evaporation, transpiration, and evapotranspiration.

2.1 Climatic factors

2.1.1 Net radiation

This is the main source of energy for the evapotranspiration process. It depends on the global solar radiation flux density and vegetation albedo. A darker vegetation absorbs more incident solar radiation and evapotranspires more. Net radiation is the primary climatic factor controlling ET when water is not limiting, especially in subhumid and humid climates. In cold humid climates, only 50 to 60% of net radiation may be converted to latent heat. In hot, arid climates, latent heat may exceed net radiation by 10 to 50% with sensible heat derived from the air and converted to latent heat. In spite of these relationships, the heat energy balance approach employed to determining or estimating ET is recognized as a reliable and conservative method. A thorough understanding of the factors controlling the energy balance of a cropped soil enables making accurate estimates or predictions of evapotranspiration and irrigation water requirements. It also facilitates more effective irrigation water management (Allen et al., 1989).

2.1.2 Temperature

Over the course of a day, an increase of the air temperature causes an increase on the saturation deficit triggering a higher evaporative demand in the air, and leading to high ET rates.

2.1.3 Relative humidity

Air relative humidity acts in conjunction with temperature. The higher relative humidity, the lesser the evaporative demand and, therefore, the lower ET will be.

2.1.4 Wind (regional advection of energy)

Advection represents the horizontal transport of energy from a drier area to another more humid, and such additional energy is utilized in the evapotranspiration process. Wind also helps remove water vapor near the plants to other regions.

2.2 Crop factors

2.2.1 Specie

This factor is related to the foliar architecture (spatial distribution of the leaves), internal resistance of the plant to water transport, and other morphological aspects (number, size, and distribution of stomata, etc.), which exert a direct influence on ET.

2.2.2 Reflection coefficient (albedo)

Radiation reflection influences directly net radiation availability for the ET process. The darker the vegetation, the lower the reflection coefficient and the higher net radiation will be.

2.2.3 Growth stage (LAI)

Such a factor is directly related to the size of transpiring foliar surface, for the larger leaf area the larger the transpiring surface, and the higher the potential for water use will be.

2.2.4 Plant height

Taller and rougher plants interact more efficiently with the atmosphere in motion, extracting more energy from the air and, therefore, increasing ET.

2.2.5 Depth of the radicular system

It is directly related to the volume of soil explored by the roots, aiming at meeting the atmospheric hydric demand. A superficial radicular system, for exploring a smaller soil volume, keeps the crop more susceptible to drying periods.

2.3 Management and environmental conditions

2.3.1 Spacing / stand

This factor determines the intraspecific competition, i.e., between plants from the same species. Small spacing results in an intensive competition for water and this causes the radicular system to deepen into the soil to enhance the volume of available water. More generous spacing allows for a more superficial radicular system, but also brings about more heating to the soil and plants and promotes a freer circulation of wind among the plants, causing as a consequence an increase on ET.

2.3.2 Orientation of the crop main line

Crops oriented perpendicularly to predominant winds tend to extract more energy from the air than those oriented in parallel. For regions with constant winds, a solution to prevent the stomata-closing would be the use of windbreaks. A windbreak reduces wind velocities and decreases the ET rate of the field directly beyond the barrier.

2.3.3 Water storage capacity

Clay soils have a higher water storage capacity than sandy soils, and are capable of maintaining a more constant ET rate for longer. However, in sandy soils the radicular system tends to be quite deeper, compensating for lower water retention.

2.3.4 Chemical / physical impediments

Impediments limit the growth of radicular system, causing the plants to explore a smaller volume of soil, resulting in negative effects both during the rainy and dry seasons. Throughout the rainy season, soil with any physical impediments gets soaking wet asphyxiating the roots. Over the dry season, the volume of available water to the roots turns out to be reduced in such a way as to preclude it from deepening into the soil in search for water.

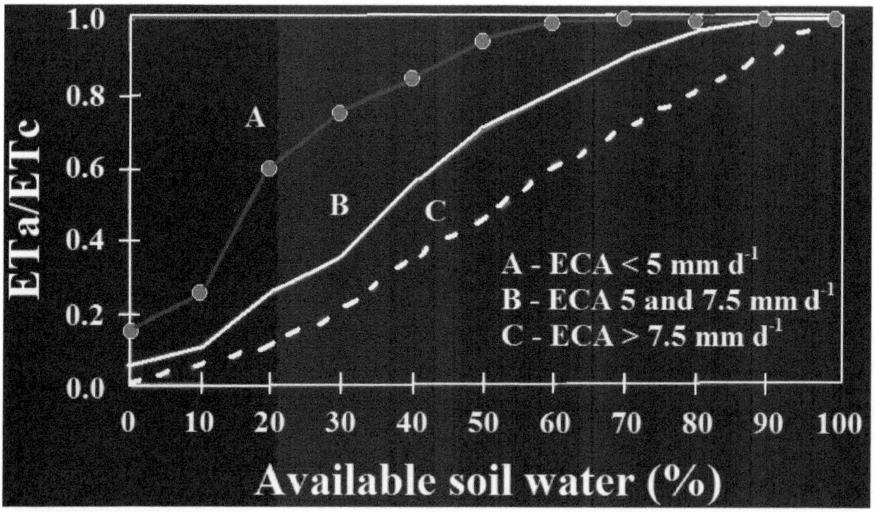

Fig. 4. Interrelationship between corn relative evapotranspiration (ETa/ETc) and soil available water and atmospheric demand expressed by ECA. Adapted by Denmead & Shaw (1962).

2.4 Interrelationship atmospheric demand – soil water supply

The soil is an active reservoir that within certain limits controls the rate of water use by the plants, always in conjunction with the atmospheric demand. The atmospheric demand depends on the availability of solar energy, relative humidity, and wind speed. Figure 4 exemplifies the interrelationship between available water in the soil (%), atmospheric demand indicated by the evaporation from a Class A pan (ECA), and relative evapotranspiration (ETa/ETc) for corn plants. Under situation A, with ECA < 5 mm day^{-1}, due to a low demand, the plant managed to extract water from the soil at potential levels (ETa/ETc ≈ 1) up to about 60% soil available water. Under situation C, in which ECA > 7.5 mm day^{-1} (high demand), even under enough amount of soil water, the plants do not manage to extract water at a rate compatible to its needs, resulting in a temporary enclosure of the stomata to avoid drying of the leaves. Such condition usually takes place at the hottest hours of the day.

3. Direct measurement of evaporation and evapotranspiration

Evaporation from an open water surface provides an index of the integrated effect of solar radiation, air temperature, relative humidity and wind speed on evapotranspiration. The Class A pan has proved its practical value and has been used successfully to estimate reference evapotranspiration by observing the evaporation loss from water surface and applying empirical coefficients to relate pan evaporation (ECA) to ETo. Virtually all ET studies are made with supporting climatic data being collected. Pan evaporation is a very common measurement and provides excellent supporting data useful for correlation and prediction. In addition, pan evaporation is commonly measured at reservoir and lake sites to be used in estimating water surface evaporation losses.

Because both pan evaporation and ET involve the same basic process, it is easy to assume that a reasonable estimate of ETo might be found by multiplying measured pan evaporation (ECA) by a factor usually less than unity. The general relationship is:

$$ETo = K_p * ECA$$

where K_p is known as a pan coefficient which is dimensionless and generally varies from zero to near unity. Doorenbos & Pruitt (1977) gave pan coefficients to estimate grass reference. Their coefficients are for Class A pan data and consider different ground covers, level of mean relative humidities, and 24-hour wind runs. The coefficients of Doorenbos & Pruitt (1977) appear in Table 1.

The direct measurement of ET is difficult and expensive, justifying its utilization only under experimental conditions. The equipment more commonly used for such a purpose is the lysimeters. Lysimeter or evapotranspirometer is equipment that consists of an impermeable box containing a soil volume which gives us data concerning the terms of water balance of the sampled soil volume. The most employed lysimeters are:

- Drainage lysimeter: This type of lysimeter works adequately for long periods of observation (± 10 days). It is based on the principle of mass conservation for water in a soil volume (Camargo, 1962):

  $$\Delta SW = P + I - ET + CR - DP$$

  Taking into account that precipitation (P) and irrigation (I) are easily measured, that change in soil water content (ΔSW) is practically null, that water transported upward by capillary rise (CR) is negligible, and that deep percolation (DP) is measured, we can determine evapotranspiration (ET) as a residue of the above equation.
- Sub-irrigation lysimeter: This kind of lysimeter adopts an automated feeding system and records of reposed water in such a way as to maintain the groundwater at a constant level, being the ET rates equal to the water volume that leaves the feeding system (Assis, 1978).
- Weighting lysimeter: This lysimeter utilizes the automated measurement of load cells set up to an impermeable box, recording its weight variation over time. Hence, faced with water consumption by the plants in the lysimeter a reduction in weight of the control volume will take place and will be proportional to ET (Gomide et al., 1996; Bergamaschi et al., 1997; Silva et al., 1999; Faria et al. 2006).

CASE A

Wind conditions*	Fetch Dist. (m)	SURROUNDED BY SHORT GREEN CROP Mean relative humidity (%)		
		<40	40-70	>70
Light	1	0.55	0.65	0.75
	10	0.65	0.75	0.85
	100	0.70	0.80	0.85
	1,000	0.75	0.85	0.85
Moderate	1	0.50	0.60	0.65
	10	0.60	0.70	0.75
	100	0.65	0.75	0.80
	1,000	0.70	0.80	0.80
Strong	1	0.45	0.50	0.60
	10	0.55	0.60	0.65
	100	0.60	0.65	0.70
	1,000	0.65	0.70	0.75
Very strong	1	0.40	0.45	0.50
	10	0.45	0.55	0.60
	100	0.50	0.60	0.65
	1,000	0.55	0.60	0.65

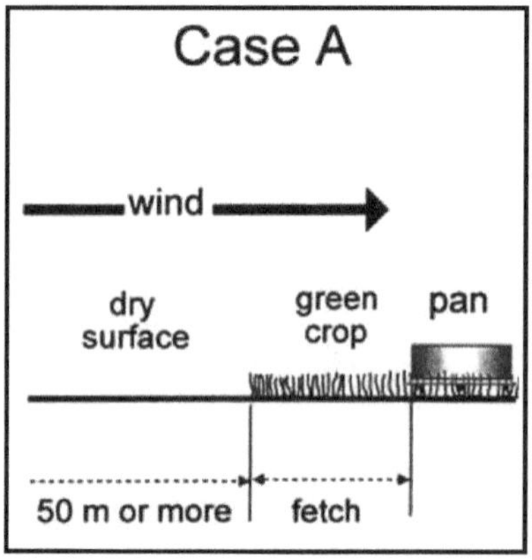

CASE B

		SURROUNDED BY SHORT GREEN CROP Mean relative humidity (%)		
		<40	40-70	>70
Light	1	0.70	0.80	0.85
	10	0.60	0.70	0.80
	100	0.55	0.65	0.75
	1,000	0.50	0.60	0.70
Moderate	1	0.65	0.75	0.80
	10	0.55	0.65	0.70
	100	0.50	0.60	0.65
	1,000	0.45	0.55	0.60
Strong	1	0.60	0.65	0.70
	10	0.50	0.55	0.65
	100	0.45	0.50	0.60
	1,000	0.40	0.45	0.55
Very strong	1	0.50	0.60	0.65
	10	0.45	0.50	0.55
	100	0.40	0.45	0.50
	1,000	0.35	0.40	0.45

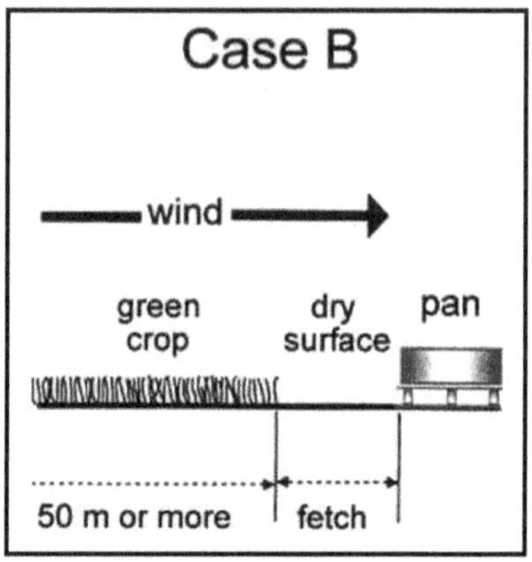

Light: <175 km/day, <2 m/s; Moderate: 175-425 km/day, 2-5 m/s; Strong: 425-700 km/day, 5-8 m/s; Very strong: >700 km/h, >8 m/s.

Table 1. Suggested values of Kp for Class A pans for the calculation of ETo for grass 8-15 cm tall. Adapted by Doorenbos & Pruitt (1977).

4. Potential evapotranspiration and crop transpiration estimation methods

Owing to the difficulty of obtaining accurate field measurements, ETo is commonly computed from weather data. A large number of empirical or semi-empirical equations have been developed for assessing ETo from meteorological information. Some of the methods are only valid under specific climatic and agronomic conditions and cannot be applied under conditions different from those under which they were originally developed.

Numerous methods taking into consideration meteorological data for calculating ETo are reported in the literature. Bernardo (1995) reports that ETo might be obtained by both direct and indirect estimation methods. Direct methods are those that make use of lysimeters and provide the highest accuracy for its determination, require installation of experimental plots in the field, control of soil moisture and a methodological procedure to assess the input and output of water in large areas. However, according to Mendonça et al. (2003), such methods due to its high costs have their use restricted to research institutions and are usually utilized for regional calibration of indirect methods.

ETo needs to be determined to provide knowledge of crop water requirements. It is desirable to have a method that estimates ETo with accuracy and from easily obtained meteorological data. Irrigation planning and decision making at a field scale are done based on calculations of maximum crop evapotranspiration (ETc).

Villa Nova et al. (2007) came up with a simplified method based on the Bowen ratio-energy balance principle to estimate ETo in Brazil. The proposed method is irrespective of monitoring wind speed data and does not require the installation of sophisticated and high-cost equipment. In order to get it validated experimental data were collected on a diurnal basis throughout the daylight period, aiming at quantifying only the daylight values of ETo, which are more representative of the water vapor transfer process to the atmosphere for a given agricultural ecosystem. The equation representative of the aforementioned method for ETo expressed in mm day^{-1} is given by:

$$ETo = 0.423 * W' * (Rn - G)$$

where W´ is the weighting factor for the effect of solar radiation on evapotranspiration that depends on air temperature, relative humidity, and psychrometric coefficient; Rn is the net radiation at surface cultivated with grass (MJ m^{-2} day^{-1}), and G is the soil heat flux in MJ m^{-2} day^{-1}. The value for W´ can be determined using the data in Tables 2 and 3.

The energy balance method simplified by Villa Nova et al. (2007) was a feasible alternative to evaluate ETo. Under local meteorological conditions of the experiment, it gave estimates practically identical to those obtained by the classical Penman-Monteith approach and added advantage of simplifying ETo calculation, leaving out information related to wind speed, making use of only net radiation, soil heat flux, mean air temperature and mean relative humidity on a daily basis. It showed high statistical accuracy when compared to ETo measurements obtained by weighing lysimeters with load cells.

Another means of calculating ETo based on the Penman approach taking into consideration only the daylight values was suggested by Villa Nova et al. (2006). In their work, the classical expression of the Bowen ratio was modified by considering the sensible heat flux emergent from the evaporative surface in conjunction with the air turbulent flux, which transports also latent heat flux. When compared to potential demand measurements obtained with weighing lysimeters, the simplified Penman approach showed a high statistical accuracy, expressed by coefficients of determination greater than 0.92, and an

T (⁰C)	Modified weighting factor (W´) Relative humidity (%)								
	45	50	55	60	65	70	75	80	85
10	0.532	0.535	0.539	0.542	0.545	0.548	0.551	0.554	0.557
11	0.546	0.549	0.552	0.556	0.559	0.562	0.565	0.568	0.571
12	0.560	0.563	0.566	0.569	0.573	0.576	0.579	0.582	0.585
13	0.573	0.576	0.580	0.583	0.586	0.589	0.593	0.596	0.599
14	0.586	0.589	0.593	0.596	0.599	0.603	0.606	0.609	0.613
15	0.599	0.602	0.606	0.609	0.612	0.616	0.619	0.622	0.626
16	0.612	0.615	0.618	0.622	0.625	0.629	0.632	0.635	0.639
17	0.624	0.628	0.631	0.634	0.638	0.641	0.645	0.648	0.651
18	0.636	0.640	0.643	0.647	0.650	0.653	0.657	0.660	0.663
19	0.648	0.652	0.655	0.659	0.662	0.665	0.669	0.672	0.675
20	0.660	0.663	0.667	0.670	0.674	0.677	0.680	0.684	0.687
21	0.671	0.675	0.678	0.682	0.685	0.688	0.692	0.695	0.698
22	0.682	0.686	0.689	0.693	0.696	0.699	0.703	0.706	0.709
23	0.693	0.697	0.700	0.704	0.707	0.710	0.714	0.717	0.720
24	0.704	0.707	0.711	0.714	0.717	0.721	0.724	0.727	0.730
25	0.714	0.717	0.721	0.724	0.728	0.731	0.734	0.737	0.740
26	0.724	0.727	0.731	0.734	0.737	0.741	0.744	0.747	0.750
27	0.734	0.737	0.740	0.744	0.747	0.750	0.753	0.756	0.760
28	0.743	0.746	0.750	0.753	0.756	0.759	0.762	0.766	0.769
29	0.752	0.756	0.759	0.762	0.765	0.768	0.771	0.774	0.777
30	0.761	0.764	0.768	0.771	0.774	0.777	0.780	0.783	0.786

Table 2. Values of the modified weighting factor W´ as a function of the observed daily mean air temperature and relative humidity for altitudes from 0 to 1000 m.

T (⁰C)	Modified weighting factor (W´) Relative humidity (%)								
	45	50	55	60	65	70	75	80	85
10	0.569	0.572	0.575	0.578	0.582	0.585	0.588	0.591	0.594
11	0.582	0.585	0.589	0.592	0.595	0.599	0.602	0.605	0.608
12	0.595	0.599	0.602	0.605	0.609	0.612	0.615	0.619	0.622
13	0.608	0.612	0.615	0.619	0.622	0.625	0.629	0.632	0.635
14	0.621	0.625	0.628	0.631	0.635	0.638	0.641	0.645	0.648
15	0.634	0.637	0.641	0.644	0.647	0.651	0.654	0.657	0.661
16	0.646	0.649	0.653	0.656	0.660	0.663	0.666	0.670	0.673
17	0.658	0.661	0.665	0.668	0.672	0.675	0.678	0.682	0.685
18	0.670	0.673	0.677	0.680	0.683	0.687	0.690	0.693	0.697
19	0.681	0.685	0.688	0.691	0.695	0.698	0.701	0.705	0.708
20	0.692	0.696	0.699	0.702	0.706	0.709	0.712	0.716	0.719
21	0.703	0.706	0.710	0.713	0.717	0.720	0.723	0.726	0.729
22	0.714	0.717	0.720	0.724	0.727	0.730	0.733	0.737	0.740
23	0.724	0.727	0.731	0.734	0.737	0.740	0.743	0.747	0.750
24	0.734	0.737	0.740	0.744	0.747	0.750	0.753	0.756	0.759
25	0.743	0.747	0.750	0.753	0.756	0.759	0.763	0.766	0.769
26	0.753	0.756	0.759	0.762	0.766	0.769	0.772	0.775	0.778

27	0.762	0.765	0.768	0.771	0.774	0.777	0.780	0.783	0.786
28	0.771	0.774	0.777	0.780	0.783	0.786	0.789	0.792	0.795
29	0.779	0.782	0.785	0.788	0.791	0.794	0.797	0.800	0.803
30	0.787	0.790	0.793	0.796	0.799	0.802	0.805	0.808	0.810

Table 3. Values of the modified weighting factor W´ as a function of the observed daily mean air temperature and relative humidity for altitudes from 1000 to 2000 m.

extremely small dispersion of the data around the 1:1 line. Therefore, given the availability of the input data required (net radiation, soil heat flux, and air temperature), it could be employed in other climatic regions besides Brazil to provide ETo estimates for irrigation scheduling.

Solar energy is the primary source for photosynthesis and transpiration in such a way as to assure the expression of the crop potential yield at a given site. A.B. Pereira et al. (2009) came up with a methodology that aims to ease the calculation of the amount of water necessary for a localized irrigation scheduling with a minimal loss possible at both citrus and apple trees orchards by means of usual available data, such as leaf area, global solar radiation flux density, net radiation, and air daily mean steam saturation deficit. In order to get the proposed methodology validated, estimated transpiration data was subjected to a regression analysis against a data set of sap flux measured by means of the heat balance approach in a citrus orchard with leaf areas of 48 and 99 m^2, as well as in apple trees with leaf areas roughly of 5, 8, 9, 11, 16 and 21 m^2. The calculated transpiration obtained as a function of the conversion efficiency of solar energy for citrus in Brazil is given by the following expression:

$$TR = (0.0923 - 0.0018 * Qg) * LA * Qg$$

where TR is the transpiration rate (l $tree^{-1}$ day^{-1}), Qg is the global solar radiation flux density (MJ m^{-2} day^{-1}) and LA is the leaf area (m^2 of leaf $tree^{-1}$).

The transpiration rate for apple trees obtained by the aforementioned methodology in France is estimated by means of the following equation:

$$TR = (0.1 + 0.0287 * \Delta e) * Rn * LA$$

where Δe is the mean vapor saturation deficit of the air (kPa) and Rn is the net radiation (MJ $m^{-2}day^{-1}$).

The results obtained by A.B. Pereira et al. (2009) revealed that it is rather feasible to estimate the amount of irrigation water throughout the whole cycle of citrus and apple trees grown under localized irrigation systems by means of a physiological model, which expresses the ability of the plants to converting solar energy into water taken up in the transpiration process at the sites in study.

A.B. Pereira et al. (2010) proposed a new methodology for the calculation of daily transpiration rates of apple trees and citrus orchards from the following meteorological data and crop parameters: mean air relative humidity, mean air temperature, photoperiod duration, and leaf area of the tree. The proposed approach dismisses the utilization of the conductance and net radiation at the dossel level and is the basis for the existing differences between water potential in the atmospheric air and within the stomachic chamber of the leaves. Such a gradient turns out to be the driving-force of the transpiration process. Its utilization as a tool for maximization of yields with a better reclamation of water resources under drip irrigation system in orchards was tested as to its viability, taking into account the data of sap flow collected by Angelocci (1996) in apple trees, as well as by Marin (2000) in citrus orchards under distinct climatic conditions.

5. Crop evapotranspiration and yield response to irrigation

The determination of a given crop´s evapotranspiration (ETc) or water demand is very important for planning water management in irrigated areas, not only from physical and biological points of view, but also from the applied engineering perspective, since the hydraulic design of an irrigation system should take into consideration ET. Work has been performed all over the world to compare values of ETc to those of ETo under different climate and soil conditions. Crop coefficients (Kc) that vary with crop type, canopy cover, and stage of growth have been experimentally calculated (Doorenbos & Kassam, 1979). Allen et al. (1998) reported extensive tables of Kc for many vegetable crops by species and stage of development.

Irrigation of high-value, water-stress-sensitive crops grown in arid environments is essential for high yield, quality, and net returns. Many of the soils most suited for high-value crops are low in organic matter and are highly susceptible to nutrient and pesticide leaching under poor irrigation scheduling. Improved irrigation and nutrient management practices are important to minimize leaching losses.

The daily ETc of potato varies according to atmospheric conditions, surface soil wetness, the stage of growth, and the amount crop cover (Wright & Stark, 1990). They observed that ETc increased as the leaf area and transpiration increased and reached near-maximum levels just before effective full cover. The leaf area index reached 3.5 by effective full cover, coincident with the highest daily ETc of 8.5 mm. Seasonal ETc was 604 mm.

Tanner (1981) reported that potato ETc measured with a lysimeter in the humid Wisconsin area for June through August ranged from 293 to 405 mm during 3 yr of study. Nkemdirim (1976), using meteorological methods, studied the ETc of a potato crop grown near Calgary, AB, Canada, and found midseason daily ETc to be about 6 mm. Daily water consumption for a potato crop grown in Botucatu, São Paulo, Brazil, during the winter season was about 3 mm and total seasonal ETc was 283 mm (A.B. Pereira et al., 1995a). Erie et al. (1965) found that the seasonal water use for potato, from February through June at Mesa, AZ, USA, averaged 617 mm.

Wright & Stark (1990) reported that seasonal water use in irrigated areas of Oregon and Washington ranged from 640 to 700 mm. For high yields at a given site, the seasonal water requirements of a potato crop with a phenological cycle varying from 120 to 150 days were within the range of 500 to 700 mm, depending on the climate (Doorenbos & Kassam, 1979).

Wright (1982) developed improved crop coefficients for various irrigated crops in the Pacific Northwest, using alfalfa to measure ETo and weighing lysimeters at an experimental field near Kimberly, ID, USA. Apart from the crop coefficient approach, potato ETc can also be estimated by means of multiple regression equations that take into consideration the LAI of the potato crop and atmospheric evaporative demand depicted by ETo or pan evaporation (A.B. Pereira et al., 1995b).

The total water requirement of onion varies considerably with location, environment, and irrigation system. De Santa Olalla et al. (1994) studied onion production with carefully managed drip irrigation and found that bulb yields of 64 to 74 Mg ha^{-1} were feasible in Spain with water applications of 100 and 120%, respectively, of onion ETc. Bulb yield increased with the addition of water up to and above ETc with the soil water tension levels at 20 or 10-kPa (Shock et al., 2000). The total amount of water applied to the 10-kPa treatment (924 mm) was higher than ETc (699 mm) and the total amount of water applied to the 20-kPa treatment (640 mm) was close to ETc. The optimal irrigation treatment, based on economic returns, would be approximately 100% ETc in 1997 and 153% ETc in 1998, a rate equivalent to pan evaporation. Bucks et al. (1981) found that the highest irrigation rate tested (100% ETc) was advantageous for drip-irrigated onion.

A linear response was found between tomato yield and ETc under surface drip irrigation in Israel (Ben-Gal & Shani, 2003). The soil texture was sandy loam. The yield-ETc relationship was similar for both spring and fall tomato crops when expressed as relative yield vs. relative ETc. A linear response between yield and ETc also was found for two cultivars at a site in Canada on sandy loam (Tan, 1993). For both studies, deficit irrigation was applied uniformly during the crop season.

Water applications of 0.4, 0.6, 0.8, 1.0, and 1.2 times the estimated ETc showed processing tomato commercial yield to increase from 9.9 Mg ha^{-1} (274 mm of water) to 109.6 Mg ha^{-1} (640 mm of water) with surface drip irrigation on sandy loam on the east side of the San Joaquin Valley (Calado et al., 1990). Deficit irrigation was applied uniformly during the crop season. Soluble solids increased from 4.6% for the highest water application (640 mm of applied water) to 6.5% for the smallest application (274 mm of applied water). These results suggested that the existing crop coefficients for tomato might underestimate the ETc needed for maximum yield because a higher yield occurred for 1.2 x ETc treatment than the 1.0 x ETc treatment.

Lettuce growth and yield depends on an adequate supply of water to replenish ETc. Gallardo et al. (1996) reported that harvested dry and fresh weight matter of lettuce grown on a sandy loam soil increased linearly as a function cumulative ETc. Yield responses to applied water of amounts equal to 150% ETc have been reported in sandy textured soils with low water capacities (Sanchez, 2000). Similarly, Bar-Yosef & Sagiv (1982) obtained highest lettuce yields on a sandy textured soil by applying 120% of evaporation from a U.S. class A pan.

Water requirements for cabbage vary from 380 to 500 mm depending on climate and the length of the growing season. The ETc increases during the growing season, with a peak toward the end of the cycle. Pawar & Firake (2003) studied the effect of three alternate-day irrigation levels (100, 75, and 50% ETc) and three irrigation methods (drip, drip-line, and microsprinkler) on cabbage yield in a middle block soil at Mahatma, Rahuri, India. They concluded that irrigation scheduled at 100% ETc produced higher yield (43.4 Mg ha^{-1}) than at 75% ETc (38.5 Mg ha^{-1}). The irrigations scheduled at 50% ETc recorded the lowest yield (30.9 Mg ha^{-1}). Yields were not influenced by the microirrigation methods. Thus, microirrigations for cabbage should be scheduled at the 100% ETc level to assure maximum yields.

Shock et al. (1998) examined the effect of three deficit irrigation treatments (100% ETc, 70% ETc, and 70% ETc during the bulking with 50% ETc thereafter) on yield and quality of four potato cultivars in three successive years (1992-1994) on a silt loam soil in eastern Oregon, USA. Yield reductions due to deficit irrigation were not as pronounced in 1993 as in 1992 or 1994. The weather pattern in 1993 was cooler and wetter during the tuber bulking period than it was in either 1992 or 1994. Both total yield and U.S. no.1 yield increased with additional water supply in each of the three years.

6. Water use efficiency

The water use efficiency (WUE) relates the biomass accumulation or commercial yield to the amount of water applied or evapotranspired by the crop (Sousa et al., 2000). In irrigated agriculture the increase of yield and determination of WUE are rather complex and require interdisciplinary knowledge. Within this context, Dinar (1993) mentions the means for enhancing the values of WUE, pointing out the importance of an adequate irrigation scheduling.

Among the means and techniques adopted to increase WUE in irrigated agriculture, drip irrigation with a water supply at a high frequency and low volume has been shown to be adequate for enhancing WUE (Srinivas et al., 1989). Such authors came up with the conclusion that the maximum WUE of a watermelon crop was obtained with drip irrigation systems, whenever water amounts were applied at an evaporation rate of 25% of class A pan (ECA), owing to the little water stress imposed and a low decrease on yield in comparison to the high reduction in water use. Similar outcomes were also obtained by Lin et al. (1983) by verifying a high WUE under a low irrigation regime for tomato.

When WUE is determined from the amount of water applied, Dinar (1993) and Letey (1993) reported its reduction, however without diminishing crop production as a way of increasing WUE. In such aspect, the irrigation system chosen (Dinar, 1993) and the reduction of the water application period throughout the crop cycle (Richards et al., 1993 e Howell et al., 1998) are important factors to be considered.

The water distribution and maintenance of optimal levels of soil moisture throughout the full cycle of the crop reduce water losses by drainage and water stress period of the crop resulting in increases in WUE. This can be attained with water applications at a high frequency and small amounts (Lin et al, 1983; Srinivas et al., 1989; Mishra et al., 1995; Saeed & El-Nadi, 1997; Sousa et al., 2000).

Gallardo et al. (1996), working on lettuce yield response to irrigation, reported that WUE was highest for the least irrigated treatment due to more effective extraction of soil water than that of the highest irrigated treatment. Similarly, Aggelides et al. (1999) obtained the highest WUE by reducing the amount of applied water on a clay loam soil. Sammis et al. (1988) describe a linear relationship between lettuce yield and applied water until maximum yield was attained on a silt loam soil in Hawaii, USA. Their model predicted similar WUE for varying amounts of applied water until maximum yield was attained, after which WUE decreased.

WUE for lettuce can also differ among irrigation methods. The use of drip irrigation can reduce evaporation from the soil surface compared to sprinkler and furrow irrigation systems. Hanson et al. (1997) found a higher WUE for drip than furrow in an unreplicated trial. Sutton & Merit (1993) calculated that drip increased WUE more than 100% compared to overhead sprinklers. However, Sammis (1980), in comparing drip, sprinkler, and furrow, reported that WUE was not significantly different among the irrigation methods when similar amounts of water were applied, except for furrow which had the highest WUE during the second season and the lowest WUE during the third season. Sammis (1980) as well as Hanson et al. (1997) noted that management was an important factor in the WUE of each irrigation method.

7. Crop potential productivity modeling

Potential productivity is the maximum possible yield of a given species or cultivar achievable under the existing conditions of solar radiation flux density with all the other environmental factors considered to be optimal. Therefore, the potential productivity is determined by the biological properties of the cultivar and the solar radiation resources available under ETc conditions. The potential yield of agronomic crops is dramatically affected by the amount of water applied during the crop-growing season at a given region.

Meteorological factors directly influence potential crop productivity, regulating its transpiration, photosynthesis, and respiration processes in such a way as to control the growth and development of the plants throughout their physiological mechanisms at a given site. The interaction of the meteorological factors with the crop responses is complex. However, by assessing physiological crop responses to environmental factors under field conditions, it is possible to derive mathematical models to estimate crop potential production as a function of climatic variables with good precision.

Research has been conducted to quantify the effects of the environment on growth, development, and yield of many agronomic crops. Among the main environmental factors that strongly govern all physiological processes of the plants are global solar radiation flux density, air temperature, and available soil water content (Coelho & Dale, 1980). Currently there is a great deal of interest in estimating crop productivity as a function of climatic factors by means of different crop weather models.

Kadaja & Tooming (2004) proposed a relatively simple model, POMOD, to calculate potato potential yield, which permits integration of the knowledge in different disciplines on the potato crop yield levels using the measured physiological, ecological, agrometeorological, and agronomical parameters of the plant. The input variables of the model can be divided into four groups: daily meteorological information, annual information, location, and cultivar. The first group includes global solar radiation, air temperature, and precipitation. The location is characterized by geographical latitude and hydrological parameters. As to cultivar, the parameters of gross and net photosynthesis, the coefficients of growth and maintenance respiration, and albedo of the crop are also needed.

The potato growth models included in DSSAT (Decision Support System for Agrotechnology Transfer) are SUBSTOR-POTATO and LINTUL-POTATO (University of Hawaii, Honolulu, Hawaii).

The SUBSTOR-POTATO crop soil weather model takes into consideration daily air temperature, photoperiod duration, intercepted solar radiation, soil water, and nitrogen supply. The model simulated fresh tuber yields ranging from 4 to 56 t ha^{-1} resulting from differences in weather patterns, soils, cultivars, and management practices (Bowen, 2003).

The LINTUL-POTATO simulation model (Kooman & Haverkort, 1995) establishes potential yield of a certain cultivar for a determining growing period and plant density and is based on incident photosynthetically active radiation (PAR), the fraction of PAR intercepted by the crop, and radiation use efficiency to produce dry matter. The potential yield established with this model was used by Caldiz & Struit (1999) to perform a preliminary yield gap analysis regarding actual and attainable potato yield in different areas of Argentina, and provided estimates ranging from 47 to 126 t ha^{-1}. Differences between actual and potential yield might be attributed to suboptimal solar radiation intercepted by the foliage, cultivar, seed management, physiological age of the seed, suboptimal management of water and fertilizer, and inadequate control measures for diseases.

A.B. Pereira & Villa Nova (2008) tested the performance of a model based on studies of maximum rates of carbon dioxide assimilation for a C_3 crop as a function of air temperature, fraction of PAR intercepted by the crop, photoperiod duration, and leaf area index to estimate the potential productivity of potato in Brazil. To assess the performance of the proposed model, the estimated values of tuber yield were compared with observed productivity data under irrigation conditions for the studied sites. Such agrometeorological model was similar to the potential productivity estimation model described by Villa Nova et al. (2001) and used by Villa Nova et al. (2005) for sugar cane in Piracicaba, State of São Paulo, Brazil. The results obtained by A.B. Pereira & Villa Nova (2008) showed that the agrometeorological model tested under the climatic conditions of the State of São Paulo, Brazil, in general underestimated irrigated potato yield by less than 10%. This justifies the recommendation to test the performance of the model in other regions for different crops and genotypes under optimal irrigation conditions in further scientific investigations.

Todorovic et al. (2009) compared the performance of AquaCrop, a crop simulation model developed by FAO, with that of two well established models CropSyst and WOFOST, in simulating sunflower growth under different water regimes in a Mediterranean environment. The models differ in the level of complexity describing crop development, in the main growth modules driving the simulation of biomass growth, and the number of input parameters. AquaCrop is exclusively based on the water-driven module, in that transpiration is converted into biomass through a water productivity parameter; CropSyst is based on both water and radiation driven modules, while WOFOST simulates crop growth using a carbon driven approach and fraction of intercepted radiation. All three models tested in this work simulated fairly well most of the situations encountered in the experimental works on sunflower growth in Southern Italy. In most of the simulation scenarios, yield was modeled with a reasonable error of ±0.5 Mg ha^{-1}. A general trend of underestimation of yield by all models was observed under severe water stress conditions.

The AquaCrop model introduces notable simplifications and requires fewer input parameters than the other two models, without affecting negatively its performances in terms of final biomass, yield, and WUE, except that CropSyst simulated WUE much better under limited water supply. However, the simplifications adopted in AquaCrop and also in CropSyst could be a limiting factor of both models when severe water stress conditions need be analyzed. The crop parameters calibrated for all three models under full irrigation in 2007 were shown to be mostly conservative enough to be used in all other simulations regardless of the water regimes and weather and soil characteristics variations in 2 yr under study. However, the predictions of biomass growth during the season were slightly better for 2007 (year of calibration) than for 2005. This means that slight modifications of crop growth parameters for 2005 could improve the simulation results by all models. This is particularly true for CropSyst and WOFOST since both models use crop growth modules that could be affected by weather characteristics (VPD and air temperature, respectively). Furthermore, it should be emphasized that the results obtained in this work depend on the calibration procedure, and for subsequent calibration/validation studies of the model(s), a parameter estimation algorithm with a well-defined goodness of fit criterion should be implemented. Moreover, for more robust model calibrations, it is usually necessary to have much more than 2 yr of experimental work under different weather and soil conditions. Also, a variation in crop growth parameters among different cultivars should be explored. Therefore, further investigations are needed to improve the model performances in predicting sunflower growth in the Mediterranean region.

There is room for improvement of simulations by all these models through more detailed measurement and elaboration of experimental data, and this could be particularly true for the WOFOST model that uses different sets of crop growth parameters as a function of development stage. Accordingly, the simulation of biomass growth during the crop time course is better by WOFOST than by the other two models. However, these more detailed input parameters connote a more complex and time consuming calibration procedure that, in some cases, could be an impediment for its extensive use. Therefore, for management purposes and in the conditions of limiting input information, the use of simpler models, such as AquaCrop, should be encouraged.

According to Steduto (2009), the aim of FAO is to have a functional canopy-level water-driven crop simulation model of yield response to water that can be used in the diverse agricultural systems that exist worldwide. It is therefore imperative that model calibration and validation, specific for each crop, are performed as extensively as possible. The current version of AquaCrop simulates several main crops (Hsiao et al., 2009 and Heng et al., 2009 for maize; García-Vila et al., 2009 and Farahani et al., 2009 for cotton; Geerts et al., 2009 for quinoa). Additionally, wheat is being calibrated with data from several locations around the world. The network of partners in this endeavor is growing and contributing to either further testing of the model calibrated already for specific crops or to parameterize and calibrate the model for additional crops (e.g., forages, oil and protein crops, tuber and root crops, and few major underutilized crops). Relative to other simulation models, AquaCrop requires a low number of parameters and input data to simulate the yield response to water, hopefully for most of the major field and vegetable crops cultivated worldwide. Its parameters are explicit and mostly intuitive, and the model has been built to maintain an adequate balance between accuracy, simplicity, and robustness. The model is aimed at a broad range of users, from engineers, economists, and extension specialists to water managers at the farm, district, and higher levels. It can be used as a planning tool or to assist in making management decisions, whether strategic, tactical or operational. AquaCrop incorporates current knowledge of crop physiological responses into a tool that can predict the attainable yield of a crop based on the water supply available.

8. Future challenges for crop production: Irrigation systems

Drip irrigation has not become a standard practice for potato production. Drip irrigation can precisely apply water and chemicals to crops at low pressure and, thus, has the potential to save water, energy, and chemicals; however, the high installation costs combined with the lack of better understanding about drip tape placement, flow rates, and efficacy of chemicals delivered through drip systems in different soil types raise growers´ concern about shifting their potato production systems to drip irrigation. Chemicals applied through a drip irrigation system could influence residue levels in the tubers or affect breakdown and movement.

Limited success was associated with deep drip tube placement (Neibling & Brooks, 1995; DeTar et al., 1996). Deep drip tape placement lowered yield and grade due to poorer water supply to the shallow root system. Shallow drip tape placement in heavy-textured soils saved water, but reduced tuber quality because of soil saturation and subsequent soil adherence on the tubers.

Drip irrigation can deliver chemicals in small doses directly to the root system of the crop; chemical use could be reduced. Irrigation scheduling and pesticide timing for more effective nematode control and control of the potato early die syndrome have the potential to increase potato yield and quality, offsetting drip system costs.

If it is possible to use drip irrigation to reduce the relative humidity of the air in potato production systems, potato late blight pressure should be reduced. Entire fields or at least the center of the filed could be converted to drip irrigation to eliminate unfavorable microclimatic conditions favoring, therefore, the development and spread of potato foliar diseases promoted by a center pivot. Fungicidal sprays might not be needed most years to control late blight where such a disease often occurs.

Changing the canopy environment to reduce duration of leaf wetness may protect the plants from pests that reduce potato yield and grade at a given site. One approach is to orient potato rows parallel to the prevailing direction of the wind (Powelson et al., 1993). Planting cultivars that do not produce extensive vines or even cultivars that have an upright growth habit may be of value in regions with a history of severe disease occurrences. An alternative is to stop irrigation in early afternoon to allow plants to dry before evening (Powelson et al., 1993). Spread of US-1 and US-8 isolates of *Phytophthora infestans* in field plots of Russet Burbank grown in Pullman, WA, USA, was favored by sprinkler irrigation during evening hours (Miller & Johnson, 2000).

Optimizing onion yields with furrow irrigation can be difficult because of low application efficiency and low distribution uniformity. Furrow irrigation can also result in excessive erosion and NO3 leaching. Different irrigation systems have been compared for onion production. Ellis et al. (1986) did not find significant differences in onion yields between sprinkler and drip irrigation systems on heavy-textured soils in Colorado, USA. When growing onion on silt loam soils in southeastern Oregon, Feibert et al. (1995) found advantages for drip and sprinkler irrigation on sites that were difficult to irrigate with furrow irrigation. Al-Jamal et al. (2000) achieved higher yields using drip than sprinkler irrigation.

Since onion has little crop canopy and a weak root system, furrow irrigation on sloping ground can lead to substantial soil and nutrient losses in runoff water (Shock et al., 1997). Mechanically applied wheat straw at 900 kg ha-1 substantially reduced sediment and nutrient losses from furrow-irrigated onion (Shock & Shock, 1998). The straw was divided into two applications to permit cultivation, 45% before the first irrigation and the remainder after the last cultivation. Straw increased onion bulb yield in replicated plots and growers fields, probably due to decreased soil water tension (Shock et al., 1999).

Statistically similar processing tomato yields were found for surface drip irrigation and furrow irrigation on loam soil at the University of California, Davis, although higher average yields occurred for the drip system compared with the furrow system (Pruitt et al., 1984). Similar ETc occurred for both irrigation methods. Lysimeter data showed furrow ETc to exceed drip ETc during irrigation, but shortly after furrow irrigation, drip ETc exceeded furrow ETc.

Furrow, sprinkler, and surface and drip irrigation of processing tomato were compared on a sandy loam for 2 years in southeast Spain (Prieto et al., 1999). Yield differences between irrigation methods were insignificant for the first year, although the highest yield occurred for sprinkler irrigation and the lowest for furrow irrigation. In the second year, the highest yield occurred for drip irrigation (statistically significant), while yields of sprinkler and furrow irrigation were similar.

Sprinkler irrigation, drip irrigation, and an unirrigated treatment were compared at a site in southwest Ontario, Canada (Tan, 1995). The soil type was sandy loam. Results were complicated by varying amounts of rainfall for each of the 4 years of experiment. Higher tomato yield occurred for drip irrigation for 2 years compared with the other irrigation treatments, while the yield of sprinkler irrigation or unirrigated treatment was higher for the other 2 years. A significant cultivar effect on yield occurred.

Several comparisons have been made among furrow, sprinkler and drip for lettuce. Sammis (1980) found that yields under furrow were comparable or higher than yields under surface and subsurface drip during the first 2 years of a 3-year field study. Flooding during stand establishment caused the furrow treatment to be the lowest yielding during the third year of the trial. Yields under sprinklers were significantly lower than the drip treatments during the first 2 years of the trial. In comparing furrow with surface and subsurface drip in large, unreplicated plots over three crop cycles, Hanson et al. (1997) reported highest yields in the furrow treatment, which received 1.3 to 2.3 times the amount of water applied by drip. Fresh weight of lettuce plants was more variable in the furrow than in the drip plots. Plant density was similar under all irrigation methods. Nitrogen content of the leaves was similar among irrigation methods for the second crop, but highest in the furrow plot of the third crop.

Sharanappa-Jangandi et al. (2001) investigated the effect of drip irrigation frequency (daily, alternate days, or every third, fourth, or fifth day) on cabbage yield in a medium black soil at Hiriyur, Karnataka, India, and compared it with irrigation every fourth day using a furrow system. Daily drip irrigation resulted in the highest yield (69.0 Mg ha^{-1}) and WUE of 54.2 g L^{-1}. Drip irrigation on alternate days resulted in good yields of 65.6 Mg ha^{-1} and WUE of 51.5 g L^{-1}. Irrigation using a furrow system produced 29.6 Mg ha^{-1}. The yield response to irrigation was found to be 2 or 4 times higher with drip irrigation than with furrow irrigation. Drip irrigation treatments used 1.27 ML ha^{-1} of water in comparison with 4.80 ML ha^{-1} for the furrow system.

9. Acknowledgments

Many thanks are owned to the Brazilian Federal Funding Agency, Conselho Nacional de Desenvolvimento Científico e Tecnológico - CNPq - for the provision of the productivity scholarship, as well as to Sandy Shelton MSc., from San Francisco, CA, USA, for the thorough revision on the English of the current book chapter.

10. References

Aggelides, S., Assimakopoulos, I., Kerkides, P. & Skondras, A. 1999. Effects of soil water potential on the nitrate content and yield of lettuce. *Commun. Soil Sci. Plant Anal.* 30:235-243.

Al-Jamal, M.S., Sammis, T.W., Ball, S. & Smeal, D. 2000. Computing the crop water production function for onions. *Agric. Water Manage.* 46:29-41.

Allen, R.G., Jensen, M.E., Wright, J.L. & Burman, R.G. 1989. Operational estimates of evapotranspiration. *Agronomy Journal* 81:650-662.

Allen, R.G., Pereira, L.S., Raes, D. & Smith, M. 1998. Crop evapotranspiration: Guidelines for computing crop water requirements. Rome: FAO. (Irrigation and Drainage Paper, 56).

Angelocci, L.R. 1996. Estimativa da transpiração máxima de macieiras (*Malus spp*) em pomares pelo método de Penman-Montheith. Tese (Livre-Docência) - Escola Superior de Agricultura "Luiz de Queiroz", Universidade de São Paulo. 95p.

Bar-Yosef, B. & Sagiv, B. 1982. Trickle irrigation and fertilization of iceberg lettuce Lactuca sativa. p.33-38. In: A. Scaife (ed.) Plant nutrition 1982: Proc. Int. Plant Nutr. Colloqu., 9th, Conventry, UK. 22-27 Aug. 1982. Commonw. Agric. Bur., Slough, UK.

Ben-Gal, A. & Shani, U. 2003. Water use and yield of tomatoes under limited water and excess boron. *Plant Soil* 256:170-186.

Bergamaschi, H., Santos, M.L.V., Medeiros, S.L.P. & Cunha, G.R. Automação de um lisímetro de pesagem através de estação meteorológica a campo. In: Congresso Brasileiro de Agrometeorologia, 10, Piracicaba, 1997, p.176-177.

Bernardo, S. 1995. Manual de irrigação. 6ª edição. Viçosa, Universidade Federal de Viçosa. 657p.

Bowen, W.T. 2003. Water productivity and potato cultivation. p.229-238. In: Kijne, J.W., R. Barker & D. Molden (eds.). Water productivity in agriculture: Limits and opportunities for improvement. CAB International Sri Lanka. Accessed on27 Nov. 2005.http://www.iwmi.cgiar.org/pubs/Book/CA_CABI_Series/Water_Productivity/Protected/0851996698ch14.pdf.

Bucks, D.A., Erie, L.J., French, O.F., Nakayama, F.S. & Pew, W.D. 1981. Subsurface trickle irrigation management with multiple cropping. *Transaction of ASAE* 1981:1482-1489.

Calado, A.M., Monzon, A., Clark, D.A., Phene, C.J., Ma, C. & Wang, Y. 1990. Monitoring and control of plant water stress in processing tomato. *Acta Horticulture* 277:129-136.

Caldiz, D.O. & Struit, P.C. 1999. Survey of potato production and possible yield constraints in Argentina. *Potato Research* 42:51-71.

Camargo, A.P. & Pereira, A.R. 1990. Prescrição de rega por método climatológico. Campinas, Fundação Cargill. 27p.

Coelho, D.T. & Dale, R.F. 1980. An energy-crop growth variable and temperature function for predicting corn growth and development. Planting to silking. *Agronomy Journal* 72:503-510.

Denmead, O.T. & Shaw, R.H. 1962. Availability of soil water to plants as affected by soil moisture content and meteorological conditions. *Agronomy Journal* 45:385-390.

De Santa Olalla, F.M., De Juan Valero, J.M. & Fabiero Cortes, C. 1994. Growth and production of onion crop (*Allium cepa* L.) under different irrigation scheduling. *European Journal of Agronomy* 3:85-92.

DeTar, W.R., Browne, G.T., Phene, C.J. & Sanden, B.L. 1996. Real-time irrigation scheduling of potatoes with sprinkler and subsurface drip systems. p.812-824. In: C.R. Camp et al. (ed.) Proc. Int. Conf. on Evapotranspiration and Irrigation Scheduling, San Antonio, TX, Am. Soc. Agric. Eng., St. Joseph, MI.

Dinar, A. 1993. Economic factors and opportunities as determinants of water use efficiency in agriculture. *Irrigation Science* 14:47-52.

Doorenbos, J. & Pruitt, W.O. 1977. Guidelines for predicting crop water requirements. Rome: FAO. 156p. (Irrigation and Drainage Paper, 24).

Doorenbos, J. & Kassam, A.H. 1979. Yield response to water. Rome: FAO. 193p. (Irrigation and Drainage Paper, 33).

Ellis, J.E., Kruse, E.G., McSay, A.E., Neale, C.M.U. & Horn, R.A. 1986. A comparison of five irrigation methods on onions. *HortScience* 21:1349-1351.

Farahani, H.J., Izzi, G., Steduto, P. & Oweis, T.Y. 2009. Parameterization and evaluation of AquaCrop for full and deficit irrigated cotton. *Agronomy Journal* 101:469-476.

Faria, R.T., Campeche, F.S.M. & Chibana, E.Y. 2006. Construction and calibration of high precision lysimeters. *Revista Brasileira de Engenharia Agrícola e Ambiental* 10:237-242.

Feibert, E.G.B., Shock, C.C. & Saunders, L.D. 1995. A comparison of sprinkler, subsurface drip, and furrow irrigation of onions. P.59-67. *In* Malheur Experiment Station annual report 1996. Spec. Rep. 947. Oregon State Univ. Agric. Exp. Stn., Corvallis.

Gallardo, M., Jackson, L.E., Schulbach, K., Snyder, R.L., Thompson, R.B. & Wyland, L.J. 1996. Production and water use in lettuce under variable water supply. *Irrigation Science* 16:125-137.

García-Vila, M., Fereres, E., Mateos L., Orgaz, F. & Steduto, P. 2009. Deficit irrigation optimization of cotton with AquaCrop. *Agronomy Journal* 101:477-487.

Geerts, S., Raes, D., Garcia, M., Miranda, R., Cusicanqui, J.A., Taboada, C., Mendoza, J., Huanca, R., Mamani, A., Condori, O., Mamani, J., Morales, B., Osco, V. & Steduto, P. 2009. Simulating Yield Response to Water of Quinoa (Chenopodium quinoa Willd.) with FAO-AquaCrop. *Agronomy Journal* 101:499-508.

Gomide, R.L., Oliveira, C.S.G. & Faccioli, G.G. Protótipo de um lisímetro de pesagem automático para estudos em casa de vegetação. In: Congresso Brasileiro de Agrometeorologia, 10, Piracicaba, 1997, p.225-227.

Hanson, B.R., Schwankl, L.J., Schulbach, K.F. & Pettygrove, G.S. 1997. A comparison of furrow, surface drip, and subsurface drip irrigation on lettuce yield and applied water. *Agric. Water Manage.* 33:139-157.

Heng, L.K., Evett, S.R., Howell, T.A. & Hsiao, T.C. 2009. Calibration and testing of FAO AquaCrop model for maize in several locations. *Agronomy Journal* 101:488:-498.

Howell, T.A., Tolk, J.A., Schneider, A.D. & Evett, S.R. 1998. Evapotranspiration, yield, and water use efficiency of corn hybrids differing in maturity. *Agronomy Journal* 90:3-9.

Hsiao, T.C., Heng, L.K., Steduto, P., Raes, D. & Fereres, E. 2009. AquaCrop—Model parameterization and testing for maize. *Agronomy Journal* 101:448-459.

Kajada, J. & Tooming, H. 2004. Potato production model based on principle of maximum plant productivity. *Agricultural and Forest Meteorology* 127:1-16.

Kooman, P.L. & Haverkort, A.J. 1995. Modelling development and growth of the potato crop influenced by temperature and daylength: LINTUL-POTATO. In: Haverkort, A.J. & D.K.L. Mac Kerron (eds.) Potato ecology and modeling of crops under conditions limiting growth. Kluwer Academic Publishers, Dordrecht, p.41-60.

Letey, J. 1993. Relationship between salinity and efficient water use. *Irrigation Science* 14:75-84.

Lin, S.S.M., Hubbel, J.N., Samson Isou, S.C.S. & Splittstoesser, W.E. 1983. Drip irrigation and tomato yield under tropical conditions. *Hortscience* 18: 460-161.

Marin, F.R. 2000. Evapotranspiração, transpiração e balanço de energia em pomar de lima-ácida "Tahiti". Dissertação (Mestrado em Irrigação e Drenagem) - Escola Superior de Agricultura "Luiz de Queiroz", Universidade de São Paulo. 73p.

Mendonça, J.C., Souza, E.F., Bernardo, S., Dias, G.P. & Grippa, S. 2003. Comparação entre métodos de estimativa da evapotranspiração de referência (ETo) na região Norte Fluminense, RJ. *Revista de Brasileira de Engenharia Agrícola e Ambiental* 7:275-279.

Miller, J.S. & Johnson, D.A. 2000. Competitive fitness of Phytophthora infestans isolates under semiarid field conditions. *Phytopathology* 90:220-227.

Mishra, H.S., Rathore, T.R. & Tomar, V.S. 1995. Water use efficiency of irrigated wheat in the Tarai Region of India. *Irrigation Science* 16:75-80.

Neibling, H. & Brooks, R. 1995. Potato production using subsurface drip irrigation - Water and nitrogen management. p.656-663. In: F.R. Lamm (ed.) Proc. Int. Microirrig. Congr., 5th, Orlando, FL. 2-6 Apr. 1995. Am. Soc. Agric. Eng., St. Joseph, MI.

Nkemdirim, L.C. 1976. Crop development and water loss - A case study over a potato crop. *Agricultural and Forest Meteorology* 16:371-388.

Pawar, S.J. & Firake, N.N. 2003. Effect of irrigation levels and micro-irrigation methods on yield of cabbage. *Journal of Maharashtra Agricultural University* 28:116-117.

Pereira, A.B, Pedras, J.F., Villa Nova, N.A. & Cury, D.M. 1995a. Water consumption and crop coefficient of potato (*Solanum tuberosum* L.) during the winter season in municipality of Botucatu-SP. *Revista Brasileira de Agrometeorologia* 3:59-62.

Pereira, A.B., Villa Nova, N.A., Tuon, R.L. & Barbieri, V. 1995b. Estimate of the maximum evapotranspiration of potato crop under the edaphoclimatic conditions of Botucatu, SP, Brazil. *Revista Brasileira de Agrometeorologia* 3:53-58.

Pereira, A.B. & Villa Nova, N.A. 2008. Potato maximum yield as affected by crop parameters and climatic factors in Brazil. *HortScience* 43:1611-1614.

Pereira, A.B., Villa Nova, N.A. & Alfaro, A.T. 2009. Water requirements of citrus and Apple trees affected by leaf area and solar energy. 2009. *Revista Brasileira de Fruticultura* 31:671-679.

Pereira, A.B., Villa Nova, N.A., Alfaro, A.T. & Pires, L.F. 2010. Transpiration of irrigated apple trees and citrus from a water potential gradient approach in the leaf-atmosphere system. *Revista Brasileira de Meteorologia* 25:xi-xf. [In print].

Pereira, A.R., Angelocci, L.R. & Sentelhas, P.C. 2002. Agrometeorologia: fundamentos e aplicações. Guaíba: Agropecuária. 478p.

Powelson, M.L., Johnson, K.B. & Rowe, R.C. 1993. Management of diseases caused by soilborne pathogens. P.149-158. In: R.C. Rowe (ed.) Potato health management. ASP Press, Wooster, OH.

Prieto, M.H., Lopez, J. & Ballesteros, R. 1999. Influence of irrigation system and strategy on the agronomic and quality parameters of the processing tomato in Extremadura. *Acta Hortic.* 487:575-579.

Pruitt, W.O, Fereres, E., Martin, P.E., Singh, H., Henderson, D.W., Hagan, R.M., Tarantino, E. & Chandio, B. 1984. Microclimate, evapotranspiration, and water-use-efficiency for drip- and furrow- irrigated tomatoes. p. 367-394. *In* Int. Congr. on Irrig. and Drain., 12th, Fort Collins, CO. 28 May-2 June 1984. Vol. 1A. U.S. Committee on Irrig. and Drain., Denver, CO.

Reichardt, K. 1985. Processos de transferência no sistema solo-planta-atmosfera. 4ª edição. Campinas, Fundação Cargill. 446p.

Richards, R.A., López-Castañeda, C., Gomez-Macpherson, H. & Condon. A.G. 1993. Improving the efficiency of water use by plant breeding and molecular biology. *Irrigation Science* 14:93-104.

Saeed, I.A.M. & El-Nadi, A.H. 1997. Irrigation effects on the growth, yield, and water use efficiency of alfalfa. *Irrigation Science* 17:63-68.

Sammis, T.W. 1980. Comparison of sprinkler, trickle, subsurface and furrow irrigation methods for row crops. *Agronomy Journal* 72:701-704.

Sammis, T.W., Kratky, B.A. & Wu, I.P. 1988. Effects of limited irrigation on lettuce and Chinese cabbage yields. *Irrigation Science* 9:187-198.

Sanchez, C.A. 2000. Response of lettuce to water and nitrogen on sand and the potential for leaching nitrate-N. *HortScience* 35:73-77.

Sharanappa-Jangandi, Shekar, B.G. & Sridhara, S. 2001. Water use efficiency and yield of cabbage as influenced by drip and furrow methods of irrigation. *Indian Agric.* 44:153-155.

Shock, C.C., Hobson, J.H., Seddigh, M., Shock, B.M., Stieber, T.D. & Saunders, L.D. 1997. Mechanical straw mulching of irrigation furrows: Soil erosion and nutrient losses. *Agronomy Journal* 89:887-893.

Shock, C.C., Feibert, E.B.G. & Saunders, L.D. 1998. Potato yield and quality response to deficit irrigation. *HortScience* 33:655-659.

Shock, C.C. & Shock, B.M. 1998. Comparative effectiveness of polyacrylamide and straw mulch to control erosion and enhance water infiltration. p. 429-444. *In* A. Wallace and R.E. Terry (ed.) Handbook of soil conditioners. Marcel Dekker, New York.

Shock, C.C., Jensen, L.B., Hobson, J.H., Seddigh, M., Shock, B.M., Saunders, L.D. & Stieber, T.D. 1999. Improving onion yield and market grade by mechanical straw application to irrigation furrows. *HortTechnology* 9:251-253.

Shock, C.C., Feibert, E.B.G. & Saunders, L.D. 2000. Irrigation criteria for drip-irrigated onions. *HortScience* 35:63-66.

Silva, F.C., Folegatti, M.V. & Maggiotto, S.R. 1999. Análise do funcionamento de um lisímetro de pesagem com célula de carga. *Revista Brasileira de Agrometeorologia* 7:53-58.

Srinivas, K., Hegede, D.M. & Havanagi, G.V. 1989. Plant water relations, canopy temperature, yield and water-use efficiency of water melon (*Citrulls lanatus* (Thamb.) Matsum et Nakai) under drip and furrow. *Journal of Horticultural Science* 64:115-124.

Sousa, V.F. de, Coelho, E.F., Fizzone, J.A., Folegatti, M.V., Andrade Júnior, A.S. & Oliveira, F. das C. 2000. Eficiência do uso da água pelo meloeiro sob diferentes freqüências de irrigação. *Revista Brasileira de Engenharia Agrícola e Ambiental* 4:183-188.

Steduto, P., Hsiao, T.C., Raes, D. & Fereres, E. 2009. AquaCrop - The FAO crop model to simulate yield response to water: I. Concepts and principles. *Agronomy Journal* 101:426-437.

Sutton, B.G. & Merit, N. 1993. Maintenance of lettuce root zone at field capacity gives best yields with drip. *Sci. Hortic.* 56:1-11.

Tan, C.S. 1993. Tomato yield-evapotranspiration relationships, seasonal canopy temperature and stomatal conductance as affected by irrigation. *Canadian Journal of Plant Science* 73:257-264.

Tan, C.S. 1995. Effect of drip and sprinkler irrigation on yield and quality of five tomato cultivars in southwestern Ontario. *Can. J. Plant Sci.* 75:225-230.

Tanner, C.B. 1981. Transpiration efficiency of potato. *Agronomy Journal* 73:257-264.

Todorovic, M., Albrizio, R., Zivotic, L., Saab, M.T.A., Stöckle, C. & Steduto, P. 2009. Assessment of AquaCrop, CropSyst, and WOFOST models in the simulation of sunflower growth under different water regimes. *Agronomy Journal* 101:509-521.

Villa Nova, N.A., Santiago, A.V. & Rezende, F.C. 2001. Energia Solar: Aspectos físicos de captura de biomassa. Departamento de Ciências Exatas: Escola Superior de Agricultura "Luiz de Queiroz", Universidade de São Paulo.

Villa Nova, N.A., Pilau, F.G., Dourado Neto, D. & Manfron, P.A. 2005. Estimativa da produtividade de cana-de-açúcar irrigada com base na fixação de CO2, radiação solar e temperatura do ar. *Revista Brasileira de Agrometeorologia* 13:405-411.

Villa Nova, N.A., Miranda, J.H., Pereira, A.B. & Silva, K.O. 2006. Estimation of potential evapotranspiration by a simplified Penman method. *Engenharia Agrícola* 26:713-721.

Villa Nova, N.A., Pereira, A.B. & Shock, C.C. 2007. Estimation of reference evapotranspiration by an energy balance approach. *Biosystems Engineering* 96:605-615.

Wright, J.L. 1982. New evapotranspiration crop coefficients. *Journal of Irrigation and Drainage ASCE* 108:57-74.

Wright, J.L. & Stark, J.C. 1990. Potato. P.859-888. In: B.A. Stewart and D.R. Nielson (ed.) Irrigation of agricultural crops. Agronomy Monograph 30. ASA, CSSA, and SSSA, Madison, WI.

10

Saving Water in Arid and Semi-Arid Countries as a Result of Optimising Crop Evapotranspiration

Salah El-Hendawy[1], Mohamed Alboghdady[1], Jun-Ichi Sakagami[2] and Urs Schmidhalter[3]
[1]*Faculty of Agriculture, Suez Canal University, 41522 Ismailia,*
[2]*Japan International Research Center for Agricultural Sciences, Tsukuba,*
[3]*Department of Plant Sciences, Technische Universität München,*
[1]*Egypt*
[2]*Japan*
[3]*Germany*

1. Introduction

Although water covers about 71 percent of our planet surface, 98 percent of it has too high salt content to be used for drinking water, for irrigation, or even for most industrial purposes. Fresh water represents one percent of all the water on the earth and is distributed unevenly on the earth surface. As a result of the dramatic increase in population, in economic activities, and a subsequent increase in water usage, the world fresh water resources became scarce during the past decades (Postel et al., 1996; Hoekstra & Chapagain, 2007).

Shortage of water currently plagues almost every country in North Africa and the Middle East (MENA) to the extent that hampers economic growth and threats social stability. Most importantly, the scenarios for global environmental change suggest a future increase in aridity and in the frequency of extreme events in many areas of the earth. Economists believe that water problem cannot be solved until water is considered as economic good. There is an urgent need to develop appropriate concepts and tools to do so.

Evapotranspiration (ET) is defined simply as the sum of the amount of water returned to the atmosphere through the processes of evaporation (moisture loss from the soil) and transpiration (biological use and release of water by vegetation). Crops are different in their response to water stress at a given growth stage. Therefore, estimating ET is an important tool to calculate the actual crop water requirements in given conditions. As a result, optimizing ET may contribute in solving water shortage problems at two levels, the farm level and the national/global level.

Firstly, at the farm level, the process of water irrigation losses has two main components: one due to evapotranspiration losses, and the other including the losses resulting from the percolation of water beneath the root zone in excess of any required leaching for salinity management. Therefore, saving water on the farm level can be achieved by using deficit irrigation strategy and irrigation scheduling. Deficit irrigation strategy, defined as the application of water below full crop-water requirement (evapotranspiration, ET), is an important tool to achieve the goal of increasing water use efficiency through their affecting on the amount of water losses by evapotranspiration (Fereres & Soriano, 2007). Irrigation scheduling, defined as determining when to irrigate and how much to apply, is another important element in improving water use efficiency through their exerting positive or negative effects on the amount of water percolating under the root zone (Bergez et al., 2002; El-Hendawy & Schmidhalter, 2010).

Full irrigation is the amount needed to achieve maximum yield; however, when irrigation water is insufficient to meet crop demand, limited irrigation management strategies should be considered. Use of limited irrigation strategies could save large amounts of water and might alleviate the issue of increasing food shortages found around the world. However, it is difficult to plan for reducing ET without a penalty in crop production, because evaporation from canopies is tightly coupled with the assimilation of carbon (Monteith, 1990; Steduto et al., 2006). For instance, El-Hendawy et al. (2008) found that average maize yield decreases under drip irrigation for 0.80 and 0.60 ET relative to 1.00 ET were 32 and 63%, respectively. Al-Kaisi & Yin (2003) also reported that average yield decreases for maize under sprinkler irrigation were 43% with 0.60 ET and 25% for 0.80 ET relative to 1.00ET. This indicates that application of water below the ET requirements requires further effective management strategies. These strategies should aim to achieve the highest possible economic return per unit of water applied. Recently, rapid and economically feasible approaches have been proposed to counteract yield losses and water use efficiency (WUE) decreases under deficit irrigation strategies. Several studies have emphasized that osmotic adjustment, which is achieved by exogenous application of osmoprotectants, is a useful approach for improving crop and water productivity under deficit irrigation (Agboma et al., 1997; Hussain et al., 2009). Of the different compatible solutes known, glycinebetaine (GB) is relatively more important as it is capable of promoting yields and WUE under water deficit. The reason for that may lie in the nature of GB as it exerts positive influence on the photosynthetic machinery (Xing & Rajashekar, 1999), and it does not inhibit enzymes even at high concentrations (Ashraf & Foolad, 2007). Therefore, it can be accumulated in the cytoplasm of plant cells to contribute to the osmotic balance between the cytoplasm and vacuole without causing any damages. This unique nature of GB has led to that foliar-applied GB is widely used for crop production throughout the world for different purposes. Exogenous application of GB to non-accumulators of GB has been taken as an alternative to improve the stress tolerance. For example, Ma et al. (2006) reported that GB-treated plants maintained a higher net photosynthetic rate during drought stress than non-GB treated.

Most importantly, foliar application of GB may be a simple and cost effective methodology to increase the net benefit per unit of water applied under limited water application. Brand et al. (2007) reported that foliar application of GB can be adapted as a management strategy to alleviate water deficit at a cost less than a US $ 2.5 per hectare. Under furrow irrigation application, Hussain et al. (2009) also found that foliar application of GB reduced water consumption by 25%, increased cost by 6% and increased net income by about 9% when the sunflower plants were exposed to water stress and treated by exogenous applications of GB simultaneously at vegetative and flowering stages. Therefore, it is important to know the optimum coupling combinations between GB levels and irrigation rate, to seek maximize yield and IWUE simultaneously under deficit irrigation strategy which is one of the objectives of this chapter.

Drip irrigation is the most effective method in terms of both maximizing yield and water conservation and also providing efficient use of limited water (Cetin & Bilgel, 2002). At the same time, several authors have shown that the water use efficiency and yield of drip-irrigated crops could be improved under limited water applications by decreasing the amount of water that moves beneath the root zone (Bergez et al., 2002; El-Hendawy et al., 2008). Thus, optimizing the coupling between irrigation frequency and water application rate could help to achieve maximum yield and water use efficiency (WUE) by exerting positive or negative effects on the amount of water percolating under the root zone. For instance, coupling very high irrigation frequency and rate will avoid stress situations, but at the cost of reduced drip irrigation efficiency and WUE as a result of the increased amount of water moving beneath the root zone because the amount of water being applied can exceed the amount extracted by roots. Coupling very low irrigation frequency and rate, by contrast, can cause water stress between successive irrigation events (especially in sandy soils) because the amount of water applied at each event is insufficient to meet the water requirement of the plants as time proceeds. Finally, coupling very low irrigation frequency and very high water application rate, particularly in sandy soils, may result in a decreased efficiency of the drip irrigation system and finally water use, because the amount of water applied at each irrigation event may be higher, and possibly excessively so, than the soil-water storage capacity, thereby increasing the amount of water that moves below the root zone so as to reduce their availability to plants as time proceeds. Therefore, it is important to know the optimum coupling combinations between irrigation frequency and water application rate, to seek maximize yield and IWUE simultaneously for drip irrigated crops which is another objective of this chapter.

Secondly, at the national/global level, while trade of real water between water-rich and water-poor regions is generally impossible due to the large distances and associated costs, it is argued that international trade moves "virtual water" from a comparatively more favorable region, where there is a surplus of soil water in soil profiles, to comparatively disadvantaged regions such as the MENA region. The virtual water is defined as the volume of water used in producing a unit of commodity, or service (Allan, 1998). Agricultural trade is by far the largest transporter to move water virtually around the world.

The virtual water concept has two types of practical uses (Hoekstra, 2003). Firstly, virtual water can be seen as an alternative source of water assisting to achieve the regional water security. Secondly, the virtual water content of a product tells something about the environmental impact of consuming this product. Raising the awareness of virtual water content of products thus providing an idea of which goods impact most on the water system and where water savings could be achieved.

The idea of actively promoting the import of virtual water in water-scarce countries is based on the idea that a nation can save its domestic water resources by importing a water-intensive product rather than produce it domestically. Import of virtual water therefore leads to a "national water saving" (Hoekstra & Chapagain, 2007). In a widely scope, Oki & Kanae (2004) argued that the virtual water trade produces "global water saving" when the agricultural product a traded from a country in which the unit requirement of water to produce a commodity (UW) is low to the country in which the UW is high. Conversely, the global water loss occurs when the trade is from a high UW country to a low UW country for a particular crop.

2. Method

2.1 Experimental design, treatments

Two field experiments of drip-irrigated maize were conducted to establish the optimal coupling combinations between ET deficit and irrigation frequency or glycinebetaine concentrations. Each experiment was conducted using a randomized complete block split-plot design with three drip irrigation rates (1.00, 0.80, and 0.60 of the estimated evapotranspiration, ET) as the main plot and four irrigation frequencies (F1, F2, F3 and F4, irrigation events once every 1, 2, 3 or 4 days, respectively) in the first experiment and five exogenous application concentrations of GB (0, 25, 50, 75 and 100 mM) in the second experiment as the split plots. The drip irrigation system was divided into four main sectors in the first experiment and three main sectors in the second experiment with the irrigation frequency treatments and irrigation rates being assigned to the four sectors and three sectors, respectively. The water application rate treatments (1.00 ET, 0.80 ET and 0.60 ET) in the first experiment and GB concentrations (0, 25, 50, 75 and 100 mM) in the second experiment were randomly nested within each main sector as a subplot, with each subplot having three replicates for subplots treatments.

2.2 Calculation of evapotranspiration

The crop evapotranspiration ET_c is given by multiplying the reference crop evapotranspiration ET_0 with the crop coefficient K_c.

$$ET_c = K_c \times ET_0 \tag{1}$$

The concept of the reference evapotranspiration was introduced to study the evaporative demand of the atmosphere independently of crop type, crop development and management practices. ET_0 is a climatic parameter and can be computed from weather data. ET_0 expresses the evaporating power of the atmosphere at a specific location and time of the year and does not consider the crop characteristics and soil factors (Allen et al., 1998).
With reference to Hoekstra & Hung (2002), crop evapotranspiration is calculated on the basis of the FAO Penman-Monteith equation:

$$ET_0 = \frac{0.408(R_n - G) + \gamma 900/(T+273)U_2(e_a - e_d)}{\Delta + \gamma(1+0.34U_2)} \tag{2}$$

where ET_0 is the reference evapotranspiration (mm day^{-1}), R_n the net radiation at the crop surface (MJ m^{-2} day^{-1}), G the soil heat flux density (MJ m^{-2} day^{-1}), T the mean daily air temperature at 2 m height (°C), u_2 the wind speed at 2 m height (m s^{-1}), e_s the saturation vapor pressure (kPa), $e_a$the actual vapor pressure (kPa), $e_s - e_a$ the saturation vapor pressure deficit (kPa), D the slope of the saturation vapor pressure curve (kPa °C^{-1}), and g is the psychrometric constant (kPa °C^{-1}).

2.3 Yield response factor (k_y)

Seasonal values of the yield response factor (k_y) for different combinations, which represent the relationship between the relative maize yield reduction $\left(1 - \frac{Y_a}{Y_m}\right)$ and the relative evapotranspiration deficit $(1-\frac{ET_a}{ET_m})$, were determined using the formula given by Doorenbos & Kassam (1979):

$$\left(1-\frac{Y_a}{Y_m}\right)=k_y\left(1-\frac{ET_a}{ET_m}\right) \tag{3}$$

Where ET_a and ET_m are the actual and maximum seasonal crop evapotranspirations (mm), respectively, and Y_a and Y_m are the corresponding actual and maximum yields (kg ha^{-1}).

2.4 Water production function

One of the more useful and widely accepted production function forms is based on the consumptive use, or evapotranspiration (ET). Grimes et al. (1969) and Gulati & Murty (1979) with wheat, barley, and sugarcane reported that the yield-evapotranspiration (Y-ET) relations for these crops were best described by quadratic functions. Many functions were tested to fit the data of the experiment. The quadratic form was the best fitting function.
Y can be described as follows:

$$Y=\alpha+\beta_1 ET+\beta_2 ET^2 \tag{4}$$

As the function is obtained from a given set of yield and ET data, it is only appropriate and reliable for the given range of the observations. In the case of the quadratic function, both α and β_2 have negative signs when the function is derived from regression analysis. The ratio Y to ET reflects the water use efficiency (WUE) for a given amount of seasonal ET. The first derivative of Y, that is, dY/dET, describes the change in Y per unit change in ET that is called marginal product of ET (MP). Corresponding to the microeconomic theory, the Y in Eq. (4) represents the total product, Y/ET the average product of the input, and dY/dET the marginal product of the input (Hexem & Heady, 1978).
The elasticity in the production function (EP) is defined as the percentage changes of Y due to one percent change in ET. Thus,

$$EP=\frac{dY}{Y}\div\frac{dET}{ET}=\frac{dY/dET}{Y/ET}=\frac{MP}{WUE} \tag{5}$$

From equation (5) the EP-ET-MP-WUE relations can be described as follows:
EP=1 if MP = WUE; that occurs at the maximum point of WUE.
EP=0 if MP = 0; that occurs at the maximum point of Y.
EP>1, when the increase in ET leads to increase in WUE.
EP<1 when the increase in ET leads to decrease in WUE.
EP<0 when the increase in ET leads to negative MP or decrease in Y.

2.5 National water saving

The national water saving NWS_{ij} as a result of trade of crop i in country j is:

$$NWS_{ij}=VWM_{ij}-VWX_{ij} \tag{6}$$

where VWX_{ij} and VWM_{ij} are the virtual water exported and imported, respectively, contained in crop i by country j. We can calculate them by multiplying the specific water demand of crop i with exported/imported quantities of the same crop in a given year. Both exported and imported virtual waters are calculated as if the mentioned crop is produced domestically.

$$VWX_{ij} = SWD_{ij} \times X_{ij} \tag{7}$$

$$VWM_{ij} = SWD_{ij} \times M_{ij} \tag{8}$$

where X_{ij} refers to the exported quantities (ton) of crop i in a given year by country j, and M_{ij} is the imported quantities (ton) of crop i by country j. SWD_{ij} refers to the specific water demand (m^3 ton^{-1}) of crop i in country j.
Intuitively, the virtual water exported ($VWX\varphi j$) or imported ($VWM\varphi j$) contained in a group of crops φ by country j is calculated by the summation of multiplying the specific water demand of each crop i in the group φ with the exported/ imported quantities of the same crop in a given year.

$$VWX_{\varphi j} = \sum_{i=1}^{n}(SWD_{ij} \times X_{ij}) \tag{9}$$

$$VWM_{\varphi j} = \sum_{i=1}^{n}(SWD_{ij} \times M_{ij}) \tag{10}$$

For crop i in country j, the specific water demand can be calculated by Hoekstra & Hung (2005)

$$SWD_{ij} = \frac{CWR_{ij}}{CY_{ij}} \tag{11}$$

CY is the crop yield (ton ha^{-1}). Although the values for crop evapotranspiration and crop water requirement are identical, crop water requirement refers to the amount of water that needs to be supplied, while crop evapotranspiration refers to the amount of water that is lost through evapotranspiration under the standard conditions (Allen et al., 1998).
It is sensible to mention that a country may not produce a specific crop and imports all the needed quantities. In this case, we have no data that are required for calculating the specific water demand such as crop yield and some factors required for calculating the crop water requirements. Thus, we compensate the domestic specific water demand of crop j by the weighted average of the regional specific water demand of the same cereal crop.
Crop water requirements are calculated by accumulation of data on daily crop evapotranspiration ETc (mm day^{-1}) over the complete growing period as follows:

$$CWR_{ij} = 10\sum_{d=1}^{L} ET_{c(ij)} \tag{12}$$

where, the factor 10 is meant to convert (mm ha^{-1}) into (m^3 ha^{-1}). The summation is done over the period from day 1 to the last day of the growing period (L) (Sallam and Abd El Nasser, 2006).

2.6 Global water saving $GWS_{i,ej}$

The global water saving $GWS_{i,ej}$ through trade of crop i from an exporting country e to an importing country j is:

$$GWS_{i,ej} = VWM_{ij} - VWX_{ie} \tag{13}$$

Where, VWM_{ij} and VWX_{ej} are the virtual water content of the importing and exporting quantities of crop i in favor of the actual specific water demand and the actual crop yield of the import and export countries.
Due to the difficulties and complications of calculating specific water demand in each exporting country as each importing country imports from many variable sources over the

given period, we compensate each specific water demand of the exporting country by the world average specific water demand for each crop.

2.7 Statistical analysis

Analysis of variance (ANOVA) appropriate for a randomized complete block split plot design was employed to test the overall significance of the data, while the least significant difference test (at P = 0.05) was used to compare the differences among treatment means. Graphical presentation of data was carried out using Microsoft Excel program (Microsoft Corporation, Los Angeles, CA, USA).

3. Results and discussion

3.1 Yield response factor (k_y)

The response of yield to water deficit is quantified through the yield response factor (k_y), which represents the decrease in yield caused by decreases in evapotranspiration (Doorenbos & Kassam, 1979). Generally, higher k_y values indicate that the crop will have a greater yield loss when the crop evapotranspiration is not met. One important finding of this study was the strong response of k_y values to the combination of evapotranspiration deficit with irrigation frequencies or glycinebetaine (GB) concentrations (Table 1). Regarding the combination between ET deficit and irrigation frequency, it is interesting to note that the lowest k_y values were recorded for the combination of 0.80 ET with irrigation frequency once every one day (k_y = 0.12) followed by 0.60 ET with the same irrigation frequency (k_y = 1.15). Both values were lower than those reported by Doorenbos & Kassam (1979; 1.25). The k_y value of 1.25 obtained by Doorenbos & Kassam (1979) represents water deficit's effect on maize crop yield for the total growing period, which means that the reduction in yield for both combinations was proportionately less than the reduction in ET. By contrast, the combination of 1.00 ET with the same irrigation frequency or 1.00, 0.80 and 0.60 ET with the irrigation frequency once every 4 days resulted in significant increases in k_y values (Table 1). These results indicate that optimal coupling combinations of ET deficit with irrigation frequency are useful for maximizing the net income per unit of water applied. Therefore, under the imposed water regime we recommended to switch from irrigation every other day to much smaller irrigations several times. This may be explained by the application of water at low irrigation frequency exceeding the soil-water storage capacity leading to excessive water percolation under the effective root zone. Furthermore, a portion of the water application was not used by the plant and the remaining available water will not meet the long term water requirements of the plants till the next irrigation event.

For the combination of ET deficit with GB concentrations, the maximum maize yield during this study was obtained from a combination between 1.00 ET and 25 mM GB. The value of evapotranspiration at 1.00 ET was assumed as maximum evapotranspiration, and evapotranspiration and grain yield at different combination of 0.80 and 0.60 ET with different GB concentrations were assumed as actual evapotranspiration and actual grain yield. According to the result in Table 1, the k_y value for the combination of 0.80 ET with 25 and 50 mM GB or 0.60 ET with 50 and 75mM GB were less than 1, which means that the reduction in yield for former combinations was proportionately less than the reduction in ET, despite the amount of water applied for 0.80 and 0.60 ET treatments was 20 and 40% lower than those for the 1.00 ET treatment, respectively. These results provided important clues to the physiological role of optimum doses of GB in possibly improving water use efficiency under ET deficit. The decreases of k_y value under ET deficit with the application of optimum dose of GB indicate that more grain yield was produced from less water use.

Treatments	Evapotranspiration deficit treatments		
	0.60 ET	0.80 ET	1.00 ET
	Irrigation frequency treatments (F)		
Once in 1 day (F1)	1.15	0.12	2.34
Once in 2 days (F2)	1.42	1.24	0.00
Once in 3 days (F3)	2.35	2.19	1.50
Once in 4 days (F4)	2.33	3.01	3.74
	Glycinebetaine concentration treatments (GB)		
0 mM	1.53	1.53	nd*
25 mM	1.32	0.87	nd
50 mM	0.97	0.51	nd
75 mM	0.93	1.49	nd
100 mM	1.50	2.07	nd

*Not determined

Table 1. Relative yield response factors (k_y) shown as the relationships between the relative yield decrease (1- Y_a/Y_m) and the corresponding relative evapotranspiration deficit (1- ET_a/ET_m) for combination of evapotranspiration deficit with irrigation frequencies or glycinebetaine concentrations.

This may be due to the effects of GB on transpiration rate. Agboma et al. (1997) reported that the transpiration rate of GB treated plants was decreased to 85% of untreated plants, strongly indicating an anti-transpirant effect of GB treatment. The slower transpiration rate allows the plant to access water for a longer period and exhibit greater photosynthesis, where it would contribute to improved yield and water use efficiency under ET deficit.

3.2 The relation of irrigation water use efficiency with evapotranspiration deficit

In recent years, several researchers have reported about "site specific management", which means applying the right amount of input in the right place at the right time to get maximum profit from per unit input. Based on this basic definition, the profitability of irrigation water applied can be maximized by determining the optimal combination of ET treatments with irrigation frequency or GB concentrations. In this study, the irrigation water use efficiency (IWUE) was significantly affected by the different combination of ET treatments with irrigation frequency or with GB concentrations (Fig. 1). It is interesting to note that the combination of 0.60 ET with irrigation frequency once every one day had IWUE values similar to those obtained for the corresponding 0.80 or 1.00 ET with irrigation frequency once every two days and was higher to those obtained for the 1.00 ET treatment with either irrigation frequency once every three or four days (Fig. 1). The combined effects of ET treatments and GB concentration on IWUE were also significant. The highest IWUE was obtained under an irrigation rate of 0.80 ET at 50 mM GB, but it was at par with an irrigation rate of 0.60 ET at 75 mM GB. By contrast, the lowest IWUE was recorded for either the combination of 1.00 ET with 75 and 100 mM GB, and 0.80 or 0.60 ET with 100 mM GB. Because of the combination effects between ET treatments and irrigation frequency on the amount of water that is percolated under the root zone or absorbed by the roots (Assouline, 2002; Wang et al., 2006; Uçan et al., 2007) and the combination effects between ET treatments and GB concentration on the amount of water that are transported from canopy (Agboma et al., 1997), the optimal combinations between ET deficit and irrigation frequency or GB concentrations were instead often crucial for maximizing net income per unit water.

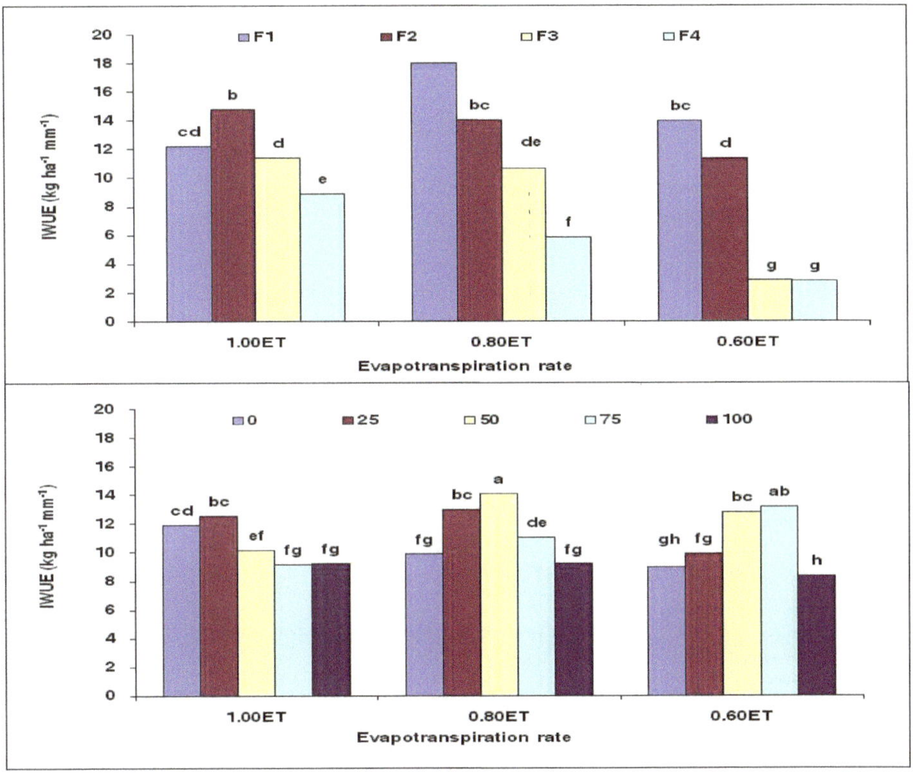

Fig. 1. Irrigation water use efficiency (IWUE) for the combination of evapotranspiration deficit (ET) with irrigation frequency (F) or with glycinebetaine concentration.

3.3 Yield-seasonal crop evapotranspiration relationship

A linear relationship has been reported between maize grain yield and seasonal crop ET (Payero et al., 2006) and was also indicated here for the combination of ET treatment with irrigation frequency or GB concentration (Fig. 2 A and B). However, when broken down

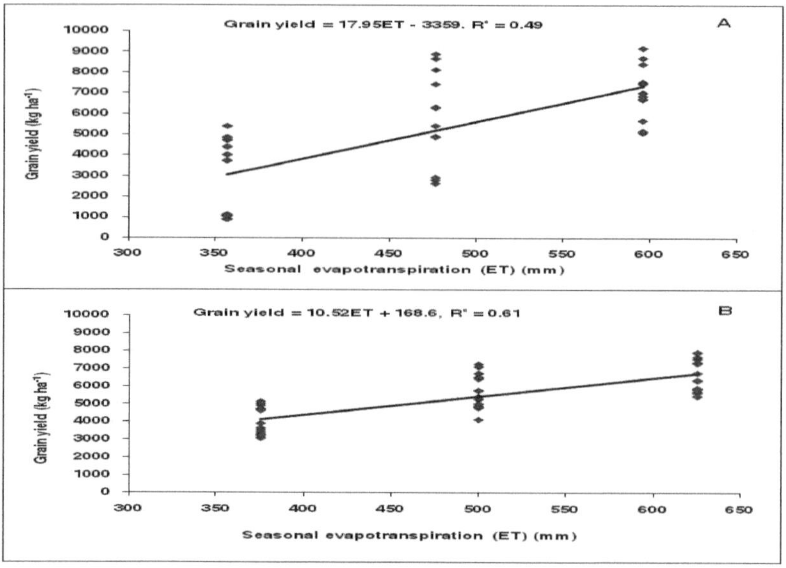

Fig. 2. Relationship between seasonal crop evapotranspiration (ET) and grain yield for the combination of ET treatment with irrigation frequency (A) or with GB concentration (B).

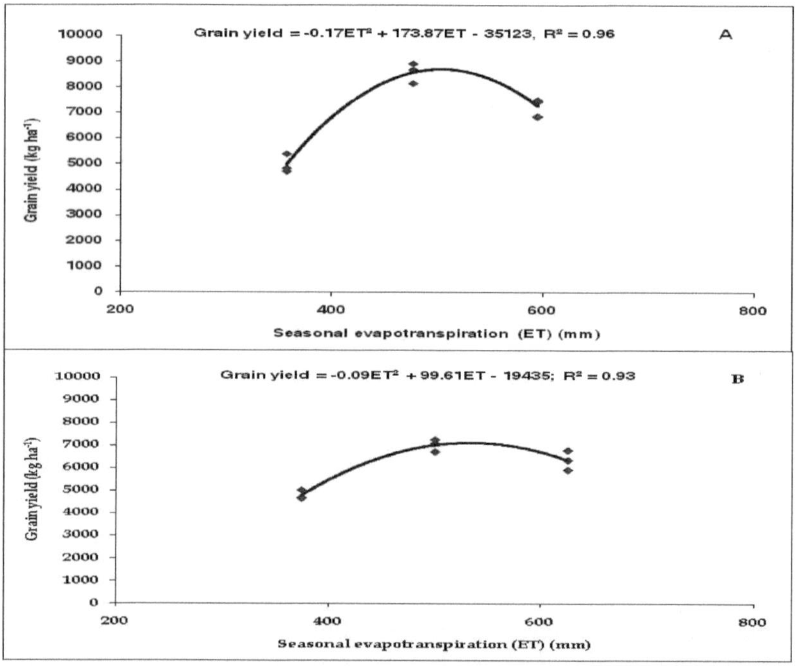

Fig. 3. Relationship between seasonal crop evapotranspiration (ET) and grain yield for irrigation frequency once every one day (A) and for 50 mM GB concentration (B).

according to the irrigation frequency treatments or GB concentration treatments, the relationships between grain yield and seasonal ET deviated from linear relationships to second order relationship for irrigation frequency once every one day or 50 mM GB concentrations (Fig. 3 A and B). Therefore, the regression of Fig. 2A and B shows that about 51 and 39% of the grain yield variation is not explained by seasonal ET under the combinations of ET treatment with irrigation frequency or with GB concentrations, respectively. The second order relationship also indicates that the increase in maize grain yield was not proportional with the increment in the amount of irrigation water. This phenomenon was especially obvious when irrigation frequency once every one day or 50 mM GB was combined with 1.00 ET because a portion of the water applied does not contribute to ET. At the same time, both former combinations resulted in significant reductions in grain yield due to the combination of 1.00 ET with the irrigation frequency once every one day resulting in a very humid region in the root zone and the combination of 1.00 ET with 50 mM GB effects on stomatal conductance and transpiration rates, which in turn both affects the crop photosynthesis. Therefore, it is suggested that the ET treatments should be matched well with other variables which directly or indirectly affects on grain yield (Paolo & Rinaldi, 2008; Farré & Faci, 2009).

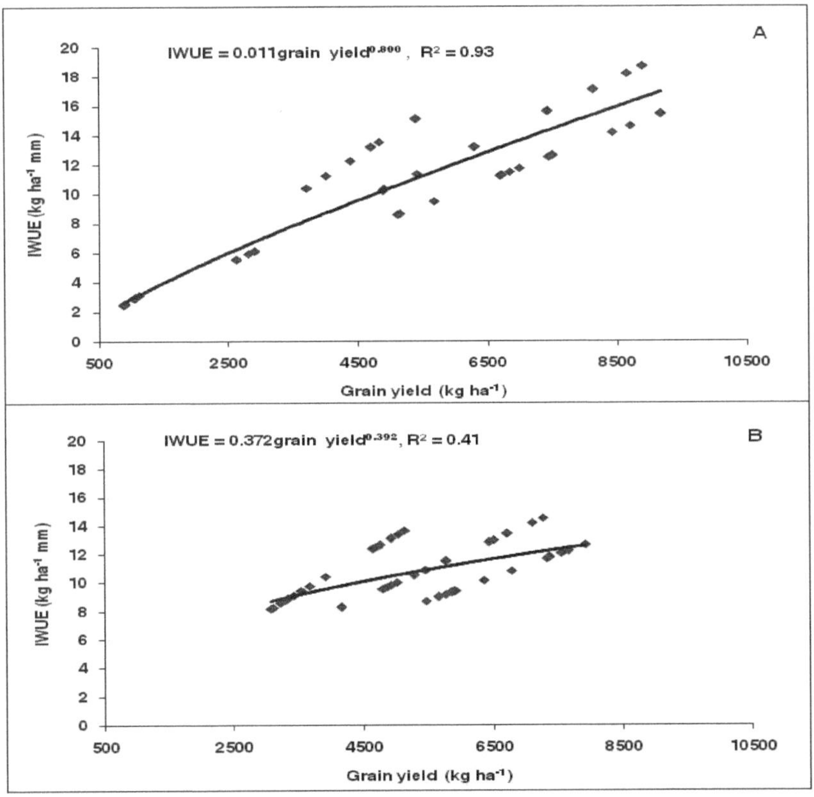

Fig. 4. Relationship between grain yield and irrigation water use efficiency (IWUE). Power regression equations; ***, * indicate significances at 0.001 and 0.05 P level, respectively.

3.4 The relation of irrigation water use efficiency with grain yield

Chen et al. (2003) mentioned that the relationship between WUE and grain yield is often used for determining the optimal irrigation strategy for arid and semiarid regions. Based on R^2 values, the power model was found to be the best fit model to describe the IWUE-yield relationship of our data, with a coefficient of determination value of 0.93 and 0.41 for the combination of ET treatments with irrigation frequency or GB concentrations, respectively (Fig. 4). In the power model, the elasticity of y with respect to x is the percentage change in y for each percentage change in x. Based on this definition, when grain yield is increased by 10%, the IWUE was increased by 8.0 and 3.9% in the combination of ET with irrigation frequency or with GB concentrations, respectively (equations in Fig 4). These equations indicated that high irrigation water use efficiency is obtained for high yield values. Therefore, it is important to know the optimum coupling combinations between ET and irrigation frequency or GB to seek maximize yield and IWUE simultaneously with saving water on the farm level.

3.5 Marginal analysis of water production function

The Y-ET function in a quadratic form is described as:

$$Y = -5921.93 + 38.6ET - 0.028ET^2 \quad R^2 = 0.78 \tag{14}$$

The relationships for EP, WUE, MP, Y, and ET based on the function of Eq. (5) are shown in Fig. 5 under a limited water supply. One of two goals can be achieved by ET applications. The first goal is maximizing the WUE, point A. Such goal can be achieved, in terms of the economic sight of the production function, at the point of which the first derivative of WUE is equal to zero, MP is equal to WUE, and EP is equal to one. The maximum WUE of 12.9 kg mm can be attained when ET was equal to 459.9 mm. The second goal is maximizing crop production, point B. At this point, the MP and EP are equal to zero as the first derivative of Y is equal to zero. ET maximizing yield is 689.3 mm at which the total yield is 7381.3 kg. Consequently, the WUE is with 1.07 kgm^{-3} less than the maximum WUE by 17.1%. It is sensible to mention that the EP is a reliable indication to recognize the range of desirable or economic production stages which fall in the range $1 \geq EP \geq 0$. With other words, the rational economic productivity must fall between the maximum WUE and maximum yield.

3.6 Virtual water as instrument to achieve water security in North Africa

The aim of this section is to analyze the consequences of North African international virtual water flows associated with cereals trade on national and global water budgets. With this aim, it quantifies and assesses national and global water savings and losses per cereal crop in the North Africa region.

Virtual water trade has a positive direct effect on water saving for the importing countries. This effect has been intensively discussed in virtual water studies since the concept of virtual water raised by Allan in the nineties. In this study, we examine the water saving at two levels. Firstly, national water saving that concentrates on the benefit gained by importing countries. Secondly the global water saving, the global net effect of virtual water trade between two nations will depend on the actual water volume that would have been required to produce a commodity in the importing and exporting countries. (Chapagain et al., 2006).

Figure 6 shows the notable variations among the specific water demand for the cereal crops production in North Africa countries and on the world average. Such highly significant variations derive mainly from two reasons. One reason is that in the North Africa region the evaporative demand is relatively high. The second reason is the low yield of the cereal crops in most NA countries. The water requirements to produce one ton of cereals in NA are equal to five times those in the world average. Highest differences were witnessed in maize and sorghum crops.

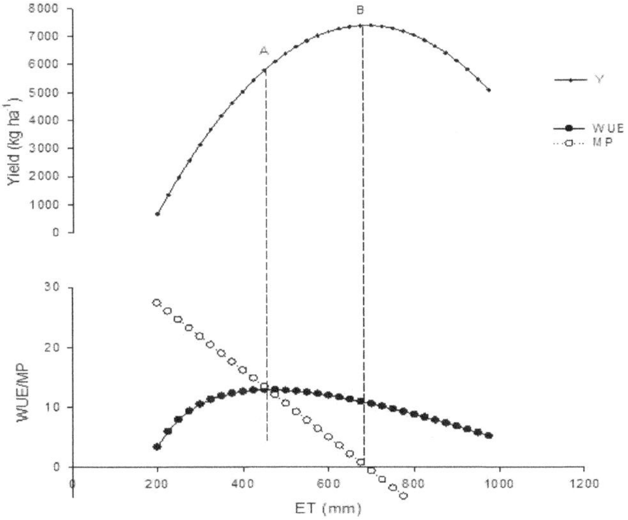

Fig. 5. Relations of EP, WUE, MP, Y, and ET for a quadratic production function for maize during 2007 -2008 in Egypt.

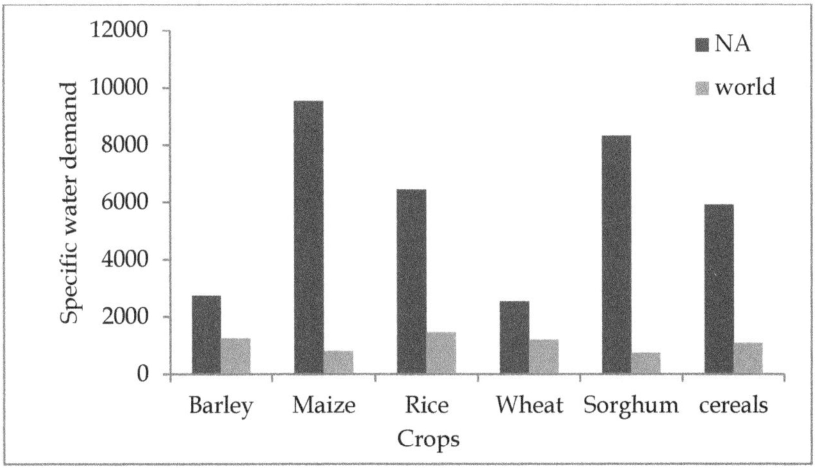

Fig. 6. Specific water demand for cereal crops in the North African and world (m^3/ton) 2003-2007

3.6.1 National water saving in NA countries through cereals trade

Table 2 gives estimates of the national water saving of the North Africa countries expressed in volumes of virtual water embedded in net cereals trade. Seeing that the cereals trade is varying significantly from year to year, the figure presents the annual average for 2003-2007.

Furthermore all respective countries are net virtual water importers, there is a wide range of national water saving from 4.93 billion m^3 in Egypt to 44.23 billion m^3 in Algeria. The percentages of the national water saving through cereals net trade, to the total fresh water resources availability in each NA countries do extremely vary from around 9 in Egypt to 2136 in Libya. One can intuitively explain such a wide span in these percentages due to the fact that Egypt and Sudan are mainly representing irrigated agriculture countries while the rest countries are mainly rain fed agricultures. In addition, Libya is one of the driest countries in the world. The annual rainfall is very low with more than 95% of the country receiving less than 100 mm y^{-1} (Wheida & Verhoeven, 2007).

	Barley	Wheat	Maize	Rice	Sorghum	Others	Total	%
Algeria	134.5	24205.7	19489.6	370.4	2.3	32.5	44234.9	378.1
Egypt	1.5	2755.5	3465.0	-1307.0	NR	15.0	4930.0	8.6
Libya	921.7	5787.0	3668.0	912.9	NR	1524.6	12814.1	2135.7
Morocco	1361.4	5597.5	17375.9	5.5	NR	101.2	24441.4	84.3
Sudan	NR	4557.4	520.6	334.4	2160.1	15.5	7588.1	11.8
Tunisia	1581.5	2506.6	6368.4	30.2	326.9	7.3	10820.7	235.2
Region	4000.5	45409.6	50887.4	346.4	2489.3	1696.0	104829.2	62.5

% is the percentage national water saving achieved by each country to the total water resources in the same country.

Table 2. National water saving (10^6 m^3) achieved by the North Africa net trade of cereals crops, average 2003-2007

Figure 7 show that maize and wheat are the prominent cereal crops in terms of net virtual water imports. Wheat representations vary from around 23% in Tunisia to around 60% in Sudan. In summation, wheat represents about 44% of the national water saving in cereal net trade in NA region. On the other hand, maize representations vary from 7% in Sudan to 71% in Morocco. On the region level, maize is the highest cereal crop in terms of virtual water imports by around 50%. From what has been said, one can conclude that maize and wheat are representing around 91% of the national water saving in the North Africa region. In addition, all crops in the region and in each individual country attain national water saving except rice in Egypt. That because Egypt is a rice net exporter.

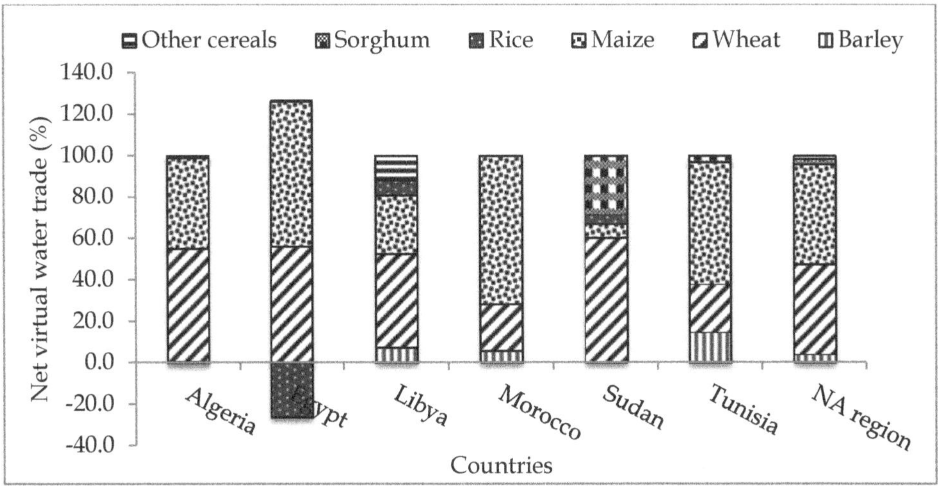

Fig. 7. Percentages of net virtual water trade embedded in each cereal crop, in the North African countries, average 2003-2007

3.6.2 Global water saving through North Africa cereals trade.

The international trade theory confirms that the global food import is approximately equal to the global food exports to achieve the global food market equilibrium over a given period. On the other hand, such equilibrium does not take place in the global virtual water trade associated with global food trade system. This imbalance is a result of the inequality of water used for producing a given amount of food between importing and exporting countries (Yang et al., 2006). Seeing that the global saving is obtained as the difference between the water productivities of the trading partners (Hoekstra & Chapagain, 2007), the positive sign on the difference, in the direction of exporting to importing countries, intuitively indicates global water saving. In contrast, the negative sign refers to global water loss.

Table 3 shows that the global saving resulting from international cereals trade between NA and the world countries is 77.2 billion cubic meter per year as on average of the period 2003 - 2007 . Approximately half of the amount of saved water originated only from Algeria. In addition, more than one quarter of the referred global water saving originated from Morocco. Thus, about 75% of the global water saving through the international trade of cereals crops in NA is derived only from Algeria and Morocco. One can see that there is notable global water saving achieved by cereal trade in all NA countries except in Egypt. The global water losing achieved by Egypt reflects a relatively high water productivity of cereal crops (Table 4). Furthermore, Egypt is the only country in the NA region with complete irrigation coverage. Consequently, the 100% irrigation coverage, adopting modern technology, and agricultural policy reforms explain its stable and significant increase of cereals production (Yang & Zehnder, 2002; El-Sadek, 2009). Considering the crop water requirements and yield, it has been noted that Egypt showed the highest cereal yield (7.5 ton ha^{-1}) among the NA countries. This yield is more than two times of the world average yield. On the other hand, the crop water requirements for cereal production in Egypt is more than the cereals water requirements of Libya, Morocco, and the world average (Table 4).

	Barley	Wheat	Maize	Rice	Sorghum	Others	Total	%
Algeria	24.1	18016.2	17853.8	286.9	2.1	16.7	36223.7	47.4
Egypt	-0.8	-3461.7	296.5	29.4	NR	26.5	-4003.8	-5.2
Libya			3360.1	707.2	NR	1251.5	10537.5	13.8
Morocco	897.3	2597.5	16206.0	-0.9	NR	85.7	19706.3	25.8
Sudan	NR	2990.0	485.3	239.3	2033.8	12.9	5859.1	7.7
Tunisia	913.4	975.9	5857.7	22.1	292.6	40.0	8133.2	10.6
Region	2554.3	25520.3	44059.5	1284.0	2328.5	1433.3	76455.9	

% is the percentage of global water saving achieved by each country to this achieved by the NA region

Table 3. Global water saving (10^6 m^3) achieved by North Africa net trade of cereals crops, average 2003-2007

	Crop water requirement (m^3/ha)	Yield (ton/ha)	Specific water demand (m^3/ton)
Algeria	8498	1.4	6007
Egypt	4495	7.5	598
Libya	3808	0.6	6123
Morocco	6617	1.2	5611
Sudan	3036	0.6	4775
Tunisia	6034	1.5	4094
World	3574	3.3	1088

Table 4. Crop water requirements, crop yields and the specific water demand of cereals in North Africa countries and in the world. Period 2003-2007.

With respect to the global water saving achieved by individual cereal crop trade in NA, it is obvious that maize trading produces 44.1 billion m^3 y^{-1} representing about 57.2% of the global water saving achieved by the region cereals trade. In the same context, wheat trading achieves 25.5 billion m^3 y^{-1} representing about 33.1%. Therefore, only maize and wheat trading are representing 90.18% of all total global water saving.

Although water scarcity is not the main driver of cereals trade on the global level, it plays an essential role in cereal trade between North Africa and the other trade partner countries in the world. At the national level, all North Africa countries achieve water saving to the extent that exceed the total fresh water resources in some cases. Trading of maize and wheat is the most important player of the national water saving among the cereal crops trade.

At the global level, reductions in global water use occur if production by the exporter is more water efficient than by the importer. All countries in the North Africa region achieve global water saving except Egypt. The global water losing achieved by Egypt reveals a relatively high water productivity of cereal crops.

4. Conclusion

The results of this chapter contribute to a better understanding of the role of evapotranspiration in saving water on the farm level and on the national and global levels through their different applications. On the farm level, the optimal coupling combinations of water application rate as ET deficit with irrigation frequency or with GB concentrations play important roles to seek maximum yield and efficient irrigation water used simultaneously under deficit irrigation conditions. At the national level, all North Africa countries achieve water savings to the extent that exceed the total fresh water resources in some cases. Trading of maize and wheat is the most important player of the national water saving among the cereal crops trade. At the global level, reductions in global water use occur if production by the exporter is more water efficient than by the importer. All countries in the North Africa region achieve a global water saving except Egypt. The global water losing achieved by Egypt reveals a relatively high water productivity of cereal crops.

5. References

Agboma, P. C., Sinclair, T. R., Jokinen, K., Peltonen-Sainio, P. & Pehu, E., (1997). An evaluation of the effect of exogenous glycinebetaine on the growth and yield of soybean: timing of application, watering regimes and cultivars. Field Crops Res. 54, 51-64.

Al-Kaisi, M.M. & Yin, X., (2003). Effects of nitrogen rate, irrigation rate, and plant population on corn yield and water use efficiency. Agron. J. 95, 1475-1482.

Allan, J.A. (1998) Virtual water: A strategic resource global solutions to regional deficits. Ground Water 36:545-546.

Allen, R.G., Pereira L.S., Raes D. & Smith M. (1998) Crop evapotranspiration - Guidelines for computing crop water requirements. FAO Irrigation and drainage paper 56.

Ashraf, M., & O'Leary, J.W., (1996). Effect of drought stress on growth, water relations and gas exchange of two lines of sunflower differing in degree of salt tolerance. Int. J. Plant Sci. 157, 729–732.

Assouline, S., (2002). The effects of micro drip and conventional drip irrigation on water distribution and uptake. Soil Sci. Soc. Am. J. 66,1630 –1636.

Bergez, J.E., Deumier, J.M., Lacroix, B., Leroy, P. & Wallach, D., (2002). Improving irrigation schedules by using a biophysical and a decisional model. Eur. J. Agron. 16, 123–135.

Brand, D., Koti, S., & Reddy, K.R., (2007). Application of glycinebetaine to alleviate drought and low temperature stress for soybean production system in Mississippi. Tri-State Soybean Forum. 5 January 2007. Vicksburg, MS, USA

Cetin, O. & Bilgel, L., (2002). Effects of irrigation methods on shedding and yield of cotton. Agric.Water Manage. 54, 1–15.

Chapagain A.K, Hoekstra A.Y. & Savenije H.H.G. (2006) Water saving through international trade of agricultural products. Hydrology and Earth System Sciences 10:455-468.

Chen, C, Payne, W.A., Smiley, R.W. & Stoltz, M.A., (2003). Yield and water-use efficiency of eight wheat cultivars planted on seven dates in northeastern Oregon. Agron J. 95,836–843.

Doorenbos, J. & Kassam, A.H., (1979). Yield response to water. FAO Irrigation and Drainage Paper 33. Rome, Italy: UN-FAO.

El-Hendawy, S.E., Hokam, E.M. & Schmidhalter, U., (2008). Drip irrigation frequency: the effects and their interaction with nitrogen fertilization on sandy soil water distribution, maize yield and water use efficiency under Egyptian conditions. J.Agron. Crop Sci. 194, 180-192.

El-Hendawy, S.E. & Schmidhalter, U., (2010). Optimal coupling combinations between irrigation frequency and rate for drip-irrigated maize grown on sandy soil. Agric. Water Manage. 97, 439- 448.

El-Sadek A. (2009) Virtual Water Trade as a Solution for Water Scarcity in Egypt. Water Resources Management.

Farre´, I.& Faci, J. M., (2009). Deficit irrigation in maize for reducing agricultural water use in a Mediterrranean environment. Agric. Water Manage. 96, 383–394.

Fereres, E. & Soriano, M.A., (2007). Deficit irrigation for reducing agricultural water use. J. Exp. Bot. 58, 147–159.

Grimes, D.W.; Yamada, H. & Dickens, W.L., (1969). Functions for cotton production from irrigation and nitrogen fertilizer variables. I. Yield and evapotranspiration. Agron. J. 61 (5), 769–773.

Gulati, H.S. & Murty, V.V.N., (1979). A model for optimal allocations of canal water based on crop production functions. Agric. Water Manage. 2 (1), 79–91.

Hexem, R.W. & Heady, E.O., (1978). Water Production Functions for Irrigated Agriculture. Iowa State University Press, Ames, IA, 215 pp.

Hoekstra, A.Y. (2003) Virtual water: An introduction. In Virtual Water Trade: Proceedings of the International Expert Meeting on Virtual Water Trade, in: A. Y. Hoekstra (Ed.),

Value of Water Research Report Series, UNESCO-IHE, Delft, The Netherlands. pp. 13-23.

Hoekstra, A.Y. & Chapagain A.K. (2007). The water footprints of Morocco and the Netherlands: Global water use as a result of domestic consumption of agricultural commodities Ecological Economics 64:143-151. DOI: 10.1016/j.ecolecon.2007.02.023.

Hoekstra, A.Y. & Hung P.Q. (2002) Virtual water Trade: A quantification of virtual water flows between nations in relation to international crop trade, UNESCO-IHE, DELFT, Netherlands.

Hoekstra, A.Y. & Hung P.Q. (2005) Globalisation of water resources: international virtual water flows in relation to crop trade. Global Environmental Change:45-56.

Hussain, M.; Malik, M. A., Farooq, M., Khan, M. B., Akram, M. & Salee, M. F. (2009). Exogenous glycinebetaine and salicylic acid application improves water relations, allometry and quality of hybrid sunflower under water deficit conditions. J.Agron. Crop Sci. 195, 98-109.

Ma, Q.Q., Wang, W.; Li, Y. H.; Li, D.Q. & Zou, Q., (2006). Alleviation of photoinhibition in drought-stressed wheat (*Triticum aestivum*) by foliar-applied glycinebetaine. J. Plant Physiol. 163, 165–175.

Monteith, J. L & Unsworth MH. (1990). Principles of environmental physics, 2nd edn. London: Edward Arnold

Oki T. & Kanae S. (2004) Virtual water trade and world water resources. Water Science and Technology 49:203-209.

Paolo, E.D. & Rinaldi, M., (2008). Yield response of corn to irrigation and nitrogen fertilization in a Mediterranean environment. Field Crop Res. 105, 202-210.

Postel, S.L., Daily G.C. & Ehrlich P.R. (1996) Human appropriation of renewable fresh water. Science 271:785-788.

Payero, J.O., Melvin, S.R., Irmak, S., & Tarkalson, D., (2006). Yield response of corn to deficit irrigation in a semiarid climate. Agric. Water. Manage. 84,101–112.

Sallam, G.A. & Abd El Nasser G. (2006) Implications of Virtual Water Concept on Water Demand Management, Riyadh- Kingdom of Saudi Arabia.

Uçan, K., Kıllı, F., Gençoğlan, H & Merdun, H., (2007). Effect of irrigation frequency and amount on water use efficiency and yield of sesame (Sesamum indicum L.) under field conditions. Field Crop Res. 101, 249-258.

Wang, F.X.; Kang, Y. & Liu, S.P., (2006). Effects of drip irrigation frequency on soil wetting pattern and potato growth in North China Plain. Agric. Water Manage. 79, 248-264.

Wheida, E. & Verhoeven R. (2007) The role of "virtual water" in the water resources management of the Libyan Jamahiriya. Desalination 205:312-316.

Xing, W., & Rajashekar, C.B. (1999). Alleviation of water stress in beans by exogenous glycinebetaine. Plant Sci. 148, 185–195.

Yang, H.; Wang L., Abbaspour K.C. & Zehnder A.J.B. (2006). Virtual water trade: an assessment of water use efficiency in the international food trade. Hydrology and Earth System Sciences 10:443-454.

Yang, H. & Zehnder, A.J.B. (2002) Water scarcity and food import: A case study for southern Mediterranean countries. World Development 30:1413-1430.

11

Evapotranspiration and Crop Water Stress Index in Mexican Husk Tomatoes (*Physalis ixocarpa* Brot)

Rutilo López- López[1], Ramón Arteaga Ramírez[2], Ignacio Sánchez-Cohen[1], Waldo Ojeda Bustamante[3] and Victor González-Lauck[1]

[1]Instituto Nacional de Investigaciones Forestales Agrícolas y Pecuarias, Km. 1 Carretera Huimanguillo-Cárdenas. 86400, Huimanguillo, Tabasco,
[2]Postgrado en Ingeniería Agrícola y Uso Integral del Agua. Universidad Autónoma Chapingo. Km 38.5 Carretera México-Texcoco. 56230, Chapingo, Texcoco,
[3]Instituto Mexicano de Tecnología del Agua, Paseo Cuauhnáhuac 8532, Progreso, Jiutepec, Morelos, 62550, México

1. Introduction

The scarcity of water availability observed in some dams in Mexico during the last years and the over-exploitation of groundwater have pushed to establish strategies for a rational and efficient use of water resources. The modernization and rehabilitation of irrigation systems stands out among these strategies, in order to improve their efficiency and profitability.

Irrigation scheduling is a decision process used to estimate the amount and timing for applying irrigations, in order to minimize deficiencies or excess in soil moisture, which could cause adverse effects in growth, yield and quality of crops; however, irrigation is usually applied without any technical advice for farmers, and rather solely on an empirical basis. Crop irrigation scheduling should consider diverse factors, such as water requirements and growth characteristics for each species and variety, atmospheric evaporation demand, and physico-chemical and biological conditions of the soil that determine its water retention capacity, since, in addition to the effective root depth, these determine the amount of water that can be used in the crop's evapotranspiration process.

Values for crop evapotranspiration and crop water requirement are identical (Allen et al., 1998); however, crop water requirement refers to the amount of water that needs to be supplied, while crop evapotranspiration refers to the amount of water that is lost through evapotranspiration (ET). Evapotranspiration includes two processes that occur simultaneously in the soil-plant-atmosphere system, and there is no easy way of separating these processes: loss of water from the soil through evaporation and from the plant through transpiration (Burman & Pochov, 1994).

In order to estimate evapotranspiration of a specific crop, it is necessary to take into account some crop characteristics and environmental conditions. Meteorological conditions determine the evaporative demand of the atmosphere, while the crop canopy and the soil humidity determine the magnitude at which the demand will be satisfied. Crop evapotranspiration can be calculated according to the FAO methodology (Allen et al., 2006) in the basic form of $ET_c = K_c ET_o$, where K_c is the crop coefficient. Then, crop evapotranspiration can be estimated if reference evapotranspiration (ET_o) measurements or estimations are available. These terms represent the meteorological demand (ET_o), and the ability of these plants and the soil to satisfy this demand (Jensen & Wright, 1978).

ET_0 is a climatic parameter expressing the atmospheric evaporation capability. According to FAO (Allen et al., 1998), reference evapotranspiration (ET_o) is defined as the maximum amount of water that a hypothetical reference crop loses; at a height of 0.12 m, a surface resistance of 70 s m^{-1} and an albedo of 0.23 m, which is similar to what happens on an extensive green grass surface of uniform height that actively grows and is well-irrigated.

Among the methods used to measure evapotranspiration on a cultivated surface, the use of a lysimeter stands out, which measures evaporation directly from the soil when it is bare, or evapotranspiration from the plants when there is a crop established. A lysimeter is a large container full of soil, generally installed in the field to represent natural environmental conditions, and where the water-soil-plant system conditions can be regulated at convenience and measured with more precision than in the natural soil profile (Hillel, 1980). This method contributes a direct measurement of the crop evapotranspiration and is frequently used to study climate effects and to evaluate estimation methods. When a weighing lysimeter is not available, the water balance method is commonly used in the field, which allows calculating the actual crop evapotranspiration (ET_c), which allows to estimate the soil moisture loss in the soil-plant-atmosphere system, of special importance for irrigation scheduling purposes (Lubana et al., 2001).

Various approaches and methods have been used for irrigation scheduling: direct and indirect measuring of soil humidity, measuring the energetic state of water in the soil, estimating the atmospheric demand and, in experimental conditions, determining plants' water potential (Buchner et al., 1994) or infrared thermometry (Giuliani et al., 2001). However, diverse experimental methods have been used to obtain the ET_0 from meteorological information and the ET_c from crop coefficients (Kc), which have generated different types of curves (Doorenbos & Pruitt, 1977; Jensen, 1981; Burman & Pochov, 1994; Allen et al., 1998; Dodds et al., 2005). In this study, the hypothesis is that with the soil's matric potential values in conditions of maximum water availability, it is possible to estimate the crop evapotranspiration and then the crop coefficients for husk tomato (*Physalis ixocarpa* Brot).

Idso et al. (1981) developed the Crop Water Stress Index (*CWSI*), an empirical method used to quantify the humidity tension in crops under arid conditions, which first depends on the determination of two baselines: with or without water stress. The baselines are specific for the crop and are influenced by the climate (Bucks et al., 1985). Jackson et al. (1981) modified the *CWSI* method by including more variables: vapor pressure deficit (*VPD*), net radiation (R_n) and aerodynamic resistance (r_a), in order to obtain a better theoretical prediction of the effects of climate on the crop temperature. This approximation is better than the empirical method, especially in humid climates (Keener & Kircher, 1983).

The *CWSI* method has had great practical use for irrigation scheduling for crops in arid and semi-arid environments (Calado et al., 1990; Itier et al., 1993; Anconelli et al., 1994; Jones, 1999; Orta et al., 2003; Yuan et al., 2004; Şimşek et al., 2005; Erdem et al., 2005). This is due primarily to the fact that the sensors required are easy to handle. In irrigated agriculture, the economic and ecological water cost is high when uncertainty in water availability is considered, something that can increase with climate change; therefore, the cost of the sensors used to quantify needed data, related to climate variables and water stress, can justify the investment associated with the method (Feldhake et al., 1997).

Water evaporated by a plant surface has the function of stabilizing the leaves' temperature in response to the atmospheric evapotranspiration. Based on this fact, Jackson et al. (1981) present the theory behind the energy balance which partitions net radiation from the sun into two components: sensible heat that heats the air and latent heat that is used for transpiration. When a crop goes through water stress, stomata close and transpiration decreases, and therefore, leaf temperature increases. When there is no water stress a plant transpires completely, and the leaf temperature fluctuates from 1 to 4° C less than the air temperature; in this case, the *CWSI* is zero. When transpiration decreases, the leaf temperature increases and can reach 4 to 6 °C more than the air temperature. In case of high water deficit, transpiration from the leaves is drastically reduced with simultaneous increase in leaf temperature; when the plant is dead or is not transpiring for a long time, the *CSWI* is one (Jackson et al., 1982).

The objectives of the study were: i) Determine the inferior and superior baselines of the *CWSI* method in husk tomato crop for irrigation scheduling; ii) Understand the effect of the irrigation depth and plastic mulching in different crop phenological stages in response to the water stress index; iii) Determine the tomato husk crop evapotranspiration from the soil matric potential and the loss of soil moisture measured by the lysimeter; iv) Propose an alternative way for Kc calculation using the foliar area index.

2. Theory of the crop water stress Index (*CWSI*) and evapotranspiration

The Crop Water Stress Index (CWSI) is a measure of the relative transpiration rate occurring from a plant at the time of measurement, using data from plant temperature (T_c) and vapor pressure deficit which is a measurement from the air dryness. The CWSI, as proposed by Idso (1981) and Jackson et al. (1981), is defined by the following relationship:

$$CWSI = \frac{\left[(T_c - T_a)_m - (T_c - T_a)_{li}\right]}{\left[(T_c - T_a)_{ls} - (T_c - T_a)_{li}\right]} \quad (1)$$

where T_c is the crop temperature and T_a is the air temperature. The "m" subscript denotes the difference between the two measured temperatures, *li* (inferior limit) denotes the non-water stress baseline expressed as the difference between the two temperatures when evapotranspiration is not restricted by water availability, and *ls* (superior limit) denotes the hypothetical non-transpiring upper baseline expressed as the difference between the two temperatures when evapotranspiration is zero.

The crop water stress index is estimated by determining the relative distance between the lower baseline representing non-stress conditions (well-irrigated condition) and the upper baseline representing no-transpiration (totally stressed condition). It is assumed for the *CSWI* to vary between 0 and 1. Since it is not normally feasible to measure crop temperature without stress and a crop with stress simultaneously, the values for the inferior and superior limit of a canopy of interest can be calculated through an energy balance analysis on the surface. This energetic balance can be expressed as:

$$R_n = G + H + \lambda E \quad (2)$$

where R_n is the net radiation (Wm-2), G is the heat flow on the soil surface (Wm-2), H is sensible heat flow in the air (Wm-2), and λE is the latent heat flow (Wm-2). The terms H and λE in equation (2) are a function of temperature gradients and vapor pressure, respectively, and can be expressed as:

$$H = \frac{\rho_a C_p (T_c - T_a)}{r_a} \tag{3}$$

and

$$\lambda E = \frac{\rho_a C_p (e_s - e_a)}{\gamma (r_a + r_c)} \tag{4}$$

where ρ_a is the water density (kg m-3), C_p is the specific heat in the air (J kg-1 °C), e_s is the saturation water vapor pressure at a temperature of T_c (kPa), e_a is the actual air water vapor pressure (e_a), γ is the psicrometric constant (kPa°C-1), r_a is the aerodynamic resistance (sm-1), and r_c is the canopy resistance to the water vapor flow (s m-1).
Equation (2) can be simplified by assuming that G = $0.1\ R_n$ (Feldhake et al., 1996), and therefore, by defining I_c as coefficient of radiation interception equal to 0.9, so that equation (2) becomes:

$$I_c R_n = H + \lambda E \tag{5}$$

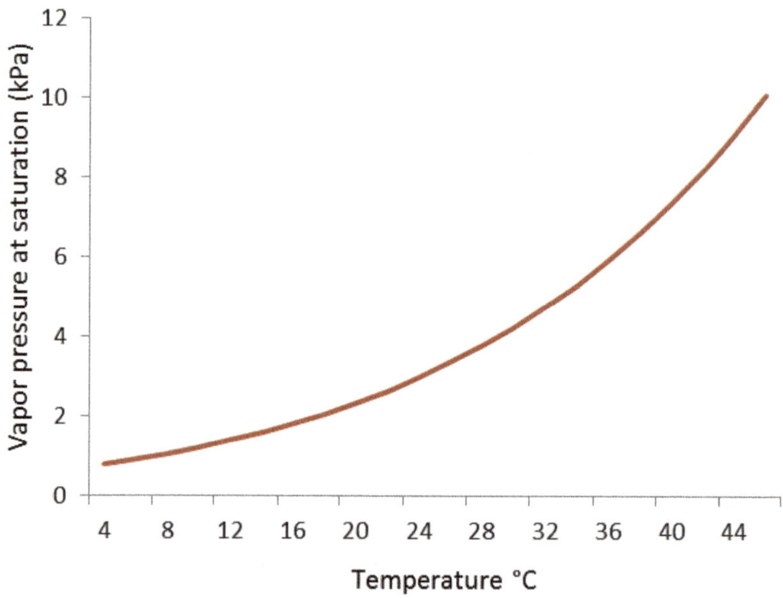

Fig. 1. Relationship between temperature and vapor pressure at saturation

When a volume of air is retained over an evaporating surface of water, equilibrium is reached between the water molecules that are incorporated into the air and those returning to the water source. At that moment, it is considered that the air is saturated, for it cannot hold any additional molecules of water. The corresponding pressure is called water vapor pressure at saturation (e_s). The amount of water molecules that can be stored in the air depends on the temperature (T). The higher the air temperature is, the higher the capacity to store water vapor and the higher the vapor pressure at saturation (Fig. 1). The curve slope changes exponentially with temperature. At low temperatures, the slope is small and varies slightly with the increase in temperature. At high temperatures, the slope is greater and slight changes in temperature produce great changes in the slope. The saturation vapor pressure curve slope (Δ) is an important parameter related with the water evaporation and is used in the calculation of reference evapotranspiration (ET_o) using weather data.
The value of Δ (kPa°C^{-1}) can be defined as:

$$\Delta = \frac{(e_s - e_a)}{(T_c - T_a)} \tag{6}$$

Jackson *et al.* (1988) found that the slope (Δ) can be calculated with:

$$\Delta = \left[45.03 + 3.014T_c - T_a + 0.05345(T_c - T_a)^2 + 0.00224(T_c - T_a)^3\right]10^{-3} \tag{7}$$

Combining equations 3, 4, 5 and 6, the temperature difference between the crop and the air can be estimated with equation (8):

$$T_c - T_a = \left[\frac{r_a I_c R_n}{\rho_a C_p}\right]\left[\frac{\gamma\left(1+\frac{r_c}{r_a}\right)}{\Delta + \gamma\left(1+\frac{r_c}{r_a}\right)}\right] - \frac{e_s - e_a}{\Delta + \gamma\left(1+\frac{r_c}{r_a}\right)} \tag{8}$$

Calculating the superior limit $(T_c - T_a)_{ls}$ when evapotranspiration is zero, r_c tends to the infinite and equation (8) is reduced to:

$$T_c - T_a = \frac{r_a I_c R_n}{\rho_a C_p} \tag{9}$$

When evapotranspiration is not limited by water availability and is equal to the reference rate, r_c approaches 0 and equation (8) is expressed as:

$$T_c - T_a = \left[\frac{r_a I_c R_n}{\rho_a C_p}\right]\left[\frac{\gamma}{\Delta + \gamma}\right] - \frac{e_s - e_a}{\Delta + \gamma} \tag{10}$$

Since the r_c does not really get to be zero in the reference evapotranspiration, the psicrometric constant γ is substituted by γ^* in equation (10):

$$\gamma^* = \gamma\left(1 + \frac{r_{cp}}{r_a}\right) \tag{11}$$

where r_{cp} (sm^{-1}) is the canopy resistance in reference evapotranspiration. The crop's resistance can be determined by the O'Toole & Real method (1986).
Aerodynamic resistance can be calculated with a semi-empirical equation, according to Thorn & Oliver (1977), which is:

$$r_a = \frac{4.72\left\{\ln\left[\frac{z-d}{z_0}\right]\right\}^2}{(1+0.54\mu)} \tag{12}$$

where z is the reference height (m), d the displacement height (m), z_0 the texture length (m), and μ the wind velocity (m s^{-1}). The terms z_0 and d can be calculated from the height of the plant (h), in crops with a full cover, these parameters are calculated with:

$$z_0 = 0.13h \tag{13}$$

and

$$d = 0.63h \tag{14}$$

Another way of developing the energy balance equation to predict the difference in temperature between the crop and the air (T_c - T_a), is by arranging the terms of the superficial energy balance (Jackson et al., 1981):

$$T_c - T_a = X_1 X_2 - X_3 \tag{15a}$$

$$X_1 = \frac{r_a(R_n - G)}{(\rho_a C_p)} \tag{15b}$$

$$X_2 = \frac{\left[\gamma\left(1+\frac{r_c}{r_a}\right)\right]}{\left[\Delta+\gamma\left(1+\frac{r_c}{r_a}\right)\right]} \tag{15c}$$

$$X_3 = \frac{(e_s - e_a)}{\left[\Delta+\gamma\left(1+\frac{r_c}{r_a}\right)\right]} \tag{15d}$$

where all the terms are previously defined. Thus, equation (15b) is equal to equation (9) and they can be used to obtain the *CWSI* superior limit (dT_{ls}), where crop resistance (r_c) approaches the infinite. Equations (15c) and (15d) are used in the case of a non-water stressed crop (inferior limit), where r_c is assumed to be equal to zero.
The *CWSI* can also be expressed in terms of evapotranspiration, based on Jackson et al. (1981):

$$CWSI = 1 - \frac{ETc}{ET_0} \tag{16}$$

where ET_c is the actual crop evapotranspiration and ET_o is the reference evapotranspiration. Substituting its values, there is:

$$CWSI = \frac{\gamma\left(1+\frac{r_c}{r_a}\right)-\gamma\left(1+\frac{r_{cp}}{r_a}\right)}{\Delta+\gamma\left(1+\frac{r_c}{r_a}\right)} \tag{17}$$

Where the r_c/r_a relation is expressed as:

$$\frac{r_c}{r_a} = \frac{\frac{r_a R}{\rho C_p}-(T_c-T_a)(\Delta+\gamma)-(e_s-e_a)}{\left[\gamma\left((T_c-T_a)-\frac{r_a R_n}{\rho C_p}\right)\right]} \tag{18}$$

r_{cp} is the canopy resistance at potential transpiration, r_c is the canopy's actual resistance, r_a is the aerodynamic resistance to the transport of sensible heat in the air; the other parameters and variables have already been defined.

The relation between actual (ET_c) and reference (ET_o) evapotranspiration approaches more the theoretical values proposed by Jackson (1982), because it does not depend as much on the conditions of wind velocity as in the classical form, according to Idso et al. (1981). When replacing the *VPD* estimations from minimum temperature (Idso, 1982) with air temperature measurements in direct radiation on irrigated plots, not only does the correlation with the actual evapotranspiration increase, but in addition, it partially corrects the influence of the superficial soil temperature in small values for the leaf area index (*LAI*). Therefore, the estimation precision for evapotranspiration through the *CWSI* is given by the relation between actual evapotranspiration (ET_c) and the leaf's water potential (ET_o) at dawn (Itier et al., 1993).

3. Methodology applied for the water stress index and evapotranspiration in crops

3.1 Crop Water Stress Index (*CWSI*) determination

Canopy surface temperature was measured with infrared thermometer and later used to estimate the crop water stress index (*CWSI*) for husk tomato (*Physalis ixocarpa* Brot.) produced under drip irrigation, to study the effect of irrigation depth and plastic mulching. A completely randomized experimental design with three repetitions was used. The effects of five irrigation depths were studied; irrigation depth reposition of 40, 60, 80, 100 and 120 % of reference evapotranspiration (ET_O), estimated with the Penman-Monteith. The *CWSI* was calculated from temperature measurements for the crop and the air, and relative humidity measured with an infrared ray gun. Then, the vapor pressure deficit (*VPD*) was estimated. The equation that defines the *CWSI* inferior limit expresses the relation between the *VDP* and the difference in crop and air temperature (T_c-T_a).

Measured or actual evapotranspiration (*E*) divided by reference evapotranspiration (E_p), defined in equation (16) when working out ETc/ET_O is:

$$\frac{ETc}{ET_0} = 1 - CWSI$$

Due to the differences in measurements of T_c-T_a vs *VPD*, the crop does not require irrigation until the *CWSI* reaches a threshold value, which can be from 0.1 to 0.2, depending on the crop. During this time, the crop is transpiring at a lower rate than the optimal and the crop yield begins to decline. The lower limit of a crop in a specific place can be determined two days after a maximum irrigation depth is applied on the crop.
Infrared thermometers or thermal seekers are used to measure the crop superficial temperature (Fig. 2). They measure the quantity of long wave radiation emitted by a surface as described by Stefan-Boltzman black body law in function of the temperature.

$$I = \varepsilon \sigma T^4 \quad (19)$$

where I is the radiation emitted by the surface (Wm^{-2}), σ is the Stefan-Boltzman constant (5.674×10^{-8} W m^{-2} K^{-4}), ε is the energy that a body emits at a given temperature, for a black body it is 1 and for others it is less than 1, T is the temperature on the surface (°K).

Fig. 2. Field measurement of parameters with infrared ray gun in husk tomato (*Physalis ixocarpa* Brot).

This CWSI method utilizes the temperature data for timing irrigation. A reduced form of expressing equation 1 is:

$$CWSI = \frac{(dT - dT_i)}{(dT_s - dT_i)} \quad (20)$$

where dT is the difference of measured air and crop temperatures; dT_s is the upper limit of the air temperatures minus the canopy temperature (crop without transpiration) and dT_i is the lower limit of temperatures in the air minus the canopy temperature (fully-irrigated crop).

In order to determine the upper and lower limits in the *CWSI* equation, the method developed by Idso et al. (1981) is used, which considers changes in both limits due to variations in the air vapor pressure deficit (*VPD*). The *VDP* is the difference between the saturation pressure and the actual vapor pressure (eq. 21) and it is a good indicator of the actual evaporating capacity of the air.

$$DPV = e_s - e_a \tag{21}$$

where e_s is the water vapor pressure at saturation at a given air temperature and e_a is the current water vapor pressure (water vapor partial pressure in the atmosphere). When the air does not become saturated, the current vapor pressure will be lower than the vapor pressure at saturation.

The water vapor pressure at saturation (e_s), in kPa, is the maximum amount of water vapor that air can hold at a given temperature (T in °C) and it is calculated with equation (22):

$$e_s(T) = 0.611\ \exp\left[\frac{17.27T}{T+237.3}\right] \tag{22}$$

The actual water vapor pressure e_a can be obtained from equation (23) if the relative humidity (*HR*) and the crop temperature are measured with the infrared ray gun.

$$HR = \frac{e_a}{e_s}100 \tag{23}$$

A *VPD* equal to zero indicates that the air holds the maximum water vapor possible (this corresponds to a relative humidity of 100%). The lower limit of the *CWSI* changes as a function of the water vapor pressure due to the *VPD*. The *CWSI* varies in the range between 0 and 1, when plants are subject to appropriate irrigation conditions and even to conditions of total water stress. Idso (1982) demonstrated that the lower limit of the *CWSI* is a linear function of the *VPD* for several crops. Once the parameters are estimated by the linear regression, the temperature difference between air and canopy can be calculated for a non-water-stressed crop (inferior limit) and a maximum stressed crop (superior limit), using the following two equations:

$$dT_i = a + b(DPV) \tag{24a}$$

$$dT_s = a + b\left[e_s(T_a) - e_s(T_a + a)\right] \tag{24b}$$

where *VPD* is expressed in kPa, $e_s(T_a)$ is the saturation vapor pressure at air temperature T_a (kPa), and $e_s\ (T_a + a)$ is the saturation vapor pressure at air temperature plus the value of the intercept for the crop. Thus, with the measurement of air humidity (relative humidity, wet bulb temperature, etc.), the air temperature and the leaf temperature, it is possible to determine the *CWSI*.

3.2 Crop evapotranspiration determination (ET_c)

In order to estimate the crop evapotranspiration, the equation must be expressed as:

$$ET_C = K_c K_s ET_o \qquad (25)$$

where K_c is the crop coefficient obtained experimentally; K_s is the soil water availability coefficient, which was assumed to be equal to one in this study; ET_o is the reference evapotranspiration, estimated through the Penman-Monteith equation (Allen et al., 1998) with the average values of ambient variables measured between 6:00 and 19:00 h with an automatic weather station. The equation for calculating ET_o is given by the following relation:

$$ET_o = \frac{0.408\Delta(R_n - G) + \gamma \frac{900}{T+273} u_2 (e_s - e_a)}{\Delta + \gamma(1 + 0.34 u_2)} \qquad (26)$$

where: R_n is the net radiation flux density on the crop surface (MJ m^{-2} d^{-1}); G is the soil heat flux density (MJ m^{-2} d^{-1}); T is the average daily air temperature (°C); u_2 is the wind speed at 2 m high (m s^{-1}); e_s is the saturation vapor pressure (kPa); e_a is the actual vapor pressure (kPa); Δ is the slope of vapor pressure-temperature curve (kPa °C^{-1}) and γ is the psychometric constant (kPa °C^{-1}).

3.2.1 Experimental site and genetic material used

The study was carried out in Chapingo, Estado de México, located geographically at 19° 29′ N and 98° 53′ W, and an altitude of 2250 m. The climate in the location corresponds to a sub-humid temperate climate with summer rains, a dry season in the winter and thermal variation between 5° and 7° C (Arteaga et al., 2006). The annual average temperature is 15.5° C, with May being the warmest month and January the coldest. The annual average precipitation is 664 mm.

Mexican husk tomato was cultivated with a drip irrigation system during the period between March and June in 2007. The "CHF1-Chapingo" variety was used, a variety released commercially by the University of Chapingo, Mexico.

3.2.2 Seedling production and transplant

The seeds were sown on February 24, 2007, Julian day (J_d) = 55, in polystyrene trays with 200 cavities filled with a substrate consistent with a mixture of peat moss and vermiculite. The transplant was performed on March 30 (J_d = 89), and the last harvest was performed on June 30 (J_d = 181). The planting arrangement was 1.5 m between lines and 0.45 m between plants, with a density of 16122 plants ha^{-1}.

3.2.3 Physical and chemical characteristics of the soil

The soil texture was clay loam with an apparent density of 1.25 and 1.36 g cm^{-3} at 0.1 and 0.3 m of depth, respectively, a real density of 2.35 at 0.1 m and 2.39 g cm^{-3} at 0.3 m of depth; soil moisture at field capacity (FC) was 29.6 % and at a permanent withering point (PWP), it was 16.5 %. The soil retention curve was determined in the lab with the pressure cell and membrane, from soil samples taken with the Uhland probe, which were later dried in the stove at 105 ° C for 24 h, for later obtaining the porous space and the saturation soil moisture. Later, the moisture content (W) was determined by weight difference, in function of successive suction changes (ψ) during a sample drying process (Figure 1). The results indicate that the soil has medium to high capacity for water retention, with a void space or volumetric saturation water content (θ s) that varies from 0.47 to 0.43 between 0.1 and 0.3 m of depth.

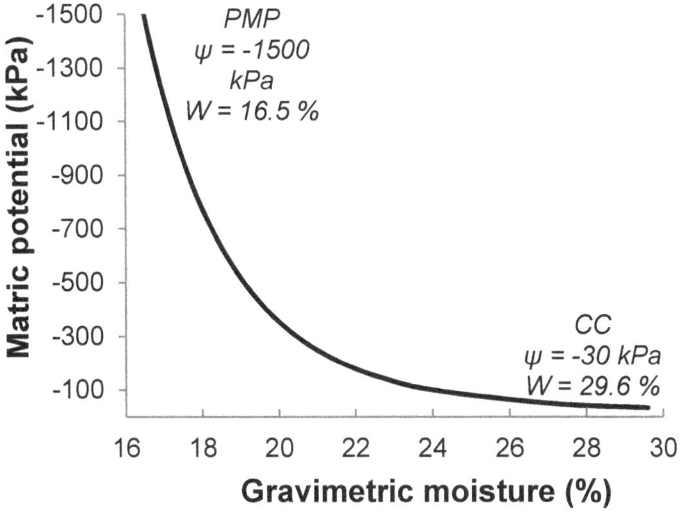

Fig. 3. Moisture retention curve in the experimental site.

The soil pH is practically neutral (6.99), moderately poor in organic material (1.48%), average in inorganic nitrogen (22.3 mg kg^{-1}), average in assimilable phosphorous (28.79 mg kg^{-1}), high in available potassium (646 mg kg^{-1}), moderately high in available calcium (2545 mg kg^{-1}), very high in available magnesium (1425 mg kg^{-1}), average in assimilable iron (8.51 mg kg^{-1}), moderately low in assimilable copper (0.79 mg kg^{-1}), average in assimilable zinc (1.44 mg kg^{-1}), moderately high in assimilable manganese (17.58 mg kg^{-1}), and very high in assimilable boron (2.63 mg kg^{-1}).

3.2.4 Treatments and experimental design

Five levels of irrigation depths were applied: 40, 60, 80, 100 and 120 % of the estimated ET_0, determined with the Penman-Monteith method (Allen et al., 1998) and two levels of silver-black plastic mulching, with and without. The treatments were distributed in a completely randomized design with three repetitions, where the depth-treatment area correspond to the area controlled by an irrigation valve, which was divided into two experimental units, one with mulching and the other without. The experimental unit for each treatment was 10 rows, 35 m long, separated at 1.5 m and 0.45 between plants.

The irrigation method was drip tape, with uniformity efficiency of 92%. The nominal characteristics are: inner diameter of 16 mm, 0.254 mm caliber, 1 L h^{-1} flow, space between transmitters of 0.3 m, and maximum pressure of 1200 kPa.

Characteristics of the plastic mulching were: 1.2 m width, for a 0.6 m bed, 2.28 mm caliber, partial perforation with a diameter of 0.063 m and 0.45 between spaces. The plastic laying was made with a mechanical mulching machine that contains devices for building the bed, fertilizing, and placing the tape and the plastic.

The watermark gypsum block sensors, with a range of measurement of 0-200 kPa, were installed at three depths (0.1, 0.3 and 0.4 m) in each treatment. A detailed description regarding the basic principles, characteristics, installation, and operating instructions can be found in Thompson et al. (2006).

3.2.5 Measuring humidity and estimating ET_c, ET_o and K_c

With the daily matric potential data measured with the watermark probes, which measure the content of water in the soil with a gravimetric or volumetric basis and the value, they are then transformed to matric potential using the soil moisture retention curve as shown on Fig. 1. The probes were installed at 0.1 and 0.3 m deep in the lysimetric container, and with the loss of measured soil moisture, a relation was obtained to estimate the ET_c as a function of matric potential. In order to estimate the crop's evapotranspiration, equation (25) was used, and the K_c was obtained experimentally; the K_s was assumed to be equal to one and the ET_0 was estimated through the Penman-Monteith equation (Eq. 26), with the average values of climatic variables measured between 6:00 and 19:00 h in an automatic station.
In order to estimate the crop evapotranspiration (ET_c), a weighing lysimeter was used, built with an undisturbed soil structure, and equipped with a mechanical-electronic system. The soil monolith had a prism-shaped, with a square base of 1.8 m per side and a depth of 1.5 m; the superior side of the monolith coincides with ground level, and the base has a draining system that allowed the exit of water. The lysimeter precision of the weighing system allowed detecting changes in weight that correspond to 0.15 mm of the water depth. The lysimeter soil water balance is indicated in Equation 27:

$$\pm\Delta S = P + R - (ET_c + I + E) \tag{27}$$

where ΔS is the change in the soil water content; P is precipitation; R is the irrigation contribution; ET_c the crop evapotranspiration; I is the deep infiltration or percolation; and E is the surface runoff.
Before establishing the experiment, the soil moisture loss was calibrated for the lysimeter with the matric potential of the first 0.3 m of depth, for which tensiometers were installed at 0.1 and 0.3 m deep. The incidence of rains was avoided and the calibration period lasted for 22 days. Based on the volumetric content at soil saturation (θ_s), which varies from 0.43 to 0.47, the irrigation depth was calculated for a specific depth (P_r) with equation 28:

$$L = (\theta_s - \theta_0)P_r \tag{28}$$

where θ_0 is the initial volumetric water content. Using the lysimeter area of 3.25 m^2, the water volume needed to saturate the top 30 cm of soil depth was calculated in 466 L.
The change in soil moisture in the lysimeter was determined daily through weight changes. The moisture loss was obtained through the weight difference between the previous hour or day and the current hour or day. The readings were registered every hour, from 8:00 to 18:00 h.
The crop evapotranspiration (ET_c) was obtained through linear models generated from experimental data from moisture loss in the weighing lysimeter (Y) and the matric potential (x). At a depth of 0.1 m: Y_1=-0.776x -1.028, R^2=0.96; at 0.3 m: Y_2=-1.362x -8.89, R^2=0.92 (Fig. 2a and 2b). Given that when the soil moisture content decreases, more energy is required to extract the water retained; these simple models are good estimators of the soil moisture loss as a function of the matric potential. They express that for each kPa of tension, there is a loss

of moisture in average of 0.78 mm in the layer 0.1 m deep; and at a depth of 0.3 m, the average loss of humidity is 1.36 mm for every kPa of tension. Finally, crop coefficients were estimated for each phenological stage, through equation (25).

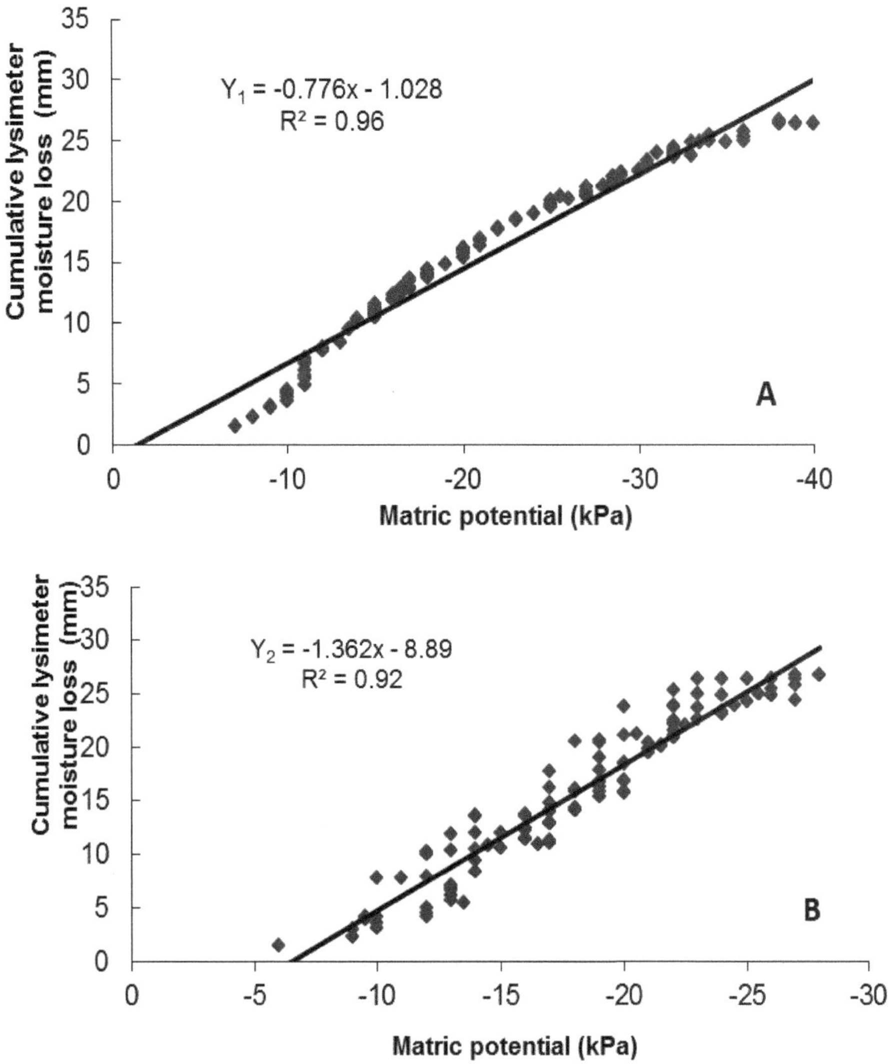

Fig. 4. Relationship between the matric potential at 0.1 m (A) and 0.3 m (B) depths and the lysimeter moisture loss in mm, obtained from experimental data during 22 continuous days. Each point of the observed values represents the water loss measured every 2 h between 8:00 and 18:00 h.

In order to relate the K_c and the leaf area index, the latter was determined in samples of 3 to 10 plants collected in the field and taken to the lab, starting on the date of transplant. Samples were taken on the following Julian days: 103, 117, 124, 138, 145, 159, 166 and 177. The leaf area was measured with a leaf area integrator, LICOR LI-3100 (LICOR, Inc. Lincoln, NE, USA).

4. Results and discussion

4.1 Upper and lower limits of the crop water stress index

Because the infrared ray gun requires sunny days to measure the water stress index, and as the method suggests that measurement must be done at the same time (from 12 to 15 pm) when the crop water demand is high. Data were taken for all treatments on Julian days: 123, 136, 145, 147, 152, 161, 162, 163, 164, 165, 166, 167 and 178. Measurements began on day 123 because it was the first day when crop water stress effects on the irrigation depth and the plastic mulching were observed.

Based on the method proposed by Idso et al. (1981), the parameters that define the lower and upper limits of the *CWSI* are presented in Figure 5. The equation that defines the lower *CWSI* baseline is: $T_c - T_a = 1.21 - 1.31VPD$ (r^2 =0.68, P<0.01, n=42), where T_c-T_a is in °C, and *VPD* in kPa. Idso (1982) reported the following relation for the lower limit in tomato crops: $T_c - T_a = 2.86 - 1.96\ VPD$. For corn, Irmak *et al.* (2000) found the following relation: $T_c - T_a = 1.39 - 0.86VPD$. It can be observed that all relations are different, which is in agreement with the results obtained by Bucks et al. (1985), who point out that the intercept and slope values vary depending on the climate, type of soil and crop being cultivated.

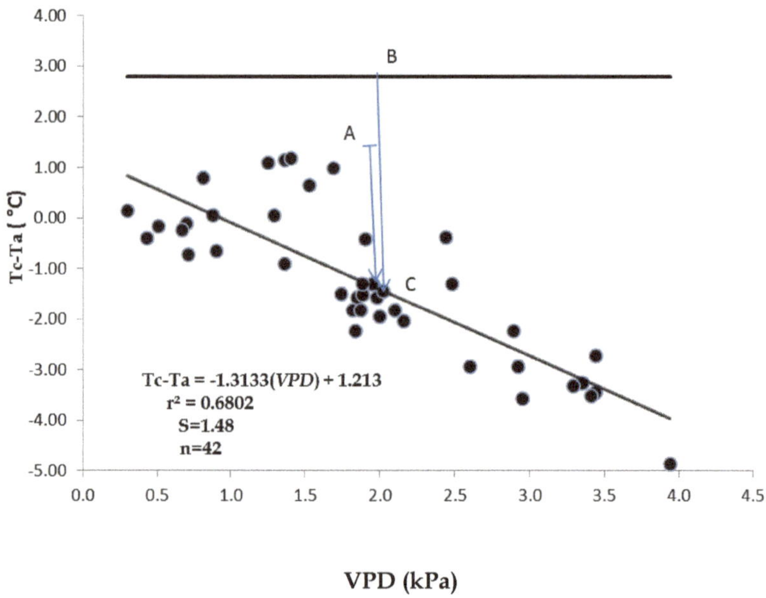

Fig. 5. The lower (A-C) and upper (B-C) baselines for husk tomato crop for determining the crop water stress index. S = Standard deviation.

The down-sloping line (Fig. 5) represents the baseline without water stress, that is, the difference between the air temperature and the crop temperature during periods with adequate water supply at different *VPD*; in this case, stomata were supposed to be open and temperature difference was a function of *VDP*, since an increase in VDP entails an increase in the drying power of the atmosphere and, therefore, in plant transpiration. The horizontal line (upper baseline, Fig. 5) is the difference between the air temperature and the crop temperature associated with periods of greater stress (with water limitations), when there is no transpiration. The average value was 2.8 °C with n = 25. For the corn crop, Irmak et al. (2000) determined an average value of 4.6 °C, a value greater than that found in this study, which means that husk tomato is more sensitive to possible water stress than the corn crop. A *VPD* equal to zero indicates that the air contains the maximum amount of water vapor possible (relative humidity = 100%). The lower limit of the *CWSI* changes as a function of vapor pressure due to the *VPD*. The *CWSI* varies between 0 and 1 when plants are subject to appropriate irrigation conditions and up to conditions of total water stress. The lower limit in this research was developed in a range of *VPD* of 0.3 to 4.0 kPa. Gardner & Shock (1989) suggest that it is necessary for the range of *VPD* to vary from 1 to 6 kPa in order to define the baseline that can be used in other locations.
Calculation of *CWSI* can be done in a graphical approach starting from the following relation: *CWSI* = *AC/BC*, where point *A* is the difference between the temperatures of the leaf minus the air at the moment of measuring, point *B* is the difference in maximum temperature between the leaf and the air (superior limit), and point *C* the minimum difference (inferior limit) in the *VPD* conditions in which temperature measuring was carried out for the leaf and the air (*A*). Therefore, the *CWSI* is determined by the relative distance between the lower line (*A-C*) that represents the conditions without stress, and the u line (*B-C*) where there is no transpiration. For example, in Figure 5, it is considered that point *A* has a value of T_c-T_a equal to 1.4 °C that corresponds to a value of *VPD* equal to 2.0 kPa. Starting from the definition by Idso (1981), the distance between point *A* and the inferior limit (*C*) is 2.8 °C, and the distance between the superior and inferior limit in 2.0 kPa is 4.2 °C. Thus, the *CWSI* is equal to the ratio of both relative distances 2.8/4.2 = 0.66. This means that a difference in temperatures of 1.4 °C between the crop and the air indicate possible problems of crop water stress.
The CWSI, estimated from infrared thermometry, can be used for crop irrigation scheduling. Various researchers have obtained the parameters to set the inferior and superior baselines for other crops (Idso, 1982, Jones et al., 1997; Orta et al., 2003 & Erdem et al., 2005).

4.2 Effect of the irrigation sheet and plastic mulching on the water stress index

The variance analysis showed that there are highly significant differences (P<0.01) of the effect of irrigation depth on the *CWSI* during the different crop phenological stages, and not so for the effect of plastic mulching, since it is only significant for the maturation stage (M), on days 161 and 165; the effect of the interaction was not significant (*P*>0.05) in the different stages of the crop development. According to the analysis of mean comparison (*P*<0.05), mulching had a mean of 0.15 to 0.2 in the vegetative stage (V), while without mulching, there was a mean value of 0.21 to 0.26 (Table 1). During the reproductive (R) and maturation stages, the mean values vary between 0.14 and 0.28 with mulching and from 0.27 to 0.33 without mulching (Table 2).

Mulching	Crop Water Stress Index (CWSI)		
	V 123 (3-05-07)[y]	*R.145 (25-05-07)*	*M.178 (27-06-07)*
No mulching (0)	0.21 a	0.21 a	0.33 a[z]
With mulching (1)	0.20 a	0.15 a	0.28 a
Mean	0.21	0.18	0.30
DSH	0.085	0.064	0.084
CME	0.012	0.007	0.012
CV(%)	54.15	45.76	36.38

[z] Values with the same letter within a column are equal according to the Tukey test, with $P<0.05$; DSH: Honestly Significant Difference; CME: Mean Square Error; and CV: Variation Coefficient.
[y] Notation V 123 (3-05-07) indicates that V is the vegetative stage, 123 the Julian day, and (3-05-07) the date corresponding to day, month and year.

Table 1. Effect of plastic mulching on the water stress index during different phenological stages in the husk tomato crop.

Irrigation depth	Crop Water Stress Index (*CWSI*)		
	V 123 (3-05-07)[y]	*R 145 (25-05-07)*	*M 178 (27-06-07)*
40 (0)	*0.40 a*	*0.37 a*	0.53 a[z]
60 (1)	0.17 b	0.22 b	0.42 ab
80 (2)	0.38 a	0.12 b	0.32 bc
100 (3)	0.08 b	0.10 b	0.11 cd
120 (4)	0.0 b	0.10 b	0.15 d
MEDIA	0.21	0.18	0.30
DSH	0.19	0.14	0.19
CME	0.012	0.007	0.012
C.V. (%)	54.15	45.76	36.38

[z] Values with the same letter within a column are equal according to the Tukey test, with $P<0.05$; DSH: Honestly Significant Difference; CME: Mean Square Error; and CV: Variation Coefficient.
[y] Notation V 123 (3-05-07) indicates that V is the vegetative stage, 123 the Julian day, and (3-05-07) the date corresponding to day, month and year.

Table 2. Effect of the irrigation depth on the water stress index during different stages of development of the husk tomato crop.

Table 2 shows the relation between irrigation depth and *CWSI*. In general, it can be noted that the treatment of 40% irrigation depth gave the highest *CWSI* values in the different stages of crop development, and is statistically significant from the other levels. The lowest *CWSI* values were obtained with the irrigation depth of 100 and 120% of the ET_0 treatments, being statistically equal to the 60 and 80% ET_0 treatments. This is because there was a normal water supply during the crop season. As water availability for the plant decreased, the CWSI value increased up to 0.7 in the treatment with severe irrigation restrictions (40%), without plastic mulching.
The functions that relate the water stress index with irrigation depth and plastic mulching were the following:
During the vegetative stage, with r^2=0.74, CME=0.025 and n=30:

$$CWSI_v = 0.73 - 0.135(a) - 0.418(b) + 0.061(b^2) + 0.037(ab) \quad (29)$$

during the reproductive stage, with r^2=0.66, CME=0.007 and n=30:

$$CWSI_r = 0.40 - 0.06(a) - 0.18(b) + 0.029(b^2) + 0.0003(ab) \quad (30)$$

and, finally, for the maturation stage with r^2=0.62, CME=0.020 and n=30:

$$CWSI_m = 0.59 - 0.267(a) - 0.232(b) + 0.028(b^2) + 0.52(ab) \quad (31)$$

where a is the effect of the plastic mulching: b represents the Penman-Monteith irrigation depth treatment; and ab is the effect of the interaction of plastic with the irrigation depth. The determination coefficients (r^2) are acceptable and indicate that the models predict the *CWSI* in an acceptable manner, and that the mean square errors are relatively small.

In figures 6a, 6b and 6c, the relations between irrigation sheets and plastic mulching with the *CWSI*, for the vegetative, reproductive and maturation stages, respectively, are presented. Equations 29, 30 and 31 are drawn, substituting the value of 0 without mulching and 1 with mulching, and the 0, 1, 2, 3 and 4 values that correspond to the irrigation sheet reposition: 40, 60, 80, 100 and 120%, respectively. The exponential models were generated from the values observed with the *CWSI* averages obtained with and without plastic padding.

The relation between the water stress index and the irrigation sheet is negative and exponential; as the irrigation sheet increases, the *CWSI* decreases until it reaches 0 when 100 or 120% of the ET_0 is applied. The differences between having mulching or not, in different phenological stages, indicate that the *CWSI* value with mulching is less than without plastic mulching. This is due primarily to the reduction of evaporation from the soil in treatments with plastic. In this regard, Şimşek et al. (2005) observed that when the irrigation sheet decreases, the rate of transpiration by the crop also decreases, producing as a result the increase in the crop's temperature and the *CWSI*; this produces a decrease in the crop yield.

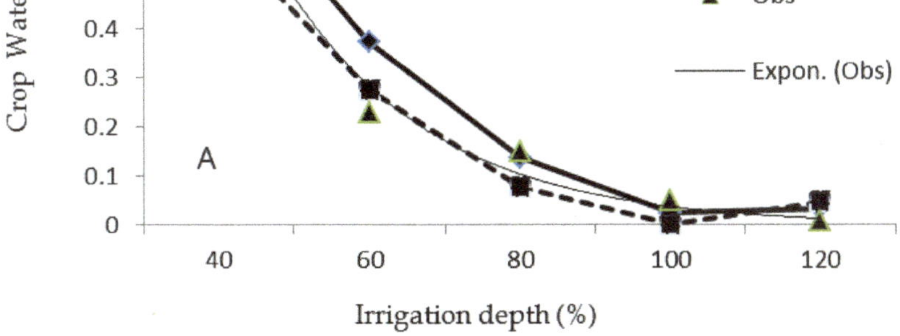

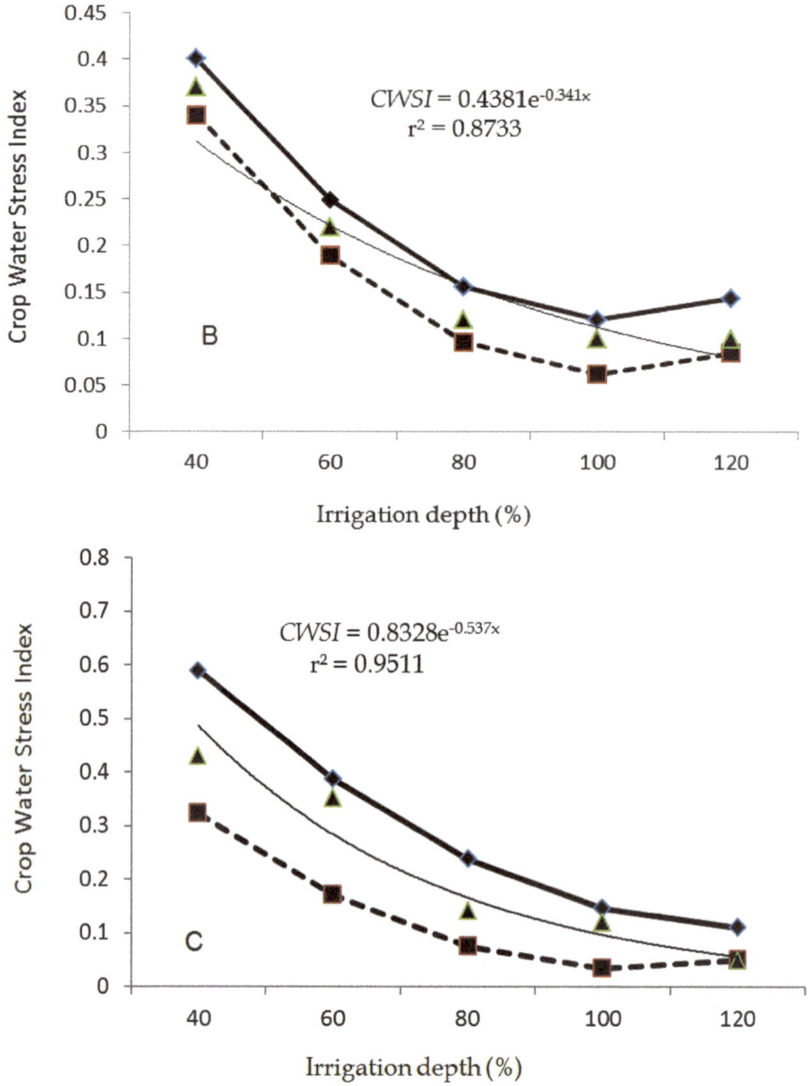

Fig. 6. Crop Water stress index estimation in husk tomato crop, during the following stages: A) vegetative, B) reproductive and C) maturation, without (SA) and with (CA) plastic mulching from the irrigation depth calculated as a fraction of the Penman-Monteith ET_o.

Prediction of the *CWSI* in the vegetative stage can be done through the exponential function: $y=2.1e^{-1.0x}$, with $r^2= 0.96$ determined from the average values associated to the irrigation depth effect on day 136; that is, taking into consideration the values obtained with and without plastic mulching. For the reproductive stage, an exponential function was obtained, $y=0.44e^{-0.34x}$, with $r^2= 0.87$, which was determined based on the average data observed on day 145, and during the stage of maturation the function $y=0.82e^{-0.53x}$, with $r^2= 0.95$, was determined according to the mean values observed on day 165 (Fig. 6a, 6b and 6c). The *CWSI* threshold varies according to the plastic mulching treatment, with and without. In general, it can be noted that with the irrigation depth calculated higher than 60% of the ET_0, the crop's water stress can be avoided, both with and without plastic mulching.

4.3 Crop coefficients per phenological stage

In this case, the crop coefficients were determined for the treatment with a depth equivalent to ET_0 = 100% and without plastic mulching. The crop coefficients were obtained through the relation between the crop evapotranspiration and the reference evapotranspiration (Eq. 26). The ET_c values were calculated daily from matric potential data of the top 0.1 m layer during the vegetative-reproductive stage and 0.3 m layer in the reproductive-fructification and maturation-senescence stages, those depths are the most active where roots take most of the water. When substituting these values in the linear equations generated in the weighing lysimeter to estimate the ET_c, the crop coefficients were obtained by phenological stage (Table 3).

Stage	Days	ET_c (mm)	ET_0 (mm)	Kc
Vegetative-Reproductive	45	50.20	165.60	0.30
Reproductive-Fructification	35	121.50	111.96	1.10
Maduration-Senescence	20	71.36	82.92	0.86

Table 3. Crop coefficients for husk tomato under drip irrigation in different phenological stages, as a function of days after transplant without plastic mulching.

The K_c values estimated here are similar to those proposed by Allen et al. (1998) for tomato (*Lycopersicon esculentum* L.) crops without plastic mulching, except during the initial stage where a higher value, of 0.6, was found.

In the cumulative curves for ET_c and ET_0 (Fig. 7), it can be noted that the ET_o has a linear behavior during the crop growth, while the ET_c does not produce a linear behavior and generally presents lower values than the ET_0. Evapotranspiration rates depend on crop management and phenological stage. The crop evapotranspiration is different of the reference evapotranspiration (ET_0), due to the differences in soil cover, vegetation properties and aerodynamic resistance, with respect to that of grass (Allen et al., 1998). The effects of the crop characteristics are incorporated into the crop coefficient. During the initial stage of growth, the main component is due to soil evaporation, particularly on its sun-exposed area, and the soil moist surface that has influence on the value of superficial resistance, which is the sum of soil resistance and root resistance. Immediately after moistening the soil from an irrigation or rain event, the rate of transference from vapor to water from the soil is high.

The crop coefficient (K_c) includes the effects of evapotranspiration on the soil and crop surfaces and depends on the water availability in the soil around the root area and the less moist soil surface exposed. These values are obtained from crops that are adequately irrigated, without water stress. Other duration time parameters can also be used, such as days of development, thermic-solar units, or cumulative ET_o (DeTar, 2004; Bandyopadhyay et al., 2005; DeTar, 2009).

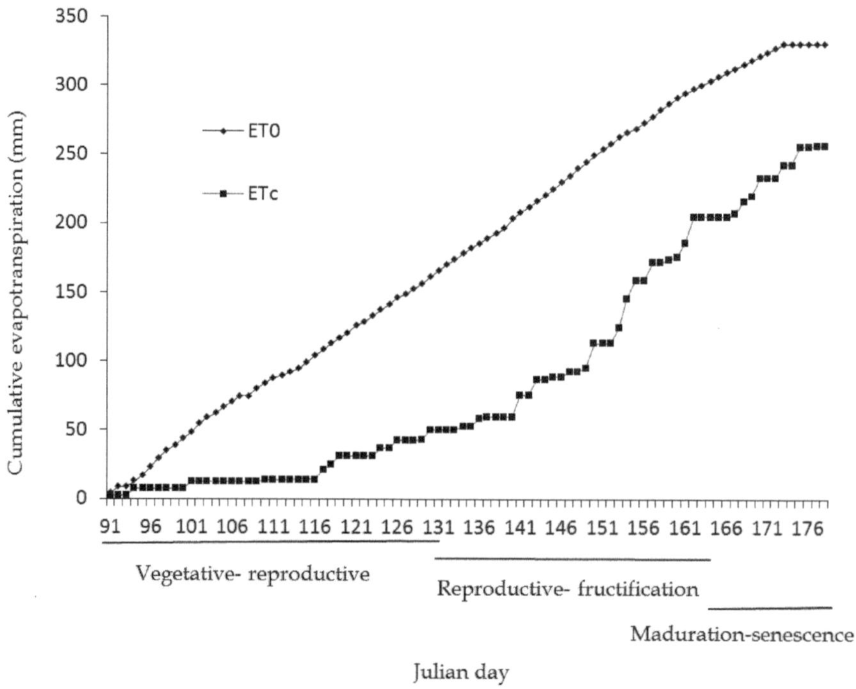

Fig. 7. Seasonal trend of cumulative ETc and ET_0 evapotranspiration in the husk tomato crop under drip irrigation,without plastic mulching and.

The unique coefficient values (temporal average) of the tomato crop (K_c) without stress, adequate management and a height of 0.6 m, in sub-humid climate with minimum relative humidity (HR_{min} = 45 %), and wind speed of u = 2 ms^{-1} for its use in the Penman-Monteith equation ET_0. Allen et al. (1998) suggest Kc values for the initial, intermediate and final stages of 0.6, 1.15 and 0.7 to 0.90, respectively; they make clear that when the crop grows from 1.5 to 2 m high, a K_c of 1.2 can be used during the intermediate stage.

4.3.1 Kc for crop with plastic mulching

Plastic mulching with drip irrigation reduces significantly water evaporation from the soil surface. However, there is a general increase in the crop transpiration due to the transference of sensible and radiative heat from the plastic cover to the crop canopy. Although the transpiration rate with plastic can increase in an average of 10-30% during the crop season, compared to no plastic cover, the value of K_c in average decreases 10-35% due to the reduction of evaporation in the soil (Allen et al., 1998).

The crop coefficients for husk tomato, under drip irrigation and plastic mulching, are reduced in 35% for those with covering according to Allen et al., 1998. In consequence, K_c values for covering can be 0.2, 0.71 and 0.56 for the vegetative-reproductive, reproductive-maturation and maturation-senescence stages, respectively.

Crop coefficients and leaf area index are important values to achieve a better use of water resources. In figures 8 and 9, the relation found between K_c and the Leaf Area Index (*LAI*) is presented, evaluated after transplant of the crop without (Eq. 32) and with plastic mulching (Eq. 33), considering well-irrigated conditions:

$$Kc = 0.6569\,LAI + 0.2404,\ R^2 = 0.96,\quad n = 8,\quad P < 0.01 \tag{32}$$

$$Kc = 0.4144\,LAI + 0.1691\quad R^2 = 0.93,\quad n = 8\quad P < 0.01 \tag{33}$$

These relations indicate that the husk tomato K_c can be estimated from LAI data. Based on the results, when LAI is equal to one, the crop coefficient approaches one in absence of plastic mulching (Figure 6), and even tends to be greater than one, which indicates that the ET_c is greater than the ET_0. However, Bandyopadhyah et al. (2005) determined for the peanut (*Arachis hypogaea* L.) crop that when the *LAI* is greater than three, K_c values are greater than one.

These results also demonstrated that plastic mulching reduces the crop coefficients and, therefore, the crop evapotranspiration, particularly during the reproductive and maturation stage, which translates into water savings during crop irrigation scheduling.

The method used for calculating *LAI* and the Penman-Monteith ET_o to evaluate the crop resistance simplifies the most complex processes of evapotranspiration; also, it is precise and reliable for evaluating crop water productivity with plastic mulching (Lovelli et al., 2008; Li et al., 2008).

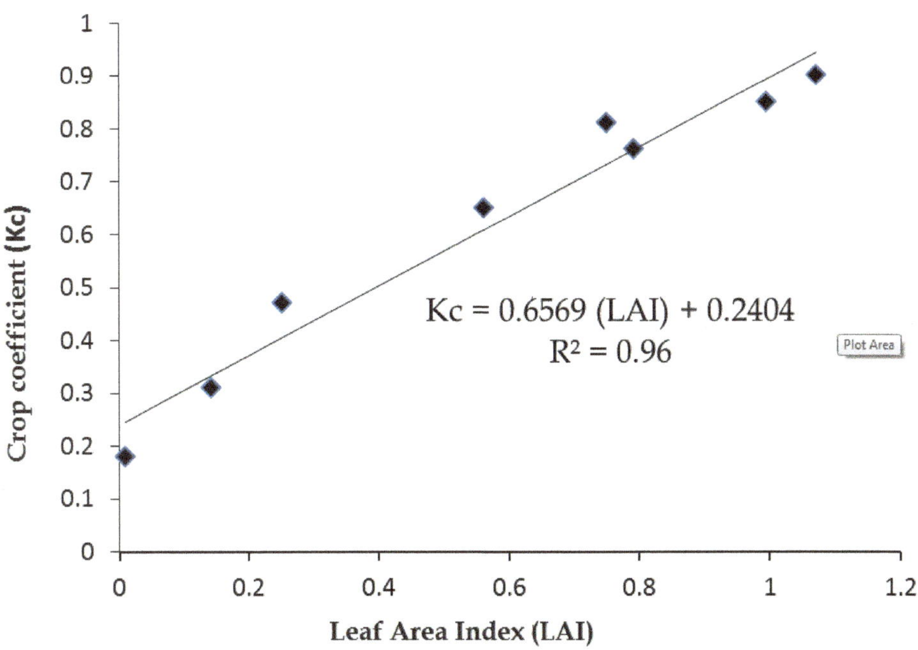

Fig. 8. Crop coefficients in husk tomato as a function of the Leaf Area Index without plastic mulching under drip irrigation.

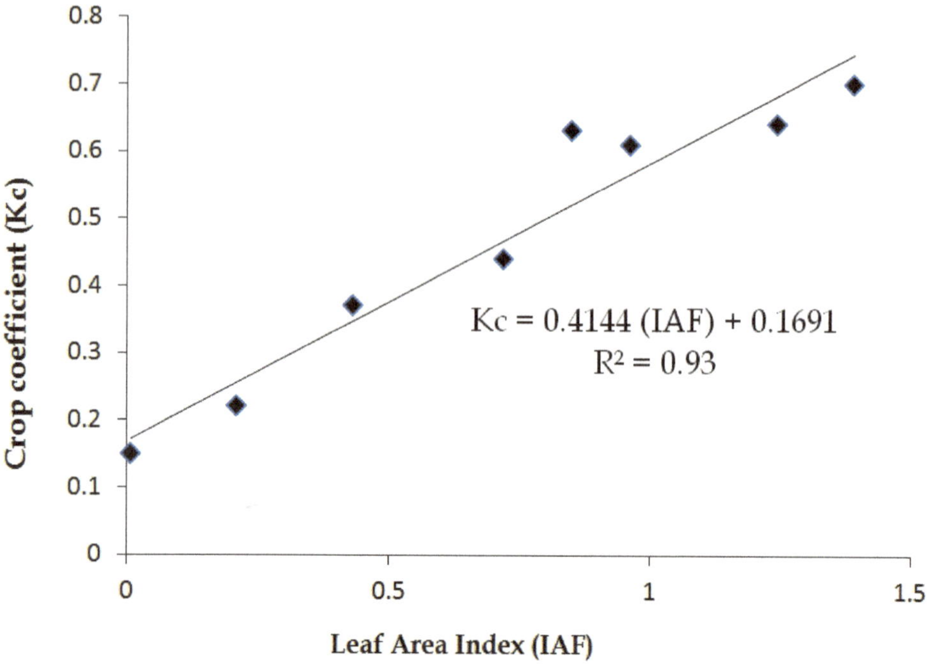

Fig. 9. Crop coefficients in husk tomato under drip irrigation as a function of the Leaf Area Index with plastic mulching.

5. Conclusions

The study's conclusions were the following: i) the use of infrared thermometry to calculate the *CWSI* is a reliable technique to schedule irrigation using the determined upper and lower *CWSI* baselines in the husk tomato crop. Its use in the initial phase of the crop is limited due to the size of the canopy. ii) The *CWSI* values close to zero corresponded to the treatments in which the totality of the irrigation depth was replaced (100 and 120 % of the ET_0) during the crop season. As the water availability decreased, the *CWSI* value increased until reaching 0.7 in the treatment with severe irrigation restrictions (40 % ET_0) without plastic mulching. The models that predict the *CWSI* from the irrigation depth are acceptably adjusted to the *CWSI* observed values. iii) Daily measurements of the matric potential level allowed estimating the losses in soil moisture in a reliable manner, as well as calculating the crop's evapotranspiration. iv) The crop coefficients (K_c) for husk tomato grown without plastic covering were estimated with a value of 0.3 in the vegetative, of 1.1 in the reproductive, and of 0.86 in maturation stage. When the crop is covered with plastic mulching, the crop coefficients were estimated with a valued of 0.2, 0.71 and 0.56 for the vegetative, reproductive and maturation stages, respectively. v) The K_c for husk tomato without mulching can be estimated from the leaf area index (*LAI*) through the equation $K_c = 0.6569(LAI) + 0.2404$, and with plastic mulching through the equation $K_c = 0.4144\ (LAI) + 0.1691$, under well irrigated conditions.

6. Acknowledgements

The authors want to express their gratitude to the Consejo Nacional de Ciencia y Tecnología (CONACYT)'s Red Temática del Agua de México, for funding for the manuscript.

7. References

Allen G.; Pereira, S.; Raes, D. & Smith, M. (1998). Crop Evapotranspiration. FAO Irrigation and Drainage Paper No. 56. 300 p. ISBN 92-5-104219-5, Rome, Italy.

Anconelli, S.; Mannini, P,: & Battilani, A. (1994). Crop Water Stress Index and baseline studies to increase quality of processing tomatoes. Acta Horticulturae (ISHS) 376:303-306. V International Symposium on the Processing Tomato. Dec.01.1994, Available from http://www.actahort.org/books/376/376_40.htm

Arteaga, R.; Vázquez, M.; Coras, P.; & Ángeles,V. (2006). Componentes de la estación de crecimiento, variación temporal y espacial en Chapingo, México. Ingeniería Hidráulica en México. Vol. 21, (Octubre 2006) pp.57-68. ISSN-0186-4076

Bandyopadhyay, P.; Mallick, S.; & Rana, S. (2005). Water balance and crop coefficients of summer-grown peanut (Arachis hypogaea L.) in a humid tropical region of India. Irrigation Science, Vol. 23, No. 4, pp. 161-169. ISSN 0342-7188

Buchner, R.; Goldhamer, D.; Shaw, D. (1994). Irrigation scheduling in kiwifruit growing and handling, J. K. Hasey, R. S. Johnson, J. A. Grant, and W. O. Reil (eds.) University of California Publication 3344: pp. 43-49.

Bucks, D.; Nakavamma, F.; French, O.; Regard, W & Alexander, W. (1985). Irrigated guayule evapotranspiration and plant water stress. Agricultural water management, Vol. 10, (Mayo 1985) pp. 61-79.

Burman R., & Pochov, L. (1994). Evaporation, evapotranspiration and climatic data. Developments in Atmospheric Science, 22. Elsevier Science B.V., 278 p. ISBN 0-444-81940-1, U.S.A.

Calado, A.; Mozon, A.; Clark, D.; Phene, C.; Ma, C. & Wang, Y. (1990). Monitoring and control of plant water stress in processing tomatoes. Acta Horticulturae. ISHS, Vol. 277, pp.129-136.

De Tar, W. (2004). Using a subsurface drip irrigation system to measure crop water use. Irrigation Science. Vol. 23, (Enero 2004), pp. 111-122.

De Tar, W. R. (2009). Crop coefficients and water use for cowpea in the San Joaquin Valley of California. Agricultural Water Manage, Vol. 96, (Enero 2009) pp.53-66.

Dodds, E; Wayne, S.; & Barton, A. (2005). A Review of Methods to Estimate Irrigated Reference Crop Evapotranspiration across Australia. CRC for Irrigation Futures Technical Report No. 04/05. CSIRO Land and Water. 04.2005. Available from www.clw.csiro.au/publications/consultancy/2005

Doorenbos, J. & Pruitt, W. (1977). Guidelines for predicting crop water requirements. FAO Irrigation and Drainage paper No. 24. F.A.O., 114 p. ISBN 92-5-104219- Rome, Italy.

Erdem, Y.; Erdem, A.; Orta,H. & Okursoy,H. (2005). Irrigation scheduling for watermelon with crop water Stress Index (IEHC). Journal of Central European Agriculture, Vol.6, No. 4, (Agosto 2005), pp. 449-460, ISSN 1332-9049

Feldhake,C.; Glenn, D.; Edwards, W. & Peterson, D. (1997). Quantifying drought for humid, temperate pastures using the Crop Water Stress Index (CWSI). New Zealand. Journal of Agricultural Research, Vol. 40, No.4 (Marzo de 1997), pp. 17-23, ISSN 0028-8233.

Gardner, B. & Shock C. (1989). Interpreting the Crop Water Stress Index. ASAE Paper 89-2642. ASAE, St. Joseph, MI.

Giuliani, R.; Magnanini, E . & Flore, J. (2001). Potential use of infrared thermometry for the detection of water deficit in apple and peach orchards. Acta Horticulturae (ISHS), Vol. 557, pp 38-43.

Hillel, D. (1980). Fundamentals of Soils Physics. Academic press, Inc. New York.

Idso, S.; Jackson, R.; Pinter, P.; Reginato, R. & Hatfield, J. (1981). Normalizing the Stress-Degree-day parameter for environmental variability. Agricultural Meteorology, Vol. 24, No.1 (Enero 1981), pp.45-55.

Idso, S. (1982). Non-water-stressed baselines: a key to measuring and interpreting plant water stress. Agricultural Meteorology Vol. 27, (Abril 1982) pp. 59-70.

Irmak, S.; Hamman, D. & Bastug, R. (2000). Determination of Crop Water Stress Index for Irrigation Timing and Yield Estimation of Corn. Agronomy Journal. Vol. 92,No.6, pp. 1221–1227, ISSN 0002-1962.

Itier, B.; Flura, D. & Belabbes, K. (1993). An Alternative Way for C.W.S.I. Calculation to Improve Relative Evapotranspiration Estimates-Results of an Experiment over Soybean. Acta Horticulturae (ISHS) Vol. 335, pp. 333-340.

Jackson, .D.; Idso, B.; Reginato, .J. & Pinter .J. (1981). Canopy temperature as a crop water stress indicator. Water Resources Research, Vol. 17 pp. 1133-1138, ISSN 1133-1138

Jackson, D. (1982). Canopy Temperature And Crop Water Stress. In: Advances in Irrigation, Vol. 1. D. I. Hillel, Editor. Academic Press. pp: 43-85.

Jensen, M. (1981). Sumary and challenges. In: Irrigation Scheduling for Water and Energy Conservation in the 80´s. Proc. ASAE, Irrigation Scheduling Conference. ASAE. P.O. Box 410. St. Joseph, Michigan. pp: 225-231.

Jensen , E; & Wright, L. (1978). The role of Evapotranspiration Models in Irrigation Scheduling. Trans. ASAE. Vol. 21, pp.82-87.

Jones, G. (1999). Use of Infrared Thermometry for Estimation of Stomatal Conductance as a Possible Aid To Irrigation Scheduling. Agricultural and Forest Meteorology. Vol. 9, No.3 (Junio 1999) pp.139-149.

Keener, E. & Kircher, L. (1983). The Use of Canopy Temperature as an Indicator of Drought Stress in Humid Regions. Agricultural Meteorology, Vol. 28, No.4, (Abril 1983), pp. 339-349.

Lovelli, M.; Perniola, M .; Arcieri,M.; Rivelli ,A. & Tommaso, T. (2008). Water use assessment in muskmelon by the Penman–Monteith "one-step" approach. Agricultural Water Manage. Vol. 95, (Abril 2008), pp.1153-1160.

Lubana, P.; Narda, N.; & Thaman, S. (2001). Performance of Summer Planted Bunch Groundnut under Different Levels Of Irrigation. Indian Journal Agricultural Science, Vol. 71, pp. 783, ISSN 0019-5022.

Orta, A.; Erdem,Y. & Erdem, T. (2003). Crop water stress index for watermelon. Scientia Horticulturae, Vol. 98, No.2 (Abril 2003), pp. 121-130, ISSN 0304-4238.

Şimşek ,M.; Tonkaz T.; Kaçira, M,; Çömlekçioğlu, N. & Doğan Z.(2005) The Effects of different Irrigation Regimes on Cucumber (Cucumbis Sativus L.) Yield and Yield Characteristics under open Field Conditions. Agricultural Water Management, Vol.73, No.3 (mayo 2005), pp. 173-191

Thompson, R.; Fernández. M.; Valdez, L.; Agüera,T. &. Gallardo, M. (2006). Evaluation of the Watermark sensor for use with drip irrigated vegetable crops. *Irrigation Science*, Vol. 24, No.3 (2006), pp. 185–202, ISSN 0342-7188

Yuan, G.; Luo, Y.; Sun, X.; & Tang, D. (2004). Evaluation of a crop water stress index for detecting water stress in winter wheat in the North China Plain. Agricultural Water Management, Vol. 64, No.1(Enero 2004), pp.29-40, ISSN 0378-3774

12

Evapotranspiration Partitioning in Surface and Subsurface Drip Irrigation Systems

Hossein Dehghanisanij[1] and Hanieh Kosari[2]
[1]*Department of Irrigation Technology, Agricultural Engineering Research Institute, Karaj,*
[2]*Department of Irrigation and Reclamation Engineering, Agricultural and Natural Resources Campus, University of Tehran, Karaj, Iran*

1. Introduction

Water transfer from the soil-plant system to the atmosphere occurs through evapotranspiration, which includes evaporation of water from the soil and other surfaces and transpiration through plant stomata. Evaporation is the process whereby liquid water is converted to water vapor (vaporization) and removed from the evaporating surface (vapor removal). Water evaporates from a variety of surfaces, such as lakes, rivers, pavements, soils and wet vegetation. Transpiration consists of the vaporization of liquid water contained in plant tissues and the vapor removal to the atmosphere. Transpiration, like direct evaporation, depends on the environmental factors including energy supply, vapor pressure gradient and wind. Hence, radiation, air temperature, air humidity and wind terms should be considered when assessing transpiration. The soil water content and the ability of the soil to conduct water to the roots also determine the transpiration rate. The transpiration rate is also influenced by crop characteristics, environmental aspects and cultivation practices. Evaporation and transpiration occur simultaneously and there is no easy way of distinguishing between the two processes (Allen et al., 1998).

Where the evaporating surface is the soil surface, the amount of water available at the soil surface is the main sources of evaporation. Accordingly any irrigation method which decreases the water availability in the soil surface will decrease the evaporation considerably. Apart from the water availability in the topsoil, the evaporation from a cropped soil is mainly determined by the fraction of the solar radiation reaching the soil surface. The degree of shading of the crop canopy is other factors that affect the evaporation process. This fraction decreases over the growing period as the crop develops and the crop canopy shades more and more of the ground area. When the crop is small, water is predominately lost by soil evaporation, but once the crop is well developed and completely covers the soil, transpiration becomes the main process.

Evapotranspiration (ET) partitioning into soil surface evaporation (E_s) and crop transpiration (T_c) is fundamental to many irrigation management studies. In many cases such as design of irrigation system, measurement of the whole ET is sufficient but when the research on water consumed by crop become more precise measurement or estimation of its two components, E_s and T_c, will be valuable. Prime attempts to partition ET include methods covering the ground surface of a plot to eliminate E_s and measure water loss by crop (Tc) and compare it with water loss by uncovered plot (ET) to reach E_s (Harrold et al., 1959; Peters and Rassel, 1959; Shaw, 1959). The researches showed, applying ground cover to partition ET changes the field and soil surface energy balance and does not estimate crop transpiration accurately in natural condition ((Fritschen and Shaw, 1961). Developing micro-lysimeter (Boast and Robertson., 1982), provided E_s measurement directly without drastic changes in soil and field condition caused by surface covers (Shawcroft and Gardner., 1983). Applying this method simultaneously with ET measurement at the same place provides ET components separately (Ham et al., 1990; Jara et al., 1998; Sepaskhah and Ilampour, 1995). Some results showed there are some limitations with using micro-lysimeter especially when E_s consisted small portion of ET (Ham et al., 1990, Jara et al., 1998). Additional researches applied ET measurement simultaneously with transpiration measurement frequently using sap flow gauges. Sakuratani (1987) was first who reported ET components separately in this way (Ham et al., 1990). In some of those researches micro-lysimeter was used for evaluating the accuracy of measured E_s with the calculated ones. Ashktorab et al. (1989) measured E_s by Bowen ratio energy balance from bare soil. Then they applied it with ET measurement using weighing lysimeter to partition ET components. Their results suggest an accurate method to measure E_s under the crop canopy (Ashktorab et al., 1994). The latest work on partitioning ET was method applying Bowen ratio energy balance (BREB) to measure both ET and E_s (Zeggaf et al., 2008). Their results showed this technique can provide a framework for partitioning ET at maize field simply and economically to previous methods (Zeggaf et al., 2008).

Applied and precise methods of ET partitioning provide useful data for farm irrigation management and water use efficiency improvement. This knowledge particularly for modern irrigation systems implementing with high costs is more important, where E_s reduction is one of the advantages of modern irrigation systems such as surface drip irrigation (DI) and subsurface drip irrigation (SDI). Accurate and efficient management should be applied to reach such advantages of these systems. Subsurface drip irrigation (SDI) is an alternative to conventional drip irrigation, which would become an attractive option to most of the farmers in arid and semi arid regions like Iran. The advantages of SDI compared to surface drip irrigation include direct application of water to the root zone, less E_s, potentially greater water use efficiency and fewer weed and disease problems (Phene et al, 1991). SDI reduced tillage using semi-permanent beds (Senn and Cornish, 2000) and removed the need for deep cultivation between the crops. SDI has been found to increase yield over surface drip (Sakellarious-Makrantonaki et al., 2002); furrow irrigation (Hanson et al., 1997); and sprinkler irrigation (De Tar et al., 2004), providing the SDI system receives good irrigation scheduling (Haman and Smajstrla, 2002).

Soil and canopy energy balances have some interactions in crop environment and irrigation systems change this environment significantly which may has influences on ET. Sprinkle irrigation increase the air humidity, surface irrigation keep the soil surface wet for at least one day after irrigation and, drip irrigation decrease the crop water stress by short irrigation interval. Toward a precise irrigation management, measuring ET component is required to have confidence on development of new and precise irrigation systems such as SDI. Besides, relation between the effective factors in ET could provide us valuable information for better farm irrigation management and water use efficiency improvement. Based on our knowledge, there is no information on proportion of E_s or T_c component under SDI, where soil surface kept dry and that may increase the soil surface temperature during the day time. In this chapter we are going to show how we could partition ET for SDI and DI systems in a maize field using a BREB method and discuss energy balance elements variation under these two irrigation systems.

2. Energy balances theories

2.1 Energy balance theory at maize field

Energy balance at field level can be expressed as:

$$R_n = G + H + \lambda E \tag{1}$$

Where R_n is net radiation reaching the field, above the maize canopy, λE is latent heat flux, H is sensible heat flux and G is soil heat flux (all units of W/m^2). In equation (1) the convention used for the signs of the energy fluxes is R_n positive downward and G is positive when it is conducted downward from the surface, λE and H are positive upward. Partitioning of energy between λE and H is determined by the BREB (Bowen., 1926, Perez et al., 1999) by the following equation:

$$\beta = \frac{H}{\lambda E} \tag{2}$$

where β is the Bowen ratio. By solving equation (1) and (2) at the same time the following expressions for λE and H are obtained:

$$\lambda E = \frac{R_n - G}{1 + \beta} \tag{3}$$

$$H = \beta \frac{R_n - G}{1 + \beta} \tag{4}$$

Assuming equality of eddy transfer coefficients for sensible heat and water vapor in the averaging period and measuring air temperature and vapor pressure gradients between the two levels, the Bowen ratio (β) is calculated by:

$$\beta = \gamma \frac{\Delta T}{\Delta e} \tag{5}$$

Where ΔT and Δe are air temperature and vapor pressure differences between the two measurement levels and γ is psychrometric constant which is calculated by the following equation:

$$\gamma = C_p P / \varepsilon L_v \tag{6}$$

Where C_p is the specific heat of air at constant pressure (1.01 kJ/kg °C), P is atmospheric pressure (kPa), ε is the ratio between the molecular weights of water vapor and air (0.622), and L_v is latent heat of vaporization (kJ/kg). Psychrometric constant for the experiment site was determined 0.058 (kPa/°C).

2.2 Energy balance theory at soil surface

Energy balance at soil surface can be expressed as:

$$R_{ns} - \lambda E_s - H_s - G = 0 \quad (7)$$

where, R_{ns} is the net radiation reaching the soil surface, λE_s is the soil surface latent heat flux, H_s is sensible heat flux from the soil surface (all units of W/m²). R_{ns} was determined by the empirical equation (8) with R_n and LAI which has been used previously by some other authors (Gardiol et al., 2003; Kato et al., 2004).

$$R_{ns} = R_n \exp(-0.622LAI + 0.055LAI^2) \quad (8)$$

Bowen ratio at soil surface was calculated similar to the energy balance computation at field level with the following equation:

$$\beta_s = \frac{H_s}{\lambda E_s} \quad (9)$$

which using equations (5) and (6) and measurement of air temperature and vapor pressure gradients by ventilated psychrometers near the soil surface, Bowen ratio at soil surface was determined. By solving equation (7) and (9) simultaneously, latent heat flux from the soil surface was determined by equation (10).

$$\lambda E_s = \frac{R_{ns} - G}{1 + \beta} \quad (10)$$

2.3 Energy balance theory at crop canopy

Energy balance at maize canopy can be expressed as below (Zeggaf et al., 2008, Ham et al., 1991):

$$R_{nc} = \lambda E_c + H_c \quad (11)$$

where R_{nc} is net radiation absorbed by crop canopy, λE_c is crop canopy latent heat flux and H_c is crop canopy sensible heat flux (all units of W/m²). Applying the principle of continuity and the definition of R_n, it can be shown that R_{nc}, equation (12), is the difference between R_n measured above and that below the maize canopy (Ham et al., 1991).

$$R_{nc} = R_n - R_{ns} \quad (12)$$

Canopy latent heat flux was calculated from equation (13):

$$\lambda E_c = \lambda E - \lambda E_s \quad (13)$$

Then H_c was calculated as a residual from equation (11).

3. Methodology development

The research was conducted in summer 2009 at experimental station of agricultural engineering research institute (AERI), Karaj-Iran (35° 21′ N, 51° 38′ E, 1312.5 m above sea level). The field soil was prepared for planting in spring. Results from soil experiments up to 80 cm below surface showed the soil type was loam texture (47 % sand, 44 % silt, 9 % clay) with ECe=1.7. Irrigation water were supplied from underground well with an quality which had no negative impact for maize (EC=0.8 dS/m and pH=7.8).

The experimental field was defined in an area of 40×60 m^2 in selected site (Fig. 1). A day before planting 50 kg potash fertilizer was added to soil and maize (Double Cross 370) was planted on 15 June 2009. The crop was planted with 0.75 m row width and north-south orientation. The field was bordered by irrigated maize field except in western side which was unplanted.

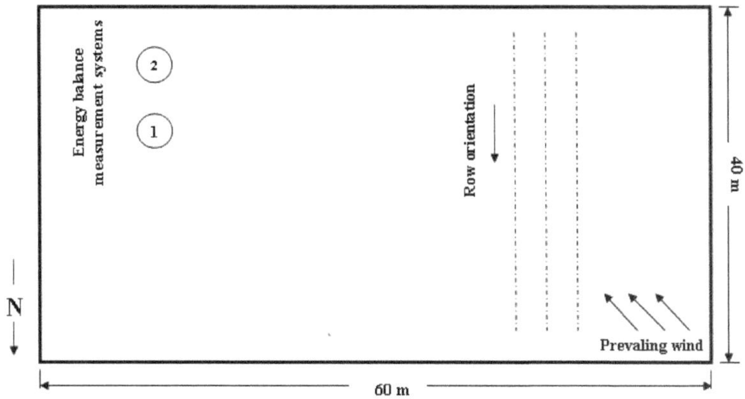

Fig. 1. Schematic diagram showing field position and location of energy balance measurement systems.

Irrigation water was supplied from the well and chemical quality analysis showed water in this region has good quality. A subsurface drip-tape irrigation system with 0.30 m dripper distance was used to apply irrigation water. Drip tapes were placed 15 cm below the soil surface in nearest place to the plant rows. Special attention was paid while positioning drip tapes to transfer water correctly. Crop water requirement was estimated based on long time meteorological data (averaging from 1988 to 2008) and calculation of crop ET by method recommended in FAO 56, Penman Monteith Method (P-M) (Allen et al., 1998). The P-M method is recommended for the Karaj area by Dehghanisanij et al. (2004). From the early crop growth period 20% over irrigation based on 3 day intervals was applied to prevent water stress. Recommended nitrate fertilizer (according to soil experiments it was 400 kg/ha) was distributed during the crop growth period and closely to crop establishment place by irrigation system (fertigation). At the period of this experiment 41-44 and 59-62 day after emergence (DAE), leaf area index (LAI) was measured in 41, 44, 59, 62 DAE. Each time 3-5 plants were selected randomly and the whole leaf area of a plant was measured with leaf area meter (Area Measurement system, DELA-T Devices, ENGLAND) in the laboratory. Then LAI was calculated from multiplying the average plant leaf area by plant density. LAI values for the days between the days of measurement obtained by linear interpolation (Gardiol et al., 2003). Automatic weather station was established in the field simultaneously with start of experiment period and hourly average values of solar radiation (Rs), air temperature, relative humidity and wind speed were measured and logged continuously.

From 41-44 and 59-62 DAE, ET and Es were determined simultaneously by measuring all energy fluxes at maize field and soil surface using two independent measurement systems.

Then by subtracting the latent heat flux at soil surface from the latent heat flux at maize field (ET), transpiration was obtained. Energy balance equipments consisted of a net radiometer (CNR1, Kipp & Zonen), two soil heat flux plates (MF-180M, EKO Japan) and four hand made thermocouple ventilated psychrometers for Bowen ratio measurement in both field level and soil surface. The details of constructed psychrometers have been described in Kosari (2010). Two independent measuring systems separated by 5 m distance, were placed 9 m from the east edge of the field as the system number 2 was positioned 5 m from the east edge to maximize fetch to height ratio when prevailing wind (north-western to south-eastern) were present (Fig. 1). That was greater than minimum adequate ratio reported by Heilman et al. (1989) for measuring Bowen ratio during our experiment period.
Measurement equipments in each measurement system were installed on a tall rod. Two ventilated psychrometers used for measuring temperature and water vapor gradients at field level above the crop. These two psychrometers were installed 1 m apart as the lowest one was positioned 0.2 m above the crop canopy (Ham et al., 1991; Jara et al., 1998). The remaining two psychrometers used for measuring temperature and water vapor gradients at soil surface. These two psychrometers were fixed 0.1 m apart on the rod as the lowest one was positioned 0.05 m above the soil surface (Ashktorab et al., 1989). Net radiation at field level was measured with net radiometer installed 1 m above crop canopy. Soil heat flux was calculated as an average of two soil heat flux plates positioned 0.02 m below the soil surface. All data were measured every minute by a CR23X data logger connected to an AM16/32 multiplexer (Campbell Scientific, Inc., UT) and averaged 30 min intervals.

4. Meteorological parameters variation

Daytime average values of meteorological parameters measured by the automatic weather station in the experiment period are shown in Table 1. Plant in days 41- 44 DAE was in developing stage and in days 59-62 DAE was in mid-season stage. Irrigation has been done on 41, 44, 59 and 63 DAE, which exceptionally because of some problems the last one irrigated with 4 days interval. In the experiment period the 42 and 60 DAE received maximum and minimum solar radiation respectively.

Growth stage	DAE	Air Temperature (°C)			Relative Humidity (%)			Wind Speed (m/s)	Solar Radiation MJ/m²day
		Min	Max	Ave	Min	Max	Ave		
Developing stage	41	16.67	34.44	24.10	18.92	78.16	53.02	2.90	45.17
	42	16.51	34.07	24.12	15.64	83.07	52.64	3.38	46.82
	43	16.10	33.30	23.39	26.80	77.82	54.09	2.95	46.21
	44	16.56	32.89	23.79	28.66	74.72	52.16	2.92	44.31
Mid-season stage	59	19.91	37.38	28.30	12.42	65.51	36.40	1.48	40.97
	60	19.96	35.64	27.42	17.78	65.98	41.67	2.33	40.14
	61	19.09	35.57	27.00	13.29	72.40	41.68	2.05	45.18
	62	16.00	34.45	24.24	13.06	59.69	38.43	1.90	45.07

Table 1. Daytime average values of meteorological values in experiment period.

5. Evapotranspiration analysis

Diurnal trend of evapotranspiration measurement by BREB method compare to evapotranspiration estimation by P-M method for 60 and 61 DAE are shown in Fig. 2. These days were selected because they are representative of cloudy and clear sky condition respectively which is believed they show all sky condition during measurement period.

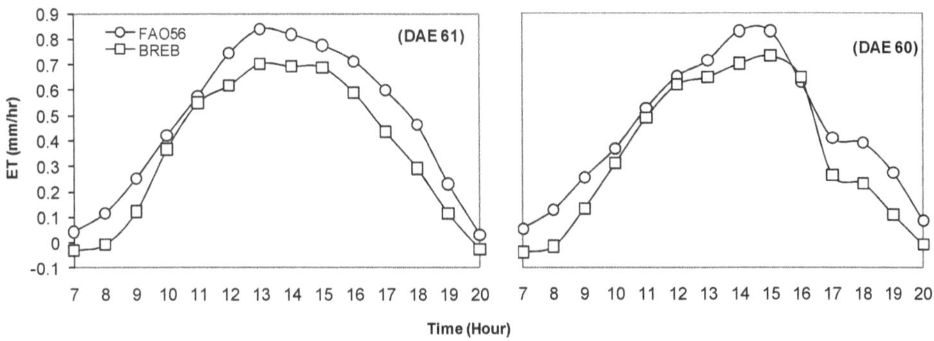

Fig. 2. Diurnal trend of evapotranspiration by BREB and P-M method for 60 and 61 DAE.

As it is shown in Fig. 2 both methods have the same trend and a good correlation (R^2=0.92 and 0.95 for DAE 60 and 61, respectively). Therefore evapotranspiration by BREB showed 9% variations compare to P-M method which can be acceptable. Positive value of MBE parameter shows overestimation of P-M method compare to BREB. These differences can be caused by different measurements of effective and required parameters in both methods. In other word, BREB evapotranspiration was obtained by direct measurement of required parameters at field level and soil surface while in P-M method maize evapotranspiration was obtained by estimation of reference evapotranspiration and crop coefficient. Ortega et al. (1995) found a good correlation between refernce evapotranspiration by BREB and Penman method on irrigated grass.

6. Energy balance at farm level

Daytime average of energy balance measurements in terms of (W/m^2) at Maize field, soil surface and crop canopy for DI and SDI are presented in tables 2 to 4. In the measurement period net radiation values ranged from 304 to 333 (W/m^2) resulted by the minimum and maximum solar radiation in corresponding days respectively (Table 2). Latent heat flux (λE) From the Maize field ranged from 207 to 267 (W/m2) for SDI and 197 to 296 (W/m2) for DI. According to the results λE ranging from 62 to 83 % of R_n under SDI and 61 to 94 % for DI, which shows maize cropping system is under non-stressed conditions (Ham et al., 1991). Herein G/R_n was ranging 6 to 8 % for SDI and 8-15 % for DI, which is close to 10 % reported by Yunusa et al. (2004).

Soil surface energy balance measurements showed R_{ns} values decreased with crop growth and canopy cover increment. Furthermore it showed R_{ns} partitioned primarily between soil heat flux and latent heat flux from the soil surface and there was very little sensible heat flux during experiment period. The G variation under SDI was 17 to 25, and it was 25 to 47

W/m² in under DI. The λEs accounted for about 40 to 73 (W/m²) in SDI, where maximum value of λE_s was 60 W/m² for surface DI. Accordingly, $\lambda E_s/R_{ns}$ and G/R_{ns} ratios ranged between 56 to 71 % and 18 to 31 % respectively (Table 3). The G under SDI was much less compared to that under DI during crop developing stage than mid-season stage. It is while, λE_s was larger under SDI compared to that for DI. These results contributed to higher possible potential for T under SDI during crop developing stage (Table 3).

DAE	R_n	Subsurface drip irrigation (SDI)				Surface drip irrigation (DI)			
		G	λE	$\lambda E/R_n$	G/R_n	G	λE	$\lambda E/R_n$	G/R_n
	W/m²			%		W/m²		%	
41	322	23	-	-	7	47	197	61	14
42	333	19	207	62	6	44	266	80	13
43	329	25	212	64	8	46	255	78	14
44	320	23	242	76	7	47	243	76	15
59	315	20	247	78	6	34	296	94	11
60	304	17	235	77	6	38	253	83	10
61	326	20	248	76	6	28	271	83	9
62	322	22	267	83	7	25	245	76	8

Table 2. Daytime average energy fluxes at maize field.

DAE	LAI	R_{ns}	Subsurface drip irrigation (SDI)				Surface drip irrigation (DI)			
			G	λE_s	$\lambda E_s/R_{ns}$	G/R_{ns}	G	λE_s	$\lambda E_s/R_{ns}$	G/R_{ns}
		W/m²			%		W/m²		%	
41	2.20	107	23	63	59	21	47	55	51	44
42	2.38	104	19	73	71	18	44	60	58	43
43	2.56	96	25	54	56	26	46	39	41	48
44	2.74	88	23	58	66	26	47	39	44	53
59	3.50	70	20	50	71	29	34	32	46	49
60	3.53	67	17	43	64	25	30	35	52	45
61	3.57	71	20	49	69	28	28	42	59	39
62	3.60	70	22	40	57	31	25	44	63	36

Table 3. Daytime average energy fluxes at soil surface.

Daytime average of energy balance measurements at Maize canopy showed λE_c increased by crop development from 134 to 227 (W/m²) which resulted $\lambda E_c/R_{nc}$ between 58 to 90% in SDI. These ratios showed canopy latent heat fluxes were often lower than available energy which resulted some sensible heat flux conducted away from canopy level. Under DI, λE_c increased by crop development from 142 to 264, which was less in average compare to that for SDI (Table 4).

7. Diurnal energy balance pattern

Diurnal trends of the energy balance components for the soil surface of maize field at 60th DAE are shown in Fig. 3. This day was representative of a day with some cloud cover in the sky. Average air temperature and relative humidity was 27.4 C° and 41.6 % respectively. Daytime average of net radiation available at maize field was 304 W/m2, which was the smallest value in the measurement period.

DAE	R_{nc}	Subsurface drip irrigation (SDI)		Surface drip irrigation (DI)	
		λE_c	$\lambda E_c/R_{nc}$	λE_c	$\lambda E_c/R_{nc}$
	W/m2		%	W/m2	%
41	215	-	-	142	66
42	230	134	58	205	89
43	233	158	68	216	93
44	232	184	79	204	88
59	245	197	80	264	108
60	237	192	81	218	92
61	255	199	78	229	90
62	252	227	90	201	80

Table 4. Daytime average energy fluxes at crop canopy.

During the day Maximum R_n and G were 674 and 70 W/m2, which occurred about 13:00 and 14:00 h respectively. As it is shown in Fig. 3 variation of R_n values is not symmetrically as a bell shape curve signifies that there were some cloud cover at sky during the day.

Most of (R_n-G) was used to drive λE and β values ranged between 0 to 0.7 while β values reported by Zeggaf et al., (2008), in maize field was lower than 0.25. It can be due to different crop growth stage in two experiments (in this research during the 60th DAE, Maize crop covered the ground relatively complete, LAI=3.53) and various climate, soil type and/or irrigation system. Sensible heat flux accounted for about 18% of available energy (R_n-G). Similar results were reported by Steduto and Hsiao, (1998), Ham et al., (1991) and Ritchie, (1971). Soil heat flux was about 17 (W/m2) which was lower than 10% of net radiation as it is found a recommended value for daytime average of soil heat flux in literature (Allen et al., 1998).

During the daytime of 60th DAE only 22% of net radiation reached the soil surface. Most of the energy was split between λE_s and G. The value of H_s was small in this balance. The λE_s was less than available energy except in the afternoon suggesting that the soil surface was absorbing energy from within-canopy (air stream) which provided energy for λE_s. Similar signification was reported in Ham et al., (1991) for soil surface energy balance relationships. Daytime average of λE_s was about 86% of R_{ns}-G and only about 14% of R_{ns}-G was used as sensible heat. During this day βs from soil surface, ranged from -0.5 to 0.5. Positive H_s values at soil surface indicates convective transport of heat away from the soil surface (Zeggaf et al., 2008), but the results from this research showed small amounts of available energy used as sensible heat flux. Field observations from subsurface drip irrigation system showed installation depth of drip tapes caused during irrigation, water rises up to the surface and makes the ground wet. Therefore the large proportion of soil surface latent heat flux from available energy can be due to the ground wetness. It can be concluded when drip tapes install in lower depth and the soil surface remains dry larger proportion of sensible heat will result.

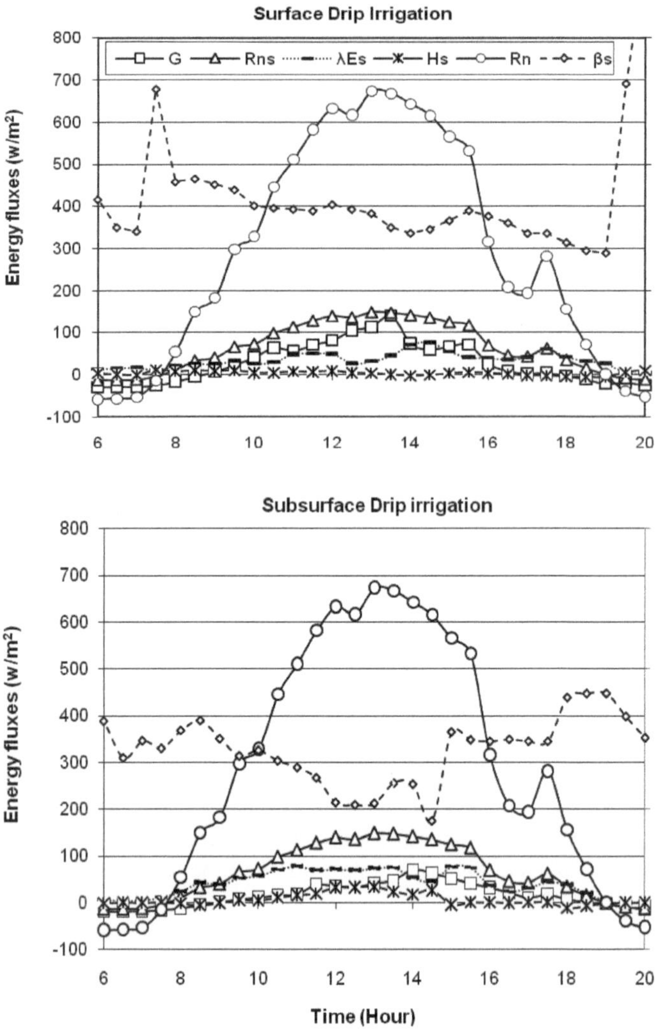

Fig. 3. Diurnal trend of energy balance components at soil surface of maize field in 60th DAE under surface and subsurface drip irrigation.

Available energy (R_n-G) and λE from maize field for the 60th DAE are shown in Fig. 4. The linear regression lines between λE and R_n-G were obtained with high values of $r^2=0.99$ for both SDI and DI. Based on the slope of the trade lines in Fig. 4, there was no a significant reduction in available energy to the maize field and soil surface between two irrigation system. Accordingly, SDI aimed at reducing soil evaporation compared to DI, is not effective when soil surface covered by canopy (LAI=3.5).

Available energy (R_{ns}-G) and λE_s from soil surface for 41th DAE are shown in Fig. 5. The 41th DAE presenting crop developing stage, when LAI was about 2.20. Accordingly, there was a wide scattering for available energy under DI which might be because soil surface is not uniformly wet under DI. However, the condition under SDI was uniformly and a linear regression lines between λE_s and R_{ns}-G were obtained.

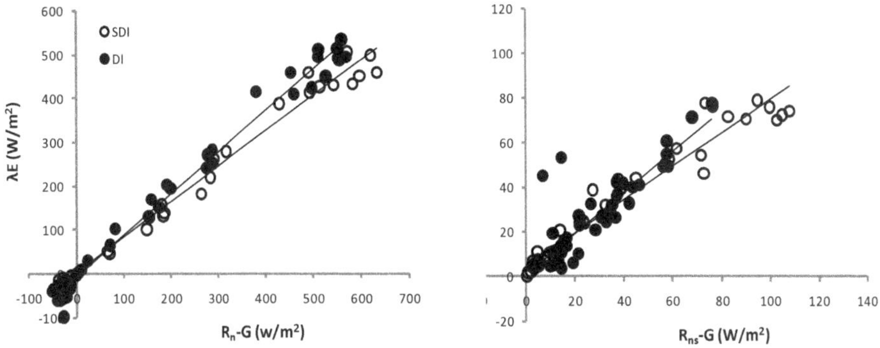

Fig. 4. Available energy (R_n-G) and latent heat flux (λE) from maize field and soil (R_{ns}-G and λE_s) in 60th DAE for surface (DI) and subsurface (SDI) drip irrigation.

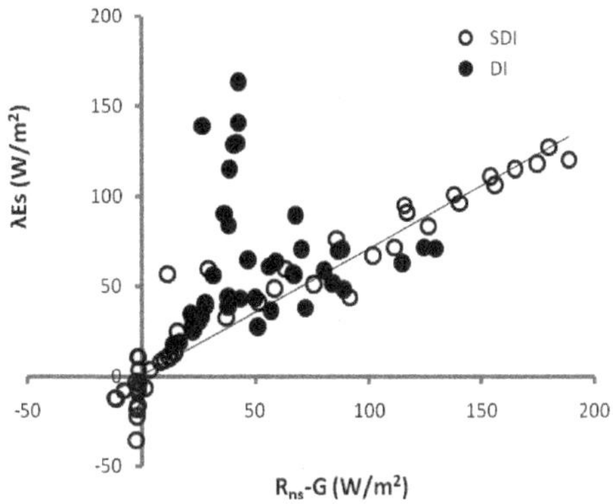

Fig. 5. Available energy (R_{ns}-G) and latent heat flux (λE_s) from soil surface in 41th DAE for surface (DI) and subsurface (SDI) drip irrigation.

8. Conclusion

The Bowen ratio method could be used for partitioning ET under surface (DI) and subsurface drip irrigation system (SDI). Partitioning ET for advance irrigation system could provide us useful information for better irrigation management during crop growth stages and development of new irrigation technique. Partitioning ET and measurement of energy balance over maize field, canopy and soil by Bowen ratio showed that soil had major impact on the energy balance between the soil and canopy when soil surface is not covered fully by crop canopy. In crop developing stage, energy balance of maize field was different under DI and SDI. This result could be contributed to more difference between the systems in early crop development stage, when soil surface is not covered fully by crop canopy.

As it was shown daytime soil heat flux values were greater under DI (25-47 W/m^2) compared to that under SDI (17-25 W/m^2). It may caused by heat convection in DI while moving down the water from the surface and higher temperature of water when drip tapes were positioned on the ground. Therefore available energy for soil evaporation, R_{ns}-G, was lower in DI. As it was shown λE_s accounted for about 41 to 63% of R_{ns} in DI while it was about 56 to 71% in SDI. It was observed the ground in both DI and SDI became wet but reverse direction of moving water in subsurface system, as may contributed to more evaporation in SDI. According to the results, more consideration should be applied using SDI systems on depth of lateral line which carrying the emitters, canopy size, crop type, and plant water stress affect soil and canopy energy balances. Those data will be useful for validation of ET models.

9. Acknowledgment

This Study was supported by Agricultural Engineering Research Institute (AERI), Iran. The authors wish to render heartfelt gratitude to Dr. Majid Liaghat and Dr. Farhad Mirzaei, Department of Irrigation and Reclamation Engineering, Agricultural and Natural Resources Campus, University of Tehran, Iran for their useful advice during the research.

10. References

Allen, R.G.; Pereira, L.S.; Raes, D. & Smith, M. (1998). Crop evapotranspiration guidelines for computing crop requirements. Irrigation and drainage paper 56. FAO. Rome, Italy. pp.300. ISBN 92-5-104219-5.

Ashktorab, H.; Pruitt, W.O.; Paw, U.K.T. & George, W.V. (1989). Energy balance determination close to the soil surface using a Micro-Bowen Ratio system. Agricultural and Forest Meteorology Journal, Vol.46, pp.259-274.

Ashktorab, H.; Pruitt, W.O. & Paw, U.K.T. (1994). Partitioning of evapotranspiration using lysimeter and Micro-Bowen Ratio system. ASCE J. of Irrigation and Drainage. Vol.120, No.2, (March/April), pp.450-464.

Boast, C.W. & Robertson T.M. (1982). A micro-lysimeter method for determining evaporation from a bare soil: Description and laboratory evaluation. Soil Sci. Soc. Am. J., Vol.46, pp.689-696.

Bowen, I. S. (1926). The ratio of heat losses by conduction and by evaporation from any water surface: In Rosenberg, N. J., Blad, B. L. and Verma, S. B., (1983). Microclimate: The biological environment. Wiley, New York, 495 PP.

De Tar, W. R. (2004). Using a subsurface drip irrigation system to measure crop water use. Irri. Sci. Vol.3. pp.111-122.

Dehghanisanij, H.; Yamamoto, T. & Rasiah, V. (2004). Assessment of evapotranspiration estimation models for use in semi-arid environment. Agricultural Water Management Journal, Vol.64, pp.91-106.

Fritschen, L.J. & Shaw, R.H. (1961). Transpiration and evaporation of corn as related to meteorological factors. Agron. J. Vol.53, pp.71-74.

Gardiol, J.M.; Serio, L.A. & Della Maggiora, A.I. (2003). Modelling evapotranspiration of corn (Zea mays) under different plant densities. Journal of Hydrology, Vol.271, pp.188-196.

Ham, J.M.; Heilman, J.L. & Lascano, R.J. (1990). Determination of soil water evaporation and transpiration from energy balance and stem flow measurements. Agric. For. Meteorol. Vol.59, pp.287-301.

Ham, J.M.; Heilman, J.L. & Lascano, R.J. (1991). Soil and canopy energy balances of a row crop at partial cover. Agron. J. Vol.83, pp.744-753.

Haman, D.Z. & Smajstral, A.G. (2002). Scheduling tips for drip irrigation of vegetable. Publication No: AE259. Florida extension service, University of Florida.

Hanson, B.R.; Schwankl, L.J.; Schulbach, K.F. & Pettygrove, G.S. (1997). A comparison of furrow, surface drip, and subsurface drip irrigation on lettuce yield and applied water. Agric. Water Manag. Vol.33, pp.139-157.

Harrold, L.L.; Peters, D.B.; Driebelbis, F.R.; & Mc-Guiness, J. L. (1959). Transpiration evaluation of corn grown on a plastic-covered lysimeter. Soil Sci. Soc. Of Am. Proc. Vol.23, pp.174-178.

Heilman, J.L.; C.L., Brittin & Neale, C.M.U. (1989). Fetch requirements of Bowen ratio measurements of latent and sensible heat fluxes. Agric. For. Meteorol. Vol.44, pp.261-273.

Jara, J.; Stockle, C.O. & Kjelgard, J. (1998). Measurement of evapotranspiration and its components in a corn (Zea Mays L.) field. Agric. For. Meteorol. Vol.92, pp.131-145.

Kato, T.; R. Kimura & Kamichica, M. (2004). Estimation of evapotranspiration, transpiration ratio and water use efficiency from a sparse canopy using a compartment model. Agric. Water Manage. Vol.65, pp.173-191.

Kosari, H.; Dehghanisanij., H.; Mirzaei, F. & Liaghat A.M. (2010). Evapotranspiration partitioning using the Bowen Ratio Energy Balance method in a sub-surface drip irrigation. Journal of Agricultural Engineering Research. Vol.11, No.3, pp.71-86.

Ortega, F.; Samuel, O.; Richard, H.; Cuenca, M. & English. (1995). Hourly grass evapotranspiration in modified maritime environment, Journal of Irrigation. and Drainage, ASCE, Vol.121, No.6. pp.369-373.

Peters, D.B. & Russell, M.B. (1959). Relative water losses by evaporation and transpiration in field corn. Soil Sci. Soc. Am. J., Vol.23, pp.170-173.

Phene, C.J.; Davis, K.R.; Hutmacher, R.B.; Bar-Yosef, B.; Meek, D. W., & Misaki, J. (1991) Effect of high frequency surface and subsurface drip irrigation on root distribution of sweet corn. Irri. Sci. Vol. 12. pp.135-140.

Ritchie, J.T. (1971). Dryland evaporation flux in a sub-humid climate: I. Micrometeorological influences. Agron. J. Vol.63, pp.51-55.

Sakellariou-Makrantonaki, M.; Kalfountzos, D. & Vyrlas, P. (2002). Water saving and yield increase of sugar beet with subsurface drip irrigation. Global Nest: The Int. J. Vol.4, No.2-3, pp.85-91.

Sakuratani, T. (1987). Studies on evapotranspiration from crops. II. Separate estimation of transpiration and evaporation from a soybean field without water shortage. In: Ham, J. M., Heilman, J. L. and Lascano, R. J. 1990. Determination of soil water evaporation and transpiration from energy balance and stem flow measurements. Agric. For. Meteorol. Vol59, pp.287-301.

Senn, A.A. & Cornish P.S. (2000). An example of adoption of reduced cultivation by Sydney, s, vegetable growers. In Soil 2000: New Horizons for a new century. Australian and New Zealand Joint Soils Conference. Volume 2: Oral papers. Eds. Adams J. A., and

Metherell, A. K. 3-8 December, Lincoln University. New Zealand Society of Soil Science. Pp: 263-264.

Sepaskhah A.R. & Ilampour S. (1995). Effects of soil moisture on evapotranspiration partitioning. Agric. Water Manage. Vol.28, pp.311-323.

Shaw, R. H. (1959). Water use from plastic-covered and uncovered corn plots. Agron .J. Vol.51, pp.172-173

Shawcroft, R.W. & Gardner, M.H. (1983). Direct evaporation from soil under a row crop canopy. Agric. Meteorol. Vol.28, pp.229-238.

Steduto, P. & Hsiao, T.C. (1998). Maize canopies under two soil water regimes: I. Diurnal patterns of energy balance, carbon dioxide flux, and canopy conductance. Agric. For. Meteorol. Vol.89, pp.169-184.

Yunusa, I.A.M.; Walker, R.R., & Lu, P. (2004). Evapotranspiration components from energy balance, sap flow and micro-lysimetry techniques for an irrigated vineyard in inland Australia. Agric. For. Meteorol. 127, 93-107.

Zeggaf, T.A.; Takeuchi, S.; Dehghanisanij, H.; Anyoji, H. & Yano, T. (2008). A Bowen Ratio Technique for partitioning energy fluxes between Maize transpiration and soil surface evaporation, Agron. J. Vol.100, pp.1-9.

13

The Impact of Seawater Salinity on Evapotranspiration and Plant Growth Under Different Meteorological Conditions

Ahmed Al-Busaidi[1] and Tahei Yamamoto[2]
[1]*College of Agricultural & Marine Sciences, Department of Soils, Water and Agricultural Engineering, Sultan Qaboos University, Muscat,*
[2]*Arid Land Research Center, Tottori University 1390 Hamasaka,*
[1]*Oman*
[2]*Japan*

1. Introduction

Conventional water resources of good quality are scarce especially in arid and semiarid regions. The salinization of soils and water places is a substantial constraint of crop productivity. The mixed application of fresh water along with seawater is becoming a common practice by the farmers near sea sides in the developing countries. Seawaters contain beneficial nutrients that can be useful for plants. The use of saline water has shown a considerable importance for crop production in the recent years. Unfortunately, the misuse of saline water for irrigation may increase soil salinity to a level higher than the tolerance of crops. The proper use of brackish water appeared to be reasonable with certain conditions. Hamdy et al. (2005) found that freshwater can be substituted to some extent with saline water without any loss in production especially in water deficit areas. Moreover, Oron et al. (2002) reported that high saline water has an agricultural potential with proper irrigation management. Petersen (1996) observed that the higher volume of irrigation water reduced soil salinity and transported salts below the root zone. Whereas, Ghadiri et al. (2005) noticed restricted water uptake by salinity due to the high osmotic potential in the soil and high concentrations of specific ions that may cause physiological disorders in the plant tissues and reduce yields.

Irrigation water, drainage salinity and evaporation are interconnected terms. Any increase or decrease in one of them will affect others. More irrigation water enhances leaching of salts, decreasing soil salinity and increasing drainage water salinity. Furthermore, evaporation and salt accumulation are also positively connected, hence increasing salinity of applied water and high evaporation lead to high concentration of salts in drainage water. Soil evaporation is mostly affected by environment factors, which includes mainly two aspects: climate aspect (net radiation, wind speed, air temperature, etc) and soil aspect (soil types, soil water content and temperature, hydrological properties, soil salinity, etc). In general, the soil water is considered as a direct factor affecting evaporation process. In fact, the surface temperature may be considered as the indirect factor on soil evaporation (Zhu & Takeo, 2010).

Under elevated temperature, plant could be water stressed due to the higher evapotranspiration. High temperatures cause an array of morpho-anatomical, physiological and biochemical changes in plants, which affect plant growth and development and may lead to a drastic reduction in economic yield. Major impact of high temperatures on shoot growth is a severe reduction in the first internode length resulting in premature death of plants (Hall, 1992). However, heat stress is a complex function of intensity, duration and rate of increase in temperature. The extent to which it occurs in specific climatic zones depends on the probability and period of high temperatures occurring during the day and/or the night. Higher air and leaf temperatures as a result of changes in climate will mean higher potential transpiration for the same stomatal conductance. Increased evaporative demand will tend to increase transpirational volume flow which will tend to increase salt damage. Any simple way out of this is confounded by the positive relationship between yield and the total quantity of water transpired (Pessarakli, 1999).

Barley (Hordeum vulgare L.) is regarded an important cereal grown in many countries of the world. It is the main cereal of arid and semiarid climates due to its lower water demand when compared to other crops. However, barley yield is also impaired by water and heat stress conditions. The productivity of barley is reported to be limited by terminal water stress and high temperatures during grain filling (Agueda et al., 1999). Matin et al. (1989) found a positive correlation between diffusion resistance in barley leaves and drought tolerance. From other side, barley is a cereal crop of salt-affected soils and has been rated as salt tolerant (Maas & Hoffman, 1977; Shannon, 1984). A considerable depression in barley grain yield after irrigation with high salinity waters was observed (Curtin et al. 1993).

There is a general consensus that productivity of plant is strongly influenced by environmental conditions. The yield potential of crops is limited by several abiotic stresses. Salinity and extreme temperatures are considered to be the most important factors (Atienza et al., 2004). Barley is widely grown in arid and semiarid areas of the Mediterranean regions for grain or forage purpose. It is well adapted to the cool and short growing season. Sharratt (1999) reported differences in growth parameters as affected by sowing date of barley due to day length and temperature. Therefore, the fundamental understanding on the effects of growth condition and seawater salinity could be useful for the quality production of barley crop in arid and semi arid countries. Moreover, evapotranspiration process is one of the main factors that is involved in irrigation scheduling, salt accumulation and other stresses. Therefore, this study was aimed to evaluate the effects of different meteorological conditions and seawater salinity on the evapotranspiration process and barley growth.

2. Materials and methods

Soil samples were collected from Sand Dune of Tottori-Japan. The samples were air-dried, and sieved (< 2 mm). Soil texture was determined by the pipette method (Gee & Bauder, 1986). Exchangeable cations were leached from the soil with neutral ammonium acetate. Their contents were determined using atomic absorption spectrophotometer (Model Z-2300 Hitachi Corp, Japan). Electrical conductivity (EC) and pH of the soil: water suspensions (1: 5) were also measured with pH and EC meters (Accumut M-10 and Horiba DS-14). Other physicochemical properties of the soil were analyzed following Klute (1986) methods (Table 1).

Property	Value
EC (1: 5) water	0.03 dS m^{-1}
pH	6.36
Exchangeable K^+	0.06 $cmol_c$ kg^{-1}
Exchangeable Ca^{2+}	0.34 $cmol_c$ kg^{-1}
Exchangeable Mg^{2+}	0.45 $cmol_c$ kg^{-1}
Exchangeable Na^+	0.10 $cmol_c$ kg^{-1}
Cation exchange capacity	2.40 $cmol_c$ kg^{-1}
Bulk density	1.50 g cm^{-3}
Infiltration rate (intake rate)	30.00 mm min^{-1}
Saturated hydraulic conductivity	0.007 cm sec^{-1}
Field capacity (pF1.8)	6 %
Permanent wilting point (pF4.2)	2 %
Texture	Sand

Table 1. Selected physicochemical characteristics of soil

Pot experiments were compared for evapotranspiration of barley and salt accumulation in soils under the effects of similar saline irrigation treatments at the Arid Land Research Center, Tottori University, Japan in the following three environments: i) a glasshouse (Gl. H.) condition from Feb to Apr, 2005; ii) a controlled growth chamber (Gr. Ch.) condition where the day / night temperature was 20 / 18 °C, relative humidity was 60 % and light intensity was 80000 lux; and iii) a greenhouse (Gr. H.) condition ranged from May to Jun, 2005. For each condition, air temperature and relative humidity were measured continuously by relative humidity and temperature meter (HOBO, Pro Series, onset, Japan) and Pyranometer (EKO, MS-601F) for solar radiation (Kw m^{-2}). Ten barley seeds were sown in each plastic pot (size- depth: 20 cm, diameter: 16 cm) filled with sand dune soil. After 18 days of sowing, plants were irrigated at a depth of 1.1 mm ET_c (crop evapotranspiration) with diluted seawater treatments of 3, 8 and 13 dS m^{-1}. The required amount of water was given daily depending on the loss of water, which was estimated by weighing the pots. Extra 10 % water was also applied as a leaching fraction. Evaporation was also monitored by placing the evaporation pans (class A) among pots. A basal dose of liquid fertilizer of NPK was added to plant in the irrigation water. The EC of drainage water was measured by a calibrated conductivity meter (Horiba DS-14). Occasionally, the pots were rotated randomly and spacing was provided in accordance with the growth of plants. Plant height and leaf area (by portable area meter LI-3000A) were monitored during the experiments. Two months old plants were harvested and their fresh and dry biomass was weighed. Stress factor, crop coefficient factor and plant water deficit were also calculated by using the formula:

$$\text{Stress factor } (K_s) = 1 - [(b / 100\ Ky)\ (EC_e - EC\ \text{threshold})]$$

where b is the percentage reduction in crop yield per 1 dS m^{-1} which is equal to 5, Ky is the yield response factor equal to 1, EC_e is the soil salinity. The threshold EC value for barley is considered as 8 dS m^{-1} (FAO, 1998).

$$\text{Crop coefficient factor } (K_c) = ET_c / ET_o$$

where ET_c is the crop evapotranspiration and ET_o is the potential evapotranspiration (FAO, 1998).

Plant water deficit (%) = [{(FWc-DWc) / FWc} - {(FWt-DWt) / FWt}] 100

where FWc and DWc indicate fresh and dry weights for control whereas FWt and DWt indicate fresh and dry weights for saline irrigation treatment respectively.
Soil water storage was also calculated with each irrigation by the formula:

Water storage = initial water + (irrigation – drainage – evapotranspiration)

Statistical analysis of data was carried out and the means were compared by LSD test at 5 % probability level.

3. Results and discussion

3.1 Irrigation water quality

Irrigation water is the most important parameter controlling plant life. Water shortage problem could stress plant and reduce its productivity. In the world, different irrigation waters were used for agriculture purposes but applications depend on soil, plant and quality of the irrigation water. Table 2 demonstrates some irrigation waters that were used by different researchers (FAO, 1998; Daoud et al., 2001; Yamamoto et al., 1988). Since fresh water is the best option for optimum plant growth but the shortage of fresh water is compelling researchers toward the use of saline irrigation. Using saline water for agriculture could add soil salinity. Whereas, under sodic conditions, saline water may improve soil physical conditions. Saline waters especially seawater could also contain some beneficial plant nutrients (Table 2). The disability of sandy soil to hold much salt as compared to clayey soil supports the idea of saline irrigation. The plant response to salt stress is complex, since it varies with the salt concentration, the type of ions, other environmental factors, and the stage of plant development depending on the growth conditions. Nutrient availability and uptake by plants in saline environments is related to i) the activity of nutrient ion in the solution, which depends upon pH, pE, concentration and composition, ii) the concentration and ratios of accompanying elements that influence the uptake and transport of this nutrient by roots, and iii) numerous environmental factors (Pessarakli, 1999).

Water type	pH	EC	Na^+	Cl^-	K^+	Ca^{2+}	Mg^{2+}	NO^{3-}	P
		dS m^{-1}				mg L^{-1}			
Fresh water 1	7.6	0.6	22.0	21.3	5.0	58.2	20.7	0.3	-
Fresh water 2	8.1	0.7	39.9	540.0	3.5	95.6	11.2	traces	1.4
Saline water 1	8.0	4.5	1150.0	560.9	3.9	48.0	21.6	-	-
Saline water 2	7.7	5.8	828.0	1366.8	15.6	720.0	108.0	-	-
Seawater	7.5	38.5	11211.0	18834.0	377.0	266.0	1976.0	0.2	2.1

Table 2. Chemical properties of different waters

3.2 Soil parameters

The meteorological data recorded during the experiment are given in Figure 1. The plants were grown during the winter season (Feb–Apr) in the glasshouse whereas greenhouse experiment was conducted in the summer season (May-Jun). The values of temperature, humidity and solar radiations remarkably differed among the places in the order of greenhouse > growth chamber > glasshouse. The greenhouse also showed the highest weather variations.

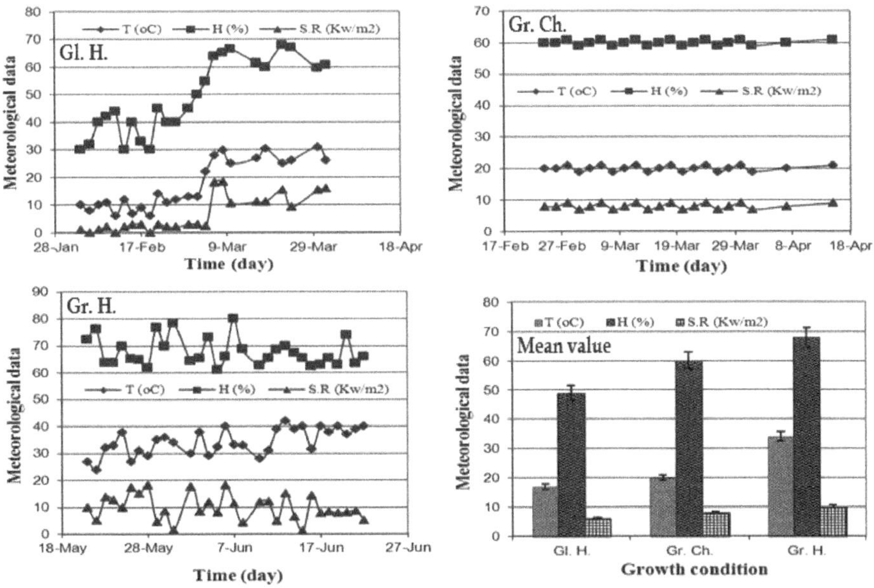

Fig. 1. Meteorological data of three experimental places

Evapotranspiration was remained the function of growing conditions and salinity treatments. Growth chamber significantly showed the highest evapotranspiration followed by the greenhouse and glasshouse (Fig. 2). This could be due to the constant temperature conditions in the growth chamber. The winter temperature of glasshouse was unable to expedite evapotranspiration rate. The weather considerably fluctuated during the day / night within the greenhouse. An averaged day temperature was recorded as 34 ^{o}C in the greenhouse. The evapotranspiration and other soils parameters were differentiated mainly due to climatic differences of the experimental sites. Plant substantially enhanced the evapotranspiration at the peak growth stage. Moreover, heating intensity along with the growing stage of plants tremendously affected the evapotranspiration and salt accumulation in the soils irrespective of the salt treatments. Plants tend to maintain stable tissue water status regardless of temperature when moisture is ample; however, high temperatures severely impair this tendency when water is limited (Machado & Paulsen, 2001).

Salts in the soil water solution can reduce evapotranspiration by making soil water less available for plant root extraction. Salts have an affinity for water and hence additional force is required for the crop to extract water from a saline soil. The presence of salts in the soil water solution reduces the total potential energy of the soil water solution. In addition, some salts cause toxic effects in plants and can reduce plant metabolism and growth (FAO, 1998).

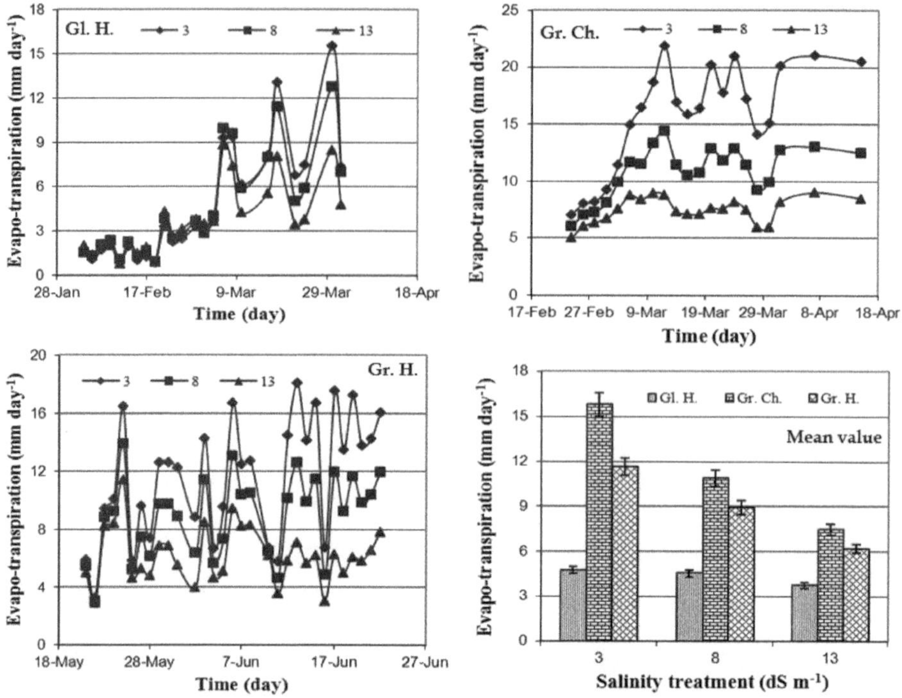

Fig. 2. Evapotranspiration as affected by temperature and salinity treatments

The evapotranspiration values were generally higher under low salt treatment regardless to the weather conditions (Fig. 3). In other words, evapotranspiration was positively related to the quality of irrigation water. A reduced water loss under high saline treatment was measured as compared to low saline water. Reduced bioavailability of water and retarded plant growth under saline irrigation produced a poor evapotranspiration in the system. The depressing effects of salinity on plant growth have been reported by various researchers (Heakal et al., 1990; Abdul et al., 1988; Koszanski & Karczmarczyk, 1985). Saline soils inhibit plant growth through reduced water absorption, reduced metabolic activities due to salt toxicity and nutrient deficiency caused by ionic interferences (Yeo, 1983). Salt concentrations in irrigation water inhibited evaporation from the soil surface (Fig. 3). This phenomenon could be related to the enhanced water density, viscosity and chemical bonds in the soil-salt system. High concentrations of salts also form salt crusts, which could reduce soil evaporation. Richards et al. (1998) reported that density, temperature and salinity affected several water characteristics e. g., evaporation etc. Abu-Awwad (2001) found that high salt concentration at the soil surface is due to high evaporation from wetted areas and the nature of soil water distribution associated with irrigation system. Moreover, Al-Busaidi & Cookson (2005) observed salt crust formation on the soil surface due to saline irrigation, which inhibited evaporation and reduced leaching efficiency.

The data showed that soil water was apparently affected by the quality of irrigation water under all conditions (Fig 4). The water lost by the drainage and evapotranspiration was

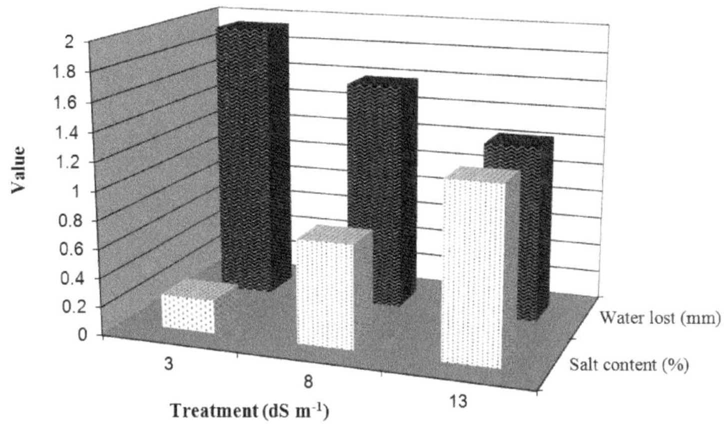

Fig. 3. Salts concentration and water loss as affected by saline water treatments

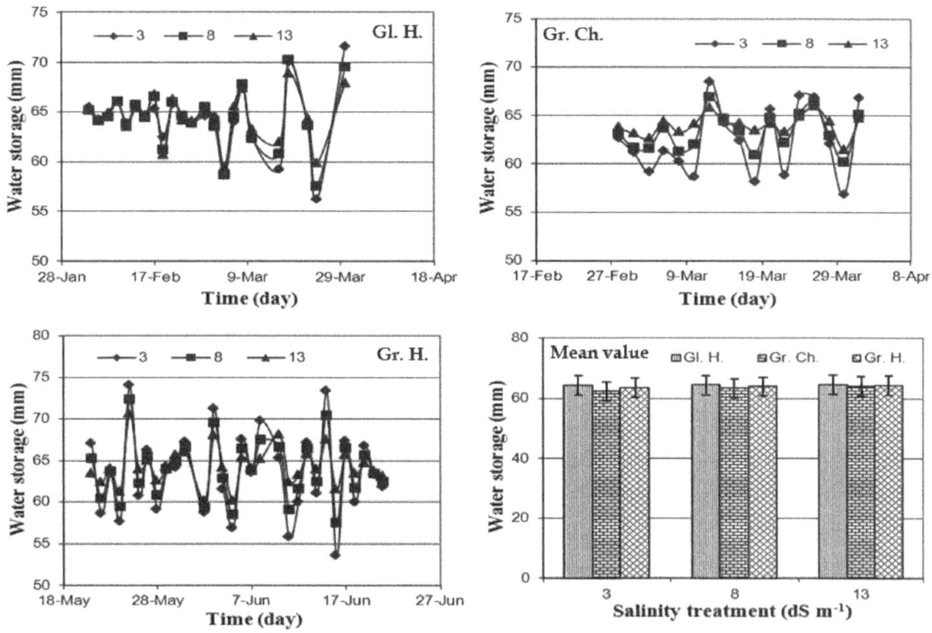

Fig. 4. Water storage as affected by different experimental conditions and saline treatments

counted to determine the water left behind in the soil. The soil water storage remained stable through out the experiment in the glasshouse. However, slight fluctuations during the last days of experiment were observed. The constant temperature of growth chamber exhibited similar trend of water contents under all salinity treatments. Water storage ranged from 54 -74 mm under greenhouse due to high variations in the air temperature and plant growth. There was a high fluctuation of soil water with low salts. The glasshouse upheld the highest magnitude of water in the pots across all saline treatments. In general the saline irrigation water, drainage water salinity and evaporation were appeared interrelated factors. Soil evaporation was positively associated to salt accumulation. High evaporation produced high concentration of salts in the drainage water under increasing salinity of water (Fig. 5). Glasshouse gave little salts increments whereas growth chamber and greenhouse induced severe salinity in drained water. The intensive temperature expedited water evaporation and accumulation salts processes under greenhouse environments. Higher plant biomass and increased evapotranspiration rate in growth chamber salinized drainage water relatively less than greenhouse. Higher drainage salinity in the salt treated pots also signified higher salts concentrations in the root zone.

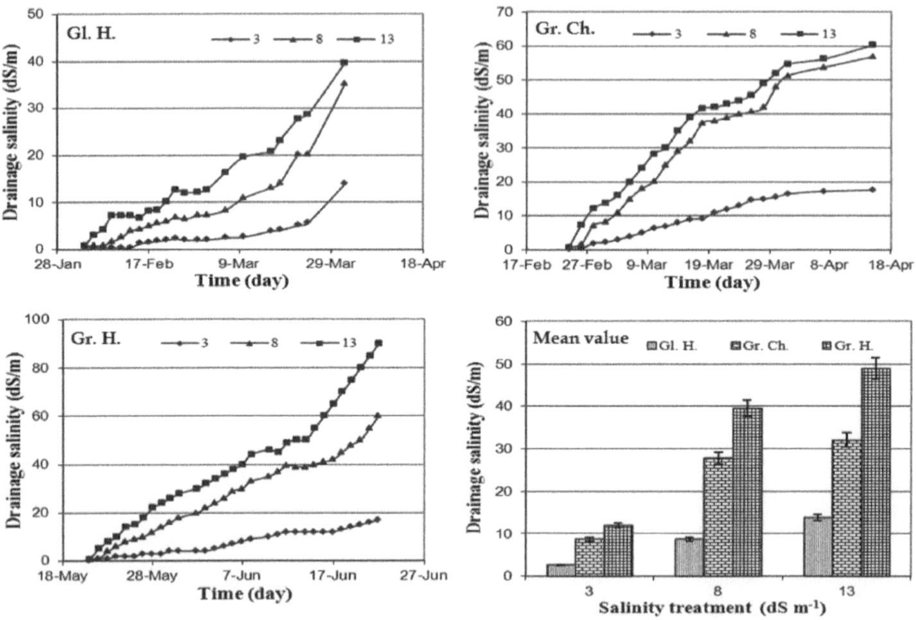

Fig. 5. Salinity status of drainage water as affected by environmental conditions and saline water

Salts accumulation in soil was highly affected by the saline irrigation water (Fig. 6). The accumulation of salts in the soils were varied in the order of glasshouse < growth chamber < greenhouse. Water uptake by plants and evaporation from the soil surface were reported the main factors for salts accumulation in the root zone (Ben-Hur et al., 2001; Bresler et al., 1982). Blanco and Folegatti (2002) found linear soil salinity with the application of saline water down the soil profile with higher salts contents near the surface. Petersen (1996) reported low soil salinity with increased volume of irrigation water due to salt transportation below the root zone.

The sustainable use of saline water in irrigated agriculture controls soil salinity in the field and make a good drainage management that may not pollute the downstream water resources (Beltran, 1999). In this experiment the leaching fraction was too low to produce sufficient drainage water. Therefore, salts were accumulated in higher strength in the soils especially under higher temperature.

3.3 Plant parameters

Soil salinity is one of the principal abiotic factors affecting crop yields in the arid and semi-arid irrigated areas. Plant growth was significantly affected by growth conditions as well as saline irrigation (Table 3). Treatment of less salinity gave the higher biomass production as compared to the high salinity. This finding was also reported by Heakal et al. (1990), Greenway & Munns (1980) and Abdul et al. (1988) when they found that dry matter yield of plant shoots decreased with increasing salinity of water. Koszanski and Karczmarczyk (1985) observed that diluted or undiluted seawater reduced plant height, grain and straw yield of barley and oats. Generally, the soils salinity affects the plants growth by producing an ionic imbalance or water deficit state in the expanded leaves. Shani and Dudley (2001) related the yield loss to reduced photosynthesis, high energy and carbohydrate expenses in osmoregulation and interference with cell functions in saline conditions.

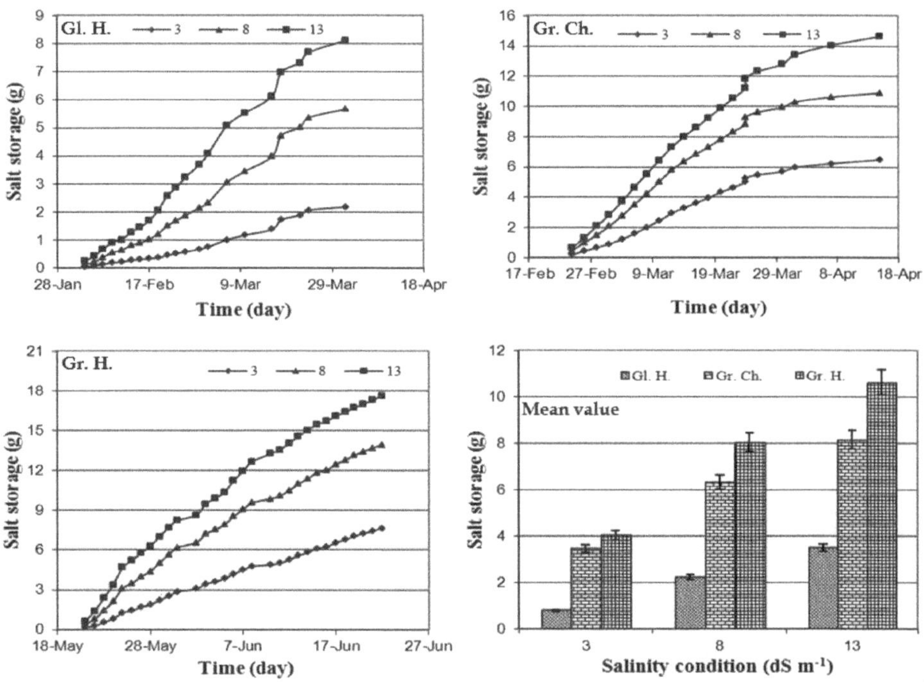

Fig. 6. Salt accumulation as affected by different environments under saline irrigation

Among growing conditions, glasshouse grown barley was the tallest and highest in biomass but relatively less in leaf area. Growth chamber produced a moderate biomass yield than other places. The biomass yield was reduced typically due to higher amount of salt depositions in the hot environment. High temperature was the major stress for plant growth. Sule et al. (2004) reported severe effects of heat stress on cereals in several countries.

The application of saline water in the presence of high temperature conditions exacerbated the process of salt accumulation and plant growth reduction. There have been several reports regarding the reduced and enhanced plant growth in various environments (Marmiroli et al., 1989; Saarikko & Carter, 1996; Sharratt, 1999).

Growing condition	Salinity	Height	Leaf area	Fresh weight	Dry weight
	$dS\ m^{-1}$	cm	cm^2	g	g
Glasshouse	3	65.5a	35.8a	401.3a	75.6a
	8	61.3b	32.0b	313.0b	61.4b
	13	52.2c	26.4c	210.9c	42.8c
Growth chamber					
	3	50.3d	31.6d	373.2d	71.9d
	8	44.5e	20.4e	253.4e	52.0e
	13	40.5f	12.6f	137.6f	33.1f
Greenhouse					
	3	54.0g	50.3g	353.0g	59.5g
	8	47.5h	43.7h	248.5h	44.5h
	13	40.0i	36.2i	144.0i	29.5i

*Means in the column with same letter indicate no difference at Duncan's Multiple Range Test at $P < 0.05$

Table 3. Plant growth parameters as affected by three experimental conditions*

Interactions between salinity and temperature were significant ($p < 0.05$) for the number of tillers, growth of tops and roots. Maximum number of tillers and the highest dry matter were produced when the root temperature was at the intermediate levels of 15 to 20 °C. Effect of salinity on most parameters tested strongly depended on the prevailing root temperature (Mozafar & Oertli, 1992). Moreover, heat stress, singly or in combination with drought, is a common constraint during anthesis and grain filling stages in many cereal crops of temperate regions (Guilioni et al., 2003). Sule et al. (2004) reported that high temperature is one of the environmental stress factors that can affect the growth and quality characteristics of barley. Moreover, it was found by Macnicol et al. (1993) that both water and heat stress reduced yield and grain size of barley. Our experiment also confirmed that growth environments differentiated the performance of barley under the same type of irrigation water. The incorporation of some salts with high temperature could lead to higher loss of plant production (Daoud et al., 2001).

The low temperature in the glasshouse exerted the lowest water deficit as compared to greenhouse and growth chamber (Fig. 7). Extended canopy and large number of tillers per plant apparently caused high transpiration rate in growth chamber. Since plants were irrigated daily at field capacity considering water loss due to evapotranspiration. The water deficit conditions under high salinity treatments could be directly attributed to the impaired water flow from soil to plant. Yeo (1999) reported that root selectivity and transpirational water flow provide the net uptake of salts whereas the salt concentration develops with the growth rate. The greater mass flow of solution through the soil-root interface or higher magnitude of evapotranspiration would increase the salt transport in plants. Thus there is a potential risk of higher salt damages in hot climate.

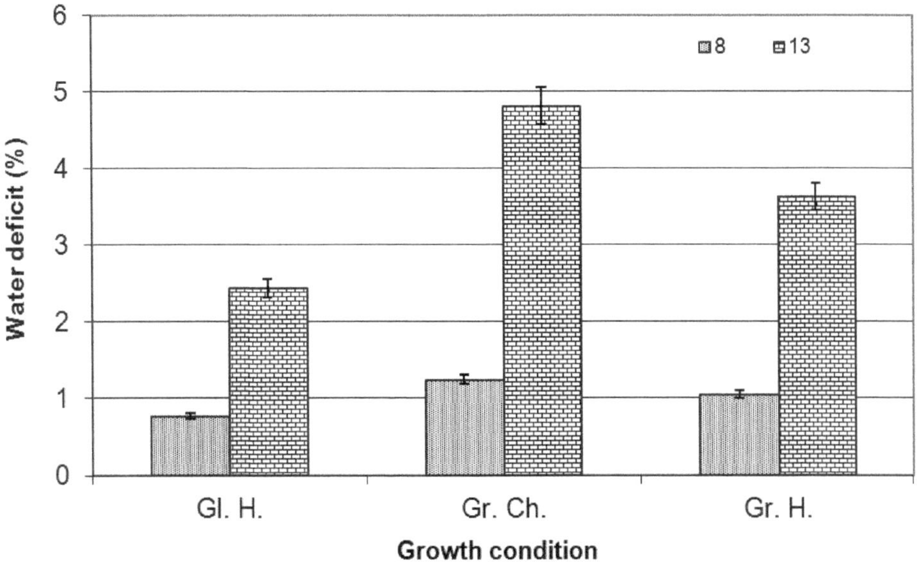

Fig. 7. Plant water deficit as affected by growth conditions

Stress factor (K_s) is an additional parameter to determine crop evapotranspiration. It is an indicator of unusual plants stress such as salinity, deficit water, disease or nutrient imbalance. It implies when its value decreases by less than 1 and smaller K_s value means higher stress. The stress co-efficient was found in the order of greenhouse > growth chamber > glasshouse (Fig. 8). The K_s values greatly decreased under high level of salinity and heat conditions. Pots irrigated with low saline water produced more biomass which did not decline K_s values. The lower K_s values depicted higher accumulation of salts in the pots under the accelerated evapotranspiration due to hot environment. Stress factor (K_s) is an indicator of salt stress problem. It can be seen from Table 4, that low salinity treatment (3 dS m^{-1}) was giving the weakest relationship between K_s and salt storage (ST). Whereas, a very strong relationship (0.90 - 0.98) was observed with high salinity treatments (8 and 13 dS m^{-1}). Table 4 is an extension of Fig. 8 and it is clearly explain how K_s value was changing as soil salinity change.

The crop coefficient factor (K_c) as affected by different conditions and salinity treatments is shown in Figure 9. The approximate standard value reported by FAO (1998) for K_c was 1.15. The K_c values obtained during our experiment were higher than FAO value. This could be attributed to the growth conditions and crop variety. There was a similar trend between K_c and ET_c under all environmental conditions. The level of K_c decreased with increasing level of salts in the water. It was reported that increased evaporation from the soil surface can counteract the reductions in K_c caused by high EC_e of the root zone (FAO, 1998). Letey et al. (1985) and Shalhevet (1994) reported that the effects of soil salinity and water stress were interactive to crop evapotranspiration.

Evapotranspiration data are needed for project planning or designing irrigation scheduling. A large number of empirical methods have been developed by several scientists over the last 50 years to estimate evapotranspiration in different climates. Relationships were often subjected to rigorous local calibrations and were proved to have certain limitations. Usually testing the accuracy of the methods under new conditions is laborious, time-consuming and costly.

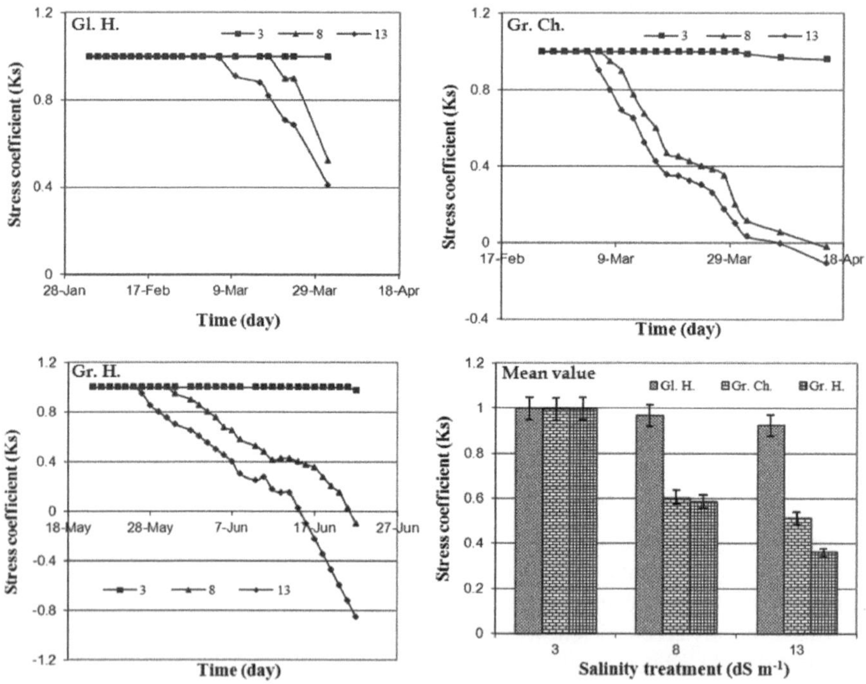

Fig. 8. Stress coefficient as affected by different environments and saline treatments

Growth condition	Salinity treatment (dS m^{-1})	Equation	R^2
Glass house	3	$K_s = 1$	-
	8	$K_s = -0.02(ST)^2 + 0.08ST + 0.95$	0.60
	13	$K_s = -0.02(ST)^2 + 0.08ST + 0.94$	0.90
Growth chamber	3	$K_s = -0.002(ST)^2 + 0.01ST + 0.99$	0.59
	8	$K_s = -0.01(ST)^2 + 0.01ST + 1.02$	0.98
	13	$K_s = -0.002(ST)^2 - 0.05ST + 1.11$	0.98
Green house	3	$K_s = -0.0003(ST)^2 + 0.002ST + 1$	0.21
	8	$K_s = -0.007(ST)^2 + 0.03ST + 0.99$	0.98
	13	$K_s = -0.007(ST)^2 + 0.04ST + 0.92$	0.96

Table 4. Salt storage (ST) relationship with stress coefficient (Ks)

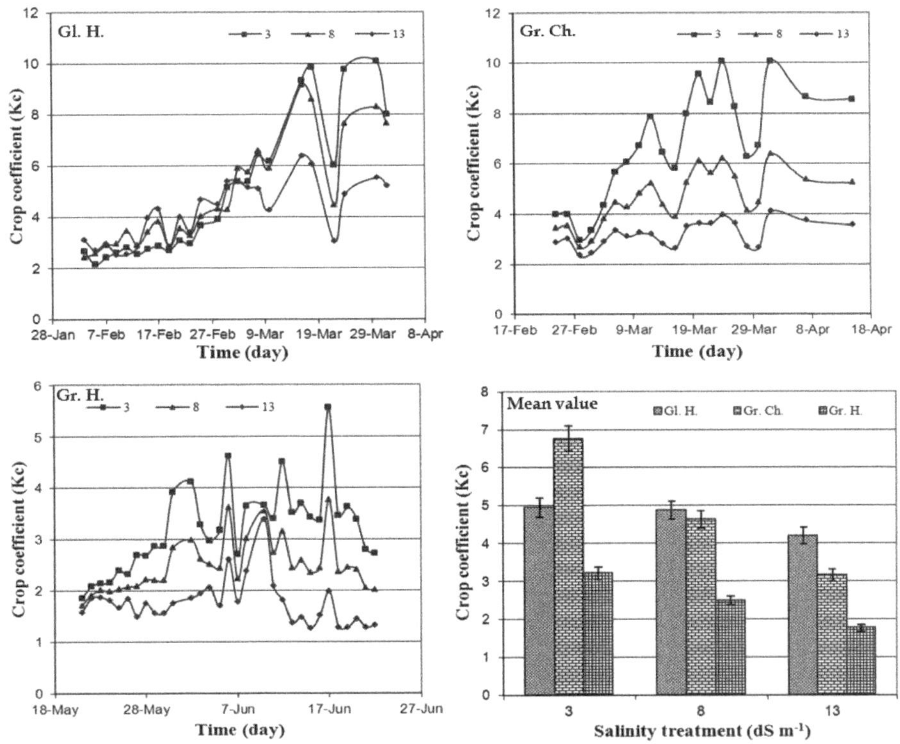

Fig. 9. Crop coefficient value as affected by different growing conditions

Crop evapotranspiration can be calculated from climatic data and crop parameters by using Blaney-Criddle, radiation, Penman and pan evaporation methods (FAO, 1998). The factors like soil salinity, soil water quality and quantity can also affect evapotranspiration value. Since evapotranspiration and crop coefficient factor (K_c) are interconnected parameters indicating plant growth statues. It can be seen that crop coefficient factor was apparently affected by interactions of soil and plant parameters (Table 5). Changes in plant development under glasshouse and growth chamber gave high values of correlation coefficient (0.80 - 0.93) for measured parameters whereas the unstable condition of greenhouse highly reduced the relationship of various parameters (0.31 - 0.60). In general, the increment in water salinity negatively impacted the correlation coefficient values. Temperature, evapotranspiration and stress coefficient controlled the value of crop coefficient factor depending on the growth conditions. It is also reported elsewhere that soil salinity, land fertility, soil management, fertilizers, soil physical condition, diseases and pests were affected crop development and evapotranspiration (FAO, 1998). In FAO report, it was mentioned that increased evaporation under high frequency irrigation of the soil surface can counteract reductions in K_c caused by high EC_e of the root zone. Under these conditions, the total K_c and ET_c are not very different from the non-saline, standard conditions under less frequent irrigation, even though crop yields and crop transpiration are reduced. Because of this, under saline conditions, the K_s reducing factor should only be applied with the dual K_c approach (FAO, 1998). In reviewed articles on impacts of salinity on crop production, Letey et al. (1985) and Shalhevet (1994) concluded that the effect of soil salinity and water stress are generally additive in their impacts on crop evapotranspiration. Therefore, the same yield-ET functions may hold for both water shortage induced stress and for salinity induced stress. Moreover, some equations used to calculate K_c or K_s are not expected to be accurate for predicting ET_c for specific days (FAO, 1998).

Growth condition	Salinity treatment ($dS\ m^{-1}$)	Dependent variable Temp (°C)	ET (mm)	WS (mm)	Ks (-)	Intersection point	Multiple correlation coefficient
Glass-house	3	0.04	0.41	-0.11	1.00	8.02	0.93
		(15.29%)	(68.03%)	(16.70%)	(-%)		
	8	0.02	0.35	-0.01	-2.96	10.90	0.91
		(10.74%)	(56.63%)	(15.36%)	(17.25%)		
	13	0.05	0.13	-0.0004	0.16	1.41	0.80
		(60.53%)	(36.10%)	(0.11%)	(3.25%)		
Growth-chamber	3	0.37	0.24	0.02	-5.43	-1.60	0.92
		(27.58%)	(65.52%)	(3.44%)	(3.41%)		
	8	0.32	0.09	0.04	-0.79	-5.25	0.92
		(42.56%)	(20.74%)	(6.57%)	(29.82%)		
	13	0.11	0.10	0.07	-0.33	-4.79	0.81
		(29.91%)	(25.47%)	(16.80%)	(27.86%)		
Green-house	3	-0.01	0.13	-0.04	45.03	-40.86	0.57
		(6.40%)	(52.59%)	(20.73%)	(20.29%)		
	8	0.01	-0.02	0.03	-0.28	0.32	0.31
		(18.60%)	(19.45%)	(30.57%)	(31.43%)		
	13	-0.06	0.08	0.04	-0.17	0.19	0.60
		(45.39%)	(22.62%)	(16.42%)	(15.56%)		

Table 5. Multiple regression analysis for crop coefficient (K_c) data

4. Conclusions

In agricultural systems, evapotranspiration process and plant productivity are strongly influenced by environmental conditions. Therefore, this study concludes that the growth conditions had profound effects on the evapotranspiration process, soil salinity and barley growth. Salt concentrations in irrigation water inhibited evaporation from the soil surface. Therefore, soil water was negatively related to the quality of irrigation water. Salt status of soils was highly increased by high temperature and evapotranspiration especially in the greenhouse and growth chamber. The barley grew well in high saline water under a favorable climate of glasshouse. Under low salinity treatment plants showed no stress as compared to high salinity treatment especially under greenhouse and growth chamber conditions. High temperature in the greenhouse and continuous heating in the growth chamber lowered the plant growth and water stress coefficient (K_s). Crop coefficient (K_c) also differed in all growing conditions with salinity treatments. The level of K_c decreased with increasing level of salts in the water and its value was controlled by temperature, evapotranspiration and stress coefficient. This experiment indicated that when saline water is used for irrigation due attention should be given to minimize root-zone salinity. However, a good management of soil and water could be a viable option for sustainable agriculture in salt affected soils. There is further need to evaluate poor water quality on different crops in arid and semi arid fields' conditions.

5. References

Abdul, K. S.; Alkam, F. M. & Jamal, M. A. (1988). Effects of Different Salinity Levels on Vegetative Growth, Yield and its Components in Barley. *ZANCO*, 1: 21-32

Abu-Awwad, A. M. (2001). Influence of Different Water Quantities and Qualities on Limon Trees and Soil Salts Distribution at the Jordan Valley. *Agricultural Water Management*, 52: 53 - 71.

Agueda, G.; Isaura, M. & Luis, A. (1999). Barley Yield in Water-Stress Conditions. The Influence of Precocity, Osmotic Adjustment and Stomatal Conductance. *Field Crops Research*, 62: 23-34.

Al-Busaidi, A. & Cookson, P. (2005). Leaching Potential of Sea Water. *Journal for Scientific Research: Agricultural and Marine Sciences (SQU)*, 9: 27-30.

Atienza, S. G.; Faccioli, P.; Perrotta, G.; Dalfino, G.; Zschiesche, W.; Humbeck, K.; Michele Stanca, A. & Cattivelli, L. (2004). Large Scale Analysis of Transcripts Abundance in Barley Subjected to Several Single and Combined A biotic Stress Conditions. *Plant Science*, 167: 1359-1365.

Beltran, J. M. (1999). Irrigation with Saline Water: Benefits and Environmental Impact. *Agricultural Water Management*, 40: 183–194.

Ben-Hur, M.; Li, F. H.; Keren, R.; Ravina, I. & Shalit, G. (2001). Water and Salt distribution in A field Irrigated with Marginal Water under High Water Table Conditions. *Soil Science Society of American Journal*, 65: 191-198.

Blanco, F. F. & Folegatti, M. V. (2002). Salt Accumulation and Distribution in A greenhouse Soil as Affected by Salinity of Irrigation Water and Leaching Management. *Revista brasileira de engenharia agricola e ambiental,* 6: 414- 419.

Bresler, E.; McNeal, B. L. & Carter, D. L. (1982). *Saline and Sodic Soils*: Principles-Dynamics-Modeling. Berlin: Springer-Verlag, p. 236.

Curtin, D.; Steppuhn, H. & Selles, F. (1993). Plant Responses to Sulfate and Chloride Salinity: Growth and Ionic Relations. *Soil Science Society of American Journal,* 57: 1304-1310.

Daoud, S.; Harrouni, M. C. & Bengueddour, R. (2001). Biomass Production and Ion Composition of some Halophytes Irrigated with Different Seawater Dilutions. *First International Conference on Saltwater Intrusion and Coastal Aquifers- Monitoring, Modeling, and Management.* Essaouira, Morocco.

FAO (1998). *Crop Evapotranspiration.* FAO Irrigation and Drainage Papers No. 56. Rome.

Gee, G. W. & Bauder, J. W. (1986). Particle-Size Analysis. In: Klute, A. (ed) *Methods of Soil Analysis, Physical and Mineralogical Methods,* Part 1. 2nd edition. Agronomy series No. 9. Am. Soc. Agronomy and Soil Sci. Soc. Am. Inc. Publ. Madison, Wisconsin USA, pp. 383-404.

Ghadiri, H.; Dordipour, I.; Bybordi, M. & Malakourti, M. J. (2005). Potential use of Caspian Sea for Supplementary Irrigation in North Iran. *Agricultural Water Management,* 79: 209–224.

Greenway, H. & Munns, R. (1980). Mechanisms of Salt Tolerance in Nonhalophytes. *Annual Review of Plant Physiology,* 31: 149–190.

Guilioni, L.; W´ery, J. & Lecoeur, J. (2003). High Temperature and Water Deficit may Reduce Seed Number in Field Pea Purely by Decreasing Plant Growth Rate. *Funct. Plant Biol.,* 30: 1151–1164.

Hall, A. E. (1992). Breeding for Heat Tolerance. Plant Breed. Rev. 10, 129–168.

Hamdy, A.; Sardo, V. & Farrage Ghanem, K. A. (2005). Saline Water in Supplemental of Wheat and Barley under Rainfed Agriculture. *Agricultural Water Management,* 78: 112–127.

Heakal, M. S.; Modaihsh, A. S.; Mashhady, A. S. & Metwally, A. I. (1990). Combined Effects of Leaching Fraction of Salinity and Potassium Content of Waters on Growth and Water Use Efficiency of Wheat and Barley. *Plant and Soil,* 125: 177-184.

Klute, A. (1986). Part I, 2nd ed, *Soil Science Society of America Book Series* No. 5. Soil Science Society of America and American Society of Agronomy.

Koszanski, Z. & Karczmarczyk, S. (1985). Use of Saline Water for Irrigation of Spring Barley and Oats. *Zeszyty-Naukowi-Akademii-Rolniczej-w-Szczecinie, -Rolnictwo,* 36: 95-105.

Letey, J.; Dinar, A. & Knapp, K. C. (1985). Crop-Water Production Function Model for Saline Irrigation Waters. *Soil Science Society of American Journal,* 49: 1005-1009.

Maas, E. V. & Hoffman, G. J. (1977). Crop Salt Tolerance-Current Assessment. *Journal of Irrigation and Drainage Division,* 103: 115-134.

Machado, S. & Paulsen, G. M. (2001). Combined Effects of Drought and High Temperature on Water Relations of Wheat and Sorghum. *Plant Soil,* 233.

Macnicol P. K.; Jacobsen J. V.; Keys M. M. & Stuart I. M. (1993). Effects of Heat and Water Stress on Malt Quality and Grain Parameters of Schooner Barley Grown in Cabinets. *Journal of Cereal Science,* 18: 61-68.

Marmiroli, N.; Lorenzoni, C.; Michele, S. A. & Terzi, V. (1989). Preliminary study of the inheritance of temperature stress proteins in barley (*Hordeum vulgare* L.). *Plant Science,* 62: 147-156.

Matin, M. A.; Brown, J. H. & Ferguson, H. (1989). Leaf Water Potencial, Relative Water Content, and Diffusive Resistance as Screening Techniques for Drought Resistance in Barley. *Agron. J.,* 81: 100-105.

Mozafar, A. & Oertli, J. J. (1992). Root-Zone Temperature and Salinity: Interacting effects on Tillering, Growth and Element Concentration in Barley. *Plant and Soil,* 139: 31–38.

Oron, G.; DeMalach, Y.; Gillerman, L.; David, I. & Lurie, S. (2002). Effect of Water Salinity and Irrigation Technology on Yield and Quality of Pears. *Biosystems Engineering,* 81: 237 - 247.

Pessarakli, M. (1999). Handbook of Plant and Crop Stress. CRC, USA.

Petersen, F. H. (1996). *Water testing and interpretation.* In: Reed, D.W. (*ed.*) Water, Media, and Nutrition for Greenhouse Crops. Batavia: Ball, pp. 31-49.

Richards, G. A.; Pereira, L. S.; Raes, D. & Smith, M. (1998). *Crop Evapotranspiration*: Guidelines for Computer Crop Water Requirements. FAO Irrigation and Drainage Paper, 65: p. 5.

Saarikko, R. A. & Carter, T. R. (1996). Phenological Development in Spring Cereals: Response to Temperature and Photoperiod under Northern Condition. *European Journal of Agronomy,* 5: 59 – 70.

Shalhevert, J. (1994). Using Water of Marginal Quality for Crop Production: major issues. *Agricultural Water Management,* 25: 233–269.

Shani, U. & Dudley, L. M. (2001). Field Studies of Crop response to Water and Salt Stress. *Soil Science Society of American Journal,* 65: 1522–1528.

Shannon, M. C. (1984). *Breeding, Selection and the Genetics of Salt Tolerance.* In: Staples, R.C., Toenniessen, A.H. (Eds.), Salinity Tolerance in Plants. John Wiley, New York, pp. 231-254.

Sharratt, B. (1999). Thermal Requirements for Barley Maturation and Leaf Development in Interior Alaska. *Field Crop Research,* 63: 179-184.

Sule, A.; Vanrobaeys, F.; Hajos, G.; Van Beeunen, J. & Devreese, B. (2004). Proteomic Analysis of Small Heat Shock Protein Isoforms in Barley Shoots. *Phytochemistry,* 65: 1853-1863.

Yamamoto, T.; Fujiyama, H.; Li, P.; Gao, Y. & Yang, Z. (1988). Studies on Irrigation Utilization of Ground Water in the Mu Us Shamo of China: A few characteristics of leaching from lysimeters grown with grass. *Sand Dune Research,* 27: 29–36.

Yeo, A. R. (1983). Salinity Resistance: Physiologies and Prices. *Plant Physiology,* 58: 214-222.

Yeo, A. R. (1999). Prediction the Interaction between the Effects of Salinity and Climate Change on Crop Plants. *Scientia Horticulture,* 78: 159–174.

Zhu, X.; & Takeo, A. (2010). Effect of Soil Water Content and Salinity on Daily Evaporation from Soil Column. *Journal of American Science*, 6: 576-580.

14

Modelling Evapotranspiration of Container Crops for Irrigation Scheduling

Laura Bacci[1*] et al.
[1]Istituto di Biometeorologia (IBIMET) – Consiglio Nazionale delle Ricerche – Firenze Italia

1. Introduction

Irrigation is now recognized as an important component in the agriculture economy of Mediterranean regions. As practiced by many growers, it is often based on traditional application methods that fail to measure the supply of water needed to satisfy the variable requirements of different crops. In order to achieve more profitable and sustainable cropping systems, it is essential to modernize existing irrigation systems and improve irrigation water use efficiency (WUE). Up-to-date methods of irrigation should likewise be based on sound principles and techniques for attaining greater control over the soil-crop-water regime and for optimizing irrigation in relation to all other essential agricultural inputs and operations. Accurate predictions of crop water requirements are necessary for an efficient use of irrigation water in container crops cultivated both outdoors and in greenhouse.

Irrigation scheduling (IS) has conventionally aimed to achieve an optimum water supply for productivity, with soil or container water content being maintained close to field capacity. Different approaches to IS have been developed, each having both advantages and disadvantages but despite the number of available systems and apparatus, not entirely satisfactory solutions have been found to rationalize IS, assuring optimal plant growth with minimal water use (Jones, 2004). Many growers, especially in the Mediterranean regions, use simple timers for automated irrigation control of containerized crops and scheduling is based on their own experience.

Evapotranspiration (*ET*) is the primary process affecting crop water requirements and, therefore, its knowledge is essential for efficient irrigation management. *ET* is the combined process of evaporation from soil or substrate and leaf transpiration. Evapotranspiration requires two essential components: a source of energy and a vapour transport mechanism. Energy is needed for phase change from liquid to vapour in sub-stomatal cavities whereas the leaf-to-air vapour pressure gradient ensures that water vapour crosses leaf stomata.

In container-grown plants, *ET* is affected by many factors, both environmental (e.g. air temperature, radiation, humidity, wind speed) and plant related characteristics (e.g. growth

---

* Piero Battista[1], Mariateresa Cardarelli[2], Giulia Carmassi[3], Youssef Rouphael[2], Luca Incrocci[3], Fernando Malorgio[3], Alberto Pardossi[3], Bernardo Rapi[1] and Giuseppe Colla[2]
[2] *Dipartimento di Scienze e Tecnologie per l'Agricoltura, le Foreste, la Natura e l'Energia – Università della Tuscia – Viterbo, Italia*
[3] *Dipartimento di Biologia delle Piante Agrarie – Università di Pisa – Pisa, Italia*

phase, leaf area), as well as type of growing substrate and container size. Any method used to accurately estimate plant water requirements must take these environmental and plant factors into account. The irrigation frequency of plants growing in container can be based on measured or calculated *ET* (Treder et al., 1997).

According to Baille (2001), the available advanced methods for controlling irrigation at the short-term decision level are based either on climate or on soil moisture status.

In the climate-based method, crop water use is computed by means of algorithms that estimate *ET* using meteorological data. The Penman-Monteith (PM) equation (Monteith, 1973; Stanghellini, 1987) and its simplified versions (e.g. Baille et al., 1994) have been used for predicting *ET* in many container-grown crops, where substrate evaporation losses are generally negligible and *ET* is determined almost exclusively by crop transpiration.

The soil-based method uses the measurement of soil water potential or content. A combination of climate and soil-based methods would be recommended, because this allows a check of the coherence and concordance between data regarding soil moisture and crop water demand thus making IS more reliable and accurate.

In this chapter, different approaches for *ET* modelling in container crops grown both in greenhouse and outdoor will be described and its application to IS is briefly discussed.

2. Modelling crop evapotranspiration in outdoor container-grown nursery stock

Among container crops, nursery stock ornamental plants typically exhibit a fast growth rate and require plenty of nutrients and water of high quality, due to the susceptibility to salinity that typically characterizes these crops. In these crops WUE is generally poor, especially in container crops, and the environmental impact may be remarkable due to the waste of water and the pollution of rivers and groundwater with fertilisers and other agrochemicals. Annual use of irrigation water ranges from 500 mm in soil-grown crops to 1,000-1,500 mm in container crops, which are increasingly used in many countries (e.g. United States, The Netherlands, Italy). For instance, in Pistoia (Tuscany, Italy) which currently is the most important district in Europe for the cultivation of landscaping ornamentals, approximately 1,400 ha, over approx. 4,500 ha of nurseries, are used for pot ornamentals with an estimated consumption of irrigation water (mostly groundwater) of more than 10,000,000 m^3/year. Even in the most advanced nurseries, simple timers are used for irrigation control and a "water-man" is in charge for adjusting timers during the growing season, whereas the estimation of crop *ET* may provide the basis for efficient IS.

In the framework of the European project Flowaid (Balendonck, 2010), between 2007 and 2009, the research group of Pisa University conducted a series of experiments to design and test a simple *ET* model for four ornamental species (*Photinia x fraseri; Viburnum tinus; Prunus laurocerasus; Forsythia intermedia*) grown in container in a peat-pumice mixture (Pardossi et al., 2009a).

Pardossi et al. (2009a) used the classical "two-step" approach to calculate of *ET* [$ET = ET0 \times Kc$ (Allen et al., 1998)], which required the knowledge of crop coefficient (K_C) and reference evapotranspiration (ET_0). The latter quantity was calculated with the California Irrigation Management Information System (CIMIS) equation (CIMIS, 2009). The CIMIS formula uses a modified PM equation with a wind function developed at the University of California, Davis, CA. Alternatively, ET_0 can be determined by evaporation pan (Brouwer & Heibloem, 1986). CIMIS equation is described in details in par. 4.2.

Main limitation to the application of "two-step" approach to calculation of ET is represented by K_C, which is crop specific and depends on leaf area, growth stage, climatic conditions and management practices. This issue is especially critical in container-grown ornamentals as the number of species/cultivars in any middle-size nursery is often in the range of hundreds and initiating time of crop production can be any time during the year for small containers. For this reason, K_c values of ornamental crops are rarely known and their behaviour and values during the growing season, as a result of cultivation in containers, are very different from those of a uniform crop canopy (Regan, 1994). Hence, it must be determined experimentally. Crop coefficient is related to the degree of soil cover by crop canopy and thus to leaf area index (*LAI*). Several authors attempted to model the evolution of K_C and *LAI* during the growing season (e.g. Orgaz et al., 2005).

In the work conducted by Pardossi et al. (2009a), the analysis of the seasonal changes of crop height (*H*), *LAI*, *Kc* and the ratio of K_C on *LAI* (*C*) in each species suggested the possibility to predict *LAI* from non-destructive measurements of *H* and then to compute K_C from *LAI* assuming a constant value for *C*:

$$ET = C \cdot LAI \cdot ET_0 \tag{1}$$

Linear regressions between the two variables *LAI* and *H* were computed for *Forsythia* (r^2=0.87), *Photinia* (r^2=0.75) and *Viburnum* (r^2=0.71), while a non-linear (exponential) regression was more adequate for *Prunus* (r^2=0.56). All these regressions were significant. In all species, *C* values oscillated during the growing season without showing any trend or possible relation with plant age and averaged 0.389, 0.352, 0.377 and 0.289 in *Forsythia, Photinia, Prunus* and *Viburnum*, respectively. The divergence of the C coefficient between *Forsythia, Photinia* and *Prunus*, respect to *Viburnum*, could be mainly due to the different crop habitus. The first three species have erect, ascending stems with large leaves, whereas *Viburnum* has a rounded, compact habit with small leaves. For model calibration, a mean value of 0.372 was used for *Forsythia, Photinia* and *Prunus*, while for *Viburnum* value was 0.289. Therefore, the following models, derived from equation 1 and implementing a crop-specific regression of *LAI* vs. *H*, were used to simulate daily *ET* of the four species under investigation (all values of r^2 were significant):

1. Forsythia: $ET = 0.372\ (2.938\ H - 0.276)\ ET_0$ ($r^2 = 0.53$)
2. Prunus: $ET = 0.372\ (0.294^{1.957\ H} - 0.276)\ ET_0$ ($r^2 = 0.43$)
3. Photinia: $ET = 0.372\ (2.938\ H - 0.276)\ ET_0$ ($r^2 = 0.42$)
4. Viburnum: $ET = 0.289\ (6.318\ H - 0.952)\ ET_0$ ($r^2 = 0.33$)

Grouping the data for all species, a significant (r^2 = 0.69) linear regression was computed between simulated and measured values, with a slope (0.96 ± 0.04) and an intercept (0.15 ± 0.09) not statistically different from 1 and 0, respectively (as assessed by using the least squares method). The mean (absolute) deviation between simulations and observations was nearly 23%; it was slightly higher for *Photinia* (27%) and *Viburnum* (25%) than for *Forsythia* (20%) and *Prunus* (18%).

In 2009, the research group of Pisa University compared two methods for IS in a nursery production scheme with four different crop species (the same used for modeling) grown in the same irrigation plot, according to the standard practice of nurserymen in Pistoia. In particular, timer-based control was compared with an irrigation control system based on *ET* modeling. In this system, irrigation was automatically activated whenever cumulated *ET* (estimated on hourly time step) of the reference crop exceeded a predetermined value corresponding to the maximum allowed substrate moisture deficit. Every week, H was measured on ten representative plants of each species and K_C was calculated accordingly. The species with the highest K_C regulated the irrigation frequency during the following week. As a matter of fact, the calculated Kc is the input value in the software integrated in the irrigation controller, which also computed ET_0 using the CIMIS PM equation. The two methods for IS did not affect crop ET and growth. However, compared to timer, the application of ET model reduced by approximately 40% the seasonal water use as a result of lower irrigation frequency.

3. Modelling evapotranspiration in greenhouse crops

Protected crops are often over-irrigated and this result in water loss and pollution due to fertilisers leaching (Vox et al., 2010). Thompson et al. (2007) reported that inappropriate IS was responsible for nitrate leaching from greenhouse tomato crops in Almeria.

Annual use of irrigation water in protected cropping system ranges from 150-200 mm (kg m^{-2}) in short-cycle, soil-grown crops, such as leafy vegetables, to 1,000-1,500 mm in soilless-grown row crops such as solanaceas and cucurbits.

ET model may be a relevant component of a decision support system (DSS) for greenhouse growing system. It could be employed for both on-line (daily management of irrigation) and off-line (simulation study to define best strategies for managing irrigation or fertigation) decisions. In addition, simulation model for ET can provide a soft-sensor for an early warning system for the grower (Elings & Voogt, 2006). A reduction of actual ET (as measured, for instance, with weighing gutter) with respect to predicted ET suggests alterations of plant water status resulting from technical failure of irrigation system, mistakes in fertigation management or the occurrence of root diseases.

Models of different complexity have been developed for predicting ET in greenhouse crops. Among the different approaches used to calculate ET, the FAO Penman-Monteith (PM) equation is currently considered a standard reference (Stanghellini, 1987; Allen et al., 1998; Walter et al., 2002):

$$\lambda \cdot ET = \frac{\Delta \cdot R_n + \left(\frac{\rho \cdot cp}{r_a} \right) \cdot \left(e_a - e_a \right)}{\Delta + \gamma \cdot \left(1 + \frac{r_c}{r_a} \right)} \tag{2}$$

where Rn is the net radiation (kJ m^{-2} s^{-1}), (es-ea) is the vapour pressure deficit of the air (kPa), ρ is the mean air density at constant pressure (kg m^{-3}), λ is the latent heat of vaporization (J kg^{-1}), cp is the specific heat of the air (kJ kg-^{1}), Δ is the slope of the saturation vapour pressure temperature relationship (kPa °C^{-1}), γ is the psychrometric constant (kPa °C^{-1}), and r_c and r_a are the canopy and aerodynamic resistances (s m^{-1}). These parameters are directly measured or calculated from weather data.

Nearly isothermal conditions generally occur under greenhouse and thus long-wave exchange can be neglected in first approximation (Bontsema et al., 2007). Some authors (Baille et al., 1994; Bailey et al., 1993) reported that in most conditions the contribution of radiation to crop transpiration depended only from the short-wave component (e.g. incident

R). Moreover, under unheated greenhouse conditions R_n generally matches R during the light period (Bailey et al., 1993). Indeed, in a study conducted with gerbera in a unheated glasshouse at the University of Pisa, R_n was closely related to R (A. Pardossi and P. Battista, unpublished):

$$Rn = 0.981 \cdot R \quad (r^2 = 0.923;\ n = 590)$$

As crop canopy is not homogenous, the radiation term in equation 2 corresponds to the radiation intercepted by the crop, which is estimated from LAI and the light interception coefficient (k) of the crop, as follows:

$$R_{\text{int}} = 1 - \exp^{-k \cdot LAI} \tag{3}$$

Along with environmental variables (net radiation intercepted by the canopy, air and leaf temperature, and vapour pressure), the r_s (stomatal resistance, s m^{-1}) and r_a have to be accurately assessed. The r_s value can be directly measured with a porometer, and the measured values can be related to the main environmental variables. For a hypostomatous crop, the exchange area for latent heat is the leaf area index (LAI). Accordingly, a mean canopy resistance, r_c, can be defined as (Bailey et al., 1993; Montero et al., 2001):

$$r_c = \frac{r_s}{LAI} \tag{4}$$

The aerodynamic resistance of the canopy to the transfer of vapour (r_a) can be obtained from the convective heat transfer coefficient, as the eddy diffusion process transports both air and water vapour. The relationship between the canopy resistance and the heat transfer coefficient for individual leaves can be assumed to be (Bailey et al., 1993):

$$r_a = \frac{\rho \cdot c_p}{2 \cdot LAI \cdot h} \tag{5}$$

where ρ is the mean air density at constant pressure (kg m^{-3}), cp is the specific heat of the air (kJ kg^{-1}), LAI is the leaf area index and h is the heat transfer coefficient (W m^{-2} K^{-1}).

Convective heat transfer is generally analyzed using the Grashof or Reynolds numbers. The numbers of Grashof and Reynolds correspond to the air flow occurring in free and forced convection, respectively. The Grashof number is a function of the temperature difference between the leaf (T_l) and the air (T_a):

$$\text{Gr} = \frac{\beta\, g\, d^3 \left| T_l - T_a \right|}{\nu} \tag{6}$$

where β is the thermal expansion coefficient of air (1/K), g is the acceleration due to gravity (m s^{-2}), d is the characteristic dimension of the leaf (m) and ν is the kinematic viscosity of air (m^2 s^{-1}).

The Reynolds number can be expressed as a function of air velocity (u, m s^{-1}):

$$\text{Re} = \frac{u \cdot d}{\nu} \tag{7}$$

Leaves are generally thin, so the temperatures of the upper and lower surfaces may be assumed to be equal. Therefore, the formula proposed by McAdams (1954) could be used to calculate the average heat transfer coefficient for free convection:

$$h = 0.37\left(\frac{k_c}{d}\right)Gr^{1/4} \tag{8}$$

where k_c is the thermal conductivity of air (W m^{-1} K^{-1}).
When moving air comes into contact with a warm body, heat is lost by forced convection. The average value of the heat transfer coefficient for forced convection is given by Grober & Erk (1961) as:

$$h = 0.60\left(\frac{k_c}{d}\right)Re^{1/2} \tag{9}$$

Under greenhouse, the air is rarely stationary. Stanghellini (1987) found that the convective heat transfer is due to forced as well as free convection (mixed convection) and proposed the following expression:

$$h = 0.37\left(\frac{k_c}{d}\right)(Gr + 6.92\,Re^2)^{1/4} \tag{10}$$

In all cases the characteristic dimension of the leaf (*d*, m), is defined by:

$$d = \frac{2}{(1/l) + (1/w)} \tag{11}$$

where *l* and *w* are, respectively, average length and width of the leaves.
Obviously, accuracy of *ET* predictions with the PM equation depends on the accuracy of the measurements or estimates of both *LAI* and r_s. Leaf area could be determined either by destructive sampling or by non-destructive measurements of leaf dimensions (e.g. Carmassi et al., 2007b; Rouphael & Colla, 2004) or whole leaf area (with hand-held ceptometer).
Generally, two sub-models for *LAI* and r_s are aggregated to the PM equation. The evolution of *LAI* is often modeled through computation of growing degree days (GDD). Accumulated thermal time was used widely to describe crop growth both in open field and in greenhouse crops (e.g. Incrocci et al., 2006; Pasian & Lieth, 1996; Xu et al., 2010).
However, since under greenhouse conditions the leaf area components (e.g. leaf appearance rate, leaf number) are not only affected by temperature, but also by photosynthetically active radiation (Heuvelink & Marcelis, 1996), the use of product of daily thermal time and daily photosynthetically active radiation instead of GDD gave a better estimation of leaf area components for greenhouse crops (Rouphael et al., 2008).
Leaf stomatal resistance could be modeled as a function of environmental variables, in particular irradiance and vapour pressure deficit (*VPD*). For instance, Rouphael & Colla (2004) reported that in pot-grown zucchini (*Cucurbita pepo* L.) the response of r_s to *R* (expressed in W m^{-2}) was explained by an exponential regression (Fig. 1).
The PM equation aggregated to sub-model for both *LAI* and r_s has been successfully applied at all scales, from single leaves to whole canopies (Stanghellini, 1987; Bailey, 1993), and for a

wide range of horticultural crops (e.g. Yang et al., 1990; Pollet et al., 1999; Montero et al., 2001; Rouphael & Colla, 2004).

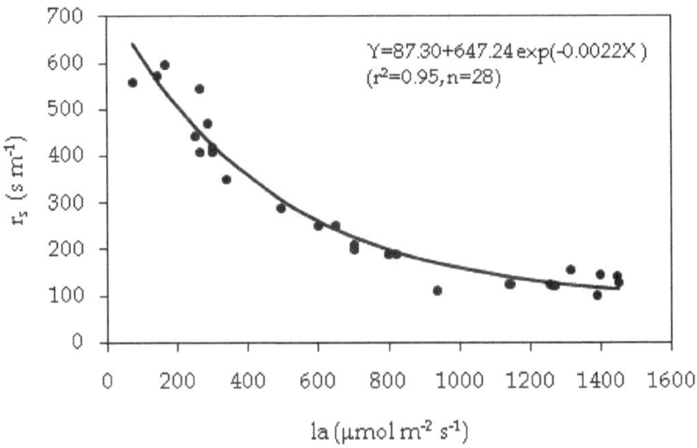

Fig. 1. Instantaneous values of leaf stomatal resistance (r_s) measured on potted zucchini squash with a porometer as a function of photosynthetically active radiation at the top of the canopy (I_a) measured during the entire growing cycle carried out at Tuscia University during spring-summer season.

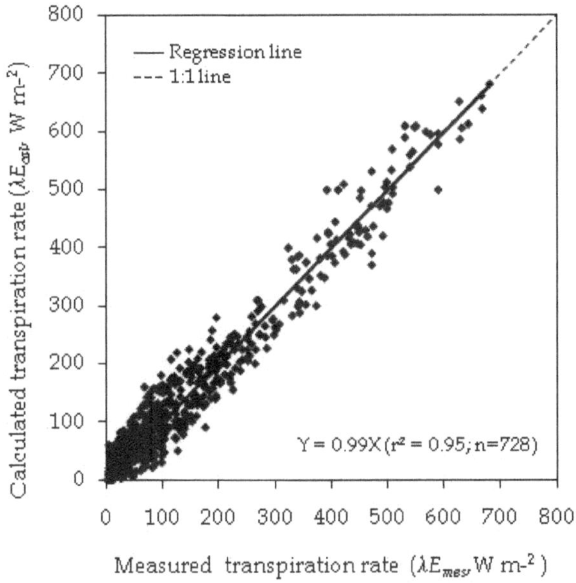

Fig. 2. Calculated daytime transpiration rates (30-min average) by the Penman-Monteith equation assuming forced convection (λE_{cal}) versus measured transpiration rates (λE_{mes}, electronic weighing balance) of potted greenhouse zucchini squash carried out at Tuscia University during spring-summer season.

Figure 2 shows the comparison of simulations and measurements of *ET* in greenhouse zucchini (Rouphael & Colla, 2004). Simulations were obtained using the PM equation where r_s and LAI were estimate as a function of irradiance (see Fig. 1) and growing degree days, respectively.

However, the application of the PM equation is not straightforward as it requires the knowledge of several inputs or parameters that are not easily available. Therefore, several authors proposed simplified equations for predicting *ET* as a function of *LAI*, R_{int} (MJ m^{-2} h^{-1}) and *VPD* (kPa) as explanatory variables:

$$ET = A \cdot \frac{R_{\text{int}}}{\lambda} + B \cdot LAI \cdot VPD \tag{12}$$

where *A* (dimensionless) and *B* (kg m^{-2} h^{-1} kPa^{-1}) are empirical coefficients.

After appropriate calibration, equation 12 predicted accurately *ET* in a variety of greenhouse crops, such as geranium (Montero et al., 2001; Colla et al., 2009), cucumber (Yang et al., 1990; de Graaf & Esmeijer, 1998; Medrano et al., 2005), rose (Baas & Rijssel, 2006; Baille et al., 1994; Kittas et al., 1999), tomato (e.g. Bailey et al., 1993; Baille et al., 1994; Carmassi et al., 2007a; Jolliet & Bailey, 1992; Okuya & Okuya, 1988; Stanghellini 1987), zucchini squash (Rouphael & Colla, 2004), and several pot ornamentals (Baille et al., 1994).

Figure 3 reports the comparison of simulations and measurements of hourly *ET* in greenhouse gerbera (*Gerbera jamesonii* Bolus ex Hook.) (A. Pardossi and G. Carmassi, unpublished); simulations were obtained using the equation 13.

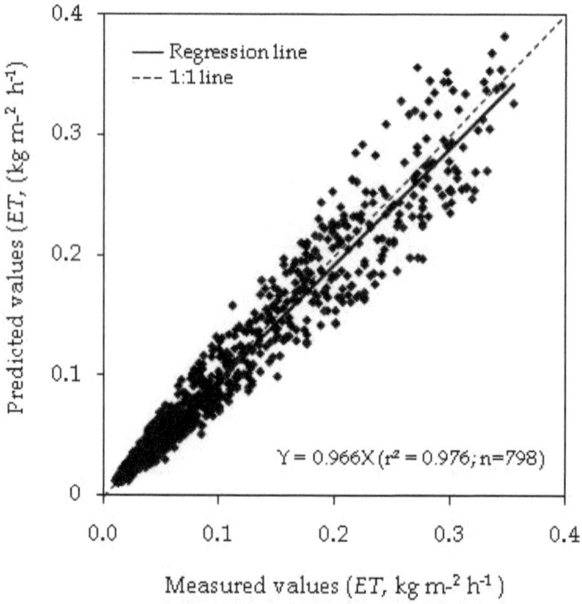

Fig. 3. Comparison between predicted and measured values of hourly *ET* in the daytime of greenhouse gerbera grown in rockwool. Measurements were taken with a weighing gutter while simulations were obtained by multiple regression equation (Eq. 13 in the text).

$$ET = 0.550\ R_{int} + 0.019\ LAI \cdot VPD \tag{13}$$

This equation is valid for *LAI* ranging from 0.9 to 2.5, approximately. R_{int} was calculated using a k value of 0.60. The regression coefficients (0.550, dimensionless; 0.019 kg m^{-2} h^{-1} kPa^{-1}) used for gerbera in the calibration procedure are similar to those reported by Baille et al. (1994) or by Montero et al. (2001) for several ornamental crops as *Gardenia, Impatiens* and *Pelargonium*.

A simple model for estimating daily *ET* of row crops in heated greenhouse was proposed by De Graaf (1988):

$$ET = \frac{h}{m} \cdot \left[a \cdot R + b \cdot \Sigma_{1440}^{1} \min_i \left(T_t - T_a \right) \right] \tag{14}$$

where h and m are actual and maximum height of the crop, respectively; R is the inside global radiation; a and b are crop-specific coefficients, the latter being attributable to heating; $\min_i$ is the successive minute during the day that the temperature of heating pipeline (T_t) is different from air temperature (T_a). The value of b is 0.22 10^{-4} for tomato and cucumber grown in Dutch growing conditions.

As solar radiation is the main climate variable influencing *ET* in protected crops, especially under unheated greenhouse conditions (Baille et al., 1994; van Kooten et al., 2008), simple linear regression of *ET* against outside or inside solar radiation has been also proposed for practical management of irrigation in greenhouse crops (Morris et al., 1957; de Villèle, 1974). In the first case, the light transmission coefficient of the covering material must be known; this coefficient typically ranges from 0.60 to 0.70.

Table 1 reports the ratio of *ET* on *R* determined in a few greenhouse crops as determined in a series of experiments conducted with soilless-grown greenhouse crops under unheated greenhouse in Central Italy. The ratio ranged from 0.65 to 0.80 and, as expected, was higher for crops with larger *LAI*.

Crop	Growing system	Season	*LAI*	*ET/R*
Tomato	Pumice; NFT*	Autumn - spring	3.0 - 3.5	0.75 - 0.80
Melon	NFT*	Autumn - spring	3.0 - 3.5	0.70 - 0.75
Strawberry	substrate	Spring	2.0 - 2.5	0.65 - 0.70
Gerbera	substrate	Year-round	2.4 – 2.8	0.65 - 0.70
Rose	substrate	Year-round	2.4 – 2.8	0.70 - 0.75

Table 1. Measured values of *LAI* and the ratio between evapotranspiration (*ET*) and incident radiation *R* (converted in kg m^{-2} day^{-1}) in different greenhouse crops grown in soilless culture. * NFT: nutrient film technique.

An aggregated model for *ET* in greenhouse tomato was designed and tested by Carmassi et al., (2007a). The model consists of two sub-models to predict the evolution of *LAI* and then daily *ET* as function of intercepted *R*. In particular, *LAI* was modeled using the Boltzmann sigmoid equation (Motulsky & Christopoulos, 2003):

$$LAI = -0.335 + \frac{(4.803 + 0.335)}{1 + \exp\left(\frac{755.3 - GDD}{134.7}\right)} \tag{15}$$

where *a*, *b*, *c* and *d* are numerical coefficients.
The second model consisted of a simple linear regression between daily *ET* and *R*:

$$ET = 0.946 \cdot [1 - \exp(-0.69\ LAI)] \cdot \frac{R}{\lambda} + 0.188 \tag{16}$$

Model's capability to predict *LAI* evolution resulted in accurate prediction of daily *ET* in both spring and autumn tomato crops.
The model proposed by Carmassi et al. (2007a) considered an appreciable night-time water consumption (the intercept of Eq. 16), which indeed reached values up to 10% of daily-cumulated *ET*. The equation did not include *VPD* as it was significantly correlated to *R* (r = 0.59). In the development of empirical regression model, the number of variables is generally limited by omitting those that are closely correlated to others.
Similarly, Colla et al., 2009 proposed a simplified model:

$$ET = 0.34\ [1\text{-}\exp(-0.6 \cdot LAI)]\ R + 30.27 \cdot LAI \cdot VPD \tag{17}$$

for predicting short-term *ET* rate in greenhouse geranium (*Pelargonium* x *hortorum*) grown in winter and spring. There was good agreement between predictions and measurements (Fig. 4).

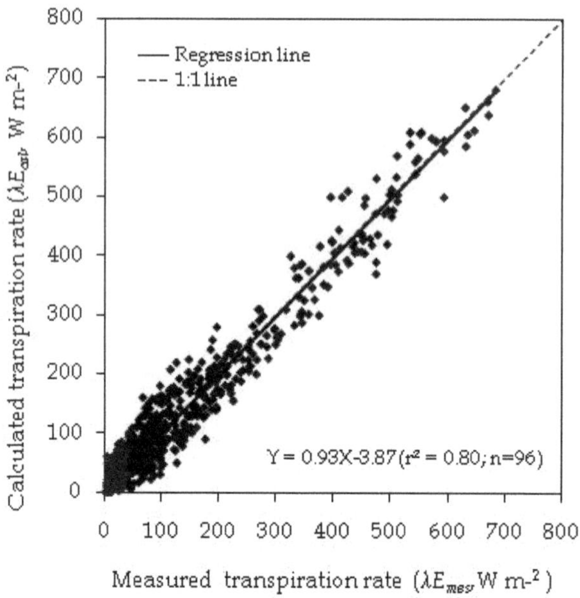

Fig. 4. Calculated daytime transpiration rates (60-min average) by the simplified Penman-Monteith equation (ET_{cal}) versus measured transpiration rates (ET_{mes} – electronic balance) of geranium plants grown in the winter and spring growing seasons.

4. Modelling evapotranspiration by integrating measurements of substrate water status and climatic parameters

According to different authors (Zazueta et al., 2002; Baille, 2001), the combined use of available advanced methods (e.g. climate and soil-based methods) for irrigation control is strongly recommended as it allows the control of data coherence and results in higher reliability of irrigation scheduling (IS). In fact, these systems are expressly designed to manage operations unattended, with only periodic human interventions, since "feed-back" and "fall-back" controls are included (Norrie et al., 1994a; Norrie et al., 1994b; Bowden et al., 2005).

Soil-based method for IS uses the measurement of soil or substrate water potential (tensiometers) or volumetric water content (e.g., dielectric sensors). Unfortunately, a series of technical problems can affect these apparatus like a wrong placement of the sensors, or a progressive decreasing of the measurement representativeness, requiring frequent and accurate controls with heavy indirect cost (Gillet, 2000, Lea-Cox et al., 2009; Pardossi et al., 2009b).

In the climate-based method, instead, ET is computed applying the simple equation $ET = ET_0 * K_c$, where ET_0 is the reference evapotranspiration and K_c is the crop coefficient. The K_c is defined as the ratio of actual evapotranspiration (ET) of a specific crop on a given condition, to the reference condition (ET_0). The methods used to estimate actual water use include direct measurement of soil moisture, by weighting or drainage lysimeters, gravimetrical method, soil moisture balance.

As already stated in par. 2 the main challenge for an effective application of the climate-method for the irrigation management, is the determination of crop coefficient (K_c). In containers, K_c typically are higher than in agronomic crops or orchard trees (Niu et al.,2006), which rarely exceed 1.3 (Doorenbos &Pruitt, 1975). They are also higher than those obtained in lysimeter experiments (García-Navarro et al., 2004).

In recent research projects, supported by Italian Ministry of Agriculture, an integrated method able to combine climate and soil-based methods, was proposed by the research group of IBIMET - Firenze (Bacci et al., 2008). In this procedure ET_0 was obtained by applying the CIMIS equation while ET was estimated first by tensiometers readings to calculate K_c values, and then by the equation ET_0*K_c. ET estimated values were compared with measured values taken with an electronic balance. Data were collected on three different container grown species, cultivated in greenhouse: *Pelargonium x hortorum*, *Callistemon viminalis* and *Petunia x hybrida*. The container volume was 1.5 L.

4.1 Estimation of ET by means of tensiometric readings (ETt)

The estimation of ET by means of tensiometer readings is valid into the range of the Available Water Content (AWC), corresponding to the amount of water retained in the substrate reservoir that can be readily used by plants. Available water is the difference between water content at matricial potential of -1 kPa (by definition, container water capacity) and that at -10 kPa (de Boodt &Verdonck, 1972).

The first step to estimate ET from tensiometer readings is the definition of a soil-specific tensiometric curve in the laboratory for the conversion of soil water potential in soil water content. One of the most simple procedure to obtain soil water retention curve was suggested by Retzlaff & South (1985). According to this methodology, laboratory soil water potential and related water content are measured by tensiometer and by gravimetric method, respectively. The relationship between water content and water potential has the following form:

$$WC = a \cdot (b - \exp(c \cdot MT)) \quad (18)$$

where WC = soil water content (g cm^{-3}); MT = absolute value of soil potential value measured by tensiometer (hPa); a, b, c = coefficient.
In table 2 a, b, c values of some substrates used in container crops are shown. Tensiometric curves of other growing media are reported by Bibbiani (1996) and Milk et al. (1989).

Substrate	*a*	*b*	*c*
Peat (60%) - Sand(10%) - Pumice (30%)	-0.1491	-3.7981	0.0196
Peat -Perlite (1:1)	-0.1825	-1.5201	0.0193
Peat -Pumice (1:1)	-0.2278	-2.0817	0.0044

Table 2. a, b, c values for equation 18 of some substrates.

Due to relatively low AWC in container crops, at least hourly measurements of soil water potential are advisable to adequately follow the changes in substrate water content by means of equation 18. The difference between two consecutive WC measurements represents the actual evapotranspiration (ETt_h):

$$ETt_h = [WC_h - WC_{h-1}] \cdot Vpot \cdot T^{-1} \quad (19)$$

where ETt_h = actual evapotranspiration (g pot^{-1} h^{-1}); WC_h = soil water content at hour h (g cm^{-3}); WC_{h-1} = soil water content at hour h-1 (g cm^{-3}); $Vpot$ = volume of pot (cm^3); T = time between two consecutive measurements (one hour).
Values of ETt_h are converted to ET per ground area based on the number of pots per m^2. Figure 5 shows the comparison between predicted (ETt) and measured (ET_b) hourly values of ET for selected days.

4.2 Estimation of reference evapotranspiration (ET_o)

ET_o can be computed at different temporal scale, but the needs to manage information at hourly step reduce the choice among a limited number of methods. The CIMIS equation (CIMIS, 2009) was chosen to calculate ET_o as appears more suitable for plants with high LAI and high density, like those grown in a greenhouse or nursery. Besides it gives a higher weight to the wind factor that has a relevant effect when container ornamentals are grown outdoor. There were no important differences between the values of ET_0 determined with CIMIS formula or the Penman-Monteith (PM) equation (Allen et al., 1998).
The CIMIS equation takes into account four meteorological parameters (net radiation, air temperature, relative humidity and wind speed) as follows:

$$ETo_h = W \cdot R_n + (1-W) \cdot VPD \cdot f(u_2) \quad (20)$$

where ETo_h= hourly reference evapotranspiration (mm h^{-1}); W = dimensionless partitioning factor; R_n = net radiation (mm h^{-1} of equivalent evaporation); VPD = vapour pressure deficit (kPa); f(u_2) = empirical wind function (mm h^{-1} kPa^{-1}).

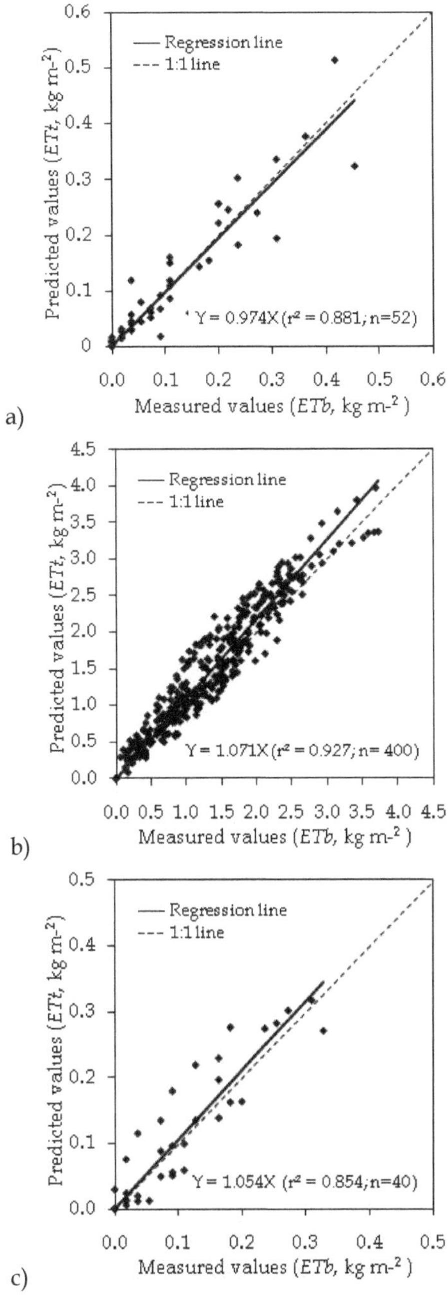

Fig. 5. Relationship between cumulated *ET* measured values (ET_b) and cumulated predicted values by tensiometric readings (*ETt*) of *Pelargonium x hortorum* (a), *Callistemon viminalis* (b) and *Petunia x hybrida hort.* (c). Values were cumulated between two successive irrigations.

The partitioning factor, *W*, was computed as follows:

$$W = s/(s+\gamma) \tag{21}$$

where *s* represents the slope of the saturation vapour pressure curve at T_a (air temperature, °C) and γ is the psychrometer constant (kPa °C^{-1}):

$$s = e_s \cdot (597.4 - 0.571\ T_a)/[0.1103\ (T_a+273.16)^2] \tag{22}$$

$$\gamma = 0.000646\ P \cdot (1+ 0.000949\ T_a) \tag{23}$$

where e_s = saturation vapour pressure at T_a (kPa); *P* = atmospheric station pressure (kPa).
The wind function, f(u_2), is:

$$f(u_2) = 0.03 + 0.0576\ u_2 \tag{24}$$

where u_2 = mean hourly wind speed at a height of 2 m (m s^{-1}).
Since a meteorological station only seldom is equipped with a net radiometer, while it is usually available a solarimeter to measure global radiation, net radiation can be estimated from global radiation data by applying simple linear relations. An example of linear relation between hourly global radiation measurements and hourly net radiation measurements is shown in equation 25. It refers to an experiment on *Hypericum hidcote* grown in container outdoor.

$$R_n = 0.7101 \cdot R - 0.3828 \quad (r^2=0.9113^{***}) \tag{25}$$

where R_n = hourly net radiation (MJ m^{-2} h^{-1}); *R* = hourly global radiation (MJ m^{-2} h^{-1}).
As already stated in par.3 in a greenhouse global radiation measurements can be used in CIMIS equation in place of net radiation data.
Solar radiation data below the threshold of 0.21 MJ m^{-2} should be excluded by the computation, assuming these values corresponding to night time period (Brown, 1998).

4.3 Estimation of *ET* by climate-method application (*ETcm*)

The values of hourly Kc_h, can be calculated by applying the following equation:

$$Kc_h = ETt_h / ETo_h \tag{26}$$

During daytime, the *Kc* usually follows solar radiation, describing theoretically a bell-shaped curve with a significant difference between early morning and half day values. For this reason, considering that plants grown in pots can transpire all available water in few hours, demanding more irrigation events during a single day, the significance of a reliable hourly *Kc* is evident.
The high variability of the hourly behavior of *Kc*, depending on meteorological and plant physiological conditions, suggested the use of an average hourly values to improve the *ETcm* estimation. In the frame of the experiments of IBIMET research group, a minimum of seven hourly values (7-day average) was used according to the following equation:

$$\overline{Kc_{hj}} = \frac{\sum_{J-1}^{J-8} Kc_{hj}}{7} \tag{27}$$

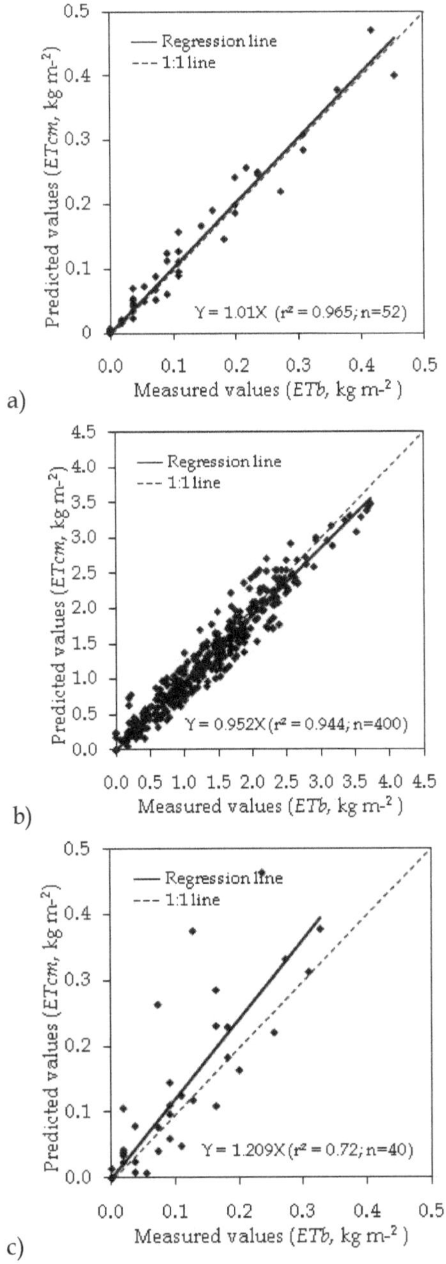

Fig. 6. Relationship between cumulated *ET* measured values (*ETb*) and cumulated predicted values by climate-method application (*ETcm)* of *Pelargonium x hortorum* (a), *Callistemon viminalis* (b) and *Petunia x hybrida hort.* (c). Values were cumulated between two successive irrigations.

where $\overline{Kc_{hj}}$ = average value of Kc_h on day j at hour h.

Obviously as new Kc_h are calculated for the successive days, new $\overline{Kc_{hj}}$ are obtained according to mobile average method. Starting from the previous data, hourly $ETcm_{hj}$ can be calculated by applying the following equation:

$$ETcm_{hj} = ETo_{hj} \cdot \overline{Kc_{hj}} \tag{28}$$

A comparison between *ET* measured and estimated values is shown in figure 6.

4.4 Data integration and irrigation management

According to the previously described method, every hour two values of *ET* are available: *ETt*, estimated by tensiometric measurements, and *ETcm*, estimated by climate-method application. The comparison between the two values allows a direct cross-control of the representativeness of the data provided by the sensors, improving the reliability of the automatic irrigation system in which this integrated method should be implemented (Fig. 7). In case of a representativeness loss of the sensor, for example, the use of average values (eq. 27) strongly reduces possible errors, while to maintain a high security level in plant watering, unrealistic differences between *ETt* and *ETcm* values can be easily detected and alternative solutions can be adopted.

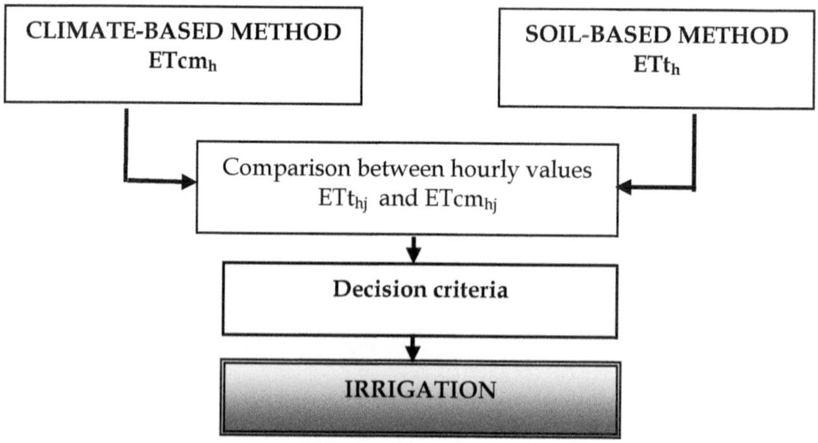

Fig. 7. Operating diagram of the integrated method.

5. Conclusions

Improved irrigation use efficiency requires appropriate irrigation methods, such trickle irrigation and subirrigation, and precise scheduling. Effective operation of irrigation systems requires a sensing system that determines crop needs in real time; this rules out manual control as an automated monitoring system is necessary.

Once implemented in a dedicated hardware/software system, *ET* models described in the previous paragraphs can be employed for daily or hourly management of irrigation or fertirrigation as they stand or integrated with the measurements of substrate moisture.

Besides *ET* model may be a relevant component for simulation studies to define best strategies for managing irrigation or fertigation.

6. Acknowledgment

Laura Bacci, Piero Battista, Bernardo Rapi, Giuseppe Colla, Mariateresa Cardarelli, and Youssef Rouphael were supported by the Italian Ministry of Agriculture projects ECOIDRIFLOR (2007-2009) "Eco-efficiency of fertigation management in floricultural and ornamental crops" and SWIFF (2009-2011) "Wireless system for the control of main greenhouse practices in floricultural field". Giulia Carmassi, Alberto Pardossi, Fernando Malorgio and Luca Incrocci were supported by a EU-FP7 (2008) project entitled "Reducing the need for external inputs in high value protected horticultural and ornamental crops (EUPHOROS)".

7. References

Allen, R.G., Pereira, L.S., Raes, D. & Smith, M. (1998). Crop evapotranspiration: Guidelines for computing crop water requirements. *Irrigation Drainage Paper 56*, FAO, Rome. Available from: <*http://www.fao.org/docrep/X0490E/X0490E00.htm*>.

Baas, R. & van Rijssel, E. (2006). Transpiration of glasshouse rose crops: evaluation of regression models, *Acta Horticulturae*, Vol.718, pp. 547-556.

Bacci, L., Battista, P. & Rapi, B. (2008). An integrated method for irrigation scheduling of potted plants, *Scientia Horticulturae*, Vol.116, No.1, pp. 89-97.

Bailey, B.J., Montero, J.I., Biel, C., Wilkinson, D.J., Anton, A. & Jolliet, O. (1993). Evapotranspiration of Ficus benjamina: comparison of measurements with predictions of the Penman-Monteith model and a simplified version, *Agricultural and Forest Meteorology*, Vol.65, pp. 229-243.

Baille, M., Baille, A. & Laury, J.C. (1994). A simplified model for predicting evapotranspiration rate of nine ornamental species vs climate factors and leaf area, *Scientia Horticulturae*, Vol.59, pp. 217-232.

Baille, A. (2001). Water management in soilless cultivation in relation to inside and outside climatic conditions and type of substrate, Italus Hortus, Vol.8, pp.16-22.

Balendonck, J. (2010). FLOW-AID Farm level optimal water management assistant for irrigation under deficit: More crop per drop. WUR, Greenhouse Horticulture, Wageningen, The Netherlands. Available from: <*http://documents.plant.wur.nl/wurglas/flowaid.pdf.*>

Bibbiani, C. (1996). Le proprietà fisiche e idrauliche dei substrati, *Flortecnica*, Vol.3, pp. 40-48.

Bontsema, J., Hemming, J., Stanghellini, C., Visser, P.H.B. de, Henten, E.J. van, Budding, J., Rieswijk, T. & Nieboer, S. (2007). On-line monitoring van transpiratie en fotosyntheseactiviteit, Wageningen UR Glastuinbouw 451, available: <*http://edepot.wur.nl/24034*>

Bowden, J.D., Bauerle, W.L., Lea-Cox, J.D. & Kantor, G. (2005). Irrigation scheduling: an overview of the potential to integrate modelling and sensing techniques in a Windows-based environment, *Proceedings of SNA Research Conference* , 50, 577-579.

Brouwer, C. & Heibloem, M. (1986). Irrigation water management: Irrigation water needs. Food and Agriculture Organization of the United Nations (FAO), Rome. 23 Dec. 2010. Available from: <*http://www.fao.org/docrep/s2022e/s2022e00.htm*>.

Brown, P.W. (1998). AZMET computation of reference crop evapotranspiration. Arizona Meteorological Network. Available from: <*http://ag.arizona.edu/azmet/et2.htm*>

California Irrigation Management Information System. (2009). PM equation. 23 Dec. 2010. Available from: <*http://wwwcimis.water.ca.gov/cimis/infoEtoPmEquation.jsp* >.

Carmassi, G., Incrocci, L., Maggini, R., Malorgio, F., Tognoni, F. & Pardossi, A. (2007a). An aggregated model for water requirements of greenhouse tomato grown in closed rockwoll culture with saline water, *Agricultural water management,* Vol.88, pp. 73-82.

Carmassi, G., Incrocci, L., Incrocci G.& Pardossi, A. (2007b). Non-destructive estimation of leaf area in (*Solanum lycopersicum* L.) and gerbera (*Gerbera jamesonii H.* BOLUS), *Agricoltura Mediterranea,* Vol.137, pp. 172-176.

Colla, G., Rouphael, Y., Cardarelli, M. , Rinaldi S.& Rea, E. (2009). Water use efficiency and transpiration of greenhouse geranium crop, *Acta Horticulturae,* Vol.807, pp. 271-276.

de Boodt, M. & Verdonck, O. (1972). The physical properties of the substrates in horticulture, *Acta Horticulturae,* Vol.26, pp. 37-44.

de Graaf, R. (1988). Automation of the water supplied of glasshouse crops by means of calculating the transpiration and measuring the amount of drainage water, *Acta Horticulturae,* Vol.229, pp. 219-231.

de Graaf, R. & Esmeijer, M.H. (1998). Comparing calculated and measured water consumption in a study of the (minimal) transpiration of cucumbers grown on rockwool, *Acta Horticulturae,* Vol.458, pp. 103-112.

de Villele, O. (1974). Besoins en eau des cultures sous serres. Essai de conduit des arrosages en function de l'ensoleillement, *Acta Horticulturae,* Vol.35, pp. 123-135.

Doorenhos, J. & Pruitt, W.O. (1975). Guidelines for predicting crop water requirements, *Irrigation Drainage Paper 24*, FAO, Rome.

Elings, A. & Voogt, W. (2006). Management of Greenhouse Crop Transpiration: the Way Forward, *Acta Horticulturae,*Vol.801, pp. 1221-1228.

García-Navarro, M.C., Evans, R.Y.& Montserrat, R.S. (2004). Estimation of relative water use among ornamental landscape species, *Scientia Horticulturae,* Vol.99, pp. 163-174.

Gillett, J. 2000. Tensiometers need periodic maintenance. Available from: <*http://www.agric.nsw.gov.au/reader/soilwater/tensio2.htm*>

Grober, H. & Erk, S. (1961). *Fundamentals of heat transfer*, McGrawHill, New York.

Heuvelink, E. & Marcelis, L.F.M. (1996). Influence of assimilate supply on leaf formation in sweet pepper and tomato, *Journal Horticulturae Science* , Vol.71, pp. 405-414.

Incrocci, L., Fila, G., Bellocchi, G., Pardossi, A., Campiotti, C.A. & Balducchi, R. (2006). Soil-less indoor-grown lettuce (Lactuca sativa L.): Approaching the modelling task, *Environmental Modelling Software,* Vol.21, pp. 121-126.

Jolliet, O. & Bailey, B.J. (1992). The effect of climate on tomato transpiration in greenhouse: measurements and models comparison, *Agricultural and Forest Meteorology,* Vol.58, pp. 43-62.

Jones, H.G. (2004). Irrigation scheduling: advantages and pitfalls of plant-based methods, *Journal of Experimental Botany,* Vol.55, pp. 2427-2436.

Kittas, C., Katsoulas, N. & Baille, A. (1999). Transpiration and canopy resistance of greenhouse soilless roses: measurements and modelling, *Acta Horticulturae,* Vol.507, pp. 61-68.

Lea-Cox, J., Ristvey, A., Ross, D. & Kantor, G. (2009). Deployment of wireless sensor networks for irrigation and nutrient management in nursery and greenhouse operations, *Proceedings of SNA Research Conference*, Vol.54, pp. 28-34.

McAdams, W.H. (1954). *Heat transmission*, 3rd edn., McGrawHill, New York.

Medrano, E., Lorenzo, P., Sanchez-Guerrero, M.C & Montero, J.I. (2005). Evaluation and modelling of greenhouse cucumber-crop transpiration under high and low radiation conditions, *Scientia Horticulturae,* Vol.105, pp. 163-175.

Milks, R.R., Fonteno, W.C. & Larson, R.A. (1989). Hydrology of horticultural substraters : I. Mathematical models for moisture characteristics of horticultural container media; II. Predicting physical properties of media in containers; III. Predicting air and water content of limited-volume plug cells, *Journal of American Society of Horticultural Science,* Vol.114, No.1, pp. 48-61.

Monteith, J.L. (1973). *Principles of environmental physics,* Arnold Ed., London, 241 pp.

Montero, J.I., Antón, A., Muñoz, P. & Lorenzo, P. (2001). Transpiration from geranium grown under high temperatures and low humidities in greenhouses, *Agricultural and Forest Meteorology,* Vol.107, pp. 323-332.

Morris, L.G., Neale, F.E. & Postlewaithe, J.D. (1957). The transpiration of glasshouse crops and its relation to the incoming solar radiation, *Journal Agricultural Engineering Resource,* Vol.2, pp. 111-122.

Motulsky, H. & Christopoulos, A. (2003). Fitting model to biological data using linear and nonlinear regression. A practical guide to curve fitting. *GraphPad Software Inc.*, San Diego, CA, Available from: *http://www.graphpad.com.*

Niu, G., Rodriguez, D.S., Cabrera, R., McKenney, C. & Mackay, W. (2006). Determining water use and crop coefficients of five woody landscape plants; *Journal of Environmental Horticulture,* Vol.24, No.3, pp. 160-165.

Norrie, J., Graham, M.E.D. & Gosselin, A. (1994a). Potential evapotranspiration as a means of predicting irrigation timing in greenhouse tomatoes grown in peat bags, *Journal of American Society of Horticultural Science,* Vol.119, pp. 163-168.

Norrie, J., Graham, M.E.D., Dubé, P.A. & Gosselin, A. (1994b). Improvement in automatic irrigation of peat-grown greenhouse tomatoes, *HortTecnology,* Vol.4, pp. 154-159.

Okuya, A. & Okuya, T. (1988). The transpiration of greenhouse tomato plants in rockwool culture and its relationship to climatic factors, *Acta Horticulturae,* Vol.230, pp. 307-311.

Orgaz, F., Fernandez, M.D. , Bonachela, S., Gallardo, M. & Fereres, E. (2005). Evapotranspiration of horticultural crops in an unheated plastic greenhouse, *Agricultural Water Management,* Vol.72, pp. 81–96.

Pasian, C.C. & Lieth, J.H. (1996). Prediction of rose shoot development: Model validation for the cultivar 'Cara Mia' and extension to the cultivars 'Royalty' and 'Sonia', *Scientia Horticulturae,* Vol.66, pp. 117-124.

Pardossi, A., Incrocci, I., Incrocci, G. & Marzialetti, P. (2009a). What limits and how to improve water and nutrient efficiency in outdoor container cultivation of ornamental nursery stocks, *Acta Horticulturae,* Vol.843, pp. 73-80.

Pardossi, A., Incrocci, L., Incrocci, G., Malorgio, F., Battista, P., Bacci, L., Rapi, B., Marzialetti, P., Hemming, J. & Balendonck, J. (2009b). Root zone sensors for irrigation management in intensive agriculture, *Sensors* 9: 2809–2835.

Pollet, S., Bleyaert, P. & Lemeur, R. (1999). Calculating the evapotranspiration of head lettuce by means of the Penman-Monteith model, *Proceedings of 3rd International Workshop: Models for plant growth and control of the shoot and root environments in greenhouses*, Bet Dagan, Israel.

Regan, R.P. (1994). Variation in water use of container-grown plants, *Comb. Proc. Intl. Plant Prop. Soc.*, Vol.44, pp.310-312.

Retzlaff, W.R. & South, D.B. (1985). A simple method for determining a partial soil water retention curve, *Tree Planters' Notes*, Vol.36, pp. 20-23.

Rouphael, Y. & Colla, G. (2004). Modelling the transpiration of a greenhouse zucchini crop grown under a Mediterranean climate using the Penman-Monteith equation and its simplified version, *Australian Journal of Agricultural Research*, Vol.55, pp. 931-937.

Rouphael, Y., Cardarelli, M., Fallovo, C., Colla, G., Salerno A., Rea, E. & Marucci, A. (2008). Predicting leaf number of greenhouse zucchini squash using degree days and photosynthetically active radiation, *Acta Horticulturae*, Vol. 801, pp.1149-1154.

Stanghellini, C. (1987). *Transpiration of greenhouse crops. An aid to climate management*. PhD Thesis, Wageningen Agricultural University, the Netherlands, 150 pp.

Thompson, R.B., Martínez-Gaitan, C., Gallardo, M., Giménez, C. & Fernández, M.D. (2007). Identification of irrigation and N management practices that contribute to nitrate leaching loss from an intensive vegetable production system by use of a comprehensive survey, *Agricultural Water Management*, Vol.89, pp. 261–274.

Treder, J., Matysiak, B., Nowak, J. & Treder, W. (1997). Evapotranspiration and potted plants water requirements as affected by environmental factors, *Acta Horticulturae*, Vol.449, pp. 235-239.

van Kooten, O., Heuvelink, E., Stanghellini, C. (2008). New development in greenhouse technology can mitigate the water shortage problem of the 21st century. *Acta Horticulturae*. Vol.767, 45-52.

Vox, G., Teitel, M., Pardossi, A., Minuto, A., Tinivella, F. & Schettini, E. (2010). Agriculture: technology, planning and management, *in* Salazar, A., Rios, I. (Eds.), *Sustainable greenhouse systems*, Nova Science Publishers, New York, pp. 1-79.

Walter, I.A., Allen, R.G., Elliott, R., Itenfisu, D., Brown, P., Jensen, M.E., Mecham, B., Howell, T.A., Snyder, R.L., Eching, S., Spofford, T., Hattendorf, M., Martin, D., Cuenca, R.H. & Wright, J.L. (2002). The ASCE standardized reference evapotranspiration equation, Report Task Com. on Standardized Reference Evapotranspiration July 9, 2002, EWRI-Am. Soc. Civil. Engr., Reston, VA, USA.

Xu, R., Dai, J., Luo, W., Yin, X., Li, Y., Tai, X., Han, L., Chen, Y., Lin, L., Li, G., Zou, C., Du, W. & Diao, M. (2010). A photothermal model of leaf area index for greenhouse crops, *Agricultural and Forest Meteorology*, Vol.150, pp. 541-552.

Yang, X., Short, T.M., Fox, R.D. & Bauerle, W.L. (1990). Transpiration, leaf temperature and stomatal resistance of a greenhouse cucumber crop, *Agricultural and Forest Meteorology*, Vol.51, pp. 197–209.

Zazueta, F.S., Bucklin, R., Jones, P.H. & Smajstrla, A.G. (2002). Basic concepts in environmental computer control of agricultural systems, Agricultural and Biological Engineering Department, Florida Cooperative Extension Service, Institute of Food and Agricultural Sciences, University of Florida. CIR1029. Available from: <*http://edis.ifas.ufl.edu/AE003*>

15

Greenhouse Crop Transpiration Modelling

Nikolaos Katsoulas[1] and Constantinos Kittas[1,2]
[1]*University of Thessaly, Dept of Agriculture, Crop Production and Rural Environment, Volos*
[2]*Centre for Research and Technology Thessaly Institute of Technology and Management of Agricultural Ecosystems, Volos, Greece*

1. Introduction

Aim of this chapter is to present the parameters affecting greenhouse crop transpiration and the existing models for greenhouse crop transpiration simulation. In the first paragraphs of the chapter, the importance of crop transpiration on greenhouse microclimate and on crop is presented and discussed. Presentation, analysis and discussion of the parameters affecting greenhouse crop transpiration and the thermal and hydrological negative feedback effects follow. Finally, the existing models for greenhouse crop transpiration simulation are presented and discussed.

Transpiration is an important component of canopy energy and water balance and thus, a major cooling mechanism of greenhouse crop canopies. Its estimation is essential for climate and irrigation control and that is why it has been given much attention in greenhouse climate research.

Air temperature and vapour pressure deficit are parameters affecting the thermal and hydrological negative feedback effects existing in a greenhouse. In addition, the main factors affecting greenhouse crop transpiration are solar radiation, vapour pressure deficit and canopy and aerodynamic conductances.

Several authors have proposed models that allow getting a more accurate estimation of the crop transpiration rate. More sophisticated transpiration models are based on leaf (canopy) transpiration and leaf energy balance models in which the transpiration is characterized by the canopy resistance, as proposed initially by Penman and modified by Monteith to account for the stomatal response of the crop (P-M formula). However, the use of the complete P-M formula requires the knowledge of several inputs or parameters that are not easily available. Particularly, the aerodynamic and stomatal leaf resistances have to be known for each crop species and possibly, for each cultivar. That is why researchers have tried to overcome the estimation of these resistances by using a simplified form of the P-M formula. Transpiration models with the greenhouse climate as a boundary condition were first developed in the northern regions of Europe and North America for horticultural crops. In these northern conditions, the glasshouse is generally poorly ventilated during a large part of the growing season. The boundary layer conductance for glasshouse crops tends to be much smaller than would be expected for similar crops growing outdoors. Thus, glasshouse crops are very strongly decoupled from the outside atmosphere by the presence of the glass, and the heat and the water released at crop surface will accumulate inside the glasshouse. Consequently, the transpiration rate will adjust until it reaches a stable equilibrium transpiration rate dictated by the net radiation received. On the contrary, greenhouse crop transpiration in Mediterranean or similar warm conditions is much more dependent on convection. As the ventilation and the turbulent mixing are vigorous, the saturation deficit at the leaf surface is closely coupled to the deficit of ambient air, and the latter is directly influenced by the outdoor saturation deficit.

2. The importance of crop transpiration on greenhouse microclimate and on crop

Transpiration is an important component of canopy energy and water balance and thus, a major cooling mechanism of greenhouse crop canopies. Its estimation is essential for climate and irrigation control and that is why it has been given much attention in greenhouse climate research. The concentration of water in plants should be kept within a narrow range so as to provide the conditions for optimum growth. A 10% reduction could affect the functioning and development of plants or even cause their death. In plants, the water is used in a wide range of functions:

- It is the component with the highest concentration and makes up about 70% to 95% of plant fresh weight. It gives shape and rigidity in plants.
- It is used as a means of dissolution and ion source, is the means for transportation of nutrients from soil to plant and is essential in many biochemical reactions.
- Finally, due to its high specific heat, the water cools the leaves through its evaporation and prevents plant overheating.

Plant water content depends on two main factors:

- the availability of water in the substrate-soil and its absorption by the roots and
- the evaporation of water from leaves, i.e. transpiration.

Crop transpiration is affected by (Figure 1):

- intercepted radiation
- the difference between the vapour pressure of air and the saturation vapour pressure at leaf temperature (crop-air vapour pressure deficit), the conductivity of the transfer of water from the interior of the leaf surface (stomatal conductance) and the conductivity of the water transfer from the surface of the leaf to the air (aerodynamic conductance).

For short periods, when the air vapour pressure deficit increases (in response to air relative humidity reduction), stomata begin to shut down gradually to reduce water stress (Choudhury and Monteith 1986). The negative effect is significantly higher for air vapour pressure deficit values higher than 1 kPa. If plants reach higher levels of water stress, the roots can not supply the aboveground part of plants with enough water, plants lose their rigidity and turgid and irreversible damage may occur on the leaves. Under drought stress, most of the water goes to the leaves and fruit growth is reduced. In addition, damage can be caused by direct 'burning' of leaves. In this case, transpiration through the process of cooling by evaporation is playing an important role in reducing the high temperature of the crop and the appropriate actions should be taken to maintain transpiration in its maximum rate.
In greenhouse conditions it is possible to control crop transpiration rate and regulate it in desired levels. To this end, shading, ventilation and cooling systems are mainly used, regulating direct radiation R_n, air temperature T_i and vapour pressure deficit D_i (= difference between the actual vapor pressure of air to saturation vapour pressure) (Figure 2), factors affecting direct (R_n) or indirectly (T_i, D_i) crop transpiration.

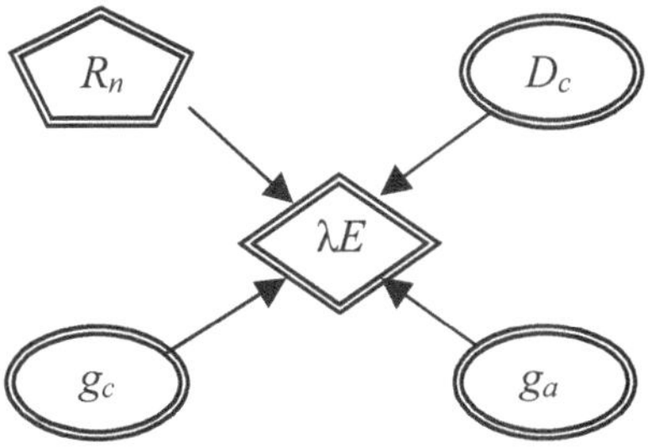

Fig. 1. Factors affecting crop transpiration. R_n = net radiation, D_c = crop-air vapour pressure deficit, g_c = crop stomatal conductance, g_a = crop aerodynamic conductance.

3. Parameters affecting greenhouse crop transpiration and the thermal and hydrological negative feedback effects

Air temperature and vapour pressure deficit are parameters affecting the thermal and hydrological negative feedback effects existing in a greenhouse. In addition, the main factors affecting greenhouse crop transpiration are solar radiation, vapour pressure deficit and canopy and aerodynamic conductances.

3.1 Natural and forced ventilation

As shown in Figure 2, ventilation directly affects the temperature and air vapour pressure deficit. At the same time affects the crop aerodynamic conductance by changes of wind velocity regime in the greenhouse. Besides crop aerodynamic conductance, ventilation affects greenhouse water vapour exchanges (greenhouse aerodynamic conductance). Experimental results on the effect of ventilation rate on greenhouse aerodynamic conductance are given in Table 1 (Fuchs et al. 1997).

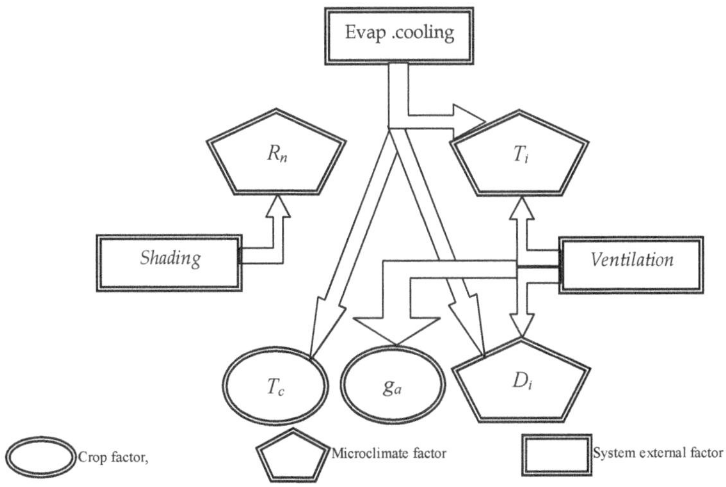

Fig. 2. Direct effects greenhouse climate control systems on greenhouse microclimate and on crop. D_i = air vapour pressure deficit, T_i = air temperature, R_n = net radiation, g_a = crop aerodynamic conductance, T_c = crop temperature.

Greenhouse ventilation effects also crop stomatal conductance through imposed effects on crop temperature and on canopy to air vapour pressure deficit crop. The last is affected by crop temperature and air vapour pressure deficit (Figure 3).
Kittas et al. (2001) observed that increasing greenhouse air exchange rate by means of forced ventilation resulted in an increase of canopy to air vapour pressure deficit and a decrease of crop stomatal conductance by 20% compared to the case of natural ventilation. Bunce (1985) observed that for the same increase in air vapour pressure deficit, total crop conductance decreased from 1.6 to 3 times more when the wind speed was 3 m s^{-1} than in the case where the wind speed was 0.5 m s^{-1}.

N h-1	ΔTi-o °C	Di kPa	ΔDi-o kPa	gv mm s-1	β
3.1	20	4.69	2.43	4.0	0.55
8.1	9.65	2.98	0.97	10.8	0.66
13.3	6.55	2.55	0.55	17.7	0.79
44.1	5.05	2.83	0.23	58.8	1.44

Table 1. Effect of ventilation rate (N h^{-1}) on greenhouse to outside air temperature difference (ΔTi^{-o} °C), greenhouse air vapour pressure deficit (Di kPa), greenhouse to outside air difference of the difference air vapour pressure deficit (ΔDi^{-o} kPa), greenhouse aerodynamic conductance (gv mm s^{-1}) and Bowen ratio (β). (Fuchs et al. 1997).

Consequently, another variable affected by air exchange rate and ventilation regime is crop aerodynamic conductance, due to changes imposed by ventilation in greenhouse wind velocity. For the case of a leaf, the aerodynamic conductance ($g_{l,a}$) is given by (Monteith 1973):

$$g_{l,a} = 6.62\,10^{-3}(u/d)^{0.5} \tag{1}$$

where u (m s^{-1}) is the mean greenhouse air velocity near the crop level and d (m) is the characteristic leaf length.
In the case of a closed greenhouse, the wind velocity inside the greenhouse is very small and a fixed value for crop aerodynamic conductance could be acceptable. This hypothesis is confirmed by experimental results (Stanghellini 1987). During periods that the greenhouse is ventilated, the wind velocity inside the greenhouse is important and its effect on crop aerodynamic conductance can not be ignored. Therefore, under these circumstances a more realistic approach should be considered which would consider crop aerodynamic conductance as a function of greenhouse ventilation rate (Seginer 1994; Kittas et al. 2001). Katsoulas (2002) states that for a rose crop with a leaf area index of 2, crop aerodynamic conductance was about 52 mm s^{-1} under natural ventilation and 123 mm s^{-1} under forced ventilation conditions.
The latent heat energy Hc (W m^{-2}) exchanged between the crop and the air is given by the following relation

$$H_c = \rho\, C_p\, g_a\, \Delta T_c \tag{2}$$

where ρ (kg m^{-3}) indicates density and C_p (J kg^{-1} °C) specific heat of air and ΔT_c indicates crop to air temperature difference. The above function shows the effect of aerodynamic conductance and accordingly of ventilation rate on energy partitioning into sensible and latent heat at the level of the crop.

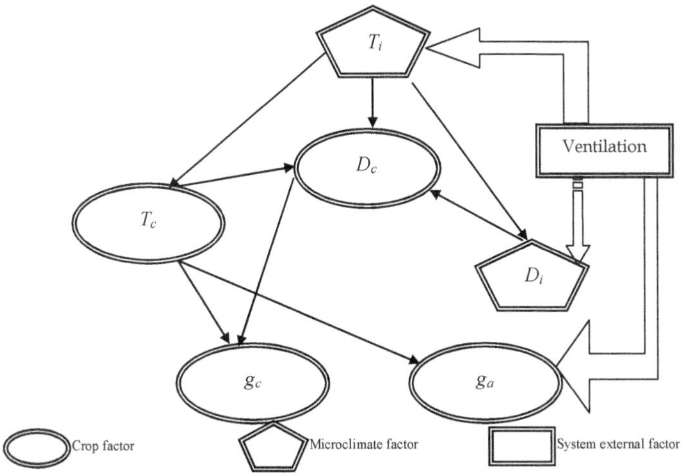

Fig. 3. Effect of greenhouse ventilation on greenhouse microclimate and on crop. T_i = air temperature, T_c = crop temperature, D_i = air vapour pressure deficit D_c = canopy to air vapour pressure deficit, g_c = crop stomatal conductance, g_a = crop aerodynamic conductance.

However, despite the significant influence of ventilation on greenhouse and crop microclimate, little is known about the effect of ventilation rate on crop energy partitioning. Boulard and Baille (1993) showed the effect of ventilation rate on greenhouse crop transpiration, indicating that transpiration is significantly affected by the air exchange rate and the air vapour pressure deficit. The above authors conclude that when vents were opened to the maximum aperture, the energy used for transpiration accounted for 45% of incoming solar radiation, while in the case of small vent openings, the transpiration accounted for 30% of incoming solar radiation.

Nevertheless, Katsoulas (2002) states that ventilation rate (with natural or forced ventilation) did not appear to significantly affect the distribution of energy into sensible and latent at the level of the crop although the average aerodynamic conductance was double under forced ventilation conditions (120 mm s^{-1}) than under natural ventilation (approximately 50 mm s^{-1}). He explains that transpiration rate remains at similar levels under the different ventilation regimes due to the interaction and feedback loops between canopy to air vapour pressure deficit and crop stomatal conductance.

Fuchs (1993) studied the transpiration rate of a tomato crop in relation to ventilation rate for different conditions of solar radiation and air relative humidity. It has to be noted that the crop aerodynamic and stomatal conductance and the greenhouse aerodynamic conductance are in series and that vapour transfer from inside the crop to outside air will be controlled by the smallest conduction. When greenhouse air exchange rate is high, transpiration will be probably controlled by crop stomatal conductance, which may not be true for low ventilation rates. Fuchs (1993), therefore, observed that for low values (10 h^{-1}) of greenhouse volume air exchange rate, transpiration was not affected by changes in crop stomatal conductance, while for large values of greenhouse air exchange rate transpiration was directly affected by changes in crop stomatal conductance.

3.2 Greenhouse shading

In the previous section it was shown that greenhouse ventilation is important for creating the necessary conditions for crop development during summer. However, natural ventilation alone is not sufficient to remove the excess sensible energy from the greenhouse during sunny summer days (Baille 1999). For this reason, ventilation should be used in conjunction with other greenhouse cooling systems.

One of the most common methods used by the growers, due to its simplicity and low cost of implementation is white washing of greenhouse roof. By greenhouse roof white washing solar radiation entering the greenhouse is reduced something that directly affects air and crop temperature due to reduction of available energy and changes induced in crop stomatal conductance. One advantage of greenhouse shading by white washing over other techniques is that shading does not affect greenhouse ventilation while internally or externally mounted curtains adversely affect ventilation efficiency. The white washing also significantly increases the proportion of diffuse radiation in the greenhouse, which is known to increase radiation use efficiency (Alados and Alados-Arboledas 1999). Shading directly affects crop temperature and stomatal conductance due to reduction of available energy in the greenhouse and the crop. Katsoulas (2002) indicates that reduction of incoming solar radiation in the greenhouse resulted in canopy to air vapour pressure deficit. At the same time, he notes that while during the period without shading the canopy to air vapour pressure deficit was high; shading of the greenhouse led the crop to non heat and water stress conditions. Thus, the same author notes that shading of the greenhouse resulted in lower crop temperature and increase of crop stomatal conductance and transpiration rate. Finally, several researchers (Boulard et al. 1991, Abreu and Meneses 2000, Dayan et al. 2000, Fernandez-Rodriguez et al. 2000) indicate that greenhouse shading causes a decrease in crop temperature and transpiration rate.

3.3 Greenhouse evaporative cooling

The effect fog cooling on greenhouse microclimate is direct through effects on temperature and air vapour pressure deficit (Cohen et al. 1983, Arbel et al. 1999) and indirect to the crop through effects on crop temperature, stomatal and aerodynamic conductance, transpiration rate, etc.).

Fog cooling affects crop performance since reduces crop temperature and air vapour pressure deficit thus helping to reduce crop heat and water stress. In addition to reducing the temperature of the air, fog cooling directly reduces crop temperature when evaporation is done on the leaves, or indirectly through the reduction of air temperature and changes in transpiration. Moreover, by reducing the temperature and air vapour pressure deficit crop stomatal and aerodynamic conductances are modified which, together with the canopy to air vapour pressure deficit affect crop transpiration. It was found that fog cooling improved rose crop production mainly due to reduction of crop water stress index (Spoelstra 1975, Plaut and Zieslin 1977). Regarding production, several researchers (Spoelstra 1975, Plaut and Zieslin 1977, Plaut et al. 1979, Javoy et al. 1990) observed an increase in quality and quantity of production. There was no effect of fog cooling on flowers life after harvest (Urban et al. 1995). Finally, it has been observed that fog cooling increased crop leaf area index (Plaut et al. 1979, Katsoulas et al. 2001). Many authors (eg: Kaufman 1982, el Sharkawy and Cock 1986, Schulze 1986, Munro 1989, Bakker 1991, Jolliet and Bailey 1992, Baille et al. 1994a) studied the effect of air vapour pressure deficit on crop stomatal conductance. Most work has been done in parts of northern Europe where low levels of radiation and vapour pressure deficit do exist. Bakker (1991) for example, observed reduction in stomatal conductance, about 65% caused by an increase of air vapour pressure deficit of 1 kPa. Montero et al. (2001) however, found no significant reduction of crop stomatal conduction when air vapour pressure deficit increased from 1.4 kPa to 3.4 kPa. indicate that there is no clear evidence that stomatal conductance is directly affected by air vapour pressure deficit. Nevertheless, Baille et al. (1994) and Katsoulas et al. (2001) that carried out measurements in Mediterranean greenhouses, noted that use of fog cooling increased crop stomatal conductance and indicated that maximum values of crop stomatal conductance where observed when solar radiation exceeded 300 W m^{-2}. Regarding the effect of fog cooling on crop transpiration, several authors (eg: Plaut and Zieslin 1977, Boulard et al. 1991, Dayan et al. 2000) reported that use of fog cooling caused a reduction in crop transpiration rate. In contrast, other researchers (eg: Boulard and Baille 1993, Urban et al. 1995) report an increase of crop transpiration under fog cooling. Furthermore, Baille et al. (1994) and Katsoulas (2002), focused on the effect of fog cooling on day lag-hysteresis observed between transpiration and vapour pressure deficit on the one hand and transpiration and solar radiation on the other hand, observing a significant effect of fog cooling in the daily course of the curves of hysteresis. The hysteresis observed was reversing counterclockwise in the case of transpiration with radiation and consistent counterclockwise in the case of transpiration and air vapour pressure deficit. Moreover, Katsoulas (2002) observed a hysteresis between stomatal conduction and solar radiation and air vapour pressure deficit.

3.4 Leaf area index

Crop transpiration, through the process of cooling by evaporation, represents the main mechanism for greenhouse and crop cooling. This explains why maintaining high levels of transpiration rate in a greenhouse crop is one of the most effective and least costly ways of cooling the greenhouse environment during the warm season with high thermal loads, as happens from April to October in the Mediterranean countries Such as Greece (Baille 1999). To achieve high rates of transpiration, it is essential that certain conditions are met:

The crop should be well developed and should have a well-irrigated root system.

The leaf area of the crop must be large enough to convert the high amounts of energy in latent heat through transpiration.

Leaf stomata should remain open, but special care must be taken to avoid environmental conditions that may cause stress to plants. Stomatal closure is usually caused by inappropriate control of ventilation. The air vapour pressure deficit can be significantly increased with the introduction by high rates of dry air coming outside the greenhouse (Seginer 1994). It is known that high levels of air vapour pressure deficit lead to water stress conditions, which could lead to partial or complete stomatal closure (Baille et al. 1994a, Monteith 1995). Accordingly, adverse effects on gas exchange (transpiration, photosynthesis) and consequently on the production and product quality should be expected.

Crop aerodynamic conductance represents the main variable to controlling crop gas exchanges because it is the main factor limiting the process of sensible energy and water transfer from crop surface to the air. A large leaf area index LAI should increase crop aerodynamic conductance ga, since, by a first approximation, can be assumed that the crop aerodynamic conductance is given by:

$$g_a = 2\ \mathrm{LAI}\ g_{l,a} \tag{3}$$

where $g_{l,a}$ is the leaf aerodynamic conductance.
The same can be applied for the crop stomatal conductance g_c

$$g_c = 2\ \mathrm{LAI}\ g_{l,c} \tag{4}$$

where $g_{l,c}$ is the leaf stomatal conductance.
Thus, maintaining a high crop leaf area index in the greenhouse, increases significantly gas exchanges, which in turn affect greenhouse microclimate, mainly the air temperature and vapour pressure deficit. Then the changes in the greenhouse microclimate affect stomatal conductance, leading to a control feedback (Figure 4), as has been discussed and formulated a number of papers (Aubinet et al. 1989, Nederhoff and Vegter 1994, Baille 1999).
Jolliet (1994), used HORTITRANS to show the effect of LAI on crop transpiration while González-Real and Baille (2001) showed the effect of leaf area index on canopy to air temperature difference and on crop transpiration rate.

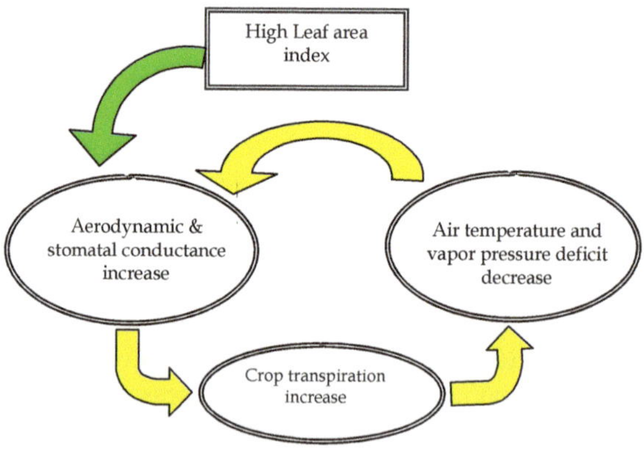

Fig. 4. Feedback control on crop transpiration.

4. Models of greenhouse crop transpiration

Transpiration is the most important part of the latent energy balance and one of the most important elements of greenhouse energy balance. As observed by many authors (eg: Morris et al. 1957, de Villèle 1972, van der Post et al. 1974 for tomato crop, Yang et al. 1990 for cucumber crop, Katsoulas et al. 2000 for rose crop) a strong correlation between crop transpiration and solar radiation exists. The air humidity inside the greenhouse is equally important but less effect on transpiration. The effect is characterized by the saturation vapour pressure deficit of air VPD (kPa), which is defined as the difference between the vapor pressure of the air at saturation e* (kPa) and the actual air vapour pressure e (kPa). The air vapour pressure deficit is variable characterising how dry the air is for a given temperature, is inversely proportional to the humidity. For constant greenhouse climate conditions VPD is correlated to solar radiation G. Therefore, in most cases, solar radiation alone can explain most of the variability of transpiration, 'hiding' the effect of VPD. However, it is necessary to consider VPD for transpiration calculation when using any heating, ventilation, cooling and dehumidification systems.

A considerable research effort has been invested in the (evapo)transpiration of greenhouse crops (e.g. Stanghellini 1987, Jolliet & Bailey 1992, Jolliet 1994, Papadakis et al. 1994, Kittas et al. 1999). Several authors have proposed models that allow getting a more accurate estimation of the crop transpiration rate. Many of these studies are based on leaf (canopy) transpiration and leaf energy balance models in which the transpiration is characterized by the crop conductance, as proposed initially by Penman and modified by Monteith to account for the stomatal response of the crop (P-M formula), including several successful attempts to combine the P–M equation with the greenhouse energy balance (Fuchs 1993, Boulard & Baille 1993, Seginer 1994, Boulard & Wang 2000).

The calculation and simulation of mass and energy exchanges between crop and air is dominated by two trends (González-Real 1995). The first assumes that the crop is divided into discrete levels and the exchange calculations require knowledge and description of the conditions in each crop level (Yang 1995). Simple sub-models are used for calculating the exchange at each level, but sophisticated models are required for calculating the required unknown parameters for each level (temperature, humidity, radiation, wind speed, CO_2 etc) and to incorporate the individual levels (eg leaf level) in a complete model (eg crop level). The second trend to calculate the mass and energy exchanges between crop and air assumes that the crop is a large leaf ('big leaf') and that all internal layers are located in the same climatic conditions (Stanghellini 1995). The transpiration of the crop in this case is given from the Penman-Monteith formula (Monteith 1973):

$$\lambda E = \frac{\delta G}{\delta + \gamma(1 + g_a/g_c)} + \frac{\rho C_p D_i g_a}{\delta + \gamma(1 + g_a/g_c)} \tag{5}$$

where G (W m^{-2}) is the solar radiation ρ and C_p are respectively the density (kg m^{-3}) and specific heat (J kg^{-1} K^{-1}) of air, and δ is the slope of the humidity-ratio (or vapour pressure) saturation curve (kPa K^{-1}).

However, the use of the complete P-M formula requires the knowledge of several inputs or parameters that are not easily available. Particularly, the aerodynamic and stomatal leaf conductances have to be known for each crop species and possibly, for each cultivar. That is why researchers have tried to overcome the estimation of these resistances by using a simplified form of the P-M formula. Transpiration models with the greenhouse climate as a boundary condition were first developed in the northern regions of Europe and North America for horticultural crops. In these northern conditions, the glasshouse is generally poorly ventilated during a large part of the growing season. The boundary layer conductance for glasshouse crops tends to be much smaller than would be expected for similar crops growing outdoors. Thus, glasshouse crops are very strongly decoupled from the outside atmosphere by the presence of the glass, and the heat and the water released at crop surface will accumulate inside the glasshouse. Consequently, the transpiration rate will adjust until it reaches a stable equilibrium transpiration rate dictated by the net radiation received. On the contrary, greenhouse crop transpiration in Mediterranean or similar warm conditions is much more dependent on convection. As the ventilation and the turbulent mixing are vigorous, the saturation deficit at the leaf surface is closely coupled to the deficit of ambient air, and the latter is directly influenced by the outdoor saturation deficit.

4.1 Aerodynamic conductance

The aerodynamic conductance, which represents the transfer of water from the surface of the leaf or crop in ambient air can be calculated in two ways:
either as a function of crop sensible energy balance and the crop to air temperature difference (Seginer 1984),

$$g_a = \frac{H_c}{\rho C_p DT_c} \tag{6}$$

or by the classical theory of heat transfer using dimensionless numbers (Stanghellini 1987). Following the non-dimensional heat transfer theory (Kreith, 1973; Monteith and Unsworth, 1990; Schuepp, 1993), the boundary layer conductance to heat transfer (g_b, m s^{-1}) from bodies of different shapes can be expressed as a function of the dimensionless Nusselt number, Nu:

$$g_b = \frac{\kappa Nu}{d} \tag{7}$$

where κ (m^2 s^{-1}) is the thermal diffusivity of air and d (m) is the characteristic dimension of the body. Nu is generally expressed as a function of the following dimensionless numbers;

- the Reynolds number:

$$Re = U\, d/\nu \tag{8}$$

where U is air velocity (m^2 s^{-1}) and ν the kinematic viscosity of the air (m2 s^{-1}),

- the Grashoff number:

$$Gr = \frac{\beta g d^3 \Delta T}{\nu^2} \tag{9}$$

where g is the gravitational acceleration (m s^{-2}), β the coefficient of volumetric expansion (K^{-1}), and ΔT (K) the temperature difference between the leaf and its environment, and

- the Prandlt number:

$$Pr = \nu/\kappa = 0.705 \text{ for air}$$

Table 2 presents some functions Nu = f(Re, Gr) in laminar regime currently used for the estimation of gb of smooth flat plates or leaf replicas, distinguishing between free, mixed or forced convection mode.

Eq. n°	Mode	Formula	References
Eq.(13)	Free	$Nu = 0.50\, Gr^{0.25}$	Monteith (1980)
Eq.(14)	Free	$Nu = 0.25\, Gr^{0.30}$	Stanghellini (1987)
Eq.(15)	Mixed	$Nu = 0.37\, (Gr + 6.92\, Re^2)^{0.25}$	Stanghellini (1987)
Eq.(16)	Forced	$Nu = 0.60\, Re^{0.5}$	Schuepp (1993)

Table 2. Formulae relating Nu to Gr and Re for a flat horizontal plate parallel to the air flow, for different convection modes. Eqs (14) and (15) were derived from measurements of tomato leaf heated replicas in greenhouse (Stanghellini, 1987).

The problem that usually arises from the use of Eqn (6) is associated with the choice of the points at which the temperature difference is calculated. There are many disagreements in the literature as to what temperature difference should be taken. If the aerodynamic conductance calculated from measurements of air temperature above and not within the crop, the result is more of an indicative value (Yang 1995). Yang (1995) considered that the assumption of the big leaf is not representative, as the wind speed above the crop is up to one order of magnitude higher than that in the crop.

4.2 Stomatal conductance

The stomatal conductance plays an important role in the division of energy into sensible and latent and is affected by a number of microclimate parameters.

The stomatal conductance at the level of leaf or crop has been associated with G, VPD, air temperature and CO2 concentration in the air, and the leaf water potential (Turner 1974, van Bavel 1974, Jarvis 1976, Takami and Uchijima 1977, Farquhar 1978, Farquhar and Sharkey 1982, Kaufmann 1982, Choudhury 1983, Zeiger 1983, Dwyer and Stewart 1984, Avissar et al. 1985, Choudhury and Idso 1985, Lindroth 1985, Simpson et al. 1985, Grantz and Zeiger 1986, Baldocchi et al. 1987, Stanghellini 1987) or physiological factors, such as photosynthesis (Ball et al. 1987, Collatz et al. 1991, Leuning 1995) and transpiration (Monteith 1973, Monteith 1995). Of these factors, as we know, the important role is played by radiation. Despite the fact that it has been found that the form of the relationship of stomatal conductance to the above factors (eg: Lange et al. 1971, Neilson and Jarvis 1975, Stanghellini 1987), a mechanistic model that could simulate stomatal conductance has not been developed and thus only empirical models are available.

The most common relationship, which reflects the influence of environmental factors on the behaviour of stomata, is that of Jarvis (1976):

$$g_c = g_M f_1(R_n) f_2(D_i) f_3(T_i) f_4(CO_2) \tag{10}$$

In this relationship, stomatal conductance is expressed as a function of maximum conductance g_M, multiplied by a number of factors. These factors are independent of each other, impose crop water stress and their result is multiplied, not added to calculate the final result. The form of mathematical functions f_1, f_2, f_3, f_4 is generally known (eg: Stanghellini 1987, Baille et al. 1994b), while the value of maximum conductance g_M, which varies from species to species, can be measured or found in the literature.

4.3 Simplified models of greenhouse crop transpiration

For greenhouse crops, the formula most currently used until now for evapotranspiration (E) prediction is based on a simple linear correlation between E and solar radiation, G (Morris et al. 1957, Stanhill and Scholte Albers 1974)

$$E = Ao\ Kc\ G + Bo \tag{11}$$

where Kc is a 'crop' coefficient depending on the crop development stage. Ao and Bo are two coefficients determined by statistical adjustment. This relation, which is mainly valid at daily or weekly time scales, presents several drawbacks, as follows:

i. A large amount of empiricism and inaccuracy in the determination of the crop coefficient Kc.

ii. The vapour pressure deficit D is not explicitly taken into account in Eq. (11). In the case of a significant correlation between G and D, Eq. (11) can give a satisfactory estimation of E. But, under greenhouses that use heating, shading screen or fog-system, such a correlation does not stand (Bakker 1991). Moreover, for most of the species, the transpiration rate depends significantly on the saturation deficit (Okuya and Okuya 1988, Baille et al. 1994).

iii. Eq. (11) assigns a constant value (Bo) to nocturnal evapotranspiration, which can rise to a significant level in the case of heated greenhouses during cold periods (De Graaf 1985), thus contributing to a large extent to the total 24 h water loss. In fact, the coefficient Bo averages in some way the influence of nocturnal heating, but cannot predict the effect of single climate variables on nocturnal values of λE. As an example, the relation between solar radiation and crop transpiration rate of a rose crop for two different seasons, winter and summer is presented in Fig. 5. In the two cases, a linear relationship $\lambda E = aG + b$ was obtained with different values of the slope a for the two cases which characterises the influence of stomatal and aerodynamic conductance and of the solar radiation to vapour pressure deficit relation on evapotranspiration rate. The offset b is significant in the two cases, and reflects the contribution of the nocturnal evapotranspiration.

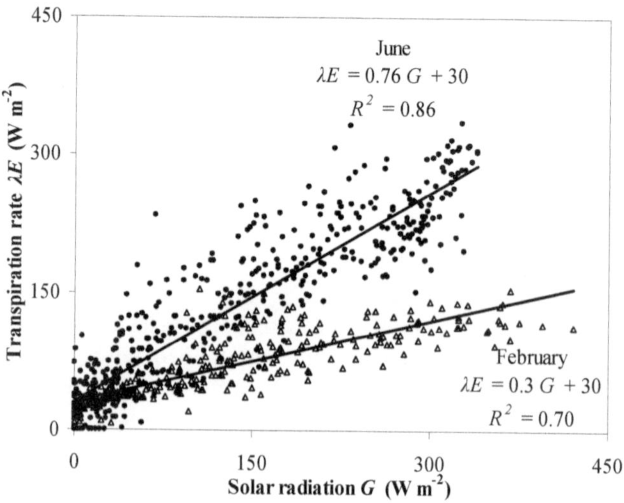

Fig. 5. Relation between solar radiation and crop transpiration rate of a rose crop for two different seasons, winter and summer.

The P-M equation (Monteith 1973) has been originally developed to calculate evapotranspiration from homogeneous vegetated surfaces. When applied to greenhouse crops it may be written as:

$$E = A\,G + B\,D_{id} \tag{12}$$

with: $A = \dfrac{\delta}{\delta + \gamma(1 + g_a/g_c)}$ and $B = \dfrac{\rho C_p\, g_a}{\delta + \gamma(1 + g_a/g_c)}$

A is referred to as the 'radiation term' and B as the 'aerodynamic term' (sometimes called 'advection term'). Hence A and B may be referred to as the 'radiation coefficient' and the 'aerodynamic coefficient'. Equation (12) may be regarded as empirical formulae, with A and B obtained by regressing measured evapotranspiration against measured solar radiation and VPD. From this point of view, A and B are often treated as constants for a given crop, or as simple functions of readily measurable quantities, such as the leaf area index (LAI). These coefficients may be corrected for changes in the environmental conditions and for water stress. The effect of light on A and B (via gc) may be ignored for high-radiation situations, and the expected increase with the ventilation rate, through its effect on ga, will also be ignored, based on previous precedents (Baille et al. 1994, Kittas et al. 1999), analysis (Seginer 1994), and experience (Seginer & Tarnopolsky, 2000). Baille et al., (1994a) suggested the following formulas for A and B as functions of the leaf area index, LAI :

$$A = a\, f_1(LAI) = A_0\,(1-e^{-k\,LAI}) \qquad (13a)$$

where k is the extinction coefficient (= 0.64 for tomato, Stanghellini, 1987). The reason for the choice of Eq. (13a) is that represents the classical relationship for radiation interception by a canopy (Varlet-Grancher et al., 1989), and

$$B = b\, f_2(LAI) = B_0\, LAI \qquad (13b)$$

The choice of Eq. (13b) is straightforward as LAI can be considered as a multiplicative factor in the 'advective' term of the Penman-Monteith equation.

The constants a and b have been identified for several greenhouse species and a summary table of A and B parameters is given by Seginer (1997).

In a comparative survey of Jolliet and Bailey (1992) between the models of: (i) Penman (FAO 1977), (ii) Stanghellini (1987), (iii) Chalabi and Bailey (1989) and (iv) Aikman and Houter (1990), the model of the Stanghellini (Stanghellini, 1987) proved to be more accurate in predicting transpiration. In conclusion, they noted that solar radiation, saturation deficit and air velocity inside the greenhouse are the most important factors affecting transpiration and should be included in each simulation was calculated.

The problem is that many of the analytical models exist, such as the Stanghellini (1987), calibrated and developed in areas and periods with low values for radiation and leaf temperature. It is characteristic that during Stanghellinis' model calibration (March and April), the air temperature was almost always less than 25 ° C, the temperature of the leaves slightly exceeded 24.5 ° C, while the radiation was usually less from 300 W m^{-2}. Moreover, in the model of Stanghellini (1987), the effect of radiation on stomatal conductance and transpiration is taken into account more indirectly than directly, by affecting the crop to air temperature difference. Thus, an increase of radiation, according to simulation, has little influence on transpiration, but reduces leaf temperature. When the leaf temperature exceeds 24.5°C, stomatal conductance decreased sharply, leading to reduced transpiration and further increase crop temperature. Stanghellinis' (1987) model is not suitable for Mediterranean areas during the summer but it is appropriate and gives accurate predictions when used in conditions similar to those calibrated. However, other authors have developed and calibration similar to Stanghellinis' (1987) model for Mediterranean conditions (eg: Papadakis et al. 1994a, Boulard et al. 1996).

Jolliet (1994) presented a relatively simple crop transpiration model, including in the calculations cover condensation. The model was verified by experimental measurements

and the results showed that it is sufficiently accurate to be able to provide transpiration, relative humidity and air saturation deficit values. Furthermore, the model is able to predict the amount of water and energy that must be removed or added to the greenhouse to achieve a specific value of air relative humidity or crop transpiration rate.
It follows that several of the proposed simplified relationships give good estimates and could therefore be used where high precision is not required (eg monitoring of irrigation). However, the validity of simplified relationships is limited because: first, can not be used in different climatic conditions and different stages of crop development than calibrated, without a new calibration for each case and, secondly, their use is limited to applications where high precision is not required. Furthermore, it is better to use correlations derived from experimental or simulated results with constant coefficients as discussed above than to use the complex equation of Penman-Monteith with wrong stomatal conductance values.

4.4 Greenhouse crop transpiration simulation based on outside climate data

The biggest problem as far it concerns greenhouse crop transpiration prediction is the interaction of humidity and transpiration. In the case of models to be used for climate control in greenhouses it is necessary to use and included sub-models for transpiration which provide transpiration simulations as a function of external climatic conditions and greenhouse characteristics.
Until recently, only complicated dynamic models were possible to achieve this (eg: Bot 1983, Kimball 1986, Chalabi and Bailey 1989). To control the humidity and estimate crop transpiration in Mediterranean countries, other researchers (eg: Fuchs 1993, Boulard and Baille 1993, Seginer 1994) presented analytical models that allow the calculation of temperature and humidity of indoor air as well and transpiration of the crop as a function of greenhouse ventilation and cooling. Boulard and Wang (2000) gave an expression that estimates crop transpiration as a function of external climatic parameters.

5. Acknowledgments

This publication was financed by the Centre for Research and Technology Thessaly.

6. References

Abreu, P. E., Meneses, J. F., 2000. Influence of soil covering, plastic ageing and roof whitening on climate and tomato crop response in an unheated plastic Mediterranean greenhouse. *Acta Horticulturae* 534: 343-350

Alados, I., Alados-Arboledas, L., 1999. Direct and diffuse photosynthetically active radiation: measurements and modelling. *Agricultural and Forest Meteorology*, 93: 27-38

Arbel, A., Yekutieli, O., Barak, M., 1999. Performance of a fog system for cooling greenhouse. *Journal of Agricultural Engineering Research*, 72: 129-136

Aubinet, M., Deltour, J., De Halleux, D., 1989. Stomatal regulation in greenhouse crops: Analysis and simulation. *Agricultural and Forest Meteorology*, 48: 21-44

Avissar, R., Avissar P., Mahrer, Y., Bravdo, B. A., 1985. A model to simulate response of plant stomata to environmental conditions. *Agricultural and Forest Meteorology*, 34: 21-29

Baille, A., 1999. Greenhouse structure and equipment for improving crop production in mild winter climates. *Acta Horticulturae*, 491: 37-47

Baille, M., Baille, A., Delmon, D., 1994a. Microclimate and transpiration of greenhouse rose crops. *Agricultural and Forest Meteorology*, 71: 83-97

Baille, M., Baille, A., Laury, J. C., 1994b. Canopy surface resistance to water vapour transfer for nine ornamental pot plants crops. *Scientia Horticulturae*, 57: 143-155.

Bakker, J. C., 1991. Leaf conductance of four glasshouse vegetable crops as affected by air humidity. *Agricultural and Forest Meteorology*, 55: 23-36

Baldocchi, D. D., Hicks, B. B., Camara, P., 1987. A canopy stomatal resistance model for gaseous deposition to vegetated surfaces. *Atmospheric Environment*, 21: 91-101

Ball, J. T., Woodrow, I. E., Berry, J. A., 1987. A model predicting stomatal conductance and its contribution to the control of photosynthesis under different environmental conditions. In: Biggings, J., (Editor). *Proceedings of the 7th international congress in 'Progress in photosynthesis research'*. Dordrecht, Martinus Nijhoff Publishers, 221-224

Boulard, T., Baille, A., 1993. A simple greenhouse climate control model incorporating effects of aeration and evaporative cooling. *Agricultural and Forest Meteorology*, 65: 145-57

Boulard, T., Baille, A., Le Gall, F., 1991. Etude de différentes méthodes de refroidissement sur le climat et la transpiration de tomates de serre. *Agronomie* 11: 543-554

Boulard, T., Wang, S., 2000. Greenhouse crop transpiration simulation from external climate conditions. *Agricultural and Forest Meteorology*, 100: 25-34

Bunce, J. A., 1985. Effect of boundary layer conductance on the response of stomata to humidity. *Plant, Cell and Environment*, 8: 55-57

Choudhury, B. J., Idso, S. B., 1985. An empirical model for stomatal resistance of field-grown wheat. *Agricultural and Forest Meteorology*, 36: 65-82

Choudhury, B., 1983. Simulating the effects of weather variables and soil water potential on a corn canopy temperature. *Agricultural and Forest Meteorology*, 29: 169-182

Choudury, B. J., Monteith, J. L., 1986. Implications of stomatal responses to saturation deficit for the heat-balance of vegetation. *Agricultural and Forest Meteorology*, 36: 215-225

Cohen, Y., Stanhill, G., Fuchs, M., 1983. An experimental comparison of evaporative cooling in a naturally ventilated greenhouse due to wetting the outer roof and inner crop soil surfaces. *Agricultural Meteorology* 28: 239-251

Collatz, G. J., Ball, J. T., Grivet, C., Berry, J. A., 1991. Physiological and environmental regulation of stomatal conductance, photosynthesis and transpiration: a model that includes a laminar boundary layer. *Agricultural and Forest Meteorology*, 54: 107-136

Dayan, E., Fuchs, M., Plaut, Z., Presnov, E., Grava, A., Matan, E., Solphoy, A., Mugira, U., Pines, N., 2000. Cooling of roses in greenhouses. *Acta Horticulturae*, 534: 351-360

De Villèle, O., 1972. Besoins en eau des cultures sous serres - Essai de conduite des arrosages en fonction de l'ensoleillement. *Acta Horticulturae* 35: 123-135

Dwyer, L. M., Stewart, D. W., 1984. Indicators of plant stress in corn (*Zea mais* L.). *Canadian Journal of Plant Science*, 64: 537-546

El Sharkawy, M. A., Cock, J. H., 1986. The humidity factor in stomatal control and its effect on crop productivity. In: Marcelle, R., Clijsters, H., van Poucke, M., (Editors), *Biological Control of Photosynthesis*. Martinus Nijhoff, Dordrecht, 187-198

FAO, 1977. Crop water requirements. *FAO irrigation and drainage,* Paper No. 24: 15-44

Farquhar, G. D., 1978. Feedforward response of stomata to humidity. *Australian Journal of Plant Physiology*, 5: 787-800

Farquhar, G. D., Sharkey, T. D., 1982. Stomatal conductance and photosynthesis. *Annual Review of Plant Physiology*, 33: 317-345

Fernandez-Rodriguez, E. J., Fernandez-Vadillos, J., Camacho-Ferre, F., Vazquez, J. J., Kenig, A., 2000. Radiative field uniformity under shading screens under greenhouse versus whitewash in Spain. *Acta Horticulturae,* 534: 125-130

Fuchs, M., 1993. Transpiration and foliage temperature in a greenhouse. In: Fuchs, M., (Editor), *Proceedings of the International Workshop on Cooling Systems for Greenhouses*, Tel Aviv, Israel, 89 - 98

Fuchs, M., Dayan, E., Shmuel, D., Zipori, I., 1997. Effects of ventilation on the energy balance of a greenhouse with bare soil. *Agricultural and Forest Meteorology*, 86: 273-282

González-Real, M. M., 1995. Estudio y modelizacion de intercambios gaseosos en cultivos de rosas bajo invernadero. *Ph D Thesis*, Universidad Politecnica de Valencia, Spain, 192 pp

González-Real, M., Baille, A., 2001. Simulating the behaviour of a greenhouse rose crop by means of a model including physical and physiological feedback loops. *Acta Horticulturae*, 559: 441-448

Grantz, D. A., Zeiger, E., 1986. Stomatal responses to light and leaf–air water VPD show similar kinetics in sugarcane and soybean. *Plant Physiology*, 81: 865-868

Jarvis, P. G., 1976. The interpretation of the variations in leaf water potential and stomatal conductance found in canopies in the field. *Philosophical Transactions of the Royal Society*, London, B, 273: 593-610

Javoy, M., Letard, M., Pelletier, B., 1990. Le concombre en culture hors sol. Les éléments de conduite. *Infos CTIFL Hos série Cultures légumières sur substrats*, 53-68

Jolliet, O., 1994. HORTITRANS, a model for predicting and optimizing humidity and transpiration in greenhouses. *Journal of Agricultural Engineering Research*, 57: 23-37

Jolliet, O., Bailey, B. J., 1992. The effect of climate on tomato transpiration in greenhouses: measurements and models comparison. *Agricultural and Forest Meteorology*, 58: 43-62

Katerji, N., Rana, G., 2006. Modelling evapotranspiration of six irrigated crops under Mediterranean climate conditions. Agric. For. Meteorol., 138:142–155.

Katsoulas N., 2002. Influence of environmental variables on greenhouse rose crop transpiration. PhD Thesis, University of Thessaly, 244 pp

Katsoulas, N. Baille, A., Kittas, C., 2002. Influence of leaf area index on canopy energy partitioning and greenhouse cooling requirements. Biosyst. Engin. 83(3), 349-359.

Katsoulas, N., Baille, A., Kittas, C., 2001. Effect of misting on transpiration and conductances of a greenhouse rose canopy. Agric. For. Meteorol. 106(3), 233-247.

Katsoulas, N., Kittas, C., Baille, A., 1999. Estimating transpiration rate and canopy resistance of a rose crop in a fan-ventilated greenhouse. Acta Horticulturae, 548: 303-309

Kaufmann, M. R., 1982. Leaf conductance as a function of photosynthetic photon flux density and absolute humidity difference from leaf to air. *Plant Physiology*, 69: 1018-1022

Kittas, C., Baille, A., Katsoulas, N., 2001. Transpiration and energy balance of a greenhouse rose crop in Mediterranean summer conditions. *Acta Horticulturae*, 559: 395-400

Kittas, C., Katsoulas, N., Baille, A., 1999. Transpiration and canopy resistance of greenhouse soilless roses in Greece. Measurements and modeling. *Acta Horticulturae*, 507: 61-68

Kittas, C., Katsoulas, N., Baille, A., 2001. Influence of greenhouse ventilation regime on microclimate and energy partitioning of a rose canopy during summer conditions. J. Agric. Eng. Res. 79(3), 349-360.

Lange, O. L., Lösch, R., Schulze, E. D., Kappen, L., 1971. Response of stomata to changes in humidity. *Planta Berlin*, 100: 76-86

Lindroth, A., 1985. Canopy conductance of coniferous forests related to climate. *Water Resource Research*, 21: 297-304

Monteith, J. L., 1973. *Principles of environmental Physics*. Contemporary Biology, Edward Arnold, London, UK, 241 pp

Monteith, J. L., 1980. Coupling of plants to the atmosphere. In: Grace, J., Ford, E. D., Jarvis, P. G. (Editors) *Plants and their Atmospheric Environment*. Blackwell Scientific Publications, Oxford, UK, 1-29.

Monteith, J. L., 1995. A reinterpretation of stomatal responses to humidity. *Plant, Cell and Environment*, 18: 357-364

Monteith, J.L., Unsworth, M.H., 1990. Principles of environmental physics, 2nd ed. Edward Arnold, London UK, 291 pp.

Morris, L. G., Neale, F. E., Postlethwaite, J. D., 1957. The transpiration of glasshouse crops and its relationship to the incoming solar radiation. *Journal of Agricultural Engineering Research*, 2(2): 111-122

Munro, D. S., 1989. Stomatal conductances and surface conductance modelling in mixed wet-land forest. *Agricultural and Forest Meteorology*, 48: 235-249

Nederhoff, E., Vegter, J. G., 1994. Canopy PS of tomato, cucumber and sweet pepper in greenhouses: measurements compared to models. *Annals of Botany*, 73(4): 421-427

Neilson, R. E., Jarvis, P. G., 1975. Photosynthesis in Sitka Spruce. VI. Response of stomata to temperature. *Journal of Applied Ecology*, 12: 721-745

Okuya A., Okuya T., 1988. The transpiration of greenhouse tomato plants in rockwool culture and its relationship to climatic factors. Acta Horticulturae, 230: 307-311

Orgaz, F., Fernandez, M.D., Bonachelac, S., Gallardoc, M. , Fereresa, E., 2005. Evapotranspiration of horticultural crops in an unheated plastic greenhouse Agr. Water Manag., 72: 81–96

Papadakis, G., Frangoudakis, A., Kyritsis, S., 1994. Experimental investigation and modelling of heat mass transfer between a tomato crop and greenhouse environment. *Journal of Agricultural Engineering Research*, 57: 217-227

Plaut, Z., Zieslin N., 1977. The effect of canopy wetting on plant water status, CO_2 fixation, ion content and growth rate of 'Baccara' roses. *Physiologia Plantarum*, 39: 317-322

Plaut, Z., Zieslin, N., Grawa, A., Gazit, M., 1979. The response of rose plants to evaporative cooling: flower production and quality. *Scientia Horticulturae*, 11: 183-190

Schuepp, P.H., 1993. Leaf boundary layers. New Phytol. 125, 477-507.

Seginer, I., 1994. Transpirational cooling of a greenhouse crop with partial ground cover. *Agricultural and Forest Meteorology*, 71: 265-281

Seginer, I., 1997. Alternative design formulae for the ventilation rate of greenhouses. *Journal of Agricultural Engineering Research*, 68(4): 355-365

Simpson, J. R., Fritshen, L. J., Walker, R. B., 1985. Estimating stomatal diffusion resistance for Douglas-Fir, Lodgepole Pine and White Oak under light saturated conditions. *Agricultural and Forest Meteorology*, 33: 299-313

Spoelstra, P. A., 1975. Comparative measurements of fan and pad cooling and plant watering. *Acta Horticulturae*, 46: 23-31

Stanghellini, C., 1987. Transpiration of greenhouse crops: an aid to climate management. *PhD Thesis*, Agricultural University of Wageningen, The Netherlands, 150 pp

Stanghellini, C., 1995. Response to comments on 'Thermal and aerodynamic conditions in greenhouses in relation to estimation of heat flux and evapotranspiration'. *Agricultural and Forest Meteorology*, 77: 137-138

Stanhill G., Scholte Albers J., 1974. Solar radiation and water loss from glasshouse roses. J. Amer. Soc. Hort, Sci., 99: 107-110.

Takami, S., Uchijima, Z., 1977. A model for the greenhouse environment as affected by the mass and energy exchange of a crop. *Agricultural Meteorology*, 33: 117-127

Urban, I., Brun, R., Urban, L., 1995. Influence of electrical conductivity, relative humidity and seasonal variations on the behaviour of cut roses produced in soilless culture. *Acta Horticulturae*, 408: 101-108

Van Bavel, C. H. M., 1974. Soil water potential and plant behaviour: a case modelling study with sunflowers. *Ecologia Plantarum*, 9: 89-109

Van der Post, C. J., van Schie, J. J., de Graaf, R., 1974. Energy balance and water supply in glasshouses in the West-Netherlands. *Acta Horticulturae*, 35: 13-22

Yang, X., 1995. Comments on 'Thermal and aerodynamic conditions in greenhouses in relation to estimation of heat flux and evapotranspiration'. *Agricultural and Forest Meteorology*, 77: 131-136

Yang, X., Short, T. H., Robert, D. F., Bauerle, W. L., 1990. Transpiration, leaf temperature and stomatal resistance of a cucumber crop. *Agricultural and Forest Meteorology*, 51:197-209

Zeiger, E., 1983. The biology of stomatal guard cells. *Annual Review of Plant Physiology*, 34: 441-475

16

Description of Two Functions I and J Characterizing the Interior Ground Inertia of a Traditional Greenhouse - A Theoretical Model Using the Green's Functions Theory

Rached Ben Younes
University of Gafsa,
Tunisia

1. Introduction

In this chapter, we are presenting a precise model translating the influence of the ground inertia in the thermal behaviour of a greenhouse without vegetation. This work takes into account all the real mechanisms of exchanges (solar conduction, convection, radiation, thermal inertia) between the various elements of the system (cover, interior air, ground), but does not take into account the mass transfers (diffusions of moisture in the ground, evapo-transpiration). We sought here to define a model constituting a core of a procedure on which new extensions will be based. We show via the Green Functions Theory (GFT) that the model's differential equations are reduced to a system of integral equations on the ground surface. These equations implicitly take into account the heat propagation in the ground. This model carefully describes in details the exchanges between the ground and the interior of the greenhouse. It aims also at defining the evolution of the greenhouse internal air temperature as well as that of the superficial temperature of its ground according to the following external data (power, exterior temperature). The mathematical study is completed by a numerical simulation on an isolated greenhouse.

One of the delicate problems in the study of thermal behaviours of the horticultural greenhouses is the modelling of their thermal inertia, which comes mainly from the ground. Indeed, the implementation of knowledge's model is an effective mean to accurately envisage the thermal behaviour of a greenhouse over long periods. Theoretical, numerical and experimental studies were the subject of many former publications. From these principal works we retain

- a simplified model which is based on a total heat balance by holding account particularly of absorbed solar radiation, and conductive losses through the wall of the greenhouse [1-6]
- another model is limited to the heat balances of the interior air and the cover, we retain of this model that the case of day and night were treated separately [7-8]
- then, another model taking into account the heterogeneity of the interior greenhouse ground, this one is subdivided into ten homogeneous layers of different conductivities [9]

- finally another model whose objective is to bring a contribution to the study of a process of setting out-freezing of a greenhouse by a technique of water sprinkling. A distinction is thus made in the choice of the system of equations according to the three following cases: summer (days and nights) and winter days, normal winter nights and winter nights with setting out-freezing.

The cover is considered at uniform temperature in the first two cases, will be subdivided into three layers in the last case to take into account of the layer of the formed ice. The ground is constituted of seven homogeneous layers in all cases [10-13]

Other modelling [14] lead to a too simplistic analytical resolution or being based on a one-dimensional approach by section of the ground's behaviour, so they do not make it possible to simulate the real behaviour [15]. The theoretical models, although often partial, contain many unknown parameters or difficult to determine experimentally.

A rigorous modelling of the interaction ground-greenhouse requires the solution of a differential equation with complex conditions of surface. The current models deal with this problem either numerically by discretizing the basement in the form of some layers [28-29], or by admitting that it behaves overall like a virtual thermal mass whose heat capacity and time-constant are given by the experiments [30].

However, the equation of propagation in the ground has an analytical solution which is obtained by GFT.

Our objective in this work is to show that this exact solution makes it possible to obtain the evolution's equation of the surface temperature $T_{Si}(t)$ of the ground's interior of the greenhouse, according to the total power absorbed by the ground and to the temperature of the exterior air of the greenhouse. Two functions characteristic of the ground's behaviour appear in this equation (hereafter in the text) and we show starting from their properties at what the approximation of the virtual thermal mass consists of. The limits of this approximation appear clearly, we thus show how to correct and compare the two results in both cases.

In section (2), we establish the evolution's equation of the greenhouse's interior air, it acts here as a simplified model (greenhouse without vegetation) where solar energy is only absorbed by the ground and where the phenomena of evapo-transpirations do not intervene Initially, we are concerned to establish a model taking into account the mechanisms of exchange by radiation, convection and conduction. In this model, we were able to control all the physical parameters in the case where it is possible to validate experimentally and quantitatively to separate the respective influences from these various modes and to determine in a reliable and univocal way the parameters suitable for each one of them: conduction, convection, radiation on the one hand and mass transfer (evapo-transpiration, condensation) on the other hand.

2. Setting in equation

2.1 Study of the heat balance of the greenhouse's internal air

Our system consists of three essential elements: the cover, internal air and the ground, the thermal behaviour of the internal air of the greenhouse, which we consider well ventilated, translating the evolution of the interior temperature $Ti(t)$, obeys to the following equation

$$\frac{dT_i(t)}{dt} = \frac{S_i H_{si}}{V_i \rho_i c_i}\left(T_{si}(t) - T_i(t)\right) + \frac{S_c H_{ci}}{V_i \rho_i c_i}\left(T_{ci}(t) - T_i(t)\right) + \frac{D_v(t)}{V_i}\left(T_e(t) - T_i(t)\right) \qquad (1)$$

The internal air exchanges its heat, by convection, with the surface of the ground of average temperature $T_{si}(t)$ and the surface of internal face of the cover, which average temperature is $T_{ci}(t)$, by two convection coefficients respectively H_{si} and H_{ci}. There is also a heat exchange between the internal and external air via openings of air renewal of volume throughput $D_v(t)$, Si: surface of the internal ground of the greenhouse, S_c: that of the cover, and V_i: the volume of the greenhouse see Figure 2.
All the temperatures are evaluated, thereafter, compared to a temperature of reference T_0 which is that of the ground taken at a depth superior to the effect of annual thermal skin. The latter is stable in a given area (evaluated at a depth of 2 meters), and practically equalizes at the annual average atmospheric temperature.

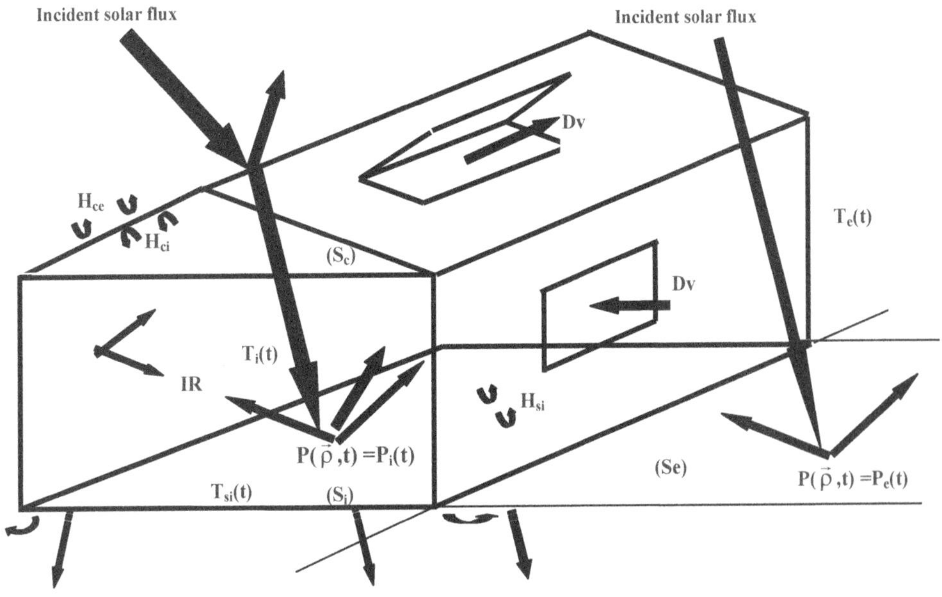

Fig. 1. Lay-out of the balance-sheet of heat exchange

2.2 Heat balance of the cover

The thickness of the cover is very low and the temperatures are slowly variable. We admit in his thickness a temperature profile constantly linear what amounts to neglecting its thermal inertia. The cover exchanges with its medium surrounding two fluxes Φ_e and Φ_i such as:
Φ_{ci}: heat flux exchanged by the internal face of the cover with the internal medium of the greenhouse.
Φ_{ce}: heat flux exchanged by the external face of the cover with the external medium of the greenhouse.

$$\Phi_{ci} = H_{ci}\left(T_i(t) - T_{ci}(t)\right) + H_{irc}\left(T_{si}(t) - T_{ci}(t)\right) \tag{2}$$

$$\Phi_{ce} = H_{ce}\left(T_e(t) - T_{ce}(t)\right) + H_{irc}\left(T_e^{'}(t) - T_{ce}(t)\right) + a_c p_r \tag{3}$$

These two fluxes are composed of a convective part and of a linearized radiative part because the temperatures oscillate slightly around the absolute annual average temperature of reference T_0. Indeed, the radiative power exchanged with the outside face of the cover is written as:

$$S_c\varphi_{ir,ce} = \varepsilon_c S_{atm} F_{atm\to ce}\varepsilon_{atm}\sigma T_{atm}^4 + \varepsilon_c S_e F_{se\to ce}\varepsilon_s\sigma T_{se}^4 - \varepsilon_c S_c F_{ce\to atm}\sigma T_{ce}^4$$

Knowing that: $\varepsilon_{atm} = 1$, $T_{atm}^4 = \varepsilon_e T_e^4$ and by neglecting the effects of the external ground on the cover, the preceding expression is reduced:

$$S_c\varphi_{ir,ce} = \varepsilon_c\sigma\left(S_{atm}F_{atm\to ce}\varepsilon_e T_e^4 - S_c F_{ce\to atm} T_{ce}^4\right)$$

The relation of reciprocity:

$$S_{atm}F_{atm\to ce} = S_c F_{ce\to atm} = S_c \text{ because } F_{ce\to atm} = 1$$

Let's make the change of variable:

$$T_e'^4 = \varepsilon_e T_e^4 \tag{4}$$

Consequently, the expression of $\phi_{ir,ce}$ becomes:

$$\varphi_{ir,ce} = \varepsilon_c\sigma\left(T_e'^4 - T_{ce}^4\right)$$

Let, $\begin{cases} T_e' = T_0 + \tilde{T}_e' \\ T_{ce} = T_0 + \tilde{T}_{ce} \end{cases}$

With $\tilde{T}_e'$, $\tilde{T}_{ce}$ are the fluctuations corresponding to each of two temperatures and T_{ce}, finally we obtain:

$\Phi_{ir,ce} = H_{irc}\left(T_e'(t) - T_{ce}(t)\right)$ with: $H_{irc} = 4\varepsilon_c\sigma T_0^4$, we took off the hats on the temperatures only to reduce the writing.

The temperature of the cover is written: $T_c(y) = a\ y + b$ with a and b two constants determined by the boundary conditions

$$T_c(y) = \frac{(T_{ce} - T_{ci})}{e} y + T_{ci} \tag{5}$$

The boundary conditions on the surface of the cover are written:

$$-k_c \left.\frac{\partial T_c}{\partial y}\right|_{y=0} = H_{ci}\left(T_i(t) - T_{ci}(t)\right) + H_{irc}\left(T_{si}(t) - T_{ci}(t)\right) \tag{6}$$

$$k_c \left.\frac{\partial T_c}{\partial y}\right|_{y=e} = H_{ce}\left(T_e(t) - T_{ce}(t)\right) + H_{irc}\left(T_e'(t) - T_{ce}(t)\right) + a_c P_r \tag{7}$$

We found previously the expression of $T_c(y)$, then

$$\frac{\partial T_c}{\partial y} = \frac{T_{ce} - T_{ci}}{e} \tag{8}$$

By injecting the expression (8), in the system of equations (6) and (7), we obtain a new system of equation ; ((6'), (7')) of unknown factors $T_{ci}(t)$ and $T_{ce}(t)$:

$$\left(\frac{k_c}{e} + H_{ci} + H_{irc}\right) T_{ci}(t) - \frac{k_c}{e} T_{ce}(t) = H_{ci} T_i(t) + H_{irc} T_{si}(t) \tag{6'}$$

$$-\frac{k_c}{e} T_{ci}(t) + \left(\frac{k_c}{e} + H_{ce} + H_{irc}\right) T_{ce}(t) = \left(H_{ce} + \varepsilon_s^{\frac{1}{4}} H_{irc}\right) T_e(t) + a_c p_r \tag{7'}$$

We deduce, starting from this system, the expressions of $T_{ci}(t)$ and $T_{ce}(t)$ taking the following forms respectively

$$T_{ci}(t) = \lambda_i T_i(t) + \lambda_{si} T_{si}(t) + \lambda_e T_e(t) + \lambda_r \tag{9}$$

$$T_{ce}(t) = \beta_i T_i(t) + \beta_{si} T_{si}(t) + \beta_e T_e(t) + \beta_r \tag{10}$$

With :

$$\lambda_i = \frac{H_{ci} \frac{e}{k_c} \left(\frac{k_c}{e} + H_{ce} + H_{irc}\right)}{Deno}, \quad \lambda_r(t) = \frac{a_c p_r}{Deno} \quad \lambda_{si} = \frac{H_{ir} \frac{e}{k_c} \left(\frac{k_c}{e} + H_{ce} + H_{irc}\right)}{Deno},$$

$$\lambda_e = \frac{\left(H_{ce} + H_{irc} \varepsilon_e^{\frac{1}{4}}\right)}{Deno}$$

$$Deno = \frac{e}{k_c}\left(\frac{k_c}{e} + H_{ci} + H_{irc}\right)\left(\frac{k_c}{e} + H_{ce} + H_{irc}\right) - \frac{k_c}{e}$$

$$\beta_i = \frac{H_{ci}}{Deno}, \quad \beta_e = \frac{\frac{e}{k_c}\left(\frac{k_c}{e} + H_{ci} + H_{irc}\right)\left(H_{ce} + \varepsilon_s^{\frac{1}{4}} H_{irc}\right)}{Deno}$$

$$\beta_{si} = \frac{H_{irc}}{Deno}, \quad \beta_r = \frac{\frac{e}{k_c}\left(\frac{k_c}{e} + H_{ci} + H_{irc}\right) a_c p_r}{Deno}$$

Let's introduce the expression of $T_{ci}(t)$ into the equation (1), this latter takes the following form

$$\frac{dT_i(t)}{dt} = G_{si} T_{si}(t) - G_i T_i(t) + G_e T_e(t) + G_r(t) \tag{11}$$

With:

$$G_{si} = \frac{S_i H_{si} + S_c H_{ci} \lambda_{si}}{V_i \rho_i C_i}, G_i = \frac{S_i H_{si} + S_c H_{ci}(1-\lambda_i) + D_v \rho_i C_i}{V_i \rho_i C_i}$$

$$G_e = \frac{S_c H_{ci} \lambda_e + D_v \rho_i C_i}{V_i \rho_i C_i} \quad G_r(t) = \frac{S_c H_{ci} \lambda_r(t)}{V_i \rho_i C_i}$$

The transformed of La place of the equation (11), leads to

$$\bar{T}_i(p) = \frac{T_i(0)}{p+G_i} + G_{si}\frac{\bar{T}_{si}(p)}{p+G_i} + G_e\frac{\bar{T}_e(p)}{p+G_i} + \frac{\bar{G}_r(p)}{p+G_i} \quad (12)$$

With, $T_i(0) \equiv T_0$: initial field of temperature, whose original is:

$$T_i(t) = T_0 e^{-G_i t} + G_{si}\int_0^t T_{si}(\tau) e^{-G_i(t-\tau)} d\tau + G_e \int_0^t T_e(\tau) e^{-G_i(t-\tau)} d\tau + \int_0^t G_r(\tau) e^{-G_i(t-\tau)} d\tau \quad (13)$$

In this integral equation, the temperature T_0 represents the field of initial temperature, $T_e(t)$ is a field of temperature which translates the influence of the exterior climatic conditions on the temperature of the internal air of the greenhouse, on the other hand one will show that $T_{si}(t)$ depends functionally on $T_i(t)$.

The hour is taken as unit of time, the temporal evolution of the internal temperature is typically during few seconds, consequently we can neglect $\frac{dT_i(t)}{dt}$, then the expression of Ti(t) takes the following form

$$T_i(t) = \frac{G_{si}}{G_i} T_{si}(t) + \frac{G_e}{G_i} T_e(t) + \frac{G_r(t)}{G_i} \quad (14)$$

2.3 Heat balance of the ground

We suppose that the basement of the greenhouse is homogeneous, isotropic and of thermal properties (voluminal heat ρc and thermal conductivity k_s) constant, we note $p(\vec{s},t)$ the absorbed power per square meter in an unspecified point $\vec{s}$ of the surface of the ground.

This power comes primarily from the absorption of the direct and indirect solar radiation, as it can include other phenomena like precipitations and evaporation. We neglect the variations coming from the phenomena of shade, variation of the surface quality, etc.

We can write:
$$p(\vec{s},t) = p_i(t) \text{ si } \vec{s} \in (S_i)$$
$$p(\vec{s},t) = p_e(t) \text{ si } \vec{s} \in (S_e)$$

The field of temperature $T(\vec{r},t)$ in a point $\vec{r}(x,y,z)$ at the moment t inside the ground obeys to the following equation

$$\frac{\partial T(\vec{r},t)}{\partial t} - a\Delta T(\vec{r},t) = \frac{p(\vec{s},t)}{\rho c}\delta(z) \text{ in } (D) \quad (15)$$

The field $T(\vec{r},t)$ must check on the surface of the ground the following conditions

2.3.1 Condition on the surface of the interior greenhouse ground (S_i)

$$-k_S \frac{\partial T(\vec{r},t)}{\partial Z}\bigg|_{Z=0} = H_{si}\left(T(\vec{r},t) - T_i(t)\right) + H_{iri}\left(T(\vec{r},t) - T_{ci}(t)\right)\Big|_{Z=0} \tag{16}$$

By replacing $T_{ci}(t)$ by his expression we obtains

$$-\frac{\partial T(\vec{r},t)}{\partial Z}\bigg|_{Z=0} = \frac{(H_{si} + H_{iri}(1-\lambda_{si}))}{k_s}\left(T(\vec{r},t) - \frac{(H_{si} + \lambda_i H_{iri})T_i + H_{iri}\lambda_e T_e}{H_{si} + H_{iri}(1-\lambda_{si})}\right)\Bigg|_{Z=0} + \frac{H_{iri}\lambda_{si}}{k_s}\left(T(\vec{r},t) - T_{si}(t)\right) - \frac{H_{iri}\lambda_r}{k_s}\bigg|_{Z=0} \tag{17}$$

Let

$$h_i' = \frac{H_{si} + H_{iri}(1-\lambda_{si})}{k_S} \tag{18}$$

$$T_2(t) = \frac{(H_{si} + \lambda_i H_{iri})T_i(t) + \lambda_e H_{iri} T_e(t)}{H_{si} + H_{iri}(1-\lambda_{si})} \tag{19}$$

$$h_i^0 = \frac{\lambda_{si} H_{iri}}{k_s} \tag{20}$$

$$R_{si}(t) = H_{iri}\lambda_r(t) \tag{21}$$

The equation (16) takes thus the following form

$$-\frac{\partial T(\vec{r},t)}{\partial Z}\bigg|_{Z=0} = h_i'\left(T(\vec{r},t) - T_2(t)\right) + h_i^0\left(T(\vec{r},t) - T_{si}(t)\right) - \frac{R_{si}(t)}{k_S} \tag{22}$$

2.3.2 Condition on the surface of the external greenhouse ground (S_e)

$$-k_S \frac{\partial T(\vec{r},t)}{\partial Z}\bigg|_{Z=0} = H_{se}\left(T(\vec{r},t) - T_e(t)\right) + H_{ire}\left(T(\vec{r},t) - T_e'(t)\right)\Big|_{Z=0} \tag{23}$$

By using the expression (4) we obtain

$$-\frac{\partial T(\vec{r},t)}{\partial Z}\bigg|_{Z=0} = \frac{H_{se} + H_{ire}}{k_s}\left(T(\vec{r},t) - \frac{\left(H_{se} + H_{ire}\varepsilon_e^{\frac{1}{4}}\right)T_e(t)}{H_{se} + H_{ire}}\right)\Bigg|_{Z=0} \tag{24}$$

Let

$$h_e' = \frac{H_{se} + H_{ire}}{k_s} \tag{25}$$

$$T_1(t) = \frac{\left(H_{se} + H_{ire}\varepsilon_e^{\frac{1}{4}}\right)T_e(t)}{H_{se} + H_{ire}} \tag{26}$$

The equation (24) is written then

$$-\frac{\partial T(\vec{r},t)}{\partial Z}\bigg|_{Z=0} = h_e^{'}\left(T(\vec{r},t) - T_1(t)\right)\Big|_{Z=0} \tag{27}$$

2.3.3 Resolution of the problem by Green's functions theory

Let's consider $G(\vec{r},\vec{r}^{'},t)$ (Green's function) translating the response in temperature of the medium to an impulse of heat into a given point, the language generally employed consists to talk about effect into $\vec{r}$ corresponding of the cause in $\vec{r}^{'}$. In addition Green's functions obey to the reciprocity's relation of the cause and the effect, if the cause is produced in $\vec{r}$, the effect will be identical in $\vec{r}^{'}$ with the proviso of respecting the same interval of time between the moment when the cause occurs t and that when the effect occurs ($t_0 = 0$) selected arbitrarily as origin of time.

The Green's function is a particular solution of the heat's equation

$$\frac{\partial G(\vec{r},\vec{r}^{'},t)}{\partial t} - a\Delta G(\vec{r},\vec{r}',t) = \delta(\vec{r}^{'} - \vec{r})\delta(z) \tag{28}$$

With the initial condition

$$G(\vec{r},\vec{r}^{'},t=0) = 0 \tag{29}$$

And checking the condition on the surface of the ground ($S_e \cup S_i$)

$$-\frac{\partial G(\vec{r},\vec{r}^{'},t)}{\partial z}\bigg|_{z=0} = h_i^{'}G(\vec{r},\vec{r}^{'},t) \tag{30}$$

The Green's function has as expression [22]

$$G(\vec{r},\vec{r}',t) = \frac{e^{\frac{-(x-x')^2-(y-y')^2}{4at}}}{4\pi at} \times \left[\frac{e^{\frac{-(z-z')^2}{4at}} + e^{\frac{-(z+z')^2}{4at}}}{\sqrt{4\pi at}} - h'_i e^{h'_i(z+z')+ah'^2_i t}\,\mathrm{erfc}\left(\frac{z+z'}{2\sqrt{at}} + h'_i\sqrt{at}\right)\right] \tag{31}$$

The Laplace's transformation of two equations (15) and (28) is

$$p\overline{T}(\vec{r},p) - T(\vec{r},0) - a\,\Delta\overline{T}(\vec{r},p) = \frac{\overline{p}(\vec{s},p)}{\rho c}\delta(z) \tag{32}$$

$$p\bar{G}(\vec{r},\vec{r}',p) \;-\; a\,\Delta\bar{G}(\vec{r},\vec{r}',p) \;=\; \delta(\vec{r}-\vec{r}')\delta(z) \qquad (33)$$

Let's multiply the equations (32) and (33) respectively by $\bar{G}(\vec{r},\vec{r}',p)$ and $\bar{T}(\vec{r},p)$, we obtain

$$p\bar{T}(\vec{r},p)\bar{G}(\vec{r},\vec{r}',p) - T(\vec{r},0)\bar{G}(\vec{r},\vec{r}',p) - a\,\bar{G}(\vec{r},\vec{r}',p)\Delta\bar{T}(\vec{r},p) = \frac{\bar{G}(\vec{r},\vec{r}',p)\bar{p}(\vec{s},p)}{\rho c}\delta(z) \qquad (34)$$

$$p\bar{T}(\vec{r},p)\bar{G}(\vec{r},\vec{r}',p) - T(\vec{r},0)\bar{G}(\vec{r},\vec{r}',p) - a\,\bar{G}(\vec{r},\vec{r}',p)\Delta\bar{T}(\vec{r},p) = \frac{\bar{G}(\vec{r},\vec{r}',p)\bar{p}(\vec{s},p)}{\rho c}\delta(z) \qquad (35)$$

Let's make the subtraction between the equations (35) and (34), we obtains

$$\bar{T}(\vec{r},p)\delta(\vec{r}-\vec{r}')\delta(z) = \bar{G}(\vec{r},\vec{r}',p)\left(T_0(\vec{r}) + \frac{\bar{p}(\vec{s},p)}{\rho c}\delta(z)\right)$$

$$-\; a\left(\bar{T}(\vec{r},p)\,\Delta\bar{G}(\vec{r},\vec{r}',p) - \bar{G}(\vec{r},\vec{r}',p)\,\Delta\bar{T}(\vec{r},p)\right) \qquad (36)$$

Let's integrate this equation on all the field (D), we obtain the field of temperature $\bar{T}(\vec{r}',p)$ then the original $T(\vec{r}',t)$, single solution of the equation (36)

$$\begin{aligned} T(\vec{r}',t) &= \int_0^t d\tau \iint_{(S=S_e\cup S_i)} \frac{p(\vec{s},t)}{\rho c} G(\vec{r},\vec{r}',t-\tau)dS + \iiint_{(D)} G(\vec{r},\vec{r}',t)T_0(\vec{r})d^3r \\ &+ a\int_0^t d\tau \iint_{(S=S_e\cup S_i)} \left(G(\vec{r},\vec{r}',t-\tau)\vec{\nabla}T(\vec{S},\tau) - T(\vec{r},\tau)\,\vec{\nabla}G(\vec{r},\vec{r}',t-\tau)\right)dS \end{aligned} \qquad (37)$$

(S) being the meeting of (S_i) and (S_e), if we take account of the boundary conditions on the surface of the ground, satisfying the conditions (22), (27) and (30) and from the initial condition we obtain

$$\begin{aligned} T(\vec{r}',t) &= \iiint_{(D)} G(\vec{r},\vec{r}',t)T_0(\vec{r})d^3r + \int_0^t d\tau \iint_{(Si)} \frac{p_i(\vec{s},t)}{\rho c} G(\vec{r},\vec{r}',t-\tau)dS_i \\ &+ \int_0^t d\tau \iint_{(Se)} \frac{p_e(\vec{s},t)}{\rho c} G(\vec{r},\vec{r}',t-\tau)dS_e \;+ \\ &+ a\int_0^t d\tau \iint_{(Si)} \left(G(\vec{r},\vec{r}',t-\tau)\frac{\partial T(\vec{r},\tau)}{\partial z} - T(\vec{r},\tau)\frac{\partial G(\vec{r},\vec{r}',t-\tau)}{\partial z}\right)dS_i \\ &+ a\int_0^t d\tau \iint_{(Se)} \left(G(\vec{r},\vec{r}',t-\tau)\frac{\partial T(\vec{r},\tau)}{\partial z} - T(\vec{r},\tau)\frac{\partial G(\vec{r},\vec{r}',t-\tau)}{\partial z}\right)dS_e \end{aligned} \qquad (38)$$

$$\begin{aligned} T(\vec{r}',t) = &\underbrace{\iiint_{(D)} G(\vec{r},\vec{r}',t)T_0(\vec{r})d^3r}_{(1)} + \underbrace{\int_0^t d\tau \iint_{(S_i)} \frac{p_i(t-\tau)}{\rho c} G(\vec{r},\vec{r}',\tau)\,dS_i}_{(2)} \\ &+ \underbrace{\int_0^t d\tau \iint_{(S_e)} \frac{p_e(t-\tau)}{\rho c} G(\vec{r},\vec{r}',\tau)\,dS_e}_{(3)} + \underbrace{a\int_0^t d\tau \iint_{(S_i)} G(\vec{r},\vec{r}',\tau)h_i'\underbrace{\left(T(\vec{r},t-\tau)-T_2(t-\tau)\right)}_{(a)}\,dS_i}_{(4)} + \end{aligned}$$

$$\underbrace{+a\int_0^t d\tau \iint_{(Si)} G(\vec{r},\vec{r}',t-\tau)h_i^0\left(T_{si}(\tau)-T(\vec{r},\tau)\right) dS_i}_{(5)} +$$

$$\underbrace{+\ a\int_0^t d\tau \iint_{(Se)} \left(\underbrace{(h_i' - h_e')}_{(b)} G(\vec{r},\vec{r}',t-\tau)T(\vec{r},\tau) + h_e' T_1(\tau) G(\vec{r},\vec{r}',t-\tau) \right) dS_e}_{(6)} \ + \qquad (39)$$

$$\underbrace{+a\int_0^t d\tau \iint_{(Si)} \frac{R_{si}(\tau)}{k_s} G(\vec{r},\vec{r}',t-\tau)\, dS_i}_{(7)}$$

With $\vec{r}' \in (D)$

- $T_0(\vec{r}',t)$ is the field of temperature in (D) due only to the initial condition given by the term (1).
- $T_{Si}(\vec{r}',t)$ is the field of temperature in the under-ground, due only to the exchanges on internal surface of the greenhouse given by the terms (2, 4, 5, 7).
- $T_{se}(\vec{r}',t)$ is the field of temperature in the under-ground, due only to the exchanges with exterior surface of the greenhouse represented by the terms (3, 6).

We can break up the field of temperature in the form
$T(\vec{r}',t) = T_0(\vec{r}',t) + T_{Si}(\vec{r}',t) + T_{Se}(\vec{r}',t)$ varies with exterior surface of the greenhouse represented by the terms (3, 6).
We note that the term (a), in (4) comes owing to the fact that the surface temperature T_{si} (t) is not completely homogeneous, the term (b) in (6) disappears since at the exterior of the greenhouse the difference $(h_i' \approx h_e')$ is negligible. These two terms, generally very weak, they could be treated as a perturbation and the expression of $T(\vec{r}',t)$ can be written as

$$\begin{aligned} T(\vec{r}',t) \;=\;& \iiint_{(D)} G(\vec{r},\vec{r}',t)T_0(\vec{r})d^3r + \int_0^t d\tau \iint_{(Si)} \frac{p_i(t-\tau)}{\rho c} G(\vec{r},\vec{r}',\tau)\, dS_i \\ &+ \int_0^t d\tau \iint_{(Se)} \frac{p_e(t-\tau)}{\rho c} G(\vec{r},\vec{r}',\tau)\, dS_e + \\ &+a\int_0^t d\tau \iint_{(Si)} G(\vec{r},\vec{r}',\tau)h_i'\left(T(\vec{r},t-\tau)-T_2(t-\tau)\right) dS_i \\ &+a\int_0^t d\tau \iint_{(Se)} h_e' T_1(\tau)G(\vec{r},\vec{r}',t-\tau)\, dS_e \;+a\int_0^t d\tau \iint_{(Si)} \frac{R_{si}(\tau)}{k_s} G(\vec{r},\vec{r}',t-\tau)\, dS_i \end{aligned} \qquad (40)$$

In fact, the knowledge of the field $T(\vec{r}',t)$ in the domain (D) is useless, what we really need is to know the surface temperature of the interior ground of the greenhouse. Indeed, let $\vec{r}'$ be close to $\vec{s} \in (S_i)$, in this case the co-ordinates z' of $\vec{r}'$ become null, and if we make the average on (S_i), the equation (40) gives us the field of surface temperature $T_{si}(t)$

$$\begin{aligned} T_{si}(t) \;&=\; \frac{1}{S_i}\iint_{S_i} T(\vec{r}',t)dS_i \\ &=\; \frac{1}{S_i}\iint_{Si} dS_i \iiint_{(D)} G(\vec{r},\vec{r}',t)T_0(\vec{r})d^3r \;+ \end{aligned}$$

$$+\int_0^t \left(\frac{p_i(t-\tau)+R_{Si}(t-\tau)}{\rho C} + ah_i' T_2(t-\tau) \right) d\tau \frac{1}{S_i} \iint_{(Si)} dS_i \iint_{(Si)} G(\vec{r},\vec{r}',\tau) dS_i$$

$$+\int_0^t \left(\frac{p_e(t-\tau)}{\rho C} + ah_e' T_1(t-\tau) \right) d\tau \frac{1}{si} \iint_{(Si)} dS_i \iint_{(Se)} G(\vec{r},\vec{r}',\tau) dS_e \quad (41)$$

We can show that

$$\frac{1}{S_i} \iint_{Si} dS_i \iint_{Si} G(\vec{S},\vec{S}',\tau) dS_i = I(\tau) J(\tau)$$

$$\frac{1}{S_i} \iint_{(Si)} dS_i \iint_{(Se)} G(\vec{S},\vec{S}',\tau) dS_e = I(\tau)(1-J(\tau))$$

with $I(\tau) = \dfrac{1}{\sqrt{\pi a \tau}} - h_i' e^{h_i'^2 a\tau} \operatorname{erfc}(h_i' \sqrt{a\tau})$

$$J(\tau) = \left[1 - \operatorname{erfc}\left(\frac{L}{2\sqrt{a\tau}}\right) + \frac{2}{L}\sqrt{\frac{a\tau}{\pi}} \left(e^{-\left(\frac{L}{2\sqrt{a\tau}}\right)^2} - 1 \right) \right]$$

and

$$* \left[1 - \operatorname{erfc}\left(\frac{\ell}{2\sqrt{a\tau}}\right) + \frac{2}{\ell}\sqrt{\frac{a\tau}{\pi}} \left(e^{-\left(\frac{\ell}{2\sqrt{a\tau}}\right)^2} - 1 \right) \right]$$

The expression of $T_{si}(t)$ becomes

$$T_{si}(t) = \frac{1}{S_i} \iint_{Si} dS_i \iiint_{(D)} G(\vec{r},\vec{r}',t) T_0(\vec{r}) d^3r + \int_0^t \left(\frac{P_i(t-\tau)}{\rho c} + ah_i' T_2(t-\tau) \right) I(\tau) J(\tau) d\tau$$

$$+ \int_0^t \left(\frac{p_e(t-\tau)}{\rho c} + ah_e' T_1(t-\tau) \right) I(\tau)(1-J(\tau)) d\tau \quad (42)$$

With

$$P_i(t-\tau) = p_i(t-\tau) + R_{si}(t-\tau) \quad (43)$$

The field of temperature can be broken up in the following form

$$T_{si}(t) = T_{si}^0(t) + T_{si,si}(t) + T_{si,se}(t) \quad (44)$$

With

$$T_{si}^0(t) = \frac{1}{S_i} \iint_{Si} dS_i \iiint_{(D)} G(\vec{r},\vec{r}',t) T_0(\vec{r}) d^3r \quad (45)$$

Representing the contribution of the initial field of temperature; it tends towards zero when t tends towards the infinite one. So, we admit that the field of temperature in all the ground is initially uniform and is equal to the temperature of reference $T_0(\vec{r})$. This approximation does not affect the precision to the beginning and over a limited time.

$$T_{si,si}(t) = \int_0^t \left(\frac{P_i(t-\tau)}{\rho c} + ah_i' T_2(t-\tau) \right) I(\tau) J(\tau) d\tau \quad (46)$$

being the field of temperature of internal greenhouse surface due to the exchanges with the basement or due to the contribution of the basement

$$T_{si,se}(t) = \int_0^t \left(\frac{p_e(t-\tau)}{\rho c} + ah_e' T_1(t-\tau) \right) I(\tau)\,(1 - J(\tau))\, d\tau \quad (47)$$

Generally, the greenhouses are of average and of large dimensions, consequently the contribution of $T_{si,se}(t)$ in the expression of $T_{si}(t)$ is negligible except in the case of the greenhouses of very small dimensions (*L* ≈ ℓ ≈ *3m*).
Consequently

$$T_{si}(t) = T_{si}^0(t) + \int_0^t \left(\frac{P_i(t-\tau)}{\rho c} + ah_i' T_2(t-\tau) \right) I(\tau) J(\tau) d\tau \quad (48)$$

By replacing $T_2(t-\tau)$ by its expression, $T_{si}(t)$ takes the following form

$$\begin{aligned} T_{si}(t) &= T_{si}^0(t) + \frac{1}{\rho c} \int_0^t P_i(t-\tau) I(\tau) J(\tau) d\tau \\ &+ \frac{H_i'}{\rho c} \int_0^t T_i(t-\tau) I(\tau) J(\tau) d\tau + \frac{H_i''}{\rho c} \int_0^t T_e(t-\tau) I(\tau) J(\tau) d\tau \end{aligned} \quad (49)$$

with $H_i' = H_{si} + H_{iri}\lambda_i$ and $H_i'' = H_{iri}\lambda_e$
We developed a fast algorithm of resolution of this type equation.

2.4 Study of the characteristic functions I(τ) et J(τ)

Equation (49) shows that the two functions I(τ) and J(τ) characterize entirely and rigorously the thermal inertia of the ground and the interaction of this one with the entire system (cover, interior ground and interior air).
These two functions are positive monotonous and tend towards zero when τ tends towards infinity, consequently they can be approximated numerically, with a good precision, by a series of exponential decreasing of time allowing a fast calculation of the product of convolution. The function $I(\tau)$ presents a singularity at the term $\tau^{-1/2}$ in the vicinity of zero fortunately this singularity can be integrated. We introduce the function
$I_1(\alpha) = e^{\alpha}\mathrm{erfc}(\sqrt{\alpha})$ with $\alpha = h_i'^2 a\tau$ in addition we saw

$$I(\tau) = \frac{1}{\sqrt{\pi a \tau}} - h_i' \exp(h_i'^2 a\tau)\,\mathrm{erfc}(h_i'\sqrt{a\tau}) \quad , \tau \in [0, +\infty[$$

We can write then for $\alpha \in [0, +\infty[$

$$I(\alpha) = h_i^{'}\left(\frac{1}{\sqrt{\pi\alpha}} - e^{\alpha}\operatorname{erfc}(\sqrt{\alpha})\right) \tag{50}$$

It is noticed finally that

$$I(\alpha) = -h_i^{'}\frac{dI_1(\alpha)}{d\alpha} \tag{51}$$

Let $u = \exp(-\alpha)$ $\quad u \in [0,1]$

$I_1(u)$ is a strictly decreasing positive function and tends towards zero when u tends towards unity; in addition its graph is not obviously linear, the numerical analysis of the graph shows that we can approach this function with a quadratic average on the interval [0,1] by polynomials of type : au^b.

The approximate expression of $I_1(u)$ is written:

$$I_1(u)_{app} = a\,u^b + c\,u^d \tag{52}$$

Calculation gives:

$a = 0{,}4269,\quad b = 4{,}676,\quad c = 0{,}499,\quad d = 0{,}1659.$

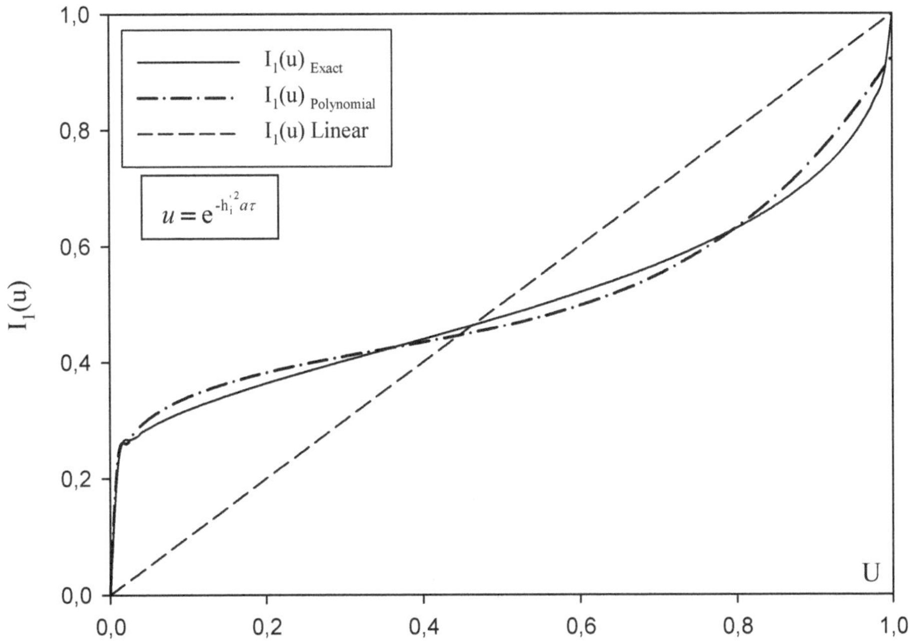

Fig. 2. Evolution of the inertia's function I_1 with its various approximations

These numbers without dimensions are thus defined once for all, we notice that exact $I_1(u)$ and its approximate expression coincide well in Figure 2 and the function I(u) can be then approximated in Figure 3, by a polynomial of following this form:

$$I(u)_{app} = h_i^{'} \left(1{,}9962\, u^{4{,}676} + 0{,}0828\, u^{0{,}1659}\right) \tag{53}$$

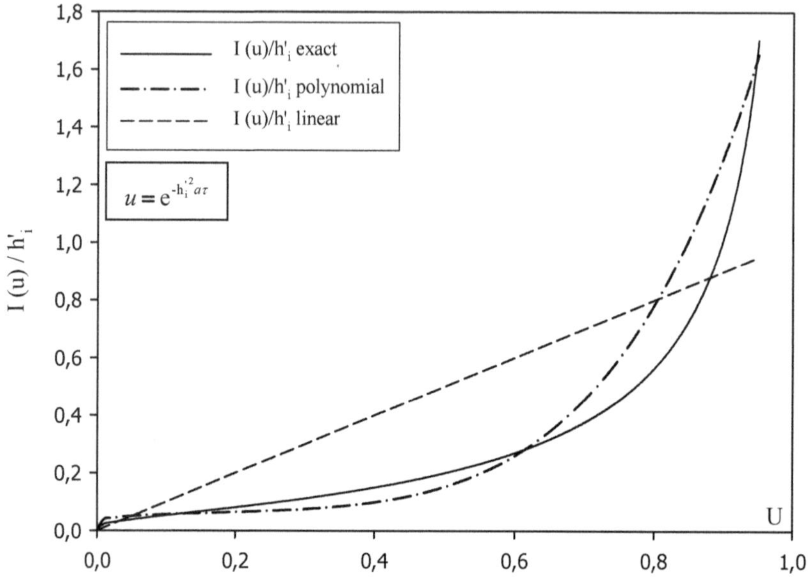

Fig. 3. Evolution of the inertia's function I(u)with its various approximations

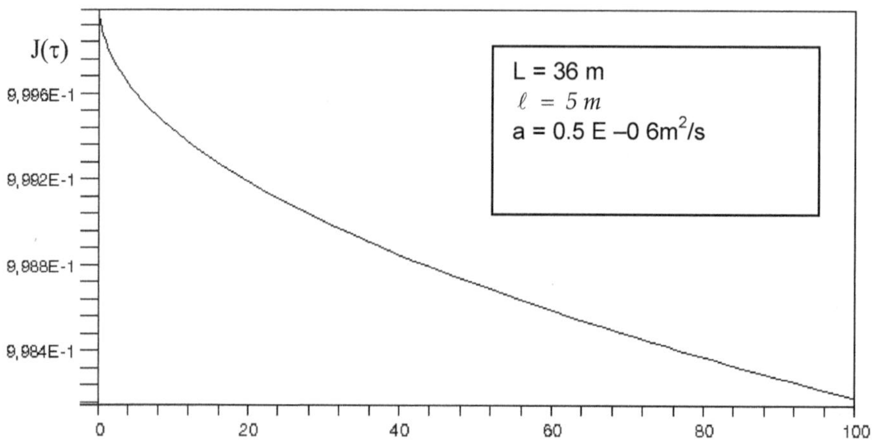

Fig. 4. Lay-out of the balance-sheet of heat exchange

The typical values of thermal conductivity (k_s) and thermal diffusivity (a) for a ground are respectively about 1 and 0,5 10^{-6}, it is obvious that for large-sized greenhouses, the effect of the surrounding ground is so negligible that we cannot measure it. It is obvious to admit that J(τ) remains practically equal to the unit Figure 4, except for greenhouses of very small sizes (L et $\ell \leq 3\,m$) where the effect can be perceptible.

3. Discussion of the numerical model

We carried out the numerical simulation on a tunnel greenhouse with a plastic cover (polyethylene) with a simple cover, isolated, of volume 354 m^3 (length 36 meters and 5 meters broad), placed on a ground of thermal diffusivity (a = 0,5 10^{-6} Wm^2/J) and of thermal conductivity (ks = 1 W/m °C), we took the function of inertia (J(τ) = 1) because the studied greenhouse is practically of great dimension.
The equations (13) and (49) that appear in their products of convolution, climactic data, exterior temperature and total solar power, contain parameters depending on the place and season.
Indeed, we took for our numerical simulation

Exterior temperature: $T_e(t) = -5\cos\left(\frac{2\pi t}{24}\right)$

Total solar power: $p_i(t) = 280\left(\cos\left(\frac{2\pi t}{24}\right)\right)^2$

In addition, the tableau1 appearing below gathers the thermo-physical constants of the air, ground and cover which we used in this simulation.

ε_{ce} =0.95	$\rho_s C_s$= 2 10^6 $JK^{-1}m^{-1}$	ρ_i = 1.117 kgm^{-3}
t_c = 0.65	ε_{si} =0.95	C_i = 1006 $Jkg^{-1} K^{-1}$
k_c = 1.5 $Wm^{-1} K^{-1}$	$T_{si\,0}$ = 9°C	T_{i0} = 279.15 K
ε_{ci} =0.95	k_s = 1 $Wm^{-1}K^{-1}$	
a_c = 0.31	a = 0,5 10^{-6} m^2s^{-1}	
h_{ce} = 0.1$Wm^{-2} K^{-1}$		
h_{ci} = 0.3 1$Wm^{-2}K^{-1}$		
L = 36 m		
$\ell = 5m$ $\ell = 5m$		

Table 1. Table of the entered parameters of the digital simulation

4. Interpretation of the results

Figure 5 shows the superimposed evolutions of incident solar flux and those exchanged with the external and internal face of the greenhouse's cover.

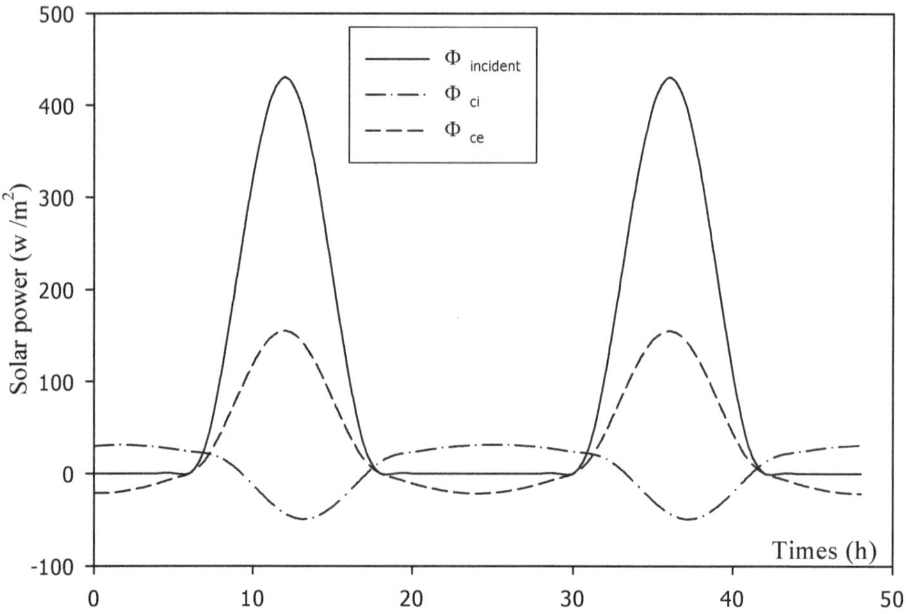

Fig. 5. Temporal evolution of the heat fluxes, incident and exchanged with internal and external faces of the greenhouse's cover

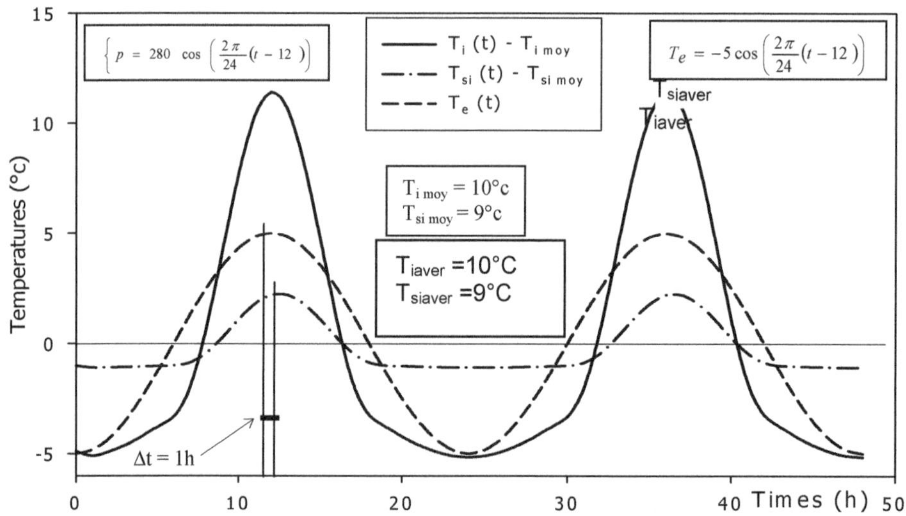

Fig. 6. Temporal evolution of temperatures, of the internal air, internal ground and the external air of the greenhouse for an air renewal' flux null

We note during the day, that the exchanged flux with the exterior face of the cover ϕ_{ce} is more important than that exchanged with the interior face ϕ_{ci}, that is foreseeable because of the absorption of a part of the incident heat flux by the cover.

During the night, the incident heat flux becomes null, consequently the interior air and ground must radiate now towards the exterior, it is the night radiation, therefore ϕ_{ci} becomes more important than ϕ_{ce}, but remain the two weak contributions.

We deferred in Figure 6, the evolutions of the exterior temperature and that of the interior ground of a closed greenhouse.

We note that the effect of inertia of the ground and the absorption of the heat of the day by its surface appears in the form of a rise in temperature of the order of 2°C and a phase shift of the order of one hour with the interior air of the greenhouse.

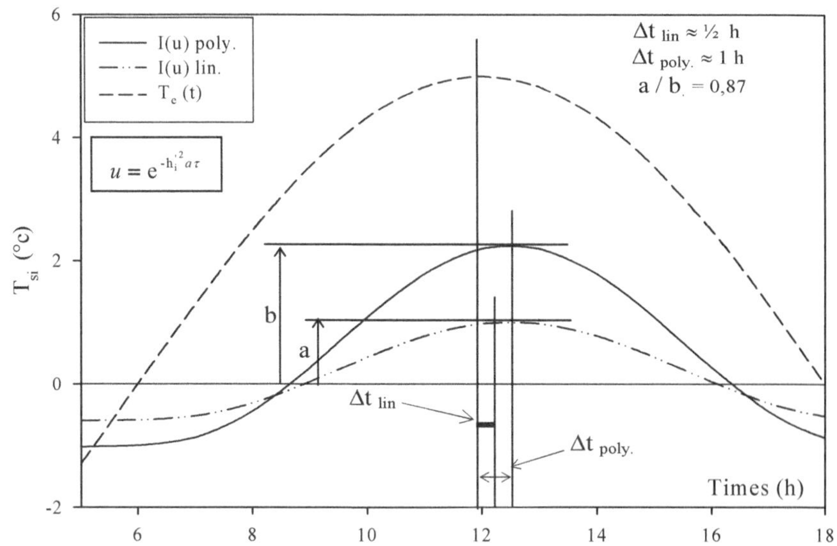

Fig. 7. Influence of the approximation's nature of the inertia's function I(τ) on the temperature's evolution of the greenhouse's internal ground

We also note the presence of the night radiation; indeed, the ground behaves like a tank of heat which was recovered and stored along the day, this heat takes part in the stabilization of the temperature of the internal air in a level higher than that of the exterior air of the greenhouse. Consequently, during the night the ground presents a thermal inertia in front of the internal air and presents also a thermal inertia compared to the exterior temperature.

The thermal inertia of the ground is characterized by the two functions I(τ) and J(τ), this latter is practically equal to unity. In order to materialize this characterization, we studied the impact of the approximation's nature of I(τ) on the temperature's evolution of the interior ground of the greenhouse compared to the exterior temperature.

For this reason, we visualized in Figure 7 the curves of the temperature's evolution of the interior ground, respectively for a polynomial and linear approximations of the function of inertia I(τ), we also deferred in the same graph the exterior ambient temperature of the air.

Going through the linear approximation which is coarse towards a more exact polynomial approximation, we announce the following remarks:

- an increase in the amplitude of the ground's temperature
- an increase in the phase shift compared to the exterior temperature
- a remarkable rise in the thermal mass (see Figure 8)

We notice for the polynomial approximation the materialisation of thermal inertia, consequently, we can affirm that the polynomial approximation is more correct because it is closer to the exact function that is proved moreover by Figures (3) and (4).

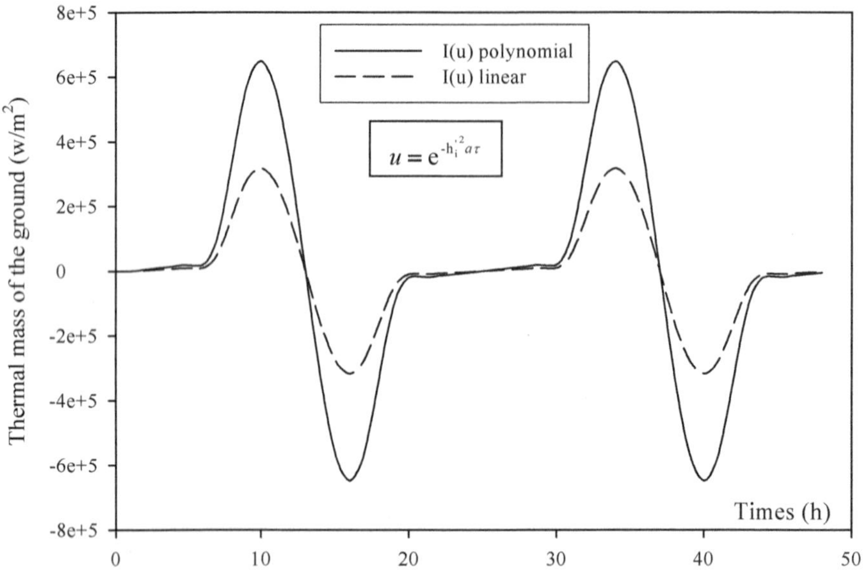

Fig. 8. Influence of the approximation's nature of the inertia's function I(τ) on the evolution of the thermal mass of the greenhouse's internal ground ($\frac{\rho_s C_s}{h_i'} \frac{dT_{si}}{dt}$)

Figure 9 shows the temperature's evolution of the greenhouse's interior air according to time for various debits of air's renewal. The continuous air's renewal obviously clearly lowers the maximum temperature of the day, as it also lowers the minimal temperature of the night. Consequently, the greenhouse's internal air becomes increasingly dependant on the exterior conditions in particular the exterior temperature.

We can say that the increase of air's renewal's debit, gradually eliminates the effect of thermal inertia of the interior air vis-à-vis to the exterior, the exterior temperature remains at a lower limit that we cannot practically exceed.

Figure 10 presents the influence of the cover's temperature on the evolution of the internal air temperature of the greenhouse. Since the cover's thickness is very low in the order of 180 μm, its thermal conductivity which is inversely proportional to the thickness is very important, consequently, the temperatures of the interior and exterior faces of the cover are finding practically the same ones.

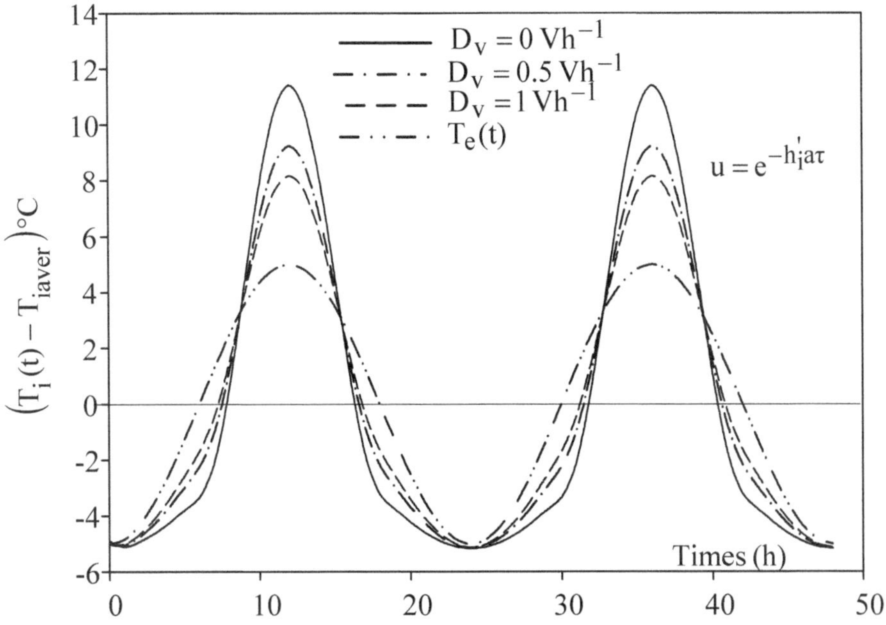

Fig. 9. Temperature's temporal evolution of the greenhouse's internal air for various renewal's flux of air

The transmissions of the external effects towards the interior of the greenhouse are carried out through the cover, what explains the important role of the latter.
During the day, the absorption of part of the incident heat flux appreciably increases the temperature of the cover beyond the temperature of the interior air of the greenhouse. During the night, when there is absence of the incident heat flux, there remains only the conduction of the exterior temperature which dominates the other modes of transfer of heat, which generates a reduction of the cover temperature under the interior air temperature. We summarize the explanation of these two observations by the fact that the cover does not have a thermal inertia.

5. Conclusion and prospects

In this study, we noticed that the ground behaves approximately like a thermal mass. We consider here a simplified model (greenhouse without vegetation) where solar energy is absorbed only by the ground where the phenomena of evaporation and transpiration do not intervene.
This model shows that it is possible to envisage the general behaviour of a naked greenhouse and can without difficulty, be supplemented to hold account in particular of the phenomena of evapo-transpiration in the case of a cultivated greenhouse.

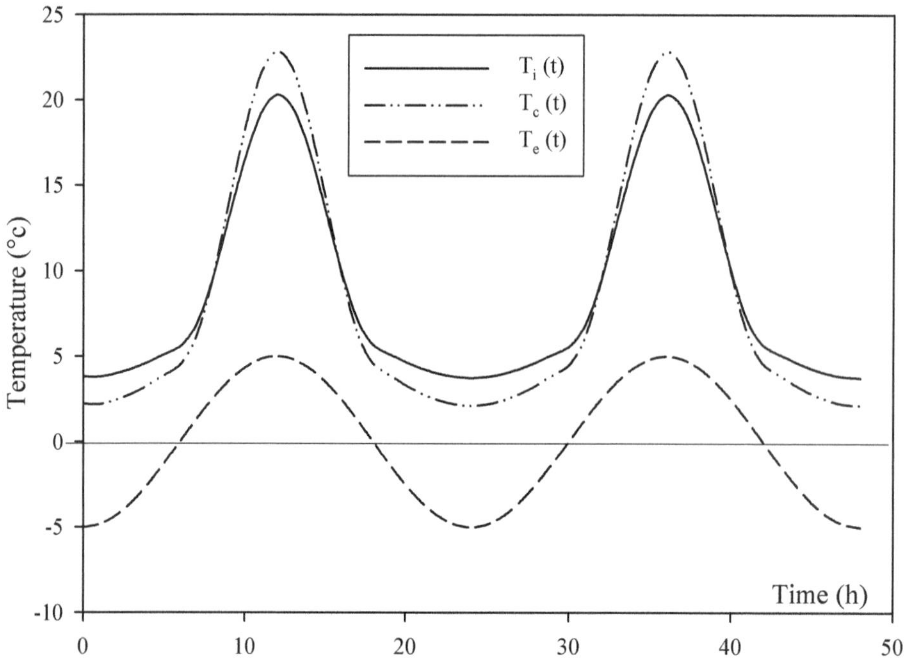

Fig. 10. Temporal evolution of the temperatures, of the internal and external air of the greenhouse and of the cover

However, the interest of the agricultural greenhouses is to increase the production period as well as the output, but requires for the periods of the unfavourable climatic conditions, the use of an expensive heating.

Consequently, the heating of the greenhouses by the integration of a significant storage unit of heat can prove to be interesting to spread out the calendar of production. This is why we highlighted theoretically and experimentally [24, 32] the interest of an underground thermal storage of short and long duration by establishing a mathematical model taking account of all the physical parameters intervening in the system.

The ground, indeed, is able to absorb the solar contributions of the greenhouse which are surplus by playing the role of a thermal wheel of inertia. But the presence of a battery of exchanger buried can play a double function, diurnal cooling of summer or nocturnal reheating of winter by providing all the year an air flow practically to the desired temperature $T_i(t)$.

Realized in the form of a battery of vertical exchangers with air buried in the internal ground of the greenhouse, this storage unit and destocking can constitute an alternative to the problem of the strong thermal amplitudes of a traditional greenhouse (considerable loss of energy during the opening).

The unification of the theory (used for this model as for the battery of exchangers) will make it possible, easy, to integrate a unit of heating in the internal atmosphere of the greenhouse.

Finally, the following stage of this work consists to confirm these numerical results on an experimental greenhouse and to find industrial partners.

6. Nomenclature

a	thermal diffusivity of the ground [m²/s]
a_c	absorption coefficient of the cover infra-red radiation
c_i	heat capacity of the air [J/kgK]
c_s	specific heat of the ground [J/kgK]
D_v	flow of air renewal [m³/h]
erfc	Error
H_{ci}	coefficient of heat exchange by convection between the cover and the internal air [W/m²K]
H_{ce}	coefficient of heat exchange by convection between the cover and the external air [W/m²K]
H_{si}	coefficient of heat exchange by convection between the ground and the internal air [W/m²K]
H_{se}	coefficient of heat exchange by convection between the ground and the external air [W/m²K]
H_{IRc}	coefficient of the linearized cover infrared exchange [|W/m²K]
H_{IRi}	coefficient of the linearized internal ground infrared exchange [W/m²K]
H_{IRe}	coefficient of the linearized external ground infrared exchange [W/m²K]
k	thermal conductivity of the ground [W/mK]
k_c	thermal conductivity of the cover [W/mK]
e	thickness of the cover [m]
L	length of the greenhouse [m]
ℓ	width of the greenhouse [m]
P	average power absorbed by the ground [W]
P_i	average power absorbed by the internal ground [W]
P_e	average power absorbed by the external groun d [W]
P_r	average power radiation [W]
T_0	annual average temperature of reference [K]
V_i	volume of the greenhouse [m³]
S_i	surface of the internal ground [m²]
S_e	surface of the external ground [m²]
S_c	surface of the cover [m²]
$T_i(t)$	temperature of the greenhouses internal air [K]
$T_e(t)$	temperature of the greenhouses external air [K]
$T_{ci}(t)$	temperature of the internal face of the cover [K]
$T_{si}(t)$	surface average temperature of the greenhouses internal ground [K]
$T_{se}(t)$	surface average temperature of the greenhouses external ground [K]
$T_{i\,aver}$	average temperature of the greenhouses internal air for the time interval of simulation [K]
$T_{si\,aver}$	average temperature of the greenhouses internal ground for the time interval of simulation [K]
Δt	time dephasing [h]
Δt_{lin}	time dephasing corresponding to a linear approximation of the function of inertia I(u) [h]
Δt_{poly}	time dephasing corresponding to a polynomial approximation of the function of inertia I(u) [h]

T'_e virtual temperature of the sky [K]

$T(\vec{r},t)$ The field of temperature inside the ground [K]

7. Indexes

c	cover
e	exterior
i	interior
s	ground
se	exterior ground
si	interior ground
ce	exterior face of the cover
ci	interior face of the cover
atm	vault of heaven
IRc	infrared exchange with the cover

Greek Symbols

ε_a	emission's total factor of the ambient air [-]
λ	dimensionless coefficient [-]
ρ	density [kg/m³]
δ	distribution of Dirac
σ	Constant of Stefan-Boatman [W/m²K⁴]

8. Appendix 1

1- The expression $\frac{1}{Si}\iint_{Si} dSi \iint_{Si} G(\vec{r},\vec{r}',\tau)dSi$ equals to $I(\tau)J(\tau)$, indeed, on the level of the ground we have $z=z'=0$ then the expression of the Green's function takes the following form:

$$G(\vec{r},\vec{r}',\tau) = \frac{e^{\frac{-(x-x')^2-(y-y')^2}{4a\tau}}}{4\pi a\tau}\left[\frac{2}{\sqrt{4\pi a\tau}} - h'_i e^{ah'^2_i\tau}\operatorname{erfc}\left(h'_i\sqrt{a\tau}\right)\right] \quad (1)$$

Let

$$I(\tau) = \frac{1}{\sqrt{\pi a\tau}} - h'_i e^{ah'^2_i\tau}\operatorname{erfc}\left(h'_i\sqrt{a\tau}\right) \quad (2)$$

$$\iint_{Si} G(\vec{r},\vec{r},\tau)\, dSi = \iint_{Si} \frac{e^{\frac{-(x-x')^2-(y-y')^2}{4a\tau}}}{4\pi a\tau} I(\tau)\, dSi = \quad (3)$$

$$\frac{I(\tau)}{4\pi\tau a}\int_0^L e^{-\frac{(x-x')^2}{4a\tau}}dx \int_0^{\ell} e^{-\frac{(y-y')^2}{4a\tau}}dy$$

Let $u = \frac{x - x'}{2\sqrt{a\tau}}$ and $v = \frac{y - y'}{2\sqrt{a\tau}}$

We can write $\begin{cases} x = x' + 2u\sqrt{a\tau} \\ y = y' + 2v\sqrt{a\tau} \end{cases}$ what gives $\begin{cases} dx = 2u\sqrt{a\tau} \\ dy = 2v\sqrt{a\tau} \end{cases}$

The equation (3) becomes:

$$\iint_{Si} G(\vec{r},\vec{r}',\tau)\, dSi = \frac{I(\tau)}{4}\left(\frac{2}{\sqrt{\Pi}}\int_{\frac{-x'}{2\sqrt{a\tau}}}^{\frac{L-x'}{2\sqrt{a\tau}}} e^{-u^2} du\right)\left(\frac{2}{\sqrt{\Pi}}\int_{\frac{-y'}{2\sqrt{a\tau}}}^{\frac{\ell-y'}{2\sqrt{a\tau}}} e^{-v^2} dv\right)$$

$$= \frac{I(\tau)}{4}\left[\mathrm{erf}\left(\frac{L-x'}{2\sqrt{a\tau}}\right) + \mathrm{erf}\left(\frac{x'}{2\sqrt{a\tau}}\right)\right]\left[\mathrm{erf}\left(\frac{\ell-y'}{2\sqrt{a\tau}}\right) + \mathrm{erf}\left(\frac{y'}{2\sqrt{a\tau}}\right)\right] \quad (4)$$

Let

$$J(x',y',\tau) = \frac{1}{4}\left[\mathrm{erf}\left(\frac{L-x'}{2\sqrt{a\tau}}\right) + \mathrm{erf}\left(\frac{x'}{2\sqrt{a\tau}}\right)\right]\left[\mathrm{erf}\left(\frac{\ell-y'}{2\sqrt{a\tau}}\right) + \mathrm{erf}\left(\frac{y'}{2\sqrt{a\tau}}\right)\right] \quad (5)$$

Finally we obtain :

$$\iint_{Si} G(\vec{r},\vec{r}',\tau)\, dSi = I(\tau)\, J(x',y',\tau) \quad (6)$$

$$\frac{1}{Si}\iint_{Si}\left(\iint_{Si} G(\vec{r},\vec{r}',\tau)\, dSi\right) dSi = \frac{1}{Si}\iint_{Si} I(\tau)\, J(x',y',\tau)\, dSi = \frac{I(\tau)}{Si}\iint_{Si} J(x',y',\tau)\, dSi \quad (7)$$

$$\frac{1}{Si}\iint_{Si} J(x',y',\tau)\, dSi = \frac{1}{4L\ell}\int_0^L \left[\mathrm{erf}\left(\frac{L-x'}{2\sqrt{a\tau}}\right) + \mathrm{erf}\left(\frac{x'}{2\sqrt{a\tau}}\right)\right] dx' \int_0^{\ell}\left[\mathrm{erf}\left(\frac{\ell-y'}{2\sqrt{a\tau}}\right) + \mathrm{erf}\left(\frac{y'}{2\sqrt{a\tau}}\right)\right] dy' \quad (8)$$

$$\int_0^L \left[\mathrm{erf}\left(\frac{L-x'}{2\sqrt{a\tau}}\right) + \mathrm{erf}\left(\frac{x'}{2\sqrt{a\tau}}\right)\right] dx' = \int_0^L \mathrm{erf}\left(\frac{L-x'}{2\sqrt{a\tau}}\right) dx' + \int_0^L \mathrm{erf}\left(\frac{x'}{2\sqrt{a\tau}}\right) dx'$$

$\begin{cases} u = \frac{L-x'}{2\sqrt{a\tau}} \\ v = \frac{x'}{2\sqrt{a\tau}} \end{cases}$ what gives $\begin{cases} dx' = -2\sqrt{a\tau}\, du \\ dx' = -2\sqrt{a\tau}\, du \end{cases}$

This leads to

$$\int_0^L \left[\operatorname{erf}\left(\frac{L-x'}{2\sqrt{a\tau}}\right) + \operatorname{erf}\left(\frac{x'}{2\sqrt{a\tau}}\right)\right] dx' = -2\sqrt{a\tau}\int_{\frac{L}{2\sqrt{a\tau}}}^{0} \operatorname{erf}(u)\, du + 2\sqrt{a\tau}\int_0^{\frac{L}{2\sqrt{a\tau}}} \operatorname{erf}(v)\, dv$$

in addition we have $\operatorname{erf}(r) = \frac{2}{\sqrt{\pi}}\int_0^r e^{-u^2} du$ and $\operatorname{erf}(u) + \operatorname{erfc}(u) = 1$

We obtain then

$$\int_0^L \left[\operatorname{erf}\left(\frac{L-x'}{2\sqrt{a\tau}}\right) + \operatorname{erf}\left(\frac{x'}{2\sqrt{a\tau}}\right)\right] dx' = 4\sqrt{a\tau}\int_0^{\frac{L}{2\sqrt{a\tau}}} \operatorname{erf}(u)\, du$$

$$= 4\sqrt{a\tau}\int_0^{\frac{L}{2\sqrt{a\tau}}} (1 - \operatorname{erfc}(u))\, du$$

$$= 4\sqrt{a\tau}\left[u - \operatorname{ierfc}(u)\right]_0^{\frac{L}{2\sqrt{a\tau}}}$$

$$= 4\sqrt{a\tau}\left[u - \left(u \operatorname{erfc}(u) - \frac{1}{\sqrt{\pi}}e^{-u^2}\right)\right]_0^{\frac{L}{2\sqrt{a\tau}}}$$

$$= 4\sqrt{a\tau}\left(\frac{L}{2\sqrt{a\tau}} - \frac{L}{2\sqrt{a\tau}}\operatorname{erfc}\left(\frac{L}{2\sqrt{a\tau}}\right) + \frac{1}{\sqrt{\pi}} e^{-\left(\frac{L}{2\sqrt{a\tau}}\right)^2} - \frac{1}{\sqrt{\pi}}\right)$$

$$= 2L\left[1 - \operatorname{erfc}\left(\frac{L}{2\sqrt{a\tau}}\right) + \frac{2}{L}\sqrt{\frac{a\tau}{\pi}}\left(e^{-\left(\frac{L}{2\sqrt{a\tau}}\right)^2} - 1\right)\right]$$

There after

$$J(\tau) = \frac{1}{s_i}\iint_{Si} J(x', y', \tau)\, dS$$

$$= \left[1 - \operatorname{erfc}\left(\frac{L}{2\sqrt{a\tau}}\right) + \frac{2}{L}\sqrt{\frac{a\tau}{\pi}}\left(e^{-\left(\frac{L}{2\sqrt{a\tau}}\right)^2} - 1\right)\right] *$$

$$* \left[1 - \operatorname{erfc}\left(\frac{\ell}{2\sqrt{a\tau}}\right) + \frac{2}{\ell}\sqrt{\frac{a\tau}{\pi}}\left(e^{-\left(\frac{\ell}{2\sqrt{a\tau}}\right)^2} - 1\right)\right] \quad (9)$$

Finally

$$\frac{1}{Si}\iint_{Si}\left(\iint_{Si} G(\vec{r},\vec{r}',\tau)\, dSi\right) dSi = \frac{I(\tau)}{Si}\iint_{Si} J(x',y',\tau)\, dS = I(\tau)\, J(\tau) \qquad (10)$$

2- The factor $\frac{1}{S_i}\iint_{(S_i)} dS_i \iint_{(S_e)} G(\vec{r},\vec{r}',\tau) dS_e = I(\tau)\left(1 - J(\tau)\right)$

Indeed, we have $(S) = (S_i) \cup (S_e)$

Then $\iint_{Se} G(\vec{r},\vec{r}',\tau) dS_e = \iint_{S} G(\vec{r},\vec{r}',\tau) dS - \iint_{Si} G(\vec{r},\vec{r}',\tau) dS_i$

in addition

$$\iint_{S} G(\vec{r},\vec{r}',\tau) dS = \iint_{S}\left(\frac{e^{-\frac{(x-x')^2}{4a\tau}}\, e^{-\frac{(y-y')^2}{4a\tau}}}{4\pi a\tau}\right. I(\tau)\, dS$$

$$= \frac{I(\tau)}{\pi}\int_{-\infty}^{+\infty} e^{-\frac{(x-x')^2}{4a\tau}} \frac{dx}{2\sqrt{a\tau}} \int_{-\infty}^{+\infty} e^{-\frac{(y-y')^2}{4a\tau}} \frac{dy}{2\sqrt{a\tau}}$$

$$= \frac{I(\tau)}{\pi}\int_{-\infty}^{+\infty} e^{-u^2}\, du \int_{-\infty}^{+\infty} e^{-v^2} dv = \frac{I(\tau)}{\pi}\left(\int_{-\infty}^{+\infty} e^{-u^2}\, du\right)^2 = I(\tau)$$

Consequently

$$\iint_{Se} G(\vec{r},\vec{r}',\tau) dS_e = I(\tau)\left(1 - J(x,y,\tau)\right) \qquad (11)$$

By using the equation (9), we obtain

$$\frac{1}{S_i}\iint_{(S_i)} dS_i \iint_{(S_e)} G(\vec{r},\vec{r}',\tau) dS_e = I(\tau)(1 - J(\tau)) \qquad (12)$$

9. Appendix 2

Approximate forms of the functions of inertia $I(\tau)$ and $J(\tau)$:
We have

$$T_{si}(t) = T_{si}^0(t) + \frac{1}{\rho c}\int_0^t P_i(t-\tau) I(\tau) J(\tau) + \frac{H_i'}{\rho c}\int_0^t T_i(t-\tau) I(\tau) J(\tau) d\tau + \frac{H_i''}{\rho c}\int_0^t T_e(t-\tau) I(\tau) J(\tau) d\tau \quad (1)$$

In order to facilitate the discretization of this expression, we must approach the two functions of inertia $I(\tau)$ and $J(\tau)$.
In addition, the expression of $I(\tau)$ can be written:

$$I(\tau) = \frac{1}{\sqrt{\pi a\tau}} - h_i' \exp(h_i'^2 a\tau)\operatorname{erfc}(h_i'\sqrt{a\tau}) \quad , \tau \in [0, +\infty[\qquad (2)$$

Let $\alpha = h_i'^2 a\tau$ then $I(\alpha) = \frac{h_i'}{\sqrt{\pi\alpha}} - h_i' e^{\alpha}\text{erfc}(\sqrt{\alpha})$

Let $I_1(\alpha) = e^{\alpha}\text{erfc}(\sqrt{\alpha})$ we find then :

$$I(\alpha) = -h_i' \frac{dI_1(\alpha)}{d\alpha} \tag{3}$$

however we have

$$I_1(\alpha)_{app} = 0.4269\, e^{-4.676\alpha} + 0.499\, e^{-0.1659\alpha} \tag{4}$$

finally

$$\begin{cases} I(\alpha) = h_i'\left(\frac{1}{\sqrt{\pi\alpha}} - e^{\alpha}\text{erfc}(\sqrt{\alpha})\right) \\ I(\alpha)_{app} = h_i'\left(1.9962\, e^{-4.676\alpha} + 0.0828\, e^{-0.1659\alpha}\right) \end{cases}$$

Or according to τ
$$\begin{cases} I(\tau) = \frac{1}{\sqrt{\pi a\tau}} - h_i' e^{h_i'^2 a\tau}\text{erfc}(h_i'\sqrt{a\tau}) \\ I(\tau)_{app} = h_i'\left(1.9962\, e^{-4.676h_i'^2 a\tau} + 0.0828\, e^{-0.1659h_i'^2 a\tau}\right) \end{cases}$$

In addition we saw that

$$J(\tau) = \left[1 - \text{erfc}\left(\frac{L}{2\sqrt{a\tau}}\right) + \frac{2}{L}\sqrt{\frac{a\tau}{\pi}}\left(e^{-\left(\frac{L}{2\sqrt{a\tau}}\right)^2} - 1\right)\right]$$

$$\left[1 - \text{erfc}\left(\frac{\ell}{2\sqrt{a\tau}}\right) + \frac{2}{\ell}\sqrt{\frac{a\tau}{\pi}}\left(e^{-\left(\frac{\ell}{2\sqrt{a\tau}}\right)^2} - 1\right)\right] \tag{5}$$

We represented J(τ), with $\tau \in [0, +\infty[$, we noted that J(τ) practically evolve in the vicinity of the unit, consequently, we can approximate the product IJ by:

$$I(\tau)J(\tau)_{app} = h_i'\left(1.9962\, e^{-4.676h_i'^2 a\tau} + 0.0828\, e^{-0.1659h_i'^2 a\tau}\right) \tag{6}$$

10. References

[1] Boulard , T. et Baille, A., " Analysis of thermal performance of a greenhouse as a solar collector", Energy in Agriculture, 6, 17, pp. 17-26, (1986).

[2] Jolliet, O., "Modélisation du comportement thermique d'une serre horticole", Thèse de docteur Sciences, école Polytechnique Fédérale de Lausanne, (1988).

[3] Comary, Y. et Nicolas, C., "La thermique des serres", Collection de la direction des études et recherches d'Electricité de France. Ed. Eyrolles (1985).

[4] Monteil, C., "Contribution informatique à l'analyse énergétique des serres agricoles", Thèse de docteur ingénieur, Institut National Polytechnique de Toulouse (1985).
[5] Chabaane, D.E., "Influence d'un écran thermique sur le microclimat nocturne d'une serre agricole", Thèse de docteur de 3ème cycle, Université de Perpignan (1986).
[6] Kabbaj, R., " Modélisation de la ventilation statique d'une serre, Rapport de Diplôme Approfondie en Energétique, Université de Perpignan (1988).
[7] Georgios K. Spanomitsios, "Temperature Control and Energy Conservation in a Plastic Greenhouse", Journal of Agricultural Engineering Research Volume 80, Issue 3 , November pp. 251-259 (2001)
[8] G. L. A. M. Swinkels, P. J. Sonneveld and G. P. A. Bot "Improvement of Greenhouse Insulation with Restricted Transmission Loss through Zigzag Covering Material", Journal of Agricultural Engineering Research Volume 79, Issue 1 , pp. 91-97 (2001).
[9] M. Trigui, S. Barrington and L. Gauthier, "A Strategy for Greenhouse Climate Control, Part II: Model Validation" , Journal of Agricultural Engineering Research ,Volume 79, Issue 1, pp.99-105 (2001).
[10] Maklouf, S., " Expérimentation et modélisation d'une serre solaire à air avec stockage par chaleur latente assisté par pompe à chaleur de déshumidification", Thèse de docteur ès-Sciences, Université de Nice (1988).
[11] De Halleux, D., "Modèle dynamique des échanges énergétiques des serres : Etude théorique et expérimentale, Thèse de doctorat de l'Université de Gembloux (1989).
[12] Walker, J.N., "*Predicting temperature in ventilated greenhouse*", Transactions of the ASAE, 8, 3, pp.445-448 (1965).
[13] Gag, A., Bartoli, R., Filelle, J.F., Bethery, J., Gérard, A. Et Feldmann, C., " *Échanges thermiques dans les serres et abris*, bulletin Technique du génie rural, n°113 (1973).
[14] Boulard, T., "Caractérisation et modélisation du climat des serres : application à la climatisation estivale", Thèse de Doctorat, E.N.S.AM., Montpellier (1996).
[15] Kimball, B.A., " Simulation of the energy balance of a greenhouse, agric.Meteorol., n°11, pp.243-260 (1973).
[16] Arinze, E.A. Schoenau, G.J. Et Besant, R.W., "*A dynamic thermal performance simulation model of an energy conserving greenhouse with thermal storage*", Transactions of the ASAE, 27, 2, pp.508-519 (1984).
[17] Bot, G.P.A., "Grenhouse climate : from physical processes to a dynamic model", Ph. D. Thesis, Agricultural University, Wageningen (1983).
[18] Kindelan, M., " Dynamic modelling of greenhouse environnement", Transaction of the ASAE, 23, 5, pp.1232-1239 (1980).
[19] Kittas, C. , "Contribution théorique et expérimentale à l'étude du bilan d'énergie des serres", Thèse de Docteur-ingénieur, Université de Perpignan (1980).
[20] Boulard, T. , Draoui, B. et Neirac, F., "Analyse du bilan thermodynamique d'une serre horticole - Application à la maîtrise du microclimat -", Colloque de la société Française des thermiciens (S.F.T.), Pau, Mai 1993, pp 632-647.
[21] Issanchou, G.,"Modélisation à la mise au point d'un logiciel de thermique appliqué à l'ingénierie des serres", Thèse de Doctorat de l'Université de Perpignan (1991).
[22] Carslaw, H.S. and Jaeger, J.P., "Conduction of heat in solids". Oxford, University Press (1959).
[23] Bbernard , R., Menguy , G. et Schwartz, M., "Le rayonnement solaire - Conversation thermique applications", éd. Technique et Documentation, Paris 1979.

[24] Ben Younes , R., "Prévision numérique et expérimentale de la réponse intrinsèque d'un échangeur bitubulaire enterré en régime continu", Thèse d'Université de Valenciennes, (1993).

[25] Ben Younes , R., Desmons , J.Y., Randriatsarafara, J.E., Khaine , D., Khaine, A. et Le Ray (M.),, "Etude Théorique et Numérique de l'intégration d'une unité de stockage à air souterrain dans une Serre Agricole". 8èmes Journées Internationales de Thermique, Marseille, (J.I.T.H.).97, Vol.2 pp 415 - 423, 7 - 10 juillet (1997).

[26] Desmons, J.Y., Ben Younes, R., Khaine, D. Et Le Ray, M. " Simulation du comportement thermique d'une serre traditionnelle - Etude de l'influence de l'inertie du sol à l'aide d'un modèle théorique utilisant les fonctions de Green". Acte du Congrès S.F.T.94, Thermique et Transports, Paris, pp. 480- 485 (1994).

[27] Ayeb, H., Ben Younes, R. Et Elouragini, S., "Mise en évidence de deux fonctions caractérisant l'inertie thermique du sol intérieur d'une serre traditionnelle - Nouvelle méthode analytique et numérique". 7ème Colloque National de Recherche en Physique de la Société Tunisienne de Physique (STP), Hammamet 21 - 24 décembre (2003).

[28] Ben Younes, R. , Desmons, J.Y. Et Randriatsarafara, J.E. " Prévision théorique et numérique du comportement thermique d'une serre en Physique de la Société Tunisienne e Physique (traditionnelle" PD20. 6ème Colloque National de Recherche STP), Hammamet 19 - 20 - 21 Mai (1999).

[29] R. Ben Younes, H. Ayeb And J.Y. Desmons. "Mathematical formulation and numerical resolution of the thermal behavior of a Greenhouse conpled with a unit of underground storage", International Journal of Heat and Technology Vol 23 N°2 december 2005.

[30] Ben Younes, R., "Numerical simulation of the thermal behaviour of a tradional greenhouse by using the theory of the green's functions", Numerical heat Transfer, Part A, 56, pp.914-929, (2009).

[31] De Brichambaut , C.P. et VAUGE, C., Le gisement solaire "évaluation de la ressource" , p.52, édition Technique et Documentation (Lavoisier), 1982.

[32] Météo France ,: Direction Régionale, Villeneuve d'Ascq, Lille (1995).

[33] Le Ray, M Cours "Energétique du Bâtiment", I.S.T.V., Université de Valenciennes, French.

[34] Desmons, J.Y. et Ben Younes, R. "Prévision à long terme de la réponse d'un stockage de chaleur sensible dans le sol", Int. J.Heat Mass Transfer Vol 40 N 13 pp. 319 - 334, (1997).

Part 3

Natural Ecosystems ET: Ecological Aspects

17

Evapotranspiration of Woody Landscape Plants

Richard C. Beeson
University of Florida
USA

1. Introduction

Landscape trees and woody shrubs are important components of residential, commercial and municipal sites, and are nearly ubiquitous in modern non-nomadic societies. Attractive landscaping can increase property values from 6 to 20% in the USA (Hardy et al., 2000; Stigarll & Elam, 2009). Proper selection and placement increases energy efficiency for heating and cooling of structures, provides storm water management, air and noise pollution abatement and carbon sequestration in urban and suburban areas (Dwyer et al., 1992; McPherson et al., 1997; Nowak et al, 2006; Sanders, 1984). An advantage of urban trees is that they are located at the source of highest CO_2 and pollution concentrations, and thus can have the greatest impact, if healthy. McPherson et al. (1999) calculated dollar benefits of the existing urban forest in Modesto, California compared to its annual tree budget cost. Including storm water and air pollution abatement, energy saving and other tangible cost, benefits outweighed cost by 2 to 1. Non-tangible aspects, such as aesthetics, positive psychological being and wildlife habitat are improved when tree and shrub quality are maintained (Hartig et al., 2010; Ulrich, 1986). Healthy maintained landscapes have a calming effect, especially for those who are under stress or depression (Ulrich, 1986).

Though most beneficial in populated areas, urban sites where woody plants are transplanted are often areas not conducive to normal growth. These sites often have constrained soil volumes (Krizek & Dubik, 1987; Lindsey & Bassuk, 1991). This is frequently due to underground utilities or street and building foundations (Ruark et al., 1983). Highly compacted soils are also a major problem (Chi, 1993). If drainage is poor or nonexistent, anaerobic conditions can lead to tree decline and death (Berrang, et al., 1985; Krizek & Dubik, 1987). Conversely, where drainage is acceptable, soil volume and/or plant available water are often inadequate to supply more than a few days of transpirational demands (Lindsey & Bassuk, 1991).

In addition to soil volumes and soil water availability, urban environments are often vastly different from natural woodlands or rural settings. Heilman et al. (1989) planted *Ligustrum japonicum* Thunb. 0.5 m from exterior walls facing the cardinal directions, actual evapotranspiration (ET_A) increased with exposure to sun due to emission of long wave radiation from walls, which increased plant temperature during and after sun exposure. Maximum sap flow was about 30% higher at the west wall than other directions. Whitlow et al. (1992) measured temperature, humidity and solar radiation at street level in a study of urban trees in Manhattan, New York City. They were compared to concurrent measurements a few blocks away in Central Park during the summer for a three year period. Duration of hours of direct sunlight was severely truncated due to shading by tall buildings at street level. The microclimate within this urban canyon had much higher daily vapor pressure deficits (vpd) than those measured at Central Park (Whitlow et al., 1992). Higher vpd was due to both drier air and higher temperature. Despite high vpd, stomata closure was rare. In a similar study conducted in Seattle, Washington, USA, solar radiation in an urban canyon was restricted to 44% of that received at a nearby park (Kjelgren and Clark, 1992). Like canyon trees in Manhattan, trees maintained an appearance of vigor. This was proposed due to acclimation to generally deep shade, resulting in larger and more leaves. In both studies, trees in the urban canyons rarely exhibited signs of water stress and stomata closure was uncommon. The limited exposure to direct sunlight lowered daily ET_A in both locations to within the volumes of water available to the trees. Additionally, reduced solar radiation would also reduce re-radiated heat from adjacent buildings and pavement. In Seattle, unlike the Manhattan canyon, vpd were similar to those at the park. In contrast, trees growing in a non-shaded paved plaza, a traffic island bordered by three major arteries in Seattle, exhibited symptoms of severe stress, such as small leaves, limited growth and low xylem water potentials. This was proposed due to chronic water stress induced by limited soil volumes and higher evaporative demand. Higher evaporative demand was principally due to two factors, both of which are common, in varying degrees, in all urban and suburban landscapes.

The first factor was wind. In the open area of the plaza surrounded by multilane highways, there were no barriers to impede natural wind. In addition, vehicles travelling at highway speeds would have augmented nature winds and created almost perpetual winds during the majority of a day. Constant winds through isolated trees would have reduced boundary layers of still air around leaves, increasing transpiration. The thinner the boundary layers, the more quickly water molecules are lost to turbulent air around a leaf (Nobel, 2009). Higher wind speeds would have also increased transpiration from shaded interior leaves in the canopy, by both lowering boundary layers and increasing vpd within the canopy by advection of drier air (Rose, 1984) arising from the heated concrete and pavement surrounding the trees.

The other environment factor contributing to severe stress was convection and re-radiation of energy absorbed by the concrete and pavement surrounding these trees. This heat energy absorbed by the leaves would have increased leaf temperature, resulting in increased density of water vapor inside a leaf, leading to a greater gradient of water vapor between inside and outside a leaf. These higher gradients would accelerate transpiration (latent heat loss). To adapt to these two forces, leaf area was substantially reduced to limit heat loading by the surroundings. Smaller leaves also have thinner boundary layers, allowing for greater conduction of heat across the boundary layer.

Whether in urban or rural areas, trees and woody shrubs usually need supplemental irrigation immediately after transplanting into a landscape to speed root growth to re-establish their natural root to leaf ratios. The duration of supplement irrigation depends on plant size and species. In central Florida (lat. 28 N), root growth can occur year round for evergreen species and 10 months for deciduous species. Here root regeneration occurs at a rate of 1 to 5 months per cm in trunk caliper (measured at 15 cm above the soil line; Gilman & Beeson, 1996). For trees with broadly spreading roots, such as *Quercus laurelifolia* and *Pinus elliotii*, establishment was 1 to 1.2 months per cm caliper whether trees were grown in above ground containers or in the ground. Root regeneration was slower for species with dense, slowly expanding root systems, such as *Ilex attenuata* 'East Palatka". Root regeneration for this species was 2.4 month per cm trunk caliper if grown in ground, or 4.8

months if transplanted from a container. Upon analysis of existing literature, Watson (2005) noted that tree establishment after transplanting was generally consistent across hardiness zones if based on trunk caliper and length of growing season when well watered. Under-irrigated trees have sparse canopies. Insufficient leaf area restricts new root generation and elongation. Such trees do not perform their intended functions in a landscape. If they survive, it can take years to regain the vigor and health they had during nursery production. To maintain aesthetically pleasing, healthy plants, supplemental irrigation is frequently required in landscapes after plant establishment. How much water woody plants use, and equally important, how much water they require, is generally unknown. Water use efficiency of most woody plants increases as soil water availability declines, up to a point. Thereafter, excessive stomata closure limits growth and eventually survival. Most research quantifying tree and woody shrub water use has been related to trees under forest conditions, or trees and shrubs during nursery production. Under these conditions, the answers have been to the question of how much water woody plants use. These answers generally establish the highest water use of a species for a given size. These values are ideal for establishing new plants into a landscape or rapidly growing plants to fulfill their place in the landscape. But once this is achieved, continuing to irrigate at production levels can result in excess pruning to maintain plant aesthetics, and is generally wasteful. It is the latter question, how much do plants require that the least is known. Going forward, the discussion will first focus on what is known of how much water plant use. Thereafter the discussion will be of how to modify known plant water use to plant water needs in landscapes and an alternative irrigation strategy that takes advantage of woody plant stomata control and perennial root systems.

2. Woody shrubs

In terms of number of plant species, there has been more research quantifying water use of woody shrubs than quantifying that of trees. However much of this research is based on small plants in 3.8 L containers, with nearly all related to irrigation needs during production. Knox (1989) grew five species of woody shrubs in north Florida (lat. 30.31 N). As plants neared market size, daily water use during late spring to early summer ranged between 0.15 to 0.30 L/day. Burger et al. (1987) grew 22 species of landscape plants in 3.8 L containers at three locations in central (38.33 N) and southern (33.56 N) California. When plants reached marketable size, daily ET_A was quantified by weighing. ET_A was then divided by the container upper surface area and reference evapotranspiration (ETo) at each location to calculate a coefficient (Kc). Actual volumes were not reported, but Kc's were. They ranged from 1.1 to 5.1. Later, Regan (1997) conducted a similar but much larger experiment, calculating Kc for 50 species of woody shrubs and trees in 3.8 L containers in northern Oregon (45.13 N). These Kc's at market size ranged from 2.3 for *Tsuga Canadensis* (l.) Carriere to 5.6 for *Hydrangea macrophylla* Thumb. 'Nikko Blue'. Since container diameters were approximately 15 cm in both studies, if one's local ETo is known, water use for these species could be estimated. However basing Kc on a fixed container diameter is only accurate if plant canopy sizes are similar between the reference plant and the plant ET_A is to be estimated for. Larger plants in the same container would have higher a Kc since larger plants would transpire more water, but the volume would be normalized by the same upper container surface area. Garcia-Navarro et al. (2004) transplanted four plant species from 3.8 L containers into 200 L drainage lysimeters, then allowed them to establish and grow for 15 months in central California (38.33 N). Daily ET_A was then quantified and averaged for three additional months. Mean ET_A was 2.3 L/day for *Spiraea x vanhouttei* (Briot.) Xabel, 1.8 L/day for *Viburnum tinus* L., 1.45 L/day for *Arctostaphylos densiflora* M.S. Baker 'Howard McMinn', and 2.0 L for *Leucophyllum frutescens* (Berl.) I.M. Johnst. Beeson (2001) reported summary results for three woody shrub species averaging marketable size in three container sizes (Table 1; 28.40 N). Mean daily ET_A ranged from 0.255 L for *Rhaphiolepis indica* (L) Lindl. Ex Ker-Gawl grown in 3.8 L containers to 2.53 L for *Viburnum odoratissimum* Ker-Gawl grown in 26.5 L containers. Winter daily water use was generally half the average, while summer water use was 50% more. In 2004, Beeson (2004) presented a graph of daily ET_A of *Ligustrum japonicum* Thumb. grown in 11.4 L containers (28.40 N). At marketable size (0.6 m tall and 0.45 m width) and spaced 43 cm on center, ET_A was about 0.80 L in the spring, with an average ETo of 4.6 mm/day. These ET_A volumes were comparable to flow rates of the same species reported by Heilman and Ham (1990; 30.36 N). Steinberg et al. (1991) reported somewhat smaller daily water use of 0.3 to 0.5 L for market size *Ligustrum japonicum* in. 7.6 L containers in early June in a shaded greenhouse in eastern Texas (30.36 N).

Container size	3.8 L	11.4 L	26.6 L
Species			
Viburnum odoratissimum	0.20 x 0.30 *** 0.40[z]	0.45 x 0.60 *** 1.14	0.90 x 1.2 *** 2.53
Ligustrum japonicum	0.20 x 0.30 *** 0.34	0.45 x 0.60 *** 1.08	0.90 x 1.2 *** 2.36
Rhaphiolepis indica	0.23 x 0.15 *** 0.25	0.57 x 0.38 *** 0.66	0.90 x 0.6 *** 1.48

Mean canopy width (m) x height (m) *** Mean L/day.

Table 1. Mean canopy dimensions and daily water lost (liters) from container grown plants. Means are based on 3 replications per species and container size. Data was collected 1 Jan. 1995 to 31 Dec. 1996 (Beeson 2001).

3. Trees

Most studies of short-term water use of both conifer and hardwood trees have occurred in forest, and thus are not accurately related to water use in landscapes. Yet the same species are among the mainstays of most landscapes. With some adjustment, detailed later, the values derived from forest trees can be used as estimates for trees in landscapes. In 1998 Wullschleger et al. (1998) reviewed forest tree water use reported from 1970 through 1998. The review covered 67 species of trees gleaned from 52 studies. It focused on contributions of the different methods of quantifying tree water use related to a holistic approach for understanding water movement within trees, between canopies and the air above. Ninety percent of the tree water use surveyed ranged from 10 to 200 L/day. Some of the studies included in the review have been included in the discussions here where specific values are relevant to landscape ET_A.

A limited number of studies have quantified water use of small landscape trees. Costello et al. (1996) reported cumulative water loss from three tree species grown in 15.6 L containers in a shaded courtyard versus an open knoll (38.33 N). The daily average over a 14 day period for the open location was 1.2, 1.18 and 0.89 L for *Liquidambar styraciflua* L. *Sequoiadendron giganteum* (lindl.) J. Bucholz and *Magnolia grandiflora* L., respectively. The average reduction in ET_A for trees in the courtyard was 54%. Tree size was not indication. Similar daily ET_A (1.3 L) for a *Taxodium distichum* (L.) Rich growing in a 15.6 L container in

July in Texas (30.36 N) was reported by Steinberg et al. (1990a). In northwest Texas (33.2 N), ET_A of *Prosopis glandulosa* Torr. with trunk diameter of 1.9 cm ranged from 0.25 to 2.5 L per day (Dugas & Mayeux, 1991). Cumulative water use of trees over a full year's growth is seldom reported. However, the annual cumulative ET_A of seedlings of 10 tropical trees transplanted into 200 L drainage lysimeters in India (26.85 N) ranged from 1,358 L for *Pongamia pinnata* (L.) Pierre to 5,324 L for a Eucalyptus hybrid (Chaturvedi et al., 1988).

3.1 Isolated trees

For large to mature trees, daily water use of only a small number of isolated trees have been quantified, and then principally for only a few days to weeks. Montague et al. (2004) measured the ET_A of five species of trees transplanted as balled and burlapped material into in ground weighting lysimeters (41.44 N) during the summer. Mean daily ET_A averaged between 3.2 L/day of *Acer platanoides* L. 'Emerald Queen' (25.2 cm^2 trunk area) to 5.1 L/day for *Fraxinus pennsylvanica* Marsh. 'Patmore' (36.3 cm^2 trunk area). Steinberg et al. (1990b) reported daily ET_A in Texas (32.12 N) of an 11cm caliper *Carya illinoensis* (Wangenh.) K.Koch 'Wichita' during the summer to range from 100 to 150 L per day over a 12 day period in August, based on weighing lysimeter measurements. The tree was 3.9 m tall, with a diameter at breast height (dbh, 1.4 m) of 7.9 cm. Green (1993) reported summer daily water use of a 10 yr old *Juglans* L. species (walnut; 3.4 x 3.1m canopy widths) in New Zealand (40.2 S) to range from 14 to 40 L, depending on environmental factors. In New Zealand pastures (40.08 S), where canopy coverage by *Populus deltoids* (Bart. Ex Marsh, Clone 178) was 66%, ET_A determined by sap flow ranged from 162 to 417 L/day for trees with projected crown areas of 97 to 275 m^2 (Guevara-Escobar et al., 2000). Part of the variability among trees was due to location on the hillside and shading by neighboring trees. Total ET_A was highly correlated with sapwood area ($r = 0.86$) and dbh ($r= 0.93$). Ruiter (1987) reported ET_A of Pinus radiata with a caliper of 13 cm (at 15 cm) to average 21 L maximum daily ET_A over a 4 week period (37.49 S) in the summer.

The largest isolated trees measured by lysimetery were reported by Edwards (1986). From these, graphs were published of the daily ET_A of four species of trees, ranging from about 8.9 to 20 cm in caliper at 15 cm (40.21 S). For the deciduous species *Populus x euramericana* (Dode) Guinier cl 'Flevo', with a trunk diameter of about 11 cm, ET_A from fall through early spring ranged from 0 to about 5 L/day. ET_A increased in spring, to maintain between 70 to 90 L/day during summer, dropping quickly to near zero in mid-autumn. A similar seasonal pattern was presented by *Salix matsudana* Koidz. When leafless, ET_A ranged from 0 to 5 L/day. During the peak of summer, ET_A ranged from 50 to 60 L/day for this 8.8 cm trunk diameter tree. An *Eucalyptus fastigata* Dean & Maid 3.3 m tall generally transpired 10 L/day during winter, but increased to a mean of 65 L/day during the summer. The other evergreen species *Pinus radiata* was about 20 cm in caliper. Winter to late spring ET_A was around 40 L/day. Throughout the summer, ET_A ranged from 85 to 125 L/day, decreasing to the 60 to 80 L range in the fall.

3.2 Forest

Within forest, daily tree water use is generally smaller than for isolated trees. In Germany (51.13 N), *Fraxinus excelsior* L. saplings 4.5 m tall and 2.84 cm dbh registered maximum transpirations of about 7 L/day in early August under optimum conditions (Stohr and Losch, 2004). In England (51.27 N), whole tree ET_A was calculated for a *Populus trichocarpa*

Toor. & A.Gray x *P. tacamahaca* L hybrid based on sap flow through branches, scaled to the whole tree level. Trees were growing on a site with a shallow water table. With trunk dbh of 4.2 to 4.6 cm and heights of 5 to 6 m, peak ET_A was 23 to 28 L/day in June (Zhang et al., 1999). In a forest near Melbourne, Australia (37.34 S), the potometer method was used to measure ET_A of a large *Eucalyptus regnans* F.J. Muell. (Vertessy et al., 2007). For a tree 55 m tall and 0.83 m dbh, flow rates through the sapwood ranged from 61 to 323 L/day. Granier (1987) found that dominant trees in a *Pseudotsuga menziesii* Carrière forest (48.44 N) averaged 22 L per day for trees about 18 cm dbh in France during the summer. Schulze et al. (1985) estimated ET_A of a *Picea abies* L. (28 cm dbh; 49.2 N) and a *Larix* hybrid (26 cm dbh; 49.95 N) growing in the Czech Republic and Germany, respectfully. In the summer, ET_A's were 63.1 L the *Picea* and 74.4 L for the *Larix*. More recently Verbeeck et al. (2007) reported maximum daily sap flows of 28 L/day in May for *Pinus sylvestris* L. growing in a forest in the Belgian Campine region (51.18 N). The tree was nearly 20 m tall with a dbh of 26.7 cm.

3.3 Tropical trees

Daily water use of tropical trees is among the highest of species quantified to date. Andrade et al. (1998) measured sap flow of five species of tropical trees in the Panama rainforest (8.58 N) during a period when soil water availability was not limited. Eighteen meter tall *Cecropia longipes* Pitt. with a dbh of 19.7 cm had a cumulative sap flow rate of 46.5 L/day. A *Spondias mombin* L. tree 23 m tall with a dbh of 33.1 cm averaged 80 L/day. *Luchea seemannii* Triana & Planch. was 129 L/day for a 29 m tall tree with a 38.2 cm dbh. A *Ficus insipida* Willd. with a 56.7 cm dbh trunk sap flow rate was 164 L/day for a 30 m tree. The highest sap flow was from an *Anacardium excelsum* (Bertero & Bald. ex Kunth) Skeels tree at 379 L/day from a 15 m tall tree with a 101.8 cm dbh. Total sap flow rate was highly correlated (r=0.99) with trunk dbh.

With the exception of isolated trees in high density urban areas, once trees approach the small forest sizes detailed above, there is little need for supplemental irrigation. On average, if unrestricted, roots systems of landscape trees are generally three times the width of their canopy (Gilman, 1988a, 1988b). At this point, unless there are substantial disruptions to the rooting volume, such as new construction, these large trees in landscapes are self-sufficient in terms of water needs.

3.4 Daily tree water use

While there have been studies quantifying short term water use of landscape trees, or trees that could be used in landscapes, until recently there have been none that track tree water use for the same tree from propagated material (seedlings or rooted cuttings) to trees of landscape stature. In 2001 a project was initiated in Central Florida (latitude 28° 40') to quantify daily water use of three landscape tree species; *Acer rubrum* L. (red maple), *Quercus virginiana* Mill. (Live oak), and *Ilex x* `Nellie R Stevens' (holly). Both the oak and holly are evergreen species. In early spring 2001 propagules were transplanted in 26 L polyethylene containers and placed in suspension lysimeters (Beeson, 2011). Each tree was surrounded by like trees, but spaced so that canopy coverage was about 50%. Tree ET_A was recorded daily for the following 5 to 6 years as trees were transplanted into ever larger containers up to 1.4 m diameter.

3.4.1 *Ilex*

In six years, the holly tree grew from a trunk caliper of 3.7 mm to 130 mm measured 15 cm above soil line. The holly cultivar, 'Nellie R. Stevens', is slow growing, with an average maximum height of 9 m (Gilman and Watson, 2006). Like most hollies, it was produced with lower branches left close to the ground. Because this holly was a true evergreen species, after the first two years the tree retained two or three years' worth of leaves. Thus leaf area was relatively constant or increasing throughout the experiment. Bell-shaped variations in ET_A over a year (Fig. 1.B - F) are therefore reflective of changes in ETo than leaf accumulation or senescence. Since ETo is principally driven by solar radiation, the bell-shape pattern reflects seasonal variability in sunlight intensity and duration of the northern hemisphere at the 28° latitude. Variability between individual points on all graphs for all species is the day-to-day variation in ET_A. On rare occasions, rainfall was all day or intermittently frequent enough such that there was no discernable mass loss from a lysimeter. These days are indicated by $ET_A = 0$.

In 2001 at the start of the experiment, holly trees averaged 0.3 m tall, with an average canopy width of 12 cm and caliper at 15 cm of a little more than 3 mm (Fig. 1). Average daily ET_A of these rooted cuttings was around 0.15 L. By early fall (Day 255) trees had more than doubled in height to 0.84 m, and increased in average canopy width and caliper 3-fold (0.35 m and 1.25 cm, respectively). Daily water use also increased 3-fold to around 0.5 L/day. During this first year growing period, maximum water use occurred during late summer in July and August. Yet it rarely exceeded 0.55 L/day. By early November (Day 305) tree water use was generally less than 0.4 L/day. Over the course of the 257 days this young tree was measured in 2001, cumulative ET_A was 86 L. Cumulative ETo was 1,074 mm.

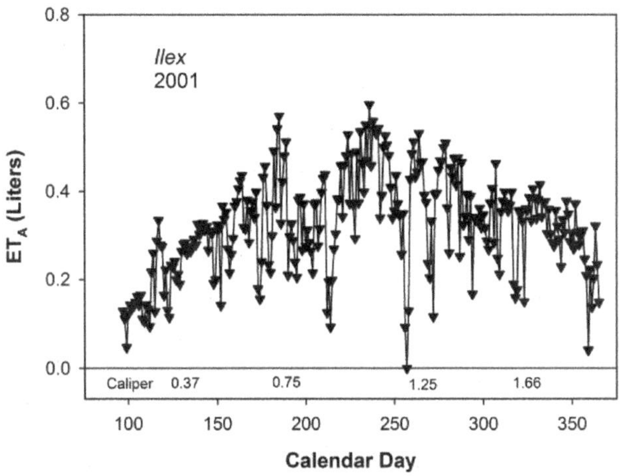

Fig. 1. Mean daily ET_A of the holly, *Ilex* x ' Nellie R. Stevens', for 2001. Trunk diameter (cm), indicated at the bottom of the graph (Caliper), measured at 15 cm inches above the substrate level.

In 2002 mean tree water use remained below 0.4 L per day through mid-March (Day 75, Fig. 2), at the beginning of shoot bud break. During the year tree height increased about 60%, to 2.1 m, with average canopy width doubling to 1.1 m between mid-March and early fall (October, Day 274). However tree water use increased 7-fold from mid-March to early August (Day 213). By mid-November (Day 320), ET_A dropped to about half the value recorded in August, without any loss of leaves, concurrent with a steep decline in daily ETo, and remained around 2.2 L/day for the remainder of the year. The cumulative ETo for 2002 was 1477 mm, associated with 566 L of cumulative ET_A.

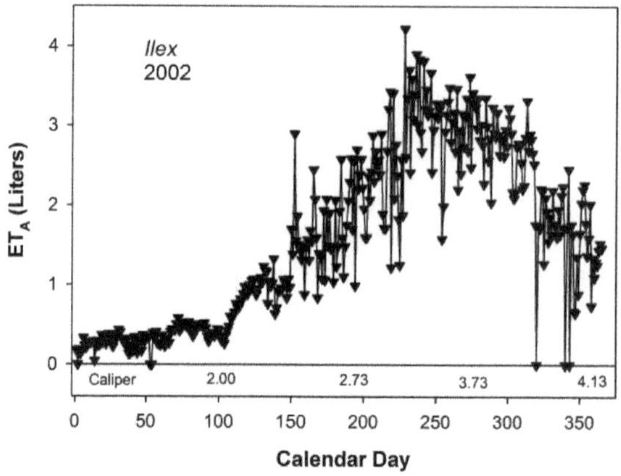

Fig. 2. Mean daily ET_A of the holly, *Ilex* x ' Nellie R. Stevens', for 2002. Trunk diameter (cm), indicated at the bottom of the graph (Caliper), measured at 15 cm inches above the substrate level.

In 2003, cumulative tree water use was 1,305 L that occurred during 1,328 mm of ETo. Increases in daily ET_A slowed to only 40% from the previous year, while tree height increased about 0.6 m for the year to 2.4 m, and canopy width increased to 1.56 m (Fig. 3). This was similar to increases in tree height in 2002 (Fig. 2), but a smaller percent increase in ET_A than the two previous years. From mid-May (Day 135) until December (Day 335) ET_A was generally consistent between 4.7 to 6.2 L/day (Fig. 3). There was little reduction in ET_A in early winter as observed in previous years.

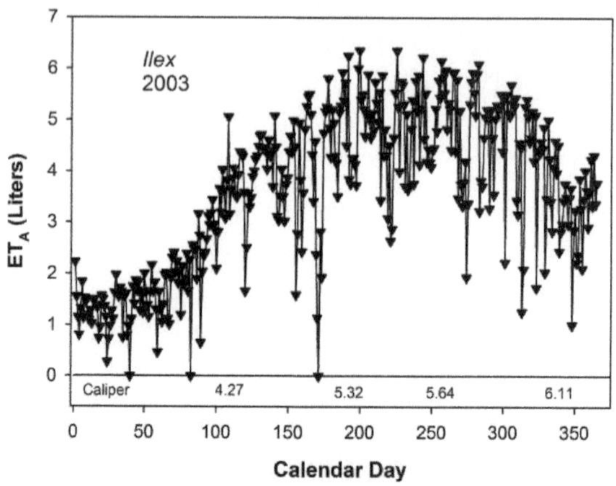

Fig. 3. Mean daily ET_A of the holly, *Ilex* x ' Nellie R. Stevens', for 2003. Trunk diameter (cm), indicated at the bottom of the graph (Caliper), measured at 15 cm inches above the substrate level.

In 2004, ET_A remained at around 4.5 L/day until early April (Day 93, Fig. 4). As in previous years, tree height increased around 0.6 m over the year. Average canopy spread increased to 2.0 m, similar to earlier years. With the increase in canopy spread, tree ET_A increased 3.5 - fold compared to the winter months. Like in previous years, with exception of 2003, ET_A declined by more than half (13 to 6.5 L/day) from early October (Day 275) to early December (Day 336). Although three hurricanes buffeted the research plot in 2004 (Days 220 to 255), they had little effect on ET_A or tree growth the holly. Cumulative ET_A, as trunk caliper increased from 5.9 to 8.0 cm was 2,976 L, with a cumulative ETo of 1,403 mm.

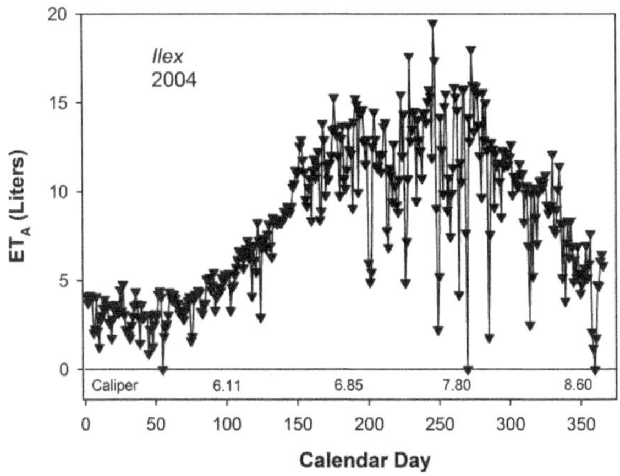

Fig. 4. Mean daily ET_A of the holly, *Ilex* x ' Nellie R. Stevens', for 2004. Trunk diameter (cm), indicated at the bottom of the graph (Caliper), measured at 15 cm inches above the substrate level.

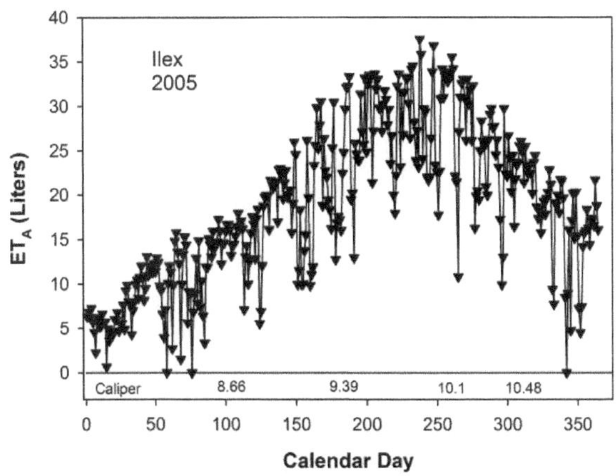

Fig. 5. Mean daily ET_A of the holly, *Ilex* x ' Nellie R. Stevens', for 2005. Trunk diameter (cm), indicated at the bottom of the graph (Caliper), measured at 15 cm inches above the substrate level.

Tree height maintained its constant pace of a 0.6 m increase in 2005. Canopy width increased at a similar constant rate of 0.5 m, while trunk caliper increase was also on pace with previous years at 19 mm. Also consistent was peak ET_A, increasing from 6 to about 34 L/day during summer (Fig. 5). Like previous years, ET_A began declining in early October (Day 275) and was down to about half of its summertime peak by December, corresponding to an approximate halving of daily ETo. Cumulative ET_A for 2005 was 6,789 L with 1,363 mm of ETo.

ET_A in 2006 peaked at around 50 L/day in late June (Fig. 6). At this point the tree was 4.2 m tall with an average canopy width of 2.78 m. From the median ET_A of 17 L/day in January, there was again a 3-fold increase in ET_A from winter to late summer. At its peak, ET_A frequently varied about 50% over periods as short as a week. Similar variability can be seen through all graphs for the holly. In 2006, mean cumulative ET_A was 10,227 L from this tree in a 1.4 m diameter container. Cumulative ETo was 1451 mm. When the experiment was terminated at the end of 2006, total ET_A to grow this tree from a rooted cutting to a tree with a trunk caliper of 13.0 cm, height and width of 4.2 m and 2.8 m, respectively, was 21,949 L.

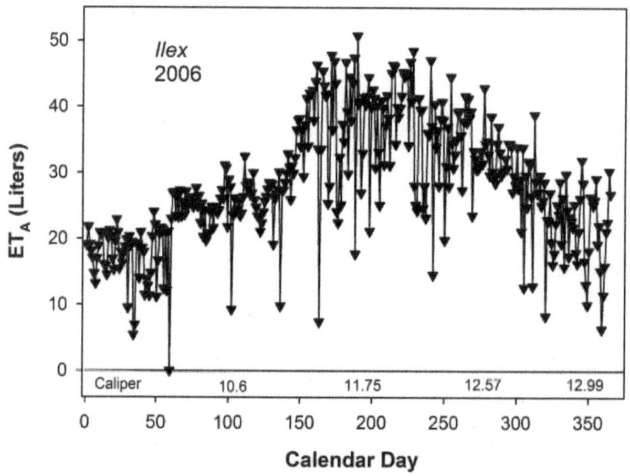

Fig. 6. Mean daily ET_A of the holly, *Ilex* x ' Nellie R. Stevens', for 2006. Trunk diameter (cm), indicated at the bottom of the graph (Caliper), measured at 15 cm inches above the substrate level.

3.4.2 *Quercus*

The Live oak seedling was germinated during the winter of 2000-2001. At the beginning in March 2001, the seedling was 0.38 m tall, with a stem caliper at 15 cm 3.04 mm (Fig. 7). During the first year of growth, tree height increased 3.6-fold, to 1.4 m, with a trunk caliper increase of 4-fold to 1.27 cm. Similar growth rates of Live oak seedlings have been reported before for trees in 11.4 L containers (Beeson and Haydu, 1995). Initial mean ET_A was similar to that of the holly, but ET_A of the oak increased more rapidly, obtaining a 4-fold increase in daily ET_A to 0.76 L/day within 120 days after transplanting (Fig. 7). Unlike holly, ET_A did not decline as much late in the year until late December. Cumulative mean ET_A for 2001 was 136 L.

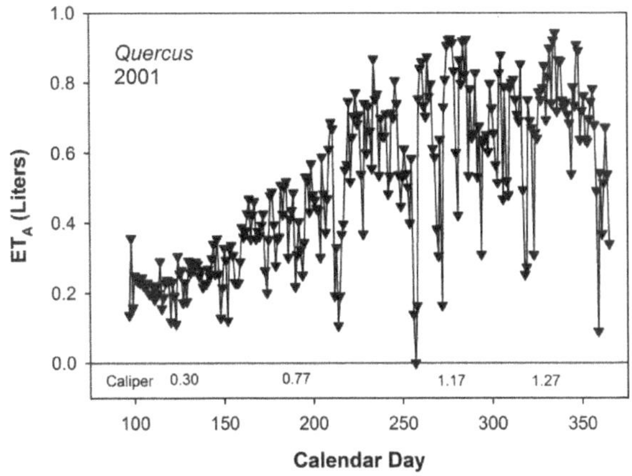

Fig. 7. Mean daily ET_A of the oak, *Quercus virginiana* for 2001. Trunk diameter (cm), indicated at the bottom of the graph (Caliper), measured at 15 cm inches above the substrate level.

Through the first portion of 2002, oak ET_A was around 0.5 L/day until early March (Day 60; Fig. 8). Though not evident in the graph, there was a dip in ET_A from mid-February (Day 45) until early March. This corresponded with annual late winter leaf drop and bud burst of Live oak. With leaf development, ET_A increased relatively gradually from early March through mid-May (Day 140), quadrupling with increasing shoot growth. By mid-August (Day 227), mean daily ET_A had increased to 9.5 L/day with a further increase in early October (Day 275) with a final cycle of shoot flush. In November, tree growth had stopped at 2.63 m in height and increased average canopy width of 2.7 m and trunk caliper of 4.82 cm. By late December, ET_A had declined to 4 L/day, and remained so until bud burst the following spring. Cumulative mean ET_A for 2002 was 1,527 L.

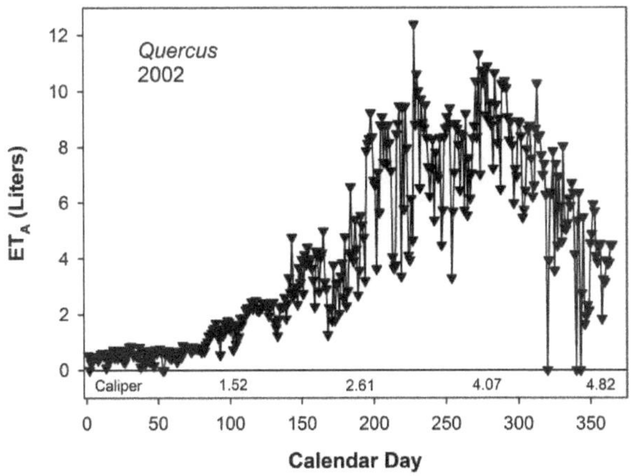

Fig. 8. Mean daily ET_A of the oak, *Quercus virginiana* for 2002. Trunk diameter (cm), indicated at the bottom of the graph (Caliper), measured at 15 cm inches above the substrate level.

In 2003, leaf drop in February briefly reduced daily ET_A to 1 L/day (Fig. 9). ET_A peaked over 36 L/day in mid-September (Day 250) as the tree grew from a height of 2.63 to 3.92 m. Daily ET_A rates between 26 and 30 L/day persisted from early July (Day 182) through mid-November (Day 315), later declining to around 16 L daily by early December. Cumulative mean ET_A for this Live oak in 2003 was 5,728 L. At the end of the year, the tree was 4.12 m tall, with an average canopy width of 2.9 m and trunk diameter of 9.42 cm.

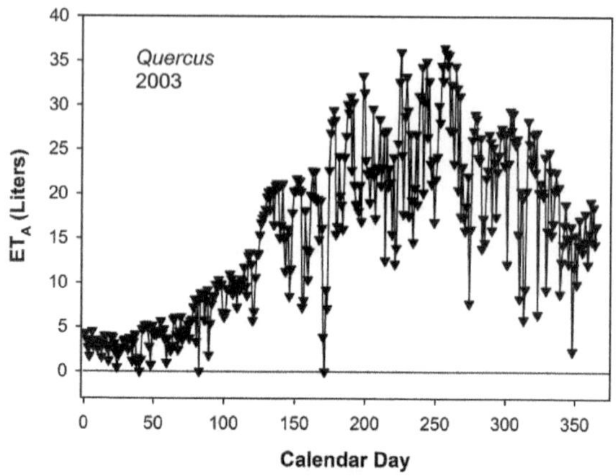

Fig. 9. Mean daily ET_A of the oak, *Quercus virginiana* for 2003. Trunk diameter (cm), indicated at the bottom of the graph (Caliper), measured at 15 cm inches above the substrate level.

In 2004, leaf change occurred closer to mid-March (Day 75), with a substantial decline in ET_A from a few weeks before (Fig. 10). The tree began transpiring at its 2003 peak rates of 30 L/day by early May (Day 125). About 60 days later in early July (Day 185), mean daily ET_A had doubled to over 65 L/day. During this time, tree height had increased by 0.47 m, with mean a mean width increase of 1.0 m. From mid-August (Day 211) through late September (Day 274), the tree was impacted by three hurricanes. The impact was minor with the first two, until the last hurricane, Jeanne, stripped leaves from southeastern side of the tree in late September (Day 273). Peak winds of 145 Km/h were recorded near the site. These leaves were replaced, resulting in a slight bulge in ET_A in late October /early November (Day 290) but with little change in canopy size. By December, ET_A declined to 30 to 35 L/day. Final dimensions of the tree for 2004 were a height of 5.28 m, mean width of 4.0 m and trunk caliper of 13.5 cm. Cumulative ET_A was 12,827 L.

In 2005, the effect of leaf drop on ET_A in the spring was more evident, occurring between late February (Day 50) and late March (Day 80; Fig. 11). By early May (Day 125), ET_A regained its summer peak of 75 L/day the year before. From late June (day 170) to mid-September (Day 255), ET_A was generally around 115 L/day or better. During this time, peak ET_A was 140 L/day with a trunk caliper of 16.7 cm. Tree height increased 0.97 m, while average tree spread increased 0.8 m during 2005. Trunk caliper had the most impressive increase, starting at 14.0 cm in spring and increasing to 16.69 cm by late September. From mid-September (Day 255) to early December, ET_A declined from 130 to 70 L/day with no loss of leaf area, due solely to decreases in ETo from an average of 5.0 mm to 1.6 mm daily. Cumulative ET_A for 2005 was 23,898 L.

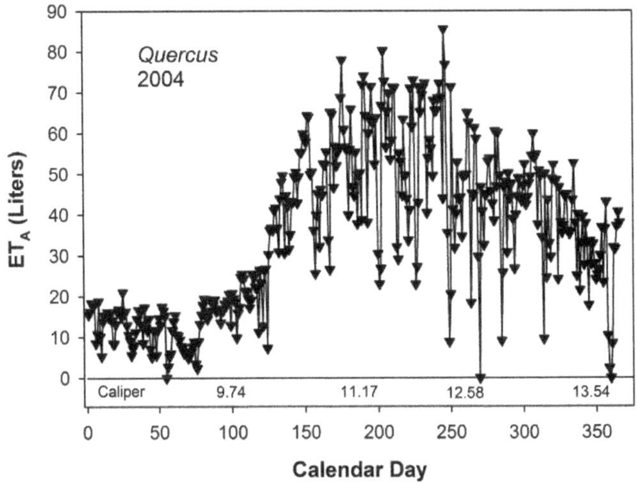

Fig. 10. Mean daily ET$_A$ of the oak, *Quercus virginiana* for 2004. Trunk diameter (cm), indicated at the bottom of the graph (Caliper), measured at 15 cm inches above the substrate level.

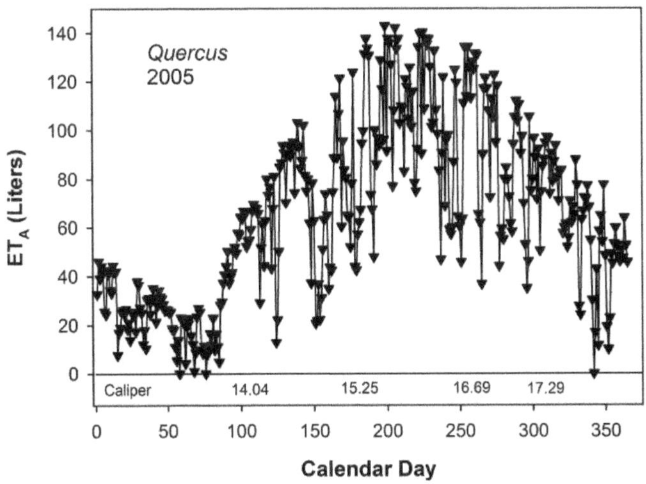

Fig. 11. Mean daily ET$_A$ of the oak, *Quercus virginiana* for 2005. Trunk diameter (cm), indicated at the bottom of the graph (Caliper), measured at 15 cm inches above the substrate level.

In 2006, leaf drop began earlier than in 2005, starting in mid-February (Day 45) and running through early March (Day 70). During this transition, ET$_A$ fluctuated between 30 and 60 L/day. The tree was 6.88 m tall and 5.05 m in average canopy with. Like 2005, trees obtained their previous summer's ET$_A$ in early May (Day 125). Mean ET$_A$ increased to 177 L/ day in late June. Shortly after the peak, the tree was blown over during a thunderstorm, destroying two of the load cells. Data collection was terminated at this time. Through the duration of data collection, tree height increased from 0.38 to 6.74 m, with a mean canopy width of 5.3 m, a trunk caliper increase from 2.3 mm to 18.73 cm. Until its demise in June, the tree transpired 13,709 L during a period of 689 mm of ETo. Mean cumulative ET_A for this live oak from seedling though mid-June 2006 was 57,825 L.

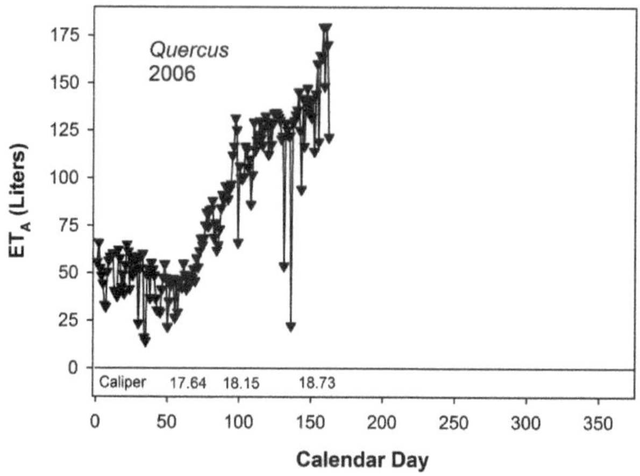

Fig. 12. Mean daily ET_A of the oak, *Quercus virginiana* for 2006. Trunk diameter (cm), indicated at the bottom of the graph (Caliper), measured at 15 cm inches above the substrate level. Data collection was terminated early due to storm damage.

3.4.3 *Acer*

Like the holly, the maple was a rooted cutting from a maple selection by Trail Ridge Nursery in north Florida. The maple was the most rapid growing of the species in terms of both shoots and root growth. Each year at transplanting, maples had completely extended their roots throughout the container volume. Being deciduous, fluctuations in ET_A were very dramatic between seasons, like those reported by Edwards (1986). From late December until early March, the tree was leafless. Thus the rise in ET_A in spring was quite rapid. This increase coincided with increasing day lengths and ETo. Conversely, leaf senescence in December lowered ET_A at the same time ETo was declining with shorter days and cooler temperatures. Thus differentiating between changes in leaf area and function, with those related to seasonal changes in ETo are difficult.

In 2001, ET_A of the maple was similar to other species initially, around 0.18 L/day (Fig. 13). As the tree grew from 0.45 to 2.05 m in height the first season, ET_A increased to around 1.1 L/day in mid-August. This ET_A was twice as much as measured for the oak and three times that of the holly. ET_A began declining in mid-September (Day 260) and was quite low by early October (Day 280). This corresponded with a disease on the leaves that induced early leaf senescence. By October, most leaves had fallen. In later years, protective sprays prevented early leaf senescence. When growth ceased for the year, mean canopy width had increased from 0.12 m to 0.65 m, while trunk caliper had increased from 4.9 to 20.0 mm. Cumulative mean ET_A for 2001 was 156 L.

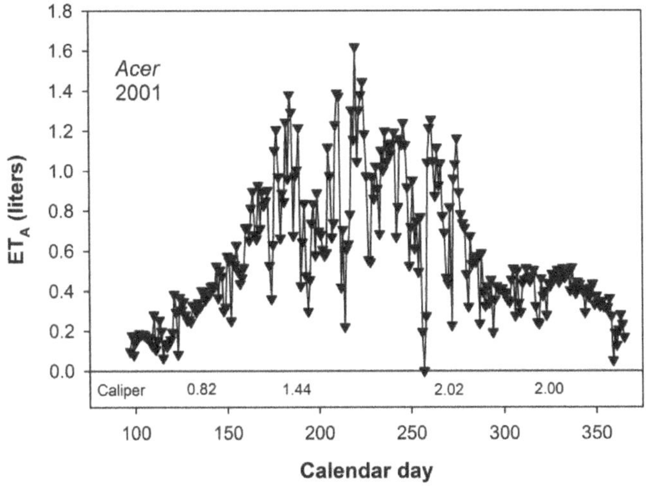

Fig. 13. Mean daily ET_A of the maple, *Acer ruburm* for 2001. Trunk diameter (cm), indicated at the bottom of the graph (Caliper), measured at 15 cm inches above the substrate level.

In 2002 bud break and leaf expansion began in late March, around Day 85 (Fig. 14). Prior to this, ET_A was equivalent to that of the young tree initially, about 0.25 L/day. ET_A increased from near nothing to 3.0 L/day over the next 60 days (Day 140, late May). ET_A was around 11 L/day from late July (Day 205) to early October (Day 275). During this time trees increased from 2.05 to 3.54 m in height and from 0.65 to 1.8 m in canopy spread (Fig. 14). Leaf senescence began in mid-December (Day 350) and was complete by month's end. Cumulative ET_A in 2002 was 1,728 L. Over the year, trunk caliper increased from 2.0 to 5.39 cm.

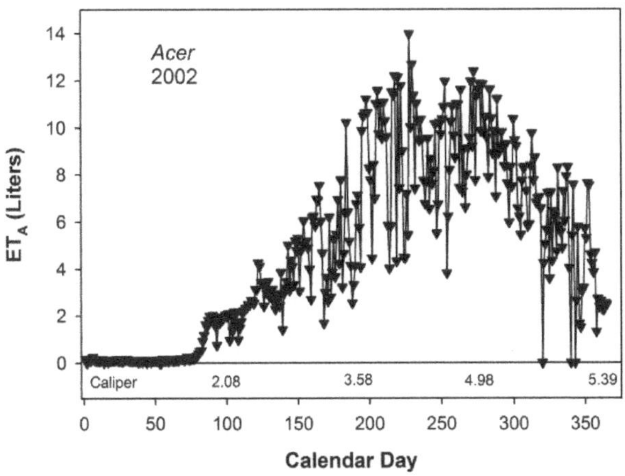

Fig. 14. Mean daily ET_A of the maple, *Acer ruburm* for 2002. Trunk diameter (cm), indicated at the bottom of the graph (Caliper), measured at 15 cm inches above the substrate level.

In 2003, the tree flowered before leaf expansion (Fig. 15). Flowering began in mid-February (Day 45), with leaf and shoot growth not beginning until late March (Day 80). Though there was likely little photosynthesis and no shoot or leaf growth, ET_A more than doubled with flowering, but was still less than 2 L/day. Increase in ET_A was more rapid with the larger tree in 2003 than in 2002 (Fig. 14), with ET_A increasing from 2.0 to 23 L/day in 45 days (Fig. 15). ET_A peaked over 30 L/day in early June (Day 155) and continued generally around 25 L/day through the first week of July. During this period, stems with red leaves, indicative of new and expanding tissue, were 30 to 45 cm in length on most major branches. The second week of July shoot elongation nearly stopped, resulting in a rapid and dramatic drop in expanding leaves. This coupled with maturation of previously expanding leaves resulted in substantial rapid decline in ET_A of 8 L/day. Shoot elongation remained limited thereafter, growing 0.58 m in height, with no measurable increase in average canopy width after the middle of July. Leaf senescence initiated in mid-December and was nearly complete by end of the year. Cumulative ET_A for 2003 was 4,121 L as the tree grew from 3.54 to 5.17 m in height and from 1.56 to 2.9 m in width, with trunk caliper increases of 3.6 cm.

Prior to flower initiation in mid-February 2004 (Day 45), ET_A was less than 4 L/day, and increased to less than 7 L/day during flowering (Fig. 16). With shoot and leaf expansion beginning in mid-March (Day 75), ET_A increased from 7 to 43 L/day over a 45 day period. From early June until mid-July (Day 200), ET_A ranged around 55 L/day. As in 2003, shoot elongation slowed dramatically in mid-July, resulting in ET_A of 38 to 45 L/day until the mid-August leaf loss due to the first hurricane (Charlie) of that year (Day 226). For the remaining of the fall, ET_A dropped following each sequential hurricane as leaves were torn or lost entirely. The 0 ET_A on Day 270 occurred during the peak of Hurricane Jeanne. Thereafter ET_A then steadily declined until leaf senescence was completed in mid-December (Day 350). Despite fall storms, the tree increased in height from 5.17 to 6.9 m and width by 2.9 m to 4.1 m. Cumulative ET_A for 2004 was 9,459 L. Trunk caliper at the end of leaf senescence was 13.25 cm.

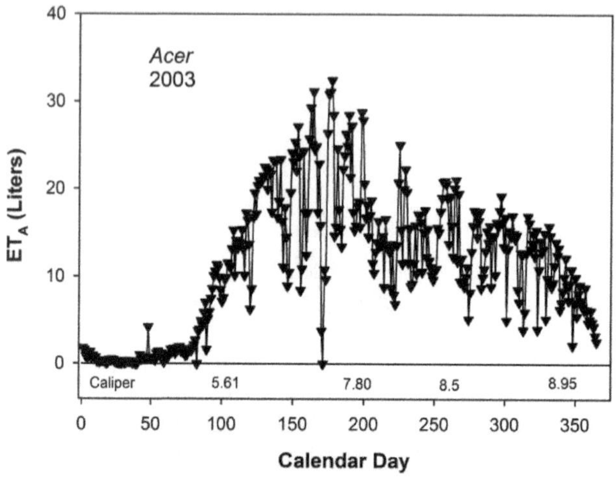

Fig. 15. Mean daily ET_A of the maple, *Acer ruburm* for 2003. Trunk diameter (cm), indicated at the bottom of the graph (Caliper), measured at 15 cm inches above the substrate level.

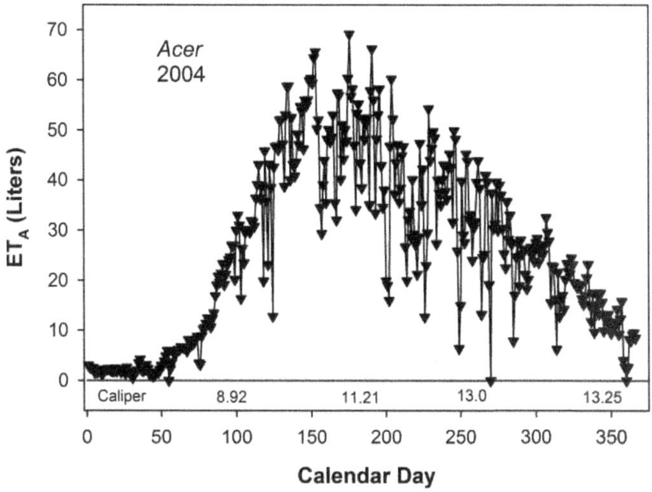

Fig. 16. Mean daily ET_A of the maple, *Acer ruburm* for 2005. Trunk diameter (cm), indicated at the bottom of the graph (Caliper), measured at 15 cm inches above the substrate level.

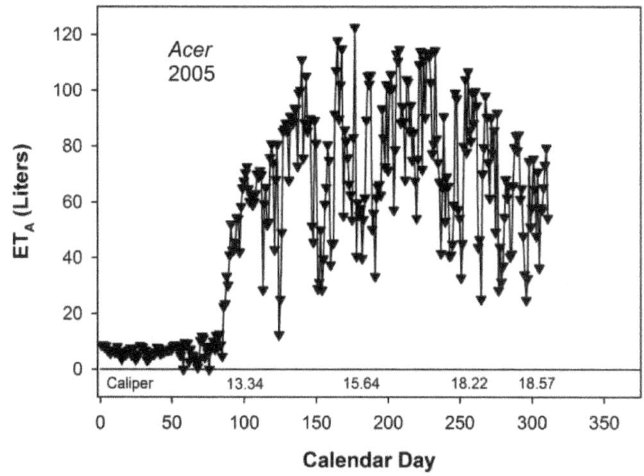

Fig. 17. Mean daily ET_A of the maple, *Acer ruburm* for 2004. Trunk diameter (cm), indicated at the bottom of the graph (Caliper), measured at 15 cm inches above the substrate level.

In 2005, ET_A was around 7 L/day until flowering in mid-February (Day 45, Fig. 17). With onset of shoot elongation (Day 90), tree water use rapidly increased from 11 to 75 L/day over a 2 week period. From first of June (Day 150) until mid-August (Day 225) ET_A was generally around 85 L/day, peaking up to 110 L/day. As in previous years, slowing of shoot elongation reduced ET_A for remainder of the fall, though not as dramatically as in 2003 (Fig. 15). This slowdown in shoot growth occurred about a month later than in previous years. Whether this was due to being a larger tree with more branches, a residual from stress from hurricanes the year before or a new pattern of grows cannot be determined. Data collection on the maple stopped the first week of November. During this last year, the tree increased 1.1 m in height growth to final height of 8.0 m, and, increased in width by 0.45 m for an average canopy spread of 4.55 m. When terminated, tree caliper was 18.6 cm. Cumulative ET_A for 2005 was 16,491 m during a period when cumulative ETo was 1,242 mm. To grow from a maple of 0.34 m tall to one of 8 m tall required 4.75 years and 31,955 L of ET_A.

4. Translating known ET_A to landscape irrigation

As noted above, ET_A of trees in forest are usually lower than that measured for smaller isolated trees. This effect is due to the degree of canopy closure, measured as projected canopy area (pca) density. In January 2003 (28.48 N), market-quality shrubs of *Viburnum odoratissiumum* in three sizes (3.8, 11.4 and 26.6 L containers) were used to verify a result obtained in 1997 (Beeson, 2010). Mean plant heights were 0.55, 0.80 and 1.23 m for the three container sizes. Plant ET_A was measured at four levels of canopy closure, ~0, 33, 67 and 100% closure with weighting lysimeters (Beeson, 2011) concurrently for nine replications of each container size. The response to canopy closure was identical for each plant size and that observed in 1997 (Fig. 18). Daily ET_A normalized by ETo was the same for isolated plants as those with 33% and 67% canopy closure, i.e. where the cumulative pca's of border and lysimeter plants was 33% and 67% of the total area on which the plants were set. At 100% canopy closure, normalized ET_A declined to 60% of that of plants at 67% closure or less. The ET_A of plants at 100% canopy closure was associated with transpiration from only the upper 40% of a canopy.

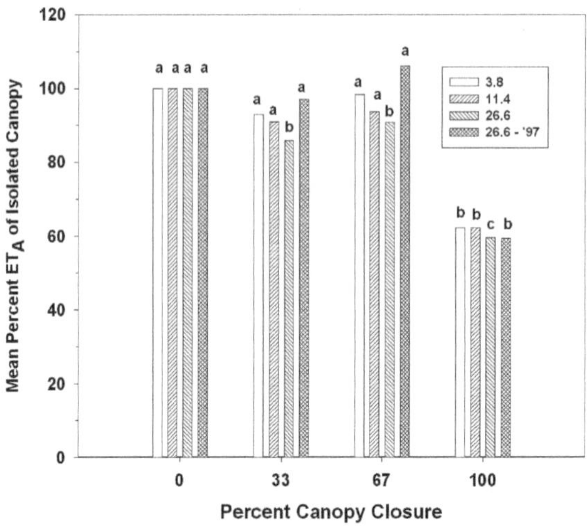

Fig. 18. Normalized ET_A of *Viburnum odoratissimum* of marketable size in three container sizes at four levels of canopy closures. Each bar is the mean of nine plants. Bars with different letters within a container size are significantly different (α=0.5) based on F-Protected LSD.

A complimentary opposite of this effect, increased ET_A, has been reported for trees that remained after forests had been partially harvested to reduce stand density (Bréda et al., 1995; Medhurst et al., 2002). Thus in forested stands, production nurseries and landscape plantings, maximum plant water use will decrease between 70 and 100% canopy closure. This occurs due to mutual leaf shading, increases in leaf boundary layers, and development of canopy boundary layers caused by reductions in canopy roughness. Water use values derived from closed forest or plant production cited above underestimate ET_A in landscapes with less than 70% canopy closure or for isolated trees. Conversely, as landscapes increase in canopy coverage to above 70% canopy closure, water use of the landscape in its entirety will decline.

After accounting for canopy closure, both in examples above and at the location where the information will be applied, utilizing this information is a two stage process. As noted earlier, for optimum establishment, newly transplanted trees and shrubs should be irrigated at their water use rates based on well-watered conditions and size. However once established, research indicates irrigation can be curtailed with woody plants, while still providing their intended aesthetic and functional expectations.

How much irrigation can be reduced has been researched mainly in the western USA. There, research has focused on the minimum level of irrigation that maintains aesthetically pleasing landscapes of established plants. Sachs et al. (1975) limited irrigation frequency of established plants of eight species in central (37.2 N) and southern (33.44 N) California. All species maintained acceptable appearance with only a bimonthly fixed-volume irrigation during the summer dry season. Paine et al. (1992) used a more flexible approach, basing irrigation rates on historical ETo rather than a fixed volume and frequency. They reported acceptable plant appearance of *Photinia fraseri* and *Rhamnus californica* when irrigated at 63.8% of historical ETo, without rain, nine months after transplanting (33.56 N). Pittenger et al. (2001) extended the versatility of using ETo for triggering irrigations (33.56 N), by multiplying daily ETo by a fraction (deficit irrigation level, DI). Irrigation occurred when the cumulative sum exceeded 4 cm, replenishing deep soil moisture. Deficit irrigation levels were 20 to 50% of measured ETo. With typical fluctuations in ETo and winter rainfall with southern California's Mediterranean climate, *Baccharis, Drosanthemum,* and *Hedera* irrigated at 20% ETo and *Vinca* at 30% ETo provided acceptable or higher aesthetic quality.

More recently, Shaw and Pittenger (2004) applied their deficit irrigation method on 30 common woody shrub species beginning two years after transplanting (33.56 N). Deficit irrigation treatments were initially 0.36, 0.24 and 0.12 times ETo, but were reduced to 0.36. 0.18 and 0 for the last two years. Aesthetic ratings were the same as used by Pittenger et al. (2001). After three years, 11 of 30 species maintained accept aesthetic quality with no (DI = 0) summer- fall irrigation. Both *Hibiscus* and *Ligustrum* exhibited continuous slow declines suggesting the 0.36 *ETo was insufficient. Three other species were not successful at the 0.36 level, three more (*Cassia, Leucophyllum* & *Galvezia*) were unacceptable during the last of summer through winter months. The other 22 species maintained acceptable aesthetical quality at the 0.12 deficit level. These included several shrubs also used in Florida landscapes, which have the opposite climate, i.e. cool dry winters and wet hot summers. These species were *Calliandra, Cassia, Lantana, Leptospermum, Otatea, Pittosporum tobira, Prunus caroliniana, Pyracantha* and *Rhaphiolepis indica.*

5. Conclusions

Little research of plant water use to date is directly applicable landscape environments. Most information must be extrapolated from production or ecological research, which often contains insufficient details of plant size or microclimate. Further research with deficit irrigations will increase the list of species that can be successfully infrequently irrigated. However major advances will require quantifying the evaporative demand in urban landscapes and linking that to plant size and algorithms that predict plant water use.

6. References

Andrade, J.L.; Meinzer, F.C.; Goldstein, G.; Holbook, H.M.; Calvelier, J; Jackson, P. & Silvera, K. (1998). Regulation of Water Flux Through Trunks, Branches and Leaves in Trees of a Lowland Tropical Forest. *Oecologia*, Vol.115 pps. 463-471.

Beeson Jr., R.C. & Haydu, J.J. (1995). Economic Feasibility of Micro-Irrigating Container-Grown Landscape Plants. *Journal of Environmental Horticulture,* Vol.15, pp. 23-29.

Beeson, Jr., R.C. (2001). Water Use of Marketable-size Shrubs in Containers - A 2-year Average. *Proceeding of the Southern Nurseryman Association Research Conference,* Vol.46, pp. 592-594.

Beeson, Jr., R. C. (2004). Modeling Actual Evapotranspiration of *Ligustrum japonicum* from Rooted Cuttings to Commercially Marketable Plants in 12 Liter Black Polyethylene Containers. *Acta Horticulturae,* Vol.664, pp. 71-77.

Beeson, Jr., R.C. (2010). Response of Evapotranspiration of *Viburnum odoratissimum* to Canopy Closure and the Implications for Water Conservation during Production and in Landscapes. *HortScience*, Vol.45, pp. 359-364.

Beeson, Jr., R.C. (2011). Weighing Lysimeter Systems for Quantifying Water use and Studies of Controlled Water Stress for Crops Grown in Low Bulk Density Substrates. *Agricultural Water Management,* Vol.98, pp. 967-976.

Bréda, N.; Granier, A. & Aussenac, G. (1995). Effects of Thinning on Soil and Tree Water Relations, Transpiration and Growth in an Oak Forest (*Quercus petraea* (Matt.) Liebl.). *Tree Physiology,* Vol.15, pp. 295-306.

Berrang, P.; Karnosky, D.F. & Stanton, B.J. (1985). Environmental Factors Affecting Tree Health in New York City. *Journal of Arboriculture,* Vol.11, pp. 185-189.

Burger, D.W.; Hartin, J.S.; Hodel, D.R.; Lukaszewski, T.A.; Tjosvoid, S.A. & Wagner, S.A. (1987). Water Use in California's Ornamental Nurseries. *California Agriculture.* Vol.41, No.9/10, pp. 7-8.

Chaturvedi, A.N.; Sharma, S.C. & Srivastava, R. (1988). Water Consumption and Biomass Production of Some Forest Tree Species. *International Tree Crops Journal,* Vol.5, pp. 71-76.

Chi Y.J. (1993). Soil Compaction as a Constraint to Tree Growth in Tropical and Subtropical Urban Habitats. *Environmental Conservation,* Vol.20, pp. 35-49.

Costello, L.R.; Thomas, D. & DeVries, J. (1996). Plant Water Loss in a Shaded Environment: A Pilot Study. *Journal of Arboriculture,* Vol.22, pp. 106-108.

Dugas, W.A.; Heuer. M.L. & Mayeux. H. (1992). Diurnal Measurements of Honey Mesquite Transpiration Using Stem Flow Gauges. *Journal of Range Management*, Vol.45, pp. 99-102.

Dwyer, J.F.; McPherson, E.G.; Schroeder, H.W. & Rowntree, R.A. (1992). Assessing the Benefits and Cost of the Urban Forest. *Journal of Arboriculture*, Vol.18, pp. 227-234.

Edwards, W.R.N. (1986). Precision Weighing Lysimetry for Trees, Using a Simplified Tared-Balance Design. *Tree Physiology* Vol.1, pp. 127-144.

Garcia-Navarro, M.C.; Evans, R.Y. & Montserrat, R.S. (2004). Estimation of Relative Water Use among Ornamental Landscape Species. *Scientica Horticulturae,* Vol.99, pp. 63-174.

Gilman, E.F. (1988a). Predicting Root Spread from Trunk Diameter and Branch Spread. *Journal of Arboriculture,* Vol.14 pp. 85-88.

Gilman, E.F. (1988b). Tree Root Spread in Relation to Branch Dripline and Harvestable Root Ball. *HortScience,* Vol. 23, pp. 351-353.

Gilman, E.F. & Beeson Jr, R.C. (1996). Production Method Affects Tree Establishment in the Landscape *Journal of Environmental Horticulture*, Vol.14, pp. 81-87.

Gilman, E.F. & Watson, D.G. (1993). Ilex x 'Nellie R. Stevens': 'Nellie R. Stevens' Holly. Environmental Horticulture Department, Florida Cooperative Extension Service, Institute of Food and Agricultural Sciences, University of Florida. EDIS EHN472.

Granier, A. (1987). Evaluation of Transpiration in a Douglas-fir Stand by Means of Sap Flow Measurements. *Tree Physiolology,* Vol.3, pp. 309-320.

Green, S.R. (1993). Radiation Balance, Transpiration and Photosynthesis of an Isolated Tree. *Agricultural Forest Meteorology,* Vol.64, pp. 201-221.

Guevara-Escobar, A.; Edwards, W.R.N.; Morton, R.H.; Kemp, P.D. & Mackay, A.D. (2000). Tree Water Use and Rainfall Partitioning in a Mature Popular-Pasture System. *Tree Physiology,* Vol.20, pp. 97-106.

Hardy, J.; Behe, B.K.; Barton, S.S.; Page, T.J.; Schutzki, R.E.; Muzii, K.; Fernandez, R.T.; Haque, M.T.; Booker, J.; Hall, C.R.; Hinson, R.; Knight, P.; McNeil, R.; Rowe, D.B.; & Safley, C (2000). Consumers' Preferences for Plant Size, Type of Plant Material and Design Sophistication in Residential Landscaping. *Journal of Environmental Horticulture*, Vol.18, pp. 224–230.

Hartig, T.; van den Berg, A.E.; Hagerhall, C.M.; Tomalak, M.; Bauer, N.; Hansmann, R.; Ojala, A. ; Syngollitou, E. & Carrus, G. (2010). Health Benefits of Nature Experience: Psychological, Social and Cultural Processes. In *Forest, Trees and Human Health.* (Nilsson, K.; de Vries, S.; Sangster, M.; Seeland, K.; Gallis, C.; Schipperijn, J. & Hartig, T. eds). Springer Science+Business Media B.V. DOI 10.1007/978-90-481-9806-1_5.

Heilman, J.L.; Brittin,C.L. & Zajicek, J.M. (1989). Water Use by Shrubs as Affected by Energy Exchange with Building Walls. *Agricultural and Forest Meteorology,* Vol.48, pp. 345-357.

Heilman, J.L. & Ham, J.M. (1990). Measurement of Mass Flow Rate of Sap in *Ligustrum japonicum*. *HortScience*, Vol.25, pp. 465-467.

Kjelgren, R.; Rupp, L & Kilgren, D. (2000). Water Conservation in Urban Landscapes. *HortScience*, Vol.35, pp. 1037-1040.

Knox, G. W. (1989). Water Use and Average Growth Index of Five Species of Container Grown Woody Landscape Plants. *Journal of Environmental Horticulture*, Vol.7, pp. 136-139.

Krizek, D.T. & Dubik, S.P. (1987). Influence of Water Stress and Restricted Root Volume on Growth and Development of Urban Trees. *Journal of Arboriculture*, Vol,13, pp. 47-55.

Lindsey, P & Bassuk, N. (1991). Specifying Soil Volumes to Meet the Water Needs of Mature Urban Street Trees and Trees in Containers. *Journal of Arboriculture*, Vol.17, pp. 141-149.

McPherson, E.G.; Nowak, D.; Heisler, G.; Grimmond, S.; Souch. C.; Grant, R. & Rowntree, R. (1997). Quantifying Urban Forest Structure, Function, and Value: The Chicago Urban Forest Climate Project. *Urban Ecosystems* 1: 49-61.

McPherson, E.G.; Simpson, J.R.; Peper, P.J. & Xiao, Q. (1999). Benefit-Cost Analysis of Modesto's Municipal Urban Forest. *Journal of Arboriculture*, Vol.25, pp. 235-248.

Medhurst, J.L.; Battaglia, M. & Beadle, C.L. (2002). Measured and Predicted Changes in Tree and Stand Water Use Following High-Intensity Thinning of an 8-year-old *Eucalyptus nitens* Plantation. *Tree Physiology*, Vol.22, pp. 775-784.

Montague, T.; Kjelgren, R.; Allen, R. & Wester, D. (2004). Water Loss Estimates for Five Recently Transplanted Landscape Tree Species in a Semi-Arid Climate *Journal of Environmental Horticulture*, Vol.22, pp. 189-196.

Nobel, P.S. Physiochemical and Environmental Plant Physiology, 4th Ed. (2009). Academic Press. Elsevier Linacre House. Oxford, UK. Pps. 582 ISBN: 978-0-12-374143-1.

Nowak, D. J.; Hoehn III, R.E.; Crane, D.E.; Stevens, J.C. & Walton, J.T. (2006). Assessing Urban Forest Effects and Values, Washington, D.C.'s urban forest. *Resource Bull. NRS-1*. Newtown Square, PA: U.S. Department of Agriculture, Forest Service, Northern Research Station. 24 p.

Paine, T.D.; Hanlon, C.C.; Pittenger, D.R.; Ferrin, D.M. & Malinoski, M.K. (1992). Consequences of Water and Nitrogen Management on Growth and Aesthetic Quality of Drought-Tolerant Woody Landscape Plants. *Journal of Environmental Horticulture*, Vol.10, pp. 94-99.

Pittenger, D.R.; Shaw, D.A.; Hodel, D.R. & Holt, D.B. (2001). Responses of Landscape Groundcovers to Minimum Irrigation. *Journal of Environmental Horticulture*, Vol.19, pp. 78-84.

Regan, R. (1997). Grouping Container-Grown Plants by Water Use. *The Digger*, Vol.41, No.6, pp. 24-27.

Rose, C.W. (1984). Modeling Evapotranspiration: An Approach to Heterogeneous Communities. *Agricultural Water Management*, Vol.8, pp. 203-221.

Ruiter, J. H. (1987). Growth, Crop Conductance and Prediction of Stem Volume Increment of Irrigated and Non-irrigated Young Radiata Pine in Non-weighing Lysimeters. *Forest Ecology,* Vol.20, pp. 79-86.

Ruark, G.; Mader, D.; Veneman, P. & Tatter, T. (1983). Soil Factors Related to Urban Sugar Maple Decline. *Journal of Arboriculture,* Vol.9, pp. 1-4.

Sach, R.M.; Kretchun, T. & Mock, T. (1975). Minimum Irrigation Requirements for Landscape Plants. *Journal of the American Society for Horticultural Science,* Vol.100 pp. 499-502.

Sanders, R.A. (1984). Urban Vegetation Impacts in the Urban Hydrology of Dayton Ohio. *Urban Ecology* 9: 361-376.

Schulze, E.D.; Cremak, J.; Matyssek, R.; Penka, M.; Zimmermann, R.; Vasicck, F.; Gries, W. & Kucera, J. (1985). Canopy Transpiration and Water Fluxes in the Xylem of the Trunk of Larix and Picea trees - A Comparison of Xylem Flow, Porometer and Cuvette Measurements. *Oecologia,* Vol.66, pp. 475-483.

Shaw, D.A. & Pittenger, D.R. (2004). Performance of Landscape Ornamentals Given Irrigation Treatments Based on Reference Evapotranspiration. *Acta Horticulturae,* Vol.664. pp. 607-614.

Steinberg, S.L.; van Bavel, C.H.M. & McFarland, M.J.. (1990a). Improved Sap Flow Gauge for Woody and Herbaceous Plants. *Agronomy Journal,* Vol.82, pp. 851-854.

Steinberg, S.L.; McFarland, M.J. & Worthington, J.W. (1990b). Comparison of Trunk and Branch Sap Flow with Canopy Transpiration in Pecan. *Journal of Experimental Botany,* Vol.41, pp. 654-659.

Steinberg, S.L.; Zajicek, J.M. & McFarland, M.J. (1991). Short-term Effect of Uniconazole on the Water Relations and Growth of *Ligustrum*. *Journal of the American Society for Horticultural Science,* Vol.116, pp. 460-464.

Stigarll, A. & Elam, E. (2009). Impact of Improved Landscape Quality and Tree Cover on the Price of Single-Family Homes. *Journal of Environmental Horticulture,* Vol.27, pp. 24-30.

Stohr, A. & Losch, R. (2004). Xylem Sap Flow and Drought Stress of *Fraxinus excelsior* saplings. *Tree Physiology,* Vol.24, pp. 169-180.

Ulrich, R.S. (1986). Human Responses to Vegetation and Landscapes. *Landscape and Urban Planning,* Vol.13, pp. 29-44.

Verbeeck, H.; Steppe, K.; Nadezhdina, N.; Op De Beeck, M.; Deckmyn,G.; Meirsoone,L; Lemeur, R.; Cermak, J.; Ceulemans, R. & Janssens, I.A. (2007). Stored Water Use and Transpiration in Scots Pine: A Modeling Analysis with ANAFORE. *Tree Physiology,* Vol.27, pp. 1671-1685.

Vertessy, R.A.; Hatton, T.J.; Reece, P.; O'Sullivan, S.K. & Benyon, R.G. (1997). Estimating Stand Water Use of Large Mountain Ash Trees and Validation of the Sap Flow Measurement Technique. *Tree Physiology,* Vol.17, pp. 747-756.

Watson, W.T. (2005). Influence of Tree Size on Transplant Establishment and Growth. *HortTechnology,* Vol.15, pp. 118-122.

Westphal, L.M. (2003). Urban Greening and Social Benefits: A Study of Empowerment Outcomes. *Journal of Arboriculture.* Vol.29, pp. 137-147.

Whitlow, T.H.; Bassuk, N.L. & Reichert, D.L. (1992). A 3-Year Study of Water Relations of Urban Street Trees. *Journal of Applied Ecology*, Vol.29, pp. 436-450.

Wullschleger, S.D.; Meinzer, F.C. & Vertessy, R.A. (1998). A Review of Whole-plant Water Use Studies in Trees. *Tree Physiology*. Vol.18, pp. 499-512.

Zhang, H.; Morison, J.I.L. & Simmonds, L. P. (1999). Transpiration and Water Relations of Poplar Trees Growing Close to the Water. *Tree Physiology*, Vol.19, pp. 563-573.

18

Interannual Variation in Transpiration Peak of a Hill Evergreen Forest in Northern Thailand in the Late Dry Season: Simulation of Evapotranspiration with a Soil-Plant-Air Continuum Model

Tanaka K.[1], Wakahara T.[2], Shiraki K.[2], Yoshifuji N.[3] and Suzuki M.[4]
[1]*Japan Agency for Marine-Earth Science and Technology*
[2]*Tokyo University of Agriculture and Technology*
[3]*Kyoto University*
[4]*The University of Tokyo*
Japan

1. Introduction

Northern Thailand, which experiences rainy and dry seasons under an Asia monsoon climate, is characterized by hilly and mountainous landscapes. The rainfall tends to increase with altitude (Kuraji et al., 2001; Dairaku et al., 2004). Forests in northern Thailand at 1000 m above sea level (a.s.l.) are classified as lower montane rain forests (Santisuk, 1988). These areas receive high amounts of precipitation and provide a stable supply of high-quality water that is crucial for irrigation and drinking water (Bruijnzeel et al., 2011). Generally, water resources or stream flow are estimated by the difference between precipitation and evapotranspiration (i.e., the sum of canopy interception, soil evaporation, and transpiration). Thus, it is important to examine how forests consume rainwater as evapotranspiration, in conjunction with hydrological and meteorological variables. Such modeling is also essential for water resource management.

This study is a continuation of previous studies of transpiration peaks in an evergreen forest in northern Thailand (18°48′N, 98°54′E, 1265–1450 m a.s.l.) in the late dry season (Tanaka et al., 2003, 2004). Tanaka et al. (2003) concluded that transpiration in evergreen forests peaked in the late dry season. They suggested that reduced canopy wetness lowered evaporation; however, stomatal conductance declined only slightly, even under the driest conditions and highest net radiation. These results counter previous reports of an evapotranspiration decline in Thailand's dry season in evergreen forests (Pinker et al., 1980) and other vegetation (Toda et al., 2002). Tanaka et al. (2004) examined the impact of rooting depth and soil hydraulic properties on forest transpiration using a newly developed soil–plant–air (SPAC) multilayer model. They found that a rooting depth of 4–5 m was needed to effectively simulate heat-pulse velocity variations corresponding to dry-season transpiration and annual discharge or stream flow. Moreover, a penetration test showed that the soil became harder at depths of 4–5 m, supporting the estimated rooting depths. Numerical simulations indicated that a late dry season transpiration peak is theoretically possible on the basis of rooting depth limitations on soil water use because the rooting depth was within the reported maximum for trees. Canadell et al. (2006) reviewed numerous reports of maximum rooting depth and calculated the average and standard deviation as 7.0 ± 1.2 m. Although these studies emphasized the late-dry-season transpiration peak and its mechanism, using combinations of modeling and observation for 2–3 years periods, the impact of interannual variation in rainfall or dry season period length on the peak was not sufficiently examined.

Our objective was to clarify the interannual variation in the late-dry-season transpiration peak in a hill evergreen forest in northern Thailand. A numerical simulation of the seasonal variation in evapotranspiration was performed using a SPAC multilayer model with hydrometeorological variables for the period 1999–2005. The heat pulse velocity corresponding to water use by individual trees was monitored and evapotranspiration was estimated from the water budget (i.e., the difference between rainfall and stream flow) for the 7-year period. These values were compared with the simulated temporal transpiration and annual evapotranspiration.

2. Materials and methods

2.1 Site

Since February 1997, the hydrological and meteorological parameters of a sub-watershed of the Kog-Ma Experimental Watershed have been measured. The sub-watershed has an area of 8.63 ha and is situated 1265–1420 m a.s.l. on Mount Pui (18°48′ N, 98°54′ E) near Chiang Mai (18°47′ N, 98°58′ E, 310 m absl.) in northern Thailand. A 50-m meteorological tower was built in the sub-watershed and equipped with instruments for measuring meteorological parameters, such as radiation, wind velocity, and air temperature. Evergreen forest covers the hills of the experimental watershed. *Fagaceae* dominates, with species including *Lithocarpus*, *Quercus*, and *Castanopsis* (Bhumibhamon & Wasuwanich, 1970). Northern Thailand is characterized by hilly and mountainous landscapes, and rainfall tends to increase with altitude (Kuraji, 2001). Dairaku et al. (2004) reported that the larger amount of high-altitude rainfall was due to duration and frequency rather than intensity, which implies that the appearance of clouds is more frequent at higher altitudes. The forest is lower montane rain forest according to the classification of Santisuk (1988). This area receives more rainfall than areas at lower altitudes such as Chiang Mai, with annual precipitation of 1183 mm for 1960–1990 (Thai Meteorological Department, 2011). It is cloudier, particularly in the rainy season, and occasionally experiences fog (N. Tanaka et al., 2011). Average annual rainfall and air temperature for a 7-year period (1999–2005) were 1881 mm and 19.8°C, respectively. The leaf area index (LAI) is approximately 4.5, with a seasonal range from 3.5 to 5.5 (Tanaka et al., 2003). Forest floor soils derive from granitic materials and are classified as Reddish Brown Lateritic (Tangtham, 1974).

Stream flow in the sub-watershed was measured at a concrete weir with a 90° triangular notch. Data were missing for 164 days in 2002, 2004, and 2005, representing 6.41% of the total days in the period 1999–2005. Stream flow on these days was estimated by data assimilation with a river flow model (Fukushima, 1988). The rainfall data and measured stream flow were used as input and output data, respectively, and the data for 10 (Sep. in 2001–Jun. in 2002), 3 (Oct.–Dec. in 2002), 8 (Jan.–Aug. in 2004), and 12 months (Jan.–Dec. in 2005) around the missing data were used to assimilate the stream flow data.

The heat pulse velocity corresponding to water use by an individual tree was monitored in three tree trunks (No. 1: *Phoebe paniculata*. Nos. 2 and 3: *Lithocarpus elegans*). The height and diameter of the three trees at 1.2 m were 28.0 m and 0.51 m, 23.0 m and 0.29 m, and 15.5 m and 0.20 m, respectively, in 1999–2005 (Tanaka et al., 2003). The observation of heat pulse velocity near a ridge, where no water table seemed to form, showed that the water use (or sap flow) of individual trees had a seasonal trend similar to those of the three trees (Tanaka et al., 2004). These trees belonged to the uppermost or second story. Therefore, the water use by these trees should reflect the transpiration over the forest as a whole (e.g., Kelliher et al., 1992; Tanaka et al., 2003) because transpiration from the upper layers is thought to contribute most of the total transpiration. Here, measured seasonal and interannual changes were used to validate the simulated transpiration.

2.2 A one-dimensional SPAC multilayer model for evapotranspiration

We used a one-dimensional SPAC multilayer model (Tanaka & Hashimoto, 2006) consisting of a soil multilayer model (Kondo & Xu, 1997) and a canopy multilayer model (Tanaka et al., 2003; Fig. 1). The soil multilayer model considers the variation in albedo and evaporation efficiency with changes in soil moisture at the top of the soil column (Kondo & Xu, 1997). The canopy multilayer model (Tanaka et al., 2003) for sensible and latent heat and CO_2 gas exchange consists of a second-order closure model for atmospheric diffusion coupled with a radiation transfer model (Tanaka, 2002), a rainfall interception model (Tanaka, 2002), a Farquhar-type photosynthesis model (Farquhar et al., 1980), and a stomatal conductance model (Ball, 1988). The rainfall interception model assumes that rainfall does not wet the lower sides of leaves with stomata, only the upper sides without stomata, while condensation wets both sides. In the photosynthesis model, the maximum potential rate of electron transport and dark respiration at 25°C (J_{MAX25} and R_{d25}) were scaled to V_{cMAX} at 25°C (V_{cMAX25}); that is, $J_{MAX25} = 2.14\ V_{cMAX25}$ after Tanaka et al. (2002) and $R_{d25} = 0.015\ V_{cMAX25}$ after Collatz et al (1991). The assumptions, functions, and procedures in the calculations of the gross CO_2 assimilation rate A, dark respiration R_d, and stomatal conductance g_s were described by Tanaka et al. (2002).

Combined, the two multilayer models by Kondo and Xu (1997) and Tanaka *et al.* (2003) consider the loss of soil moisture by water uptake (or transpiration) and the effect of soil water content on stomatal closure (Tanaka et al., 2004). The water uptake at depth z was assumed to be proportional to the ratio of the extractable to the entire extractable soilwater content (W_e; Tanaka et al., 2006). The sum of the water uptakes corresponds to the temporal canopy transpiration. When canopy transpiration can be supplied by the entire extractable soilwater content at 0–1m depths (see Case A in Fig. 1), where the major plant nutrients C, N, P, and K appear to be concentrated (Jackson et al., 2000), W_e is calculated between the depths of 0–1m from which the water uptake is supplied. In the other case, W_e is calculated as the extractable soilwater content between 1m and the maximum rooting depth Z_{ROOT} (here, = 4 m) (see Case B in Fig. 1), and the water uptake is supplied from soil layers at 1m to Z_{ROOT}. The water uptake at depth z, regardless of the vertical root distribution, is expressed as

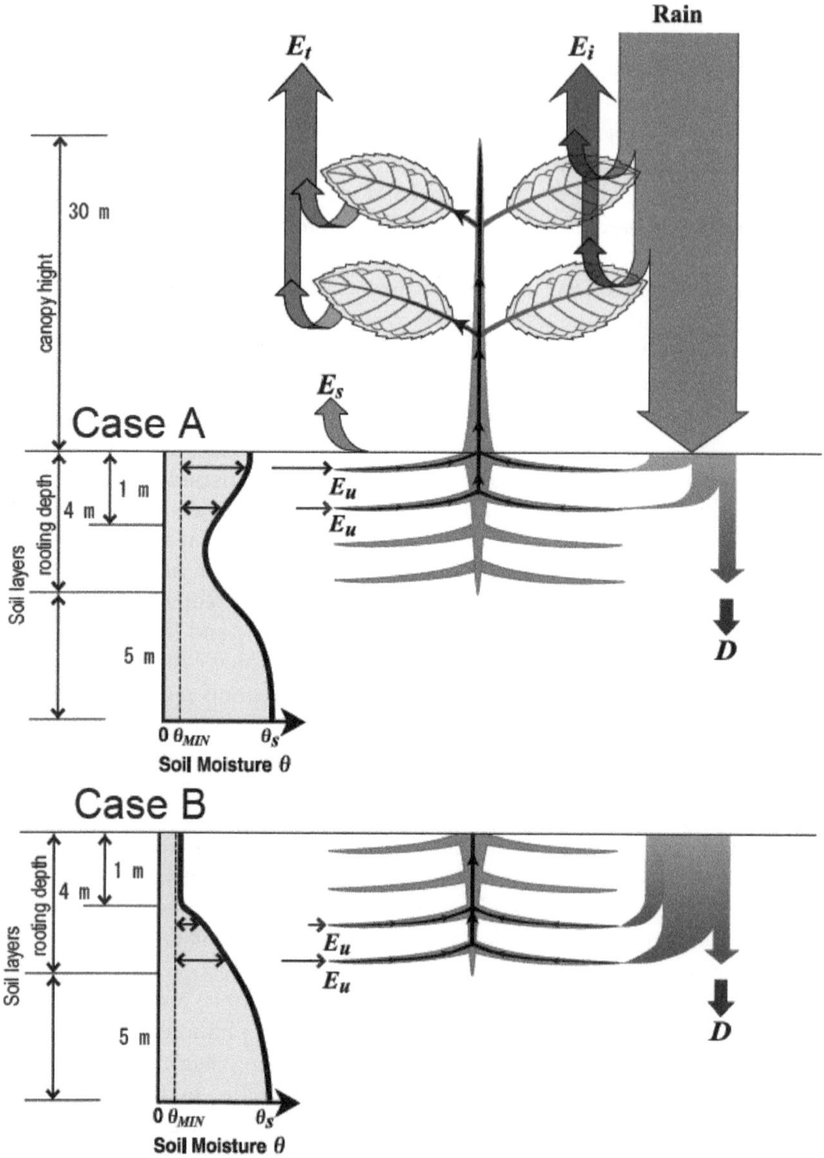

Fig. 1. A one-dimensional soil–plant–air continuum multilayer model for evapotranspiration (i.e., the sum of canopy transpiration E_t, canopy interception E_i, and soil evaporation E_s). Discharge D was calculated as the downward water flux passing through the rooting depth (4 m), and volumetric soil moisture at the bottom of the soil layer (9 m) was set to the saturated volumetric soil moisture θ_s. E_u and θ_{MIN} are the water uptake by roots and soil moisture, respectively, at the lower limit of soil water potential ψ_{LL} (–100 m), where trees cannot take up water. Case A shows water uptake when canopy transpiration can be

supplied by the entire extractable soilwater content at 0–1m depths, while Case B shows water uptake in the other case. E_u at depth z was assumed to be proportional to the ratio of the extractable soil water content (i.e., $\theta(z)$ - θ_{MIN}) to the entire extractable soil water content W_e at 0–1m depths (Case A) or from 1m to the maximum rooting depth Z_{Root} (= 4 m) (Case B) (Tanaka et al., 2006). The canopy height was set to 30 m (Tanaka et al., 2003).

$$E_u(z) = E_t \frac{(\theta(z) - \theta_{MIN})}{W_e} \tag{1}$$

Here θ_{MIN} is the volumetric soil water content at which trees cannot take up water. This corresponds to the value at the upper limit of the soil water potential (ψ_{UL} = -100 m). W_e in Case A or B is expressed as

$$W_e = \int_0^1 (\theta(z) - \theta_{MIN})dz \text{ , or } W_e = \int_1^{Z_{ROOT}} (\theta(z) - \theta_{MIN})dz \tag{2}$$

This assumption of water uptake is simple compared to another frequently used weighting scheme (e.g., Dickinson et al., 1993; Desborough, 1997) based on the assumption that the root length density distribution is proportional to water extraction throughout the profile. Radersma and Ong (2004) did not find a clear relationship between root length density and water extraction. Other researchers have questioned the various proposed relationships between root length density and water uptake (Dardanelli et al., 2004). These findings suggest that the process of water uptake by roots is not entirely clear. Therefore, we used a simpler assumption.

Stomatal conductance was assumed to decrease with the ratio R_{We} of integrated extractable water content W_e from the surface to the rooting depth (i.e., $W_e = \int_0^{Z_{ROOT}} (\theta(z) - \theta_{MIN})dz$) to the integrated extractable water content at saturation from the surface to the rooting depth $W_{es} = (\theta_s - \theta_{MIN})Z_{ROOT}$, given as

$$g_s = f(R_{We}) \cdot g_{sW} \tag{3}$$

where g_{sW} is the stomatal conductance in well-watered soil and $f(R_{We})$ is a function of the ratio $R_{We} = W_e/W_{es}$ ranging from 0 to 1. The function $f(R_{We})$ was calculated as

$$f(R_{We}) = \min\left[1.6R_{We} + 0.2;1\right] \text{ at } R_{We} > 0 \tag{4}$$

$$f(R_{We}) = 0 \text{ at } R_{We} = 0 \tag{5}$$

Equation (4), including the values of the slope and intercept, is close to the relationship between the extractable water content and stomatal conductance shown by Gollan et al. (1985).

In the canopy multilayer model, the evapotranspiration depends on the LAI, the slope m in Ball's stomatal conductance model (Ball, 1988), and V_{cmax} at 25°C (V_{cmax25}) in a Farquhar-type photosynthesis model (Farquhar et al., 1980). These parameters are based on the estimated LAI and determined by referring to the measured net photosynthesis rate and stomatal conductance for single leaves (Tanaka et al., 2003). The values were set at 4.5, 10, and 25 μmol m^{-2} s^{-1}, respectively. The vertical profile of the LAI is also a required parameter. It was assumed to obey a beta distribution, with the greatest leaf area density at 0.7 times the canopy height (B-type canopy; see Figure 6 of Tanaka et al. 2003). Tanaka et al. (2003) investigated the impact of each parameter on evapotranspiration. Kondo and Xu (1997) verified the method by comparing observed and calculated results for four soil textures (i.e., volcanic ash, clay loam, silty sand, and sand). Silty sand was selected as the sub-watershed soil texture, whose observed relationship between volumetric soil water content θ and soil water potential ψ was close to that in the model (Tanaka et al., 2004). The soil and rooting depth were set at 9 and 4 m, respectively (Tanaka et al., 2004). Kondo and Xu (1997) and Tanaka et al. (2003) detailed the other parameter values used in the simulation.

The canopy (height = 30 m) was divided into 50 layers. Each soil layer was 0.1 m thick. The time interval was set at 3 min in the soil multilayer model because of the thin soil layers (0.1 m), but it was set at 15 min in the canopy multilayer model. The model simulated soil evaporation E_s, canopy interception (i.e., evaporation from a wet canopy) E_i, transpiration E_t, discharge D, and soil moisture. The profiles of all the meteorological elements were calculated repeatedly among all sub-models until the differences between the previous and new values of leaf temperature, air temperature, humidity, ambient CO_2 concentration, downward and upward longwave radiation, and water storage on both upper and lower sides of the leaves were within 1% (Tanaka, 2002). The maximum number of repetitions was set at 100 (Tanaka, 2002). Here, D was calculated as the downward water flux passing through the rooting depth (Fig. 1). The initial soil moisture condition at the beginning of 1998 calculated by Tanaka et al. (2004) was used here. The initial soil moisture condition was calculated repeatedly until it corresponded to the value at the end of 2000 in the study by Tanaka et al. (2004). This implies that the total rainfall was used as E_t, E_i, and E_s, and discharged without changing into stored soil water between the beginning and end of the 3-year period (1998–2000). Soil moisture at the bottom of the soil layer (= 9 m; Fig. 1) was set to the saturated soil moisture θs. This initial condition did not consider the impact of the decrease in rainfall in the rainy season in 1997 caused by the 1997–1998 El Niño (Wang & Weisberg, 2000). The initial soil moisture appeared greater because of the impact of more rainfall in 2001. Heat pulse velocity was not monitored in the late dry season in 1998. Therefore, simulation results for 1999-2005 are shown here. The initial soil moisture at the beginning of 1999 was calculated using hydrometeorological variables in 1998.

3. Results

Figure 2 shows seasonal and interannual temporal variations in hydrometeorological variables in 1999–2005. The study area has three seasons in terms of air temperature and rainfall changes: a rainy season and early and late dry seasons (Tanaka et al., 2003). The light gray, gray, and black bars in Fig. 2a indicate the point 30 days before the day when the rainfall amount exceeded 150 mm (i.e., the wet period; WP) (Fig. 2b), the days excluding those in the WP whose following 5 days had mean air temperatures below 21°C (i.e., the cool dry period; CDP), and the days excluding those in the WP whose following 5 days had average values of air temperature over 21°C (i.e., hot dry period; HDP), respectively. The horizontal bars in Fig. 2b show the points at which the 90 previous days had less than 50 mm of total rainfall (i.e., a drought condition; DC). The CDP was concentrated in the early dry season, while the HDP was concentrated in the late dry season. The HDP occasionally appeared in the early dry seasons, in much shorter periods than in the late dry seasons. The

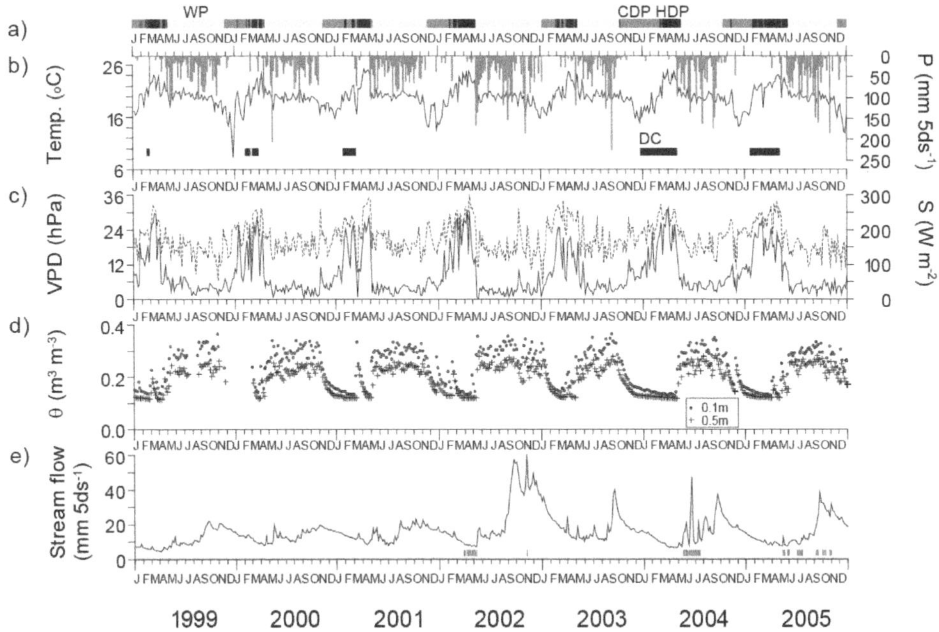

Fig. 2. (a) Hydrometeorological periods in 1999–2005. Light gray, gray, and black bars show the days on which the amount of rainfall 30 days previously was over 150 mm (i.e., the wet period; WP), the days excluding those in the WP for which the following 5 days had average air temperature values below 21°C (i.e., the cool dry period; CDP), and the days excluding those in the WP for which the following 5 days had average air temperature values over 21°C (i.e., the hot dry period; HDP), respectively. Seasonal changes in air temperature and rainfall *P* (b), vapor pressure deficit and downward solar radiation *S* (c), volumetric soil moisture θ at depths of 0.1 and 0.5 m (d), and stream flow (e) are shown for 1999–2005. The horizontal bars in (b) show the days for which the previous 90 days had less than 50 mm of total rainfall (i.e., drought conditions; DC). The horizontal bars in (e) show the days without measurements of stream flow. During the days with missing data, stream flows were estimated by data assimilation using a river flow model (Fukushima, 1988). The shaded areas in (e) correspond to the data assimilation periods.

lengths of both the CDP and HDP changed interannually, being longest in 2004–2005 and shortest in 2002–2003. In the longest dry season, the DC period was longest. The DC period did not appear in the 2001–2002 and 2002–2003 dry seasons. The annual amount of rainfall was smallest in 2003 (= 1504 mm) and largest in 2002 (= 2458 mm). The vapor pressure deficit (VPD) and downward solar radiation S peaked in the HDP (Fig. 2c), indicating the strongest atmospheric evaporative demand. These quantities were lower in the WP. The VPD peak was the lowest in the HDP of 2003 because the shortest dry season was in 2002–2003. The solar elevation at noon and the day length were higher and longer, respectively, in the rainy season, but the frequent appearance of clouds modified the less intense solar radiation. The volumetric soil moisture values at 0.1 and 0.5 m were also lowest in the late dry season (Fig. 2d). The duration of the driest conditions increased with the DC period, particularly in 2004. The stream flow was never interrupted, even in the late dry season (Fig. 2e). This implies that the deeper soil portion was still moist even though the soil surface layer at depths of 0.1-0.5 m was dry. The peak stream flow appeared in the late rainy season or at the end of WP (e.g., September-November). The peak value was the largest in 2002 due to the rainfall amount. The horizontal bars in Fig. 2e show the days without stream flow measurements. For days with missing data, stream flows were complemented by data assimilation with a model for river flow (Fukushima, 1988). The shaded areas in Fig. 2e correspond to the data assimilation periods. The total measured and modeled stream flows were 3025 mm and 3040 mm, respectively, for 835 days during the periods with measured stream flow. The total amount of complemented stream flow was 498 mm for the 164 days with missing data, and the estimated error was extremely small (several millimeters).

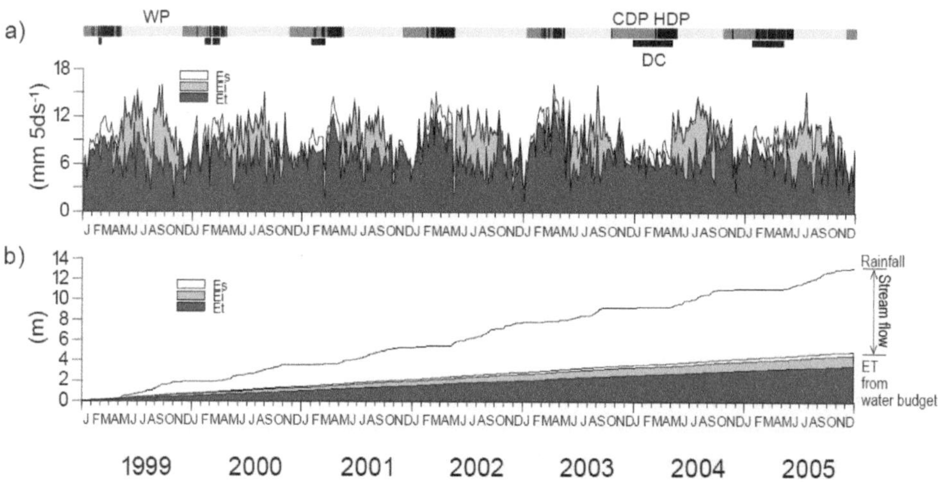

Fig. 3. (a) Simulation of the seasonal variation in evapotranspiration (i.e., the sum of soil evaporation E_s, canopy interception or evaporation from the wet canopy E_i, and transpiration E_t) in 1999-2005. (b) The cumulative result. Light gray (wet period; WP), gray (cool dry period; CDP), and black bars (hot dry period; HDP) in (a) correspond to those in Fig. 2a. The shorter black bar (drought condition; DC) also corresponds to that in Fig. 2b.

Figure 3a shows the simulation results of evapotranspiration ET (i.e., the sum of soil evaporation E_s, canopy interception E_i, and transpiration E_t) in 1999-2005. The E_i appeared to increase with rainfall in the rainy seasons. In particular, the ratio of E_i to ET was close to half of the ET in 2002, the year with the largest rainfall. The E_t peaked in the HDP in 1999-2003, particularly in 2002 and 2003, but the E_t in the HDP almost equaled that in the WP in 2004 and 2005. The simulated ET maintained large values during the HDP and WP in 1993-2003, although the ET was smaller in the HDP in 2004 and 2005. The simulated ET values were smaller in the CDP due to weaker atmospheric conditions (lower temperature, lower VPD, and less intensive solar radiation due to the decline in solar elevation), even though the soil was wetter. The simulated E_s was small under humid conditions within a canopy due to both the lower VPD and the wetter soil, while E_s increased in the HDP and/or the

DC. Figure 3b shows the cumulative results from Fig. 3a. The annual amounts of simulated ET, E_s, E_i, and E_t were 707, 49, 151, and 507 mm yr^{-1}, respectively, for the 7-year period. The closed canopy reduced E_s, and almost all E_i disappeared outside the rainy season. Thus, the percentages of E_s, E_i, and E_t within ET were 7, 21, and 72% in the 7-year period, respectively. Assuming a negligible difference in the storage of soil moisture between the beginning of 1999 and the end of 2005, the annual ET was 694 mm yr^{-1}, the difference between rainfall (1881 mm yr^{-1}) and stream flow (1187 mm yr^{-1}). The ET value from the water budget was very close to the simulated ET. The error in stream flow estimated by data assimilation appeared to be negligible (several millimeters over 7 years), as the above-mentioned.

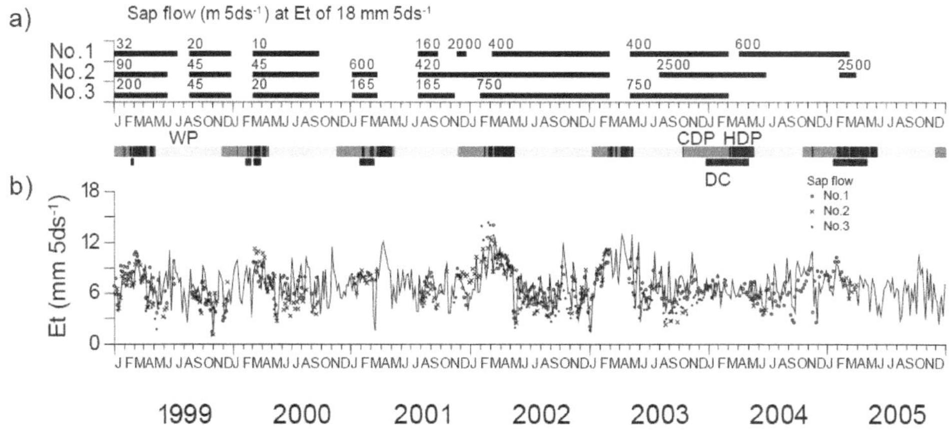

Fig. 4. (a) Continuous period of measurement of heat pulse velocity of three trees (number 1, *Phoebe paniculata*; numbers 2 and 3, *Lithocarpus elegans*) using the same sensor. The values in the figure correspond to the heat pulse velocity (m 5ds^{-1}) at 18 mm per 5 days of transpiration. (b) Simulation of the seasonal variation in transpiration E_t and heat pulse velocity of the three trees in 1999–2005. Light gray (wet period; WP), gray (cool dry period; CDP), and black bars (hot dry period; HDP) in (b) correspond to those in Fig. 2a. The shorter black bar (drought condition; DC) also corresponds to that in Fig. 2b.

Figure 4 shows seasonal and interannual variations in the simulated E_t and heat pulse velocity of the three trees (number 1, *Phoebe paniculata*; numbers 2 and 3, *Lithocarpus elegans*). The horizontal bars in Fig. 4a indicate the duration of the measurement of heat pulse velocity for each tree with the same sensor and position. The values above the bars correspond to the heat pulse velocity (m s^{-1}) at 18 mm per 5 days of transpiration. These changes depended on where and how deep the probe and sensor were placed in the trunks (Phillips et al., 1996; Takizawa et al., 1996; James et al., 2002). Probes and sensors were inserted in the trunks. The simulated E_t captured the variation in heat pulse velocity. Both E_t and the heat pulse velocities exhibited late-dry-season transpiration peaks in 1999, 2000, and particularly 2002. However, the peaks were smaller in 2003 and 2004 and almost the same as E_t in the rainy season.

Figure 5 shows the simulated mean±1 standard deviation values of daily transpiration during the HDP (see Fig. 4b) in the dry season from 1999–2000 to 2004–2005. The gray, black, and narrower black bars denote the number of days in the CDP, HDP, and DC, respectively, in the dry season from 1999–2000 to 2004–2005. Daily transpiration in the HDP was the smallest and second smallest in the 2003–2004 and 2004–2005 periods, respectively. These periods had longer dry seasons; in particular, the longest CDP and HDP were in 2004–2005. There were also 54 and 78 days of DC in the 2003–2004 and 2004–2005 periods, respectively. The longer period of atmospheric strong evaporative demand and insufficient soil moisture decreased the transpiration in the HDP. The transpiration in the HDP in 2004–2005 was larger than that in 2003–2004, although both the HDP and DC were longer in 2004–2005 than in 2003–2004. In the shortest dry season without DC (for which the sum of CDP and HDP was 118 days; 2002–2003), the daily transpiration was the largest.

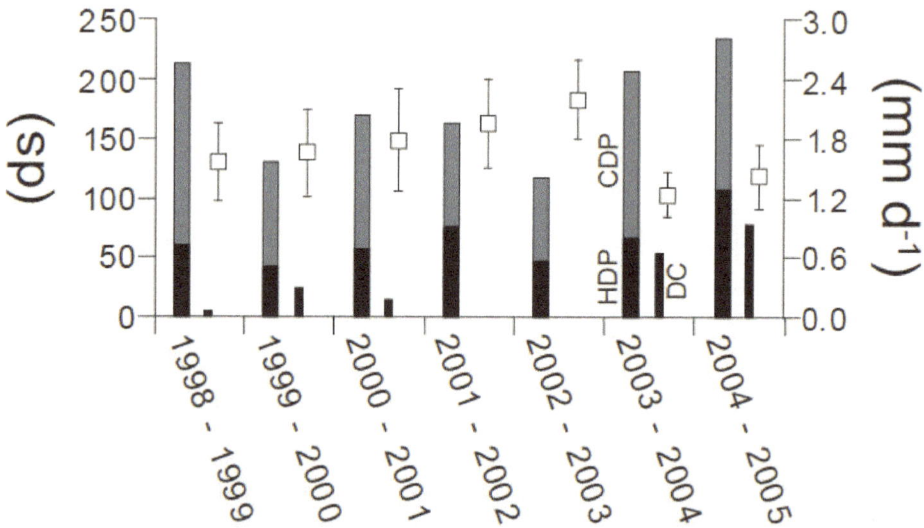

Fig. 5. Simulated mean±1 standard deviation values of daily transpiration E_t during the hot dry period (HDP, see Fig. 4b) in the dry season from 1999–2000 to 2004–2005. Gray, black, and narrow black bars show the number of days of cool dry period (CDP), HDP, and drought condition (DC), respectively, in the dry season from 1999–2000 to 2004–2005.

4. Discussion

The SPAC multilayer model (Tanaka & Hashimoto, 2006) was used to simulate the interannual variation in *ET* at a hill evergreen forest in northern Thailand between 1999 and 2005. The simulated annual *ET* was close to the difference between rainfall and stream flow (i.e., the *ET* from the water budget) during the 7-year period. The simulated transpiration E_t captured the measured heat pulse velocity corresponding to water use by an individual tree, particularly the decrease in the late-dry-season E_t in 2004 and 2005. The assumption that the decrease in extractable soil moisture had reduced impact on stomatal closure (i.e., equation 4) thus appeared to be appropriate for the estimation of transpiration at the forest canopy level. To confirm this assumption, the model should be applied to data for different vegetation under drought conditions. Tanaka et al. (2009) found that this assumption was reasonable for modeling deciduous teak plantations with shallower rooting depth under drought conditions in northern Thailand. Although the simulated canopy interception E_i was not validated here, Tanaka et al. (2003) simulated the E_i of the forest in 1998–1999 using a canopy multilayer model (Tanaka, 2002) but no soil multilayer model (Kondo & Xu, 1997). They compared the simulated E_i with that estimated from the difference between rainfall and the sum of throughfall and stemflow and found that the annual values were close. Although the simulated soil evaporation E_s was also not validated, E_s likely occupied a small portion of the ET due to the decline in downward solar radiation and wind velocity because of the closed canopy, as in the simulation (Fig. 3; the ratio of E_i to ET is annually 7%). Thus, this study is the first to reveal features of the interannual variation in ET (i.e., the sum of E_s, E_i, and E_t).

The simulated annual ET (= 707 mm yr^{-1}) was smaller than that estimated from the water budget (= 694 mm yr^{-1}) by 13 mm. The smaller LAI decreased E_t and E_i and increased E_s (Tanaka et al., 2009). Smaller values of V_{cMAX25} or shallower rooting decreased E_t and changed little both E_i and E_s within the model (Tanaka et al., 2004, 2009). These impacts on ET were larger in the late dry season and the HDP, when atmospheric evaporation demand was stronger than in the rainy season because of the higher temperatures and VPD and the intensive solar radiation (Tanaka et al., 2003, 2004). Therefore smaller values of V_{cMAX25} or shallower rooting depths are required for simulated ET to become closer to that estimated from measurements. Then the seasonal and interannual variation in ET would show slightly smaller E_t values in the HDP, and neither E_s nor E_i would change significantly. In tuning with smaller LAI values, the decrease in E_i and E_t and increase in E_s must be considered.

The simulated E_t and monitored heat pulse velocity showed that the late-dry season E_t peak was lower in 2004 and 2005 (Fig. 4). The rooting depth was set at 4 m. This is deeper than the 1 m depth often used in land surface models within general circulation models (Kleidon & Heimann, 1998). Tanaka et al. (2004) noted that, during dry periods, rainwater remaining from the previous rainy season may be sufficient for trees with greater water use capacities. The longest DC periods, however, as well as the longest dry season lengths in 2003–2004 and 2004–2005, largely limited E_t in the late dry season beyond this larger water use capacity. Kume et al. (2007) compared heat pulse velocities (or sap flow) among two large trees corresponding to Nos. 1 and 2 in this study (Fig. 4b) and two smaller trees (4.8 m and 1.4 m height) in the study forest in 2003 and 2004. They found that the reduced impact of soil drought on sap flow was clearer in the smaller trees in the late dry season in 2004, likely due to their shallower rooting depths. They suggested that the larger trees might avoid water uptake limitations with their deeper roots. Furthermore, transpiration over the whole forest canopy could also have been limited in the late dry season in 2004 and 2005 (Figs. 4b and 5).

The annual ET (= 694 mm) was small compared with values reported for other tropical and sub-tropical forests (e.g., Doley 1981); in the latter, values often exceed 1000 mm, with maxima of 1750 mm, and the ratio of annual discharge to annual rainfall exceeds that of the annual ET. Air temperature decreases with altitude, while rainfall tends to increase with altitude (Kuraji, 2001; Dairaku et al., 2004) in northern Thailand, and the downward solar radiation decreases due to the frequent appearance of clouds during the rainy season (Fig. 2c). Under the weaker evaporative demand in the rainy season, most rainwater probably infiltrates the soil layers with relatively smaller *ET* values. This rainwater is likely used by evergreen trees with deeper roots, even in the following dry season. Such trees can continuously transpire, control their leaf temperature, and assimilate carbon, although late-dry-season transpiration peaks likely decreased in 2004 and 2005. Hence the trees can maintain leaves all year round. The population of deciduous trees increases as altitude decreases below 1000 m a.s.l. in northern Thailand (Santisuk, 1998). Tanaka et al. (2009) numerically simulated canopy net assimilation A_n, *ET*, and soil moisture in a deciduous teak plantation with shallow rooting depth (< 1 m) in a dry tropical climate in northern Thailand (18°25′N, 99°43′; 380 m a.s.l.) using the SPAC multilayer model. That site had annual rainfall of 1361 mm for 2001–2008 and higher annual temperature of 25.4°C (K. Tanaka et al., 2011). The first experiment in that study involved seasonally varying LAI estimates based on time-series of radiative transmittance through the canopy, and the second experiment applied an annually constant LAI. The first simulation captured the measured seasonal changes in soil surface moisture; the simulated transpiration agreed with seasonal changes in heat pulse velocity, corresponding to the water use of individual trees. The simulated A_n almost always became positive during leafy seasons. The simulated annual *ET* was ~900–1200 mm. However, in the second simulation in the dry season, A_n and E_t became negative and small, respectively, because the decline in stomatal conductance due to severe soil drought limited the assimilation. The simultaneous increase in leaf temperature increased dark respiration. These experiments revealed that leaflessness in the dry season is reasonable for carbon gain, and trees cannot maintain leaves year round at the site. Therefore, it may be more difficult (easier) for trees to maintain leaves in the late dry season as altitudes decrease (increase) in northern Thailand.

5. Conclusion

The E_t simulated with the SPAC multilayer model and heat pulse velocities indicated that the late-dry-season transpiration peak weakened in 2004 and 2005, even with an assumed rooting depth of 4 m. The 2003–2004 and 2004–2005 dry seasons were relatively longer, and they had the second longest (= 67 days) and longest (= 108 days) DC days. The soil moisture likely became insufficient beyond the rooting depth limitations on soil water use because of the duration of drought conditions along with the stronger atmospheric evaporative demand.

6. Acknowledgments

We are grateful to Nipon Tangtham and Chatchai Tantasirin of Kasetsart University for their helpful suggestions. We also thank Nobuaki Tanaka of the University of Tokyo, Tomonori Kume of the National Taiwan University, and Hideki Takizawa and Izumi Kosaka of Nihon University for their help with the measurements. We would also like to thank Fujio Kimura of Japan Agency for Marine-Earth Science and Technology (JAMSTEC), Tetsuzo Yasunari of Nagoya University/JAMSTEC, and our colleagues at JAMSTEC, in particular those in the Advanced Atmosphere-Ocean-Land Modeling Program, who gave K. Tanaka such valuable comments.

7. References

Ball, J. T. (1988), An Analysis of Stomatal Conductance. Ph.D. Thesis, Stanford University, CA, 89 pp.

Bhumibhamon, S. & Wasuwanich, P. (1970), The Structural Characteristics of Huay Kog-Ma Forest, *Kog-Ma Watershed Research Bulletin*, pp. 20, Department of Conservation, Faculty of Forestry, Kasetsart University, (in Thai with an English abstract)

Bruijnzeel, L.A., Mulligan, M. & Scatena, F.N. (2011), Hydrometeorology of Tropical Montane Cloud Forests: Emerging Patterns, *Hydrological Processes*, Vol. 25, No. 3, (January 2011), pp. 319-326, ISSN 1099-1085

Canadell, J., Jackson, R.B., Ehleringer, J.R., Mooney, H. A., Sala, O. E. & Schulze, E.-D. (1996), Maximum Rooting Depth for Vegetation Types on Global Scales, *Oecologia*, Vol. 108, No. 4, (November 1996), pp. 583-595, ISSN 0029-8549

Collatz, G.J., Ball, J.T., Grivet, C. & Berry, J.A. (1991), Physiological and Environmental Regulation of Stomatal Conductance, Photosynthesis and Transpiration: A Model that Includes a Laminar Boundary Layer, *Agricultural and Forest Meteorology*, Vol.54, No. 2-4, (April 1991), pp.107-136, ISSN 0168-1923

Dairaku, K., Emori, S. & Oki, T. (2004). Rainfall Amount, Intensity, Duration, and Frequency Relationships in the Mae Chaem Watershed in Southeast Asia. *Journal of Hydrometeorology*, Vol. 5, No. 3, (June 2004), pp. 458-470, ISSN 1525-755X

Dardanelli, J. L., Ritchie, J. T., Calmon, M., Andriani, J. M. & Collino, D. J. (2004) An Empirical Model for Root Water Uptake, *Field Crops Research*, Vol. 87, No. 1, (April 2004), pp. 59-71, ISSN 0378-4290

Desborough, C. E. (1997), The Impact of Root Weighting on the Response of Transpiration to Moisture Stress in Land Surface Schemes, *Monthly Weather Review*, Vol. 125, No. 8, (August 1997), pp.1920-1930, ISSN 0027-0644

Dickinson, R. E., Henderson-Sellers, A., & Kennedy, P.J. (1993), Biosphere-Atmosphere Transfer Scheme (BATS) Version 1e as Coupled to the NCAR Community Climate model, *Tech Note/TN-387+STR*, The National Center for Atmospheric Research, Boulder, CO. (August 1993), 72 pp.

Doley, D. (1981), Tropical and Subtropical Forests and Woodlands, In *Water Deficits and Plant Growth*, Vol. VI, Kozlowski, T. T., Academic Press, pp. 209-323, ISBN 0-12-424156-5

Farquhar, G. D., von Caemmerer, S. & Berry,J. A. (1980), A Biochemical Model f Photosynthetic CO2 Assimilation in Leaves of C3 Species, *Planta*, Vol. 149, No. 1, (January 1980) pp. 78-90, ISSN 0032-0935

Fukushima, Y. (1988), A Model of River Flow Forecasting for a Small Forested Mountain Catchment, *Hydrological Processes*, Vol. 2, No. 2, (April/June 1988), pp. 167-185, ISSN 1099-1085

Gollan, T., Turner, N. C. & Schulze, E.-D. (1985) The Responses of Stomata and Leaf Gas exchange to Vapour Pressure Deficits and Soilwater Content, *Oecologia*, Vol. 65, No. 3, (February 1985), pp. 356-362, ISSN 0029-8549

Jackson, R.B., Schenk, H.J., Jobbágy, E.G., Canadell, J., Colello, G.D., Dickinson, R.E., Field, C.B., Friedlingstein, P., Heimann, M., Hibbard, K., Kicklighter, D.W., Kleidon, A., Neilson, R.P., Parton, W.J., Sala, O.E. & Sykes, M.T., 2000. Belowground Consequences of Vegetation Change and their Treatment in Models. *Ecological Applications*, Vol. 10, No. 2 (April 2000), pp.470–483., ISSN 1051-0761

James, S.A., Clearwater, M.J., Meinzer, F.C. & Goldstein,G.. (2002), Heat dissipation Sensors of Variable Length for the Measurement of Sap Flow in Trees with Deep Sapwood, *Tree Physiology*, Vol.22, No.4, (March 2002), pp.277-283, ISSN 0829-318X

Kelliher, F.M., Kostner, B.M.M., Hollinger, D.Y., Byers, J.N., Hunt, J.E., McSeveny, T.M., Meserth, R., Weir, P.L., & Schulze, E.-D. (1992), Evaporation, Xylem Sap Flow, and Tree Transpiration in a New Zealand Broad-Leaved Forest, *Agricultural and Forest Meteorology*, Vol. 62, No. 1-2, (December 1992), pp. 53-73, ISSN 0168-1923

Kleidon, A. & Heimann, M. (1998), A Method of Determining Rooting Depth from a Terrestrial Biosphere Model and its Impacts on the Global Water and Carbon Cycle, *Global Change Biology*, Vol. 4, No. 3, (March 1998), pp. 275-286, ISSN 1354-1013

Kondo, J., & Xu, J. (1997), Seasonal Variations in the Heat and Water Balances for Nonvegetated Surfaces, *Journal of Applied Meteorology*, Vol. 36, No. 12, (December 1997), pp. 1676-1695, ISSN 0894-8763

Kume, T., Takizawa, H., Yoshifuji, N., Tanaka, K., Tantasirin, C., Tanaka, N. & Suzuki, M. (2007), Impact of Soil Drought on Sap Flow and Water Status of Evergreen Trees in a Tropical Monsoon Forest in Northern Thailand, *Forest Ecology and Management*, Vol. 238, No. 1-3, (January 2007), pp. 220-230 , ISSN 0378-1127

Kuraji, K., Punyatrong, K. & Suzuki, M. (2001). Altitudinal Increases in Rainfall in the Mae Chaem Watershed, Thailand. *Journal of the Meteorological Society of Japan*, Vol. 79, No. 1B, (March 2001). pp. 353-363, ISSN 0026-1165

Phillips, N., Oren, R. & Zimmermann,R. (1996), Radial Patterns of Xylem Sap Flow in Non-, Diffuse- and Ring-Porous Tree Species, *Plant, Cell and Environment*,Vol.19, no. 8 (August, 1996), pp.983-990, ISSN 0140-7791

Pinker, R.T., Thompson, O.E., & Eck, T.F. (1980), The Energy Balance of a tropical evergreen forest, *Journal of Applied Meteorology*, Vol. 19, No. 12, (December 1980), pp. 1341-1350, ISSN 0021-8952

Radersma, S. & Ong,C. K., (2004), Spatial Distribution of Root Length Density and Soil Water of Linear Agroforestry Systems in Sub-Humid Kenya: Implications for Agroforesty Models, *Forest Ecology and Management*, Vol. 188, No. 1-3, (February 2004), pp. 77-89, ISSN 0378-1127

Santisuk, T. (1988), An account of the Vegetation of Northern Thailand, In: *Geoecological Research*, Vol. 5, Flohn, H., Jusatz, H.J. & Legris, J.J., Franz Steiner Verlag Wiesbaden GmbH Stuttgart, ISBN 3-515-0511-8

Takizawa, H., Kubota, J. & Tsukamoto, Y. (1996), Distribution of water flow velocity at a stem cross-section (in Japanese with English abstract), *Journal of the Japanese Forest Society*, Vol.78, No.2, (May 1996), pp.190-194, ISSN 0021-485X

Tanaka, K. (2002), Multi-Layer Model of CO2 Exchange in a Plant Community Coupled with the Water Budget of Leaf Surfaces, *Ecological Modelling*, Vol. 147, No. 1, (January 2002), pp. 85-104, ISSN 0304-3800

Tanaka, K. & Hashimoto, S. (2006), Plant Canopy Effects on Soil Thermal and Hydrological Properties and Soil Respiration, *Ecological Modelling*, Vol. 196, No. 1-2, (July 2006), pp. 32-44, ISSN 0304-3800

Tanaka, K., Kosugi, Y. & Nakamura, A. (2002), Impact of Leaf Physiological Characteristics on Seasonal Variation in CO2, Latent and Sensible Heat Exchanges over a Tree Plantation, *Agricultural and Forest Meteorology*, Vol. 114, No. 1-2, (December 2002) pp 103-122, ISSN 0168-1923

Tanaka, K., Takizawa, H., Tanaka, N., Kosaka, I., Yoshifuji, N., Tantasirin, C., Piman, S., Suzuki, M., & Tangtham, N. (2003), Transpiration Peak over a Hill Evergreen Forest in Northern Thailand in the Late Dry Season: Assessing the Seasonal Changes in Evapotranspiration Using a Multilayer Model, *Journal of Geophysical Research*, Vol. 108, No. D17, (September 2003), 4533, doi:10.1029/2002JD003028, ISSN 0148-0227

Tanaka, K., Takizawa, H., Kume, T., Xu, J., Tantasirin, C. & Suzuki, M. (2004), Impact of Rooting Depth and Soil Hydraulic Properties on the Transpiration Peak of an Evergreen Forest in Northern Thailand in the Late Dry Season, *Journal of Geophysical Research*, Vol. 109, No. D23107, (December 2004), doi:10.1029/2004JD004865, ISSN 0148-0227

Tanaka, K., Yoshifuji, N., Tanaka, N., Shiraki, K., Tantasirin, C. & Suzuki, M. (2009), Water Budget and the Consequent Duration of Canopy Carbon Gain in a Teak Plantation in a Dry Tropical Region: Analysis Using a Soil-Plant-Air Continuum Multilayer Model, *Ecological Modelling*, Vol. 220, No. 12, (June 2009), pp.1534-1543, ISSN 0304-3800

Tanaka, K., Tantasirin, C. & Suzuki, M. (2011), Interannual Variation in Leaf expansion and Outbreak of Teak Defoliator at a Teak Stand in Northern Thailand, *Ecological Applications*, Vol. 21, No. 5 (July 2011), pp.1792-1801, ISSN 1051-0761

Tanaka, N., Kuraji, K.,Tantasirin, C., Takizawa, H., Tangtham, N. & Suzuki, M. (2011). Relationships between Rainfall, Fog and Throughfall at a Hill Evergreen Forest Site in Northern Thailand. *Hydrological Processes*, Vol. 25, No. 3, (January 2011), pp. 384-391, ISSN 1099-1085

Tangtham, N. (1974), Preliminary Study of the Ecosystem of a Hill Evergreen Forest in Northern Thailand, Ph.D. Thesis, Pennsylvania State University, pp. 65.

Thai Meteorological Department (April 2011), 30 year Average (1961-1990) - Chiang Mai, In: Thai Meteorological Department, 15.04.2011, Available at http://www.tmd.go.th/EN/province_stat.php?StationNumber=48327

Toda, M., Nishida, K., Ohte, N., Tani, M., & Musiake, K. (2002), Observations of Energy Fluxes and Evapotranspiration over Complex Terrestrial Land Covers in the

Tropical Monsoon Environment, *Journal of the Meteorological Society of Japan*, Vol. 80, No. 3, (June 2002), pp. 465-484, ISSN 0026-1165

Wang, C. & Weisberg, R. H., (2000), The 1997-98 El Nino Evolution Relative to Previous El Nino Events, *Journal of Climate*, Vol. 13, No. 2, (January 2000), pp. 488-501, ISSN 0894-8755

Part 4

ET and Groundwaters

19

The Role of the Evapotranspiration in the Aquifer Recharge Processes of Mediterranean Areas

Francesco Fiorillo
Dipartimento di Science per la Biologia, Geologia e l'Ambiente, University of Sannio Benevento, Italy

1. Introduction

Evapotranspiration is an important natural process controlling the amount of recharge into the aquifers, and relative groundwater fluctuations.

The main factors controlling the evapotranspiration processes can be distinguished in: (i) temperature and humidity of the air, (ii) water condition of the soil, and (iii) land-use and pedological characteristic of the catchment.

Air temperature and humidity are the main climate variables controlling the evapotranspiration processes; based on theoretical formula, evapotranspiration is an non-linear function of the temperature and humidity, depending also on the radiation and wind velocity.

The evapotranspiration processes increase in function of the soil water availability; generally the maximum rate of the evapotranspiration occur when soil is at field capacity and decrease rapidly when soil water tends to be near wilting point, independently from other variables.

Land-use has a great importance in evapotranspiration processes, and in particular the type and distribution of the vegetable cover. Generally, in the Mediterranean climate area evapotranspiration processes reached the maximum stage during spring time, and can lead the soil to wilting point for some vegetable species. Only endemic and autochthonous species are able to resist over the dry season without any water supply.

Usually the actual evapotranspiration is difficult to evaluate, and generally is determined indirectly from water balance of the soil cover. In many practical cases, the evapotranspiration can be estimate from the comparison of rainfall with potential evapotranspiration.

Based on monthly scale, the difference between rainfall and potential evapotranspiration is a first useful tool to evaluate the recharge processes of aquifers. However, a detailed approach needs an evaluation of the hydrological soil balance at daily scale, which allows to identify the amount of the rainfall which is free to percolate into the aquifer, recharging the water table, or run out from the aquifer.

An application on two nearby karst systems, the Terminio-Tuoro and Cervialto system, belonging to Picentini mountain, Southern Italy, have been carried out. These karst systems

feed important karst springs, which supply with water several million of people in Campania and Puglia regions. They are monitored by several climate stations, and spring discharges are measured since the beginning of last century.
In the karst systems the runoff amount can be very small due to both high permeability of the rock outcropping which favours infiltration processes and the presence of wide endorheic areas. These characteristics allow to compare the output (spring discharge) with input (afflux), evaluating the role of the evapotranspiration on the recharge processes.

2. Geological and hydrogeological features

The Picentini Mountains constitute a large karst system in the Campania region of Italy, over 600 km^2 wide (Fig.1).
Outcropping rocks in the region primarily belong to the calcareous and calcareous-dolomite series (Late Triassic-Miocene), are 2500 m thick, heavily fractured and faulted and frequently reduced to breccias. These karstic rocks are mantled by pyroclastic deposits of Vesuvius activity, which play an important role in the infiltration of water into the karst substratum below. The calcareous-dolomite series are tectonically bounded by terrigenous and impermeable deposits, constituting complex argillaceous (Paleocene) and flysch sequences (Miocene). Quaternary continental deposits, including slope breccias and debris and alluvial and lacustrine deposits, cover the marine substratum. General geological features of the Southern Apennine can be found in Parotto and Praturlon (2004) and ISPRA (2009).
All the slopes are covered by pyroclastic deposits, due to late Pleistocene-Holocene volcanic activity of Vesuvius. As a consequence of volcanic eruptions and weathering processes between deposition events, the pyroclastic mantle is generally composed of several irregular, ashy pumiceous layers, alternating with buried soil (Fig.2). Generally the thickness varies from steep slopes to flat/gentle slopes, from few decimetres to several metres. However, along the Cervialto slopes, rarely the pyroclastic mantle reaches 1 m thick, due to higher distance from Vesuvius. Some geotechnical characteristics of the pyroclastic mantle are shown in Table 1. These slopes are covered mainly by trees of beech and chestnut, and subordinately by trees of pine, fir, birk and oak.
From a hydrogeological point of view, the northern sector the Picentini mountain is formed by two main karst systems: the Terminio-Tuoro and Cervialto. These karst systems feed powerful springs, generally located along the tectonic contact between the carbonate rocks and flysch sequences.
The main karst springs fed by the Termino-Tuoro system (Civita, 1969) are the Serino springs (Acquaro-Pelosi, 377-380 m a.s.l., and Urciuoli, 330 m a.s.l.) located on the western side, along the Sabato river; the Sorbo Serpico springs located on the northern side; the Cassano springs (Bagno della Regina, Peschiera, Pollentina and Prete springs, 473-476 m a.s.l.) located on the eastern side, along the Calore river.
The Cervialto system (Celico & Civita, 1976) feeds the Caposele spring (Sanità, 417 m a.s.l.) located on the eastern side, which constitutes the main outflow from the aquifer.
The two karst systems present a different distribution of the ground-elevation, as can be seen in figure 3. In particular, over 70% of the Cervialto catchment lies above 1000 m a.s.l., with the peak at 1200-1300 m a.s.l, whereas only 30% of the Terminio-Tuoro catchment lies above 1000 m a.s.l., with a peak at 800-900 m a.s.l. These different features have an important role in the snow accumulation during the winter period and have consequences on the spring regimes.

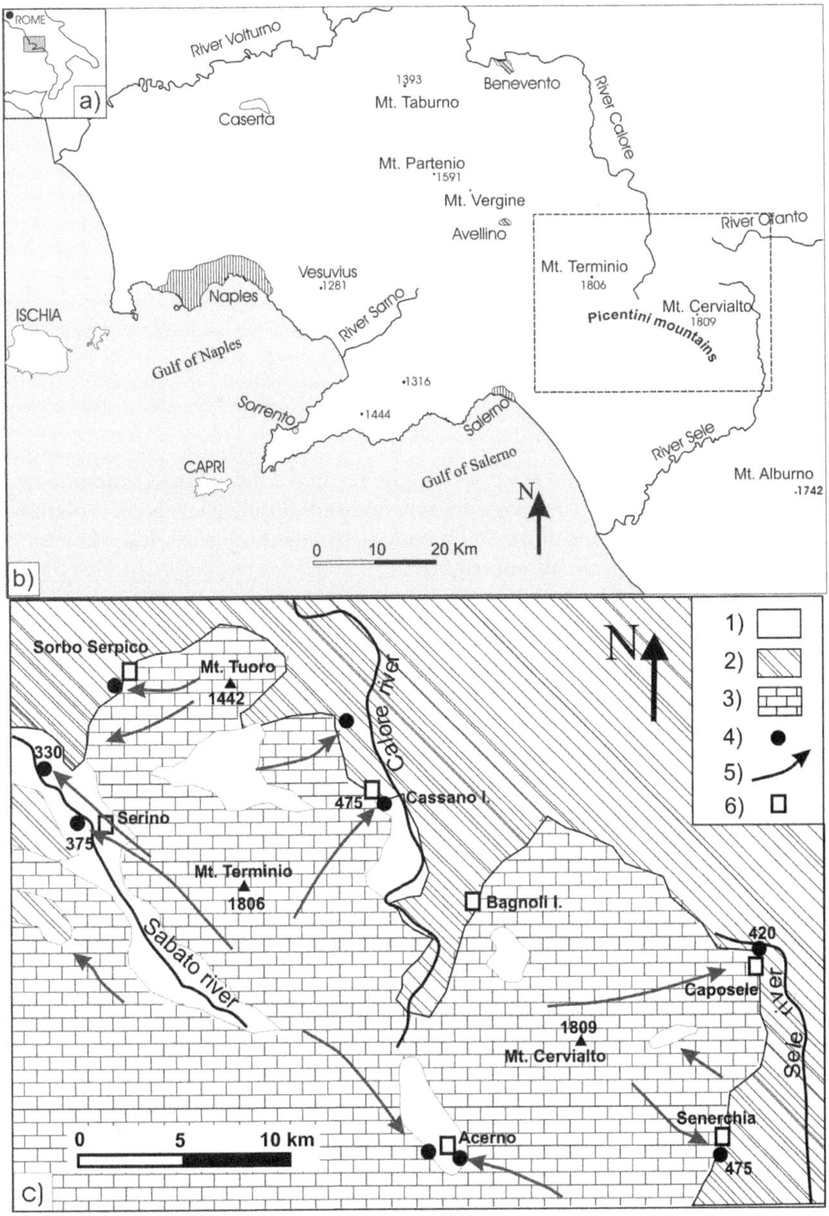

Fig. 1. a) Southern Italian peninsula. b) Map of the Western Campania region. Delimited rectangular area is detailed in c). c) Hydrogeological map of the Northern Picentini Mountains. 1) Slope breccias and debris, pyroclastic, alluvial and lacustrine deposits (Quaternary); 2) argillaceous complex and Flysch sequences (Paleogene–Miocene); 3) calcareous-dolomite series (Jurassic–Miocene); 4) main spring; 5) groundwater flow direction; 6) main village.

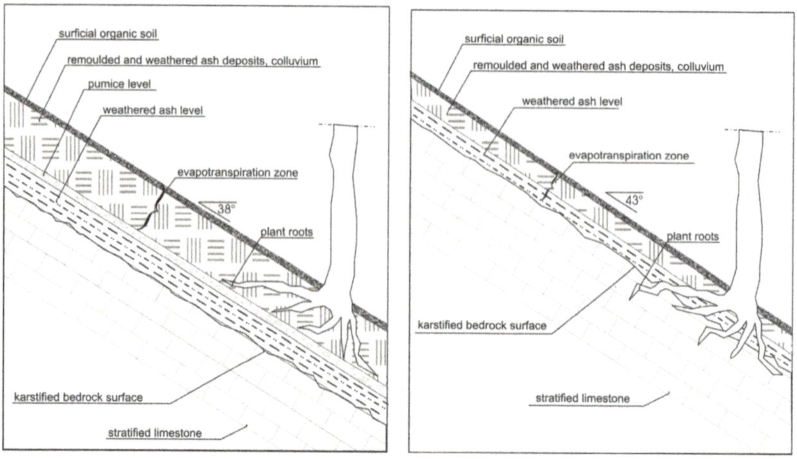

Fig. 2. Schematic profile of a typical slope of high-elevated zones of the Campania region (overlapping and thickness of the levels are schematised). Pumice levels limit plant roots development in the upper zone of the soil mantle (left); plant roots reach the karst substratum if pumice levels are absent (right).

Level	Gs (-)	clay (%)	silt (%)	sand (%)	gravel (%)	γ_d (kN/m^3)	n (-)	n_{eff} (-)
B_w	2,55	5	20	54	21	10,8	0,57	0,05÷0,07
C	2,55	-	2	52	46	8,1	0,68	0,33÷0,37
B_t	2,55	15	25	60	-	8,3	0,67	0,05÷0,06

Table 1. Some geotechnical characteristics of the pyroclastic soil (extracted from Fiorillo & Wilson, 2004). B_w, weathered and remoulded ash deposits; C, pumice level; B_t, weathered ash level; G_s, specific gravity; γ_d, dry bulk density; n, total porosity; n_{eff}, effective porosity.

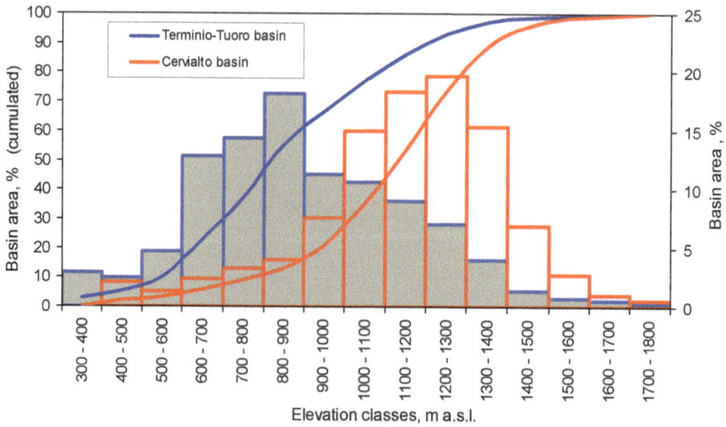

Fig. 3. Elevation classes distribution of the Terminio-Tuoro and Cervialto basin.

3. Climate and hydrological features

The region is located in an area characterised by typical Mediterranean climate, with dry and warm summer and a wet period that occurs during autumn, winter and spring. Table 2 reports some climate parameters of four local stations, with exception of the high-elevation Montevergine climate station (20 km NW from Mt. Terminio, figure 1a), which records data since 1884.

Station, m a.s.l.	autumn		winter		spring		summer		year	
	P (mm)	T (°C)	P (mm)	T (°C)	P (mm)	T (°C)	P (mm)	T (°C)	P; σ (mm)	T (°C)
Caposele, 426	390	10,9	471	5,5	298	14,3	123	19,3	1250; *255*	12,2
Cassano, 456	434	11,2	490	5,4	297	11,1	117	20,0	1344; *320*	11,9
Serino, 374	432	12,1	502	5,1	301	10,8	115	19,7	1352; *280*	11,9
M.Vergine, 1270	662	9,5	705	1,0	500	6,2	193	16,1	2174; *587*	8,2

Table 2. Seasonal (autumn, winter, spring and summer) and annual mean values of the Caposele, Cassano, Serino and Montevergine climate stations. P, precipitation; T, temperature; σ, standard deviation. Mean of precipitation period 1930-2006 (1884-2006 for Montevergine rain gauge) and mean of temperature period 1950-2000.

Rainfall spatial distribution depends strongly on the ground-elevation as can see in figure 4. Figure 5 shows the annual mean rainfall distribution along the northern sector of the Picentini mountains, based on the available rain gauge stations, and adding further fictitious rain gauges in the high-elevated zones. Rainfall distribution rapidly decrease along the northern and eastern sector of the Picentini mountains, as consequence of the Atlantic origin of the main meteorological storms.

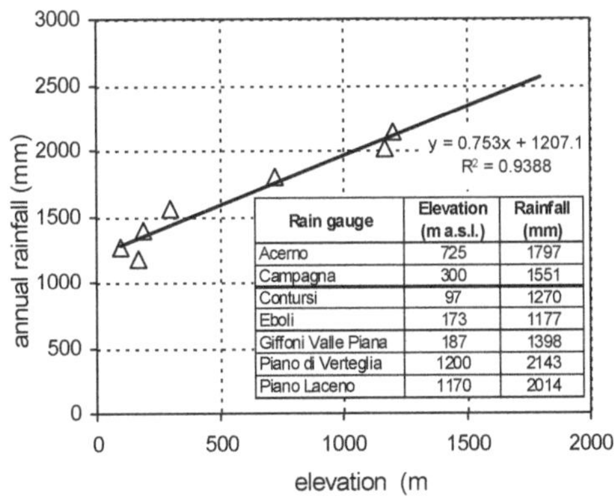

Fig. 4. Annual mean rainfall in relation to elevation. A high-significance regression line is obtained using rain gauges located along the western side of the Picentini mountains, as the main precipitations come from Atlantic zone.

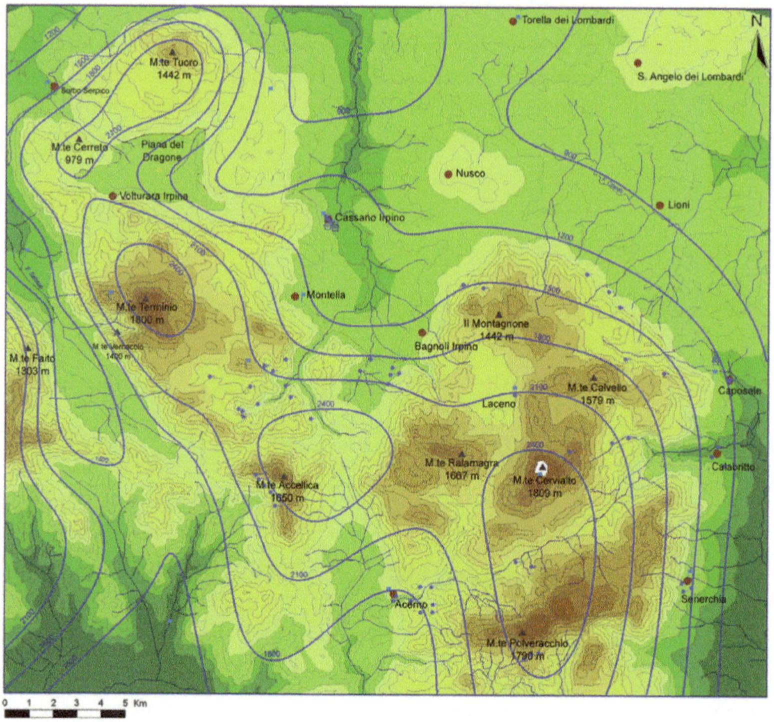

Fig. 5. Annual mean rainfall along the northern sector of the Picentini mountain, based on available rain gauges and other added following the relation in Fig.4. IDW method has been used to plot isohyets.

Monthly rainfall distribution, evapotranspiration and recharge have been evaluated for an high and low elevation zone (Figs. 6-7). Figures 6a and 7a show the monthly rainfall distribution, P_i; to consider the effect of the evapotranspiration, the effective rainfall, P_{eff}, has been also plotted as difference P_i-E_p=P_{eff}, where E_p is the potential evapotranspiration computed by the method of Thornthwaite (1948); for values of P_i<E_p, effective rainfall has been fixed null (P_{eff}=0). The effective rainfall expresses the amount of rainfall which is free to charge the soil moisture, to percolate and charge the groundwater, or runs off. Generally during the spring and mainly the summer, due to high rate of evapotranspiration, the effective rainfall, P_{eff}, is null; however, high-elevation areas (Figure 6a) present higher precipitation and lower temperature, and are characterised by shorter period of null effective rainfall.

As consequence of the rainfall increase and evapotranspiration processes decrease with ground-elevation, high-elevated zones are characterised by value of recharge (R=1582, figure 6) more than twice respect to low-elevated zone (R=755 mm, figure 7). However, the recharge values of figures 6 and 7 should be reduced, as unknown amount of rainfall runs off. A further evaluation of the recharge and of the evapotranspiration processes can be deducted from the ratio between output from the catchment (spring discharge) and the annul mean rainfall of each hydrogeological catchment.

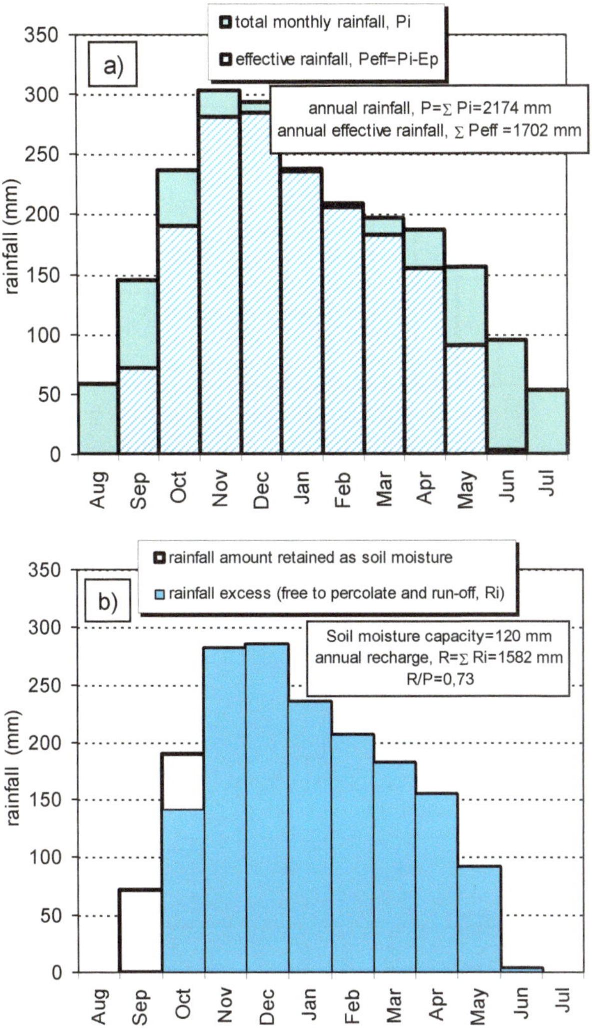

Fig. 6. High-elevated rain gauge (Montevergine, 1270 m a.s.l.). a) monthly rainfall, P_i, and effective rainfall, P_{eff}, through the hydrological year. Monthly potential evapotranspiration, E_p, has been computed by the Torthwaite (1948) method. b) rainfall free to infiltrate and percolate, R_i, though the hydrological year, obtained subtracting the soil moisture capacity (c=120 mm) from the effective rainfall, P_{eff}. The annual recharge, R, is obtained by difference between annual effective rainfall and soil moisture capacity (R=1702-120=1582 mm).

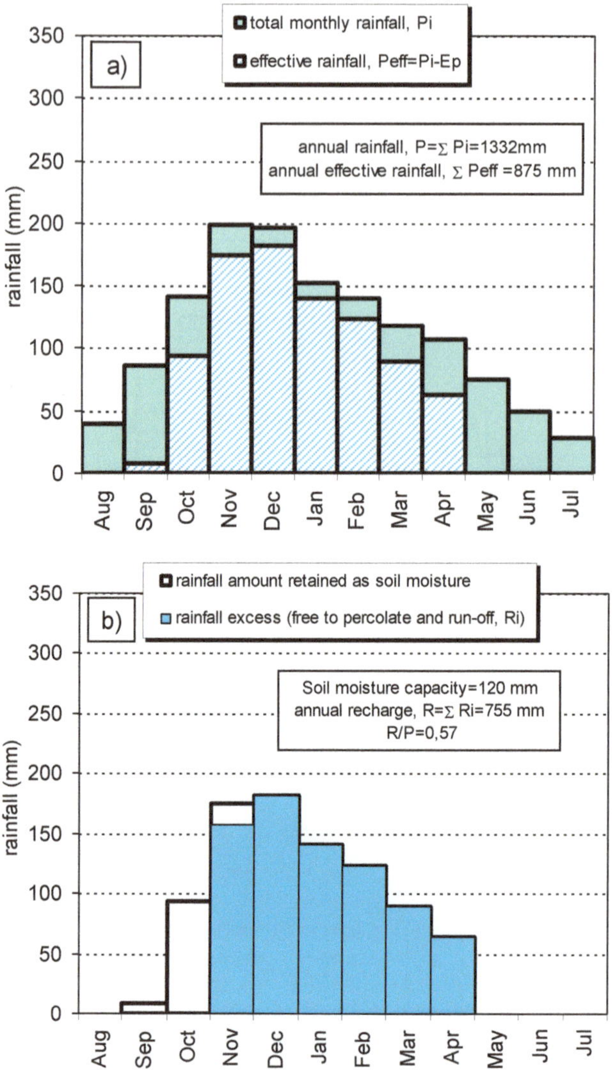

Fig. 7. Low-elevated rain gauge (Serino, 374 m a.s.l.). a) monthly rainfall, P_i, and effective rainfall, P_{eff}, trough the hydrological year. Monthly potential evapotranspiration, E_p, has been computed by the Torthwaite (1948) method. b) rainfall free to infiltrate and percolate, R_i, though the hydrological year, obtained subtracting the soil moisture capacity (c=120 mm) from the effective rainfall, P_{eff}. The total recharge, R, is obtained from the difference between the total effective rainfall and the soil moisture capacity (R=875-120=755 mm).

The annual mean rainfall of each hydrogeological catchment (afflux) has been evaluated in *Arc*GIS environment, using the regression found in figure 4a. The results are shown in Table 3.

Elevation classes (m a.s.l.)	Computed rainfall (mm)	Terminio-Tuoro basin		Cervialto basin	
		Area (Km^2)	Afflux (m^3 x10^6)	Area (Km^2)	Afflux (m^3 x10^6)
300 - 400	1455	5.14	7. 48	0	0
400 - 500	1545	4.31	6.66	2.43	3.75
500 - 600	1621	8.36	13.55	1.42	2.30
600 - 700	1696	22.68	38.47	2.71	4.59
700 - 800	1771	25.42	45.03	3.71	6.57
800 - 900	1847	32.18	59.43	4.59	8.48
900 - 1000	1922	19.98	38.41	8.7	16.72
1000 - 1100	1997	18.83	37.61	17.3	34.56
1100 - 1200	2072	15.97	33.10	21.09	43.72
1200 - 1300	2148	12.44	26.72	22.66	48.68
1300 - 1400	2223	7.03	15.63	17.7	39.36
1400 - 1500	2298	2.46	5.6	7.9	18.16
1500 - 1600	2374	1.33	3.16	3.13	7.43
1600 - 1700	2449	0.9	2.20	1.16	2.84
1700 - 1800	2524	0.46	1.16	0.63	1.59
Total		*177.49*	*334.31*	*115.13*	*238.77*

Table 3. Annual mean rainfall distribution of different elevation classes, computed by the regression line of figure 4. Afflux is the product of the area and rainfall height.

The outputs from the catchments have been computed by the annual mean discharge from the springs, and are shown in Table 4. The recharge of the Cervialto basin is higher than that of Terminio-Tuoro basin, according to higher ground-elevation of the basin which favours an higher afflux. Besides, the recharge values computed in table 4 are lower than the mean of the recharge computed in figures 6 and 7, confirming that an amount of rainfall leaves the catchments as run off.

However, each procedure presents some doubts: in the first case (Figs.6-7), the main uncertainty is the evaluation of the effective rainfall, as any method to compute potential evapotranspiration cannot be considered foolproof (Ward and Robinson, 2000).

The second procedure presents the main uncertainty in the evaluation of the rainfall distribution with ground-elevation, which can only roughly determined. A recent method to estimate the recharge has been proposed by Andreo et al. (2008) by a parametric method developed in GIS environment, which considers the altitude, slope angle, lithology, infiltration and soil nature (APLIS).

Basin	Springs group	Annual mean discharge (m^3/s)	Annual mean discharged volume (m^3x10^6)	*Total afflux* (m^3x10^6)	Recharge (%)
Terminio-Tuoro	Serino	2.25	69.98	*334.31*	50.1
	Cassano I.	2.65	82.43		
	Sorbo S.	0.43	13.37		
	Montella	0.15	4.67		
	Total	*5.48*	*170.45*		
Cervialto	Caposele	4.00	124.42	*238.77*	54.3
	Bagnoli I.	0.07	2.18		
	Others	0.10	3.11		
	Total	*4.17*	*129.70*		

Table 4. Output from Terminio-Tuoro and Cervialto basins by the annual mean spring discharge. Recharge has been computed by the ratio between Total output and Total afflux.

4. Daily scale analysis of recharge processes and evapotranspiration role

The soil mantle of the mountains here considered has a fundamental role in the evapotranspiration processes, as it retains an amount of rainfall at the beginning of the rainy season, to satisfy the water deficit accumulated during the previous dry period. The nature, physical characteristics and thickness of each soil stratum, together with the vegetation cover, control the soil storage capacity.

Below, some variables have been fixed on the basis of in situ observations, and allow to carry out a preliminary simulation of the evapotranspiration and recharge processes.

In the study area the soil undergoing evapotranspiration can be fixed at h=50 cm, as plant roots reach depth between 40 and 60 cm along the slopes of the Picentini mountains. Plan roots can cross the pyroclastic mantle and anchor directly into karst substratum. However, pant roots are unable to cross the pumice layers (Fig.2); these levels act as a drainage sheet, and break the root grooving deeply. As consequence, the evapotranspiration processes occur prevalently in the upper part of the pyroclastic mantle, deeply limited by the first pumice level. In situ suction measurements highlight the increasing of the suction toward summer, the decreasing towards autumn; below pumice level the suction tends to be constant during the year. Fiorillo and Wilson (2004) estimated the following parameters for the pyroclastic mantle of Campania slopes:

- (volumetric) water content at field capacity, θ_{max}, 51%;
- minimum water content reached at the end of dry season (summer), θ_{min}, 27%.

If the evapotranspiration processes occur similarly into all soil thickness (50 cm), and based on the technical characteristics of the soil mantle (table 1), the soil storage capacity, *m*, is 120 mm. The soil generally reaches the field capacity (θ_{max}) by October and November, and remains at or near that level until April. During this period, any additional rainfall percolates down, recharging the water table of carbonate aquifers, or runs off as surface stream flow. Between July and August, the moisture content of the pyroclastic cover is reduced to its minimum level (θ_{min}). To increase the water volumetric content from θ_{min} up to θ_{max}, an amount of 120 mm of water is needed, corresponding to the fixed soil storage capacity, *m*. Due to evapotranspiration processes, this amount of water corresponds to an higher amount of rainfall, up to 200-250 mm, depending on the temporal distribution of the daily rainfall.

In order to evaluate the soil moisture conditions and periods of the recharge periods, a hydrological balance of the pyroclastic soil has been carried out, based on the method of Thornthwaite (1948), adapted at daily scale (Fiorillo & Wilson, 2004). A wet and dry hydrological year are shown in figure 8 and 9, respectively.

Figure 8a shows daily rainfall of 2005-06; the total cumulative rainfall reaches the value of 1812 mm, which is higher than annual historical mean (1352 mm). In figure 8b, rainfall excess occurs during October and April, when soil moisture reaches the field capacity (Θ_{max}=51%); the rainfall excess is able to percolate deeply into the aquifer and recharge the water table, and a minor part is drained as runoff. Fig.8c shows the hydrograph of Bagno della Regina spring (Cassano Irpino group). Note that single rain event (daily rainfall) has no direct influence on the spring discharge, which generally is characterised by one or several smoothed peak during spring season (Fiorillo, 2009). These behaviour is connected to wide catchment area and to thickness of the vadose zone, which can reach value higher than 1000 m, specially for the Caposele spring catchment. The pyroclastic mantle reduces rapid infiltration into the karst system, which may also contribute to the smoothed shape of the spring hydrographs. The cumulative rainfall excess (Fig.8c) increases up late April; after that, due to rainfall decrease and evapotranspiration increase, no groundwater recharge occur. During this period, spring discharge decreases up to the next recharge, and the hydrograph generally shows a typical concave shape.

Note that at beginning and at the and of the hydrological year, the spring discharge is 1,07 and 1,16 m^3/s, respectively; this means that the recharge of wet years can influence positively the spring discharge of the following year (Fiorillo, 2009).

Figure 9a shows daily rainfall of 2001-02; the total cumulative rainfall reaches the value of 902,5 mm, which is well below than annual historical mean (1352 mm). The 2001-2002 was the driest year recorded in many rain gauges of southern Italy (Fiorillo & Guadagno, 2010), and caused an extreme hydrological drought (Tallaksen and Van Lanen, 2004) in a wide area of the Mediterranean basin. In figure 9b, rainfall excess occurs between late November and May; this amount of rainfall was unable to increase the spring discharge, which shows a continuous decreasing trend in Fig.9c. Besides, at the beginning and at the and of the hydrological year, the spring discharge is 0,67 and 0,19 m^3/s, respectively; this means that the recharge of the dry years influence negatively the spring discharge of the following year (Fiorillo, 2009).

During the 2001-02, about half of the total rainfall is lost by evapotranspiration processes (432 mm). This amount is comparable to that of 2005-06 wet year (485 mm), indicating that the amount of the evapotranspiration appears to be almost the same during dry or wet year in this area. Thus, the main factor influencing the recharge in this area seems to be the annual rainfall, as evapotranspiration processes waste almost the same amount in any hydrological year in this area.

However, the amount of the evapotranspiration water depends strongly on the soil moisture capacity, *m*. In order to evaluate the different role of the soil moisture capacity on the evapotranspiration water amount, the rainfall excess has been computed considering different value of *m*. Figure 10 shows the results obtained for the 2001-02 and 2005-06 years, where is possible to observe from the higher to smaller value of the soil moisture capacity

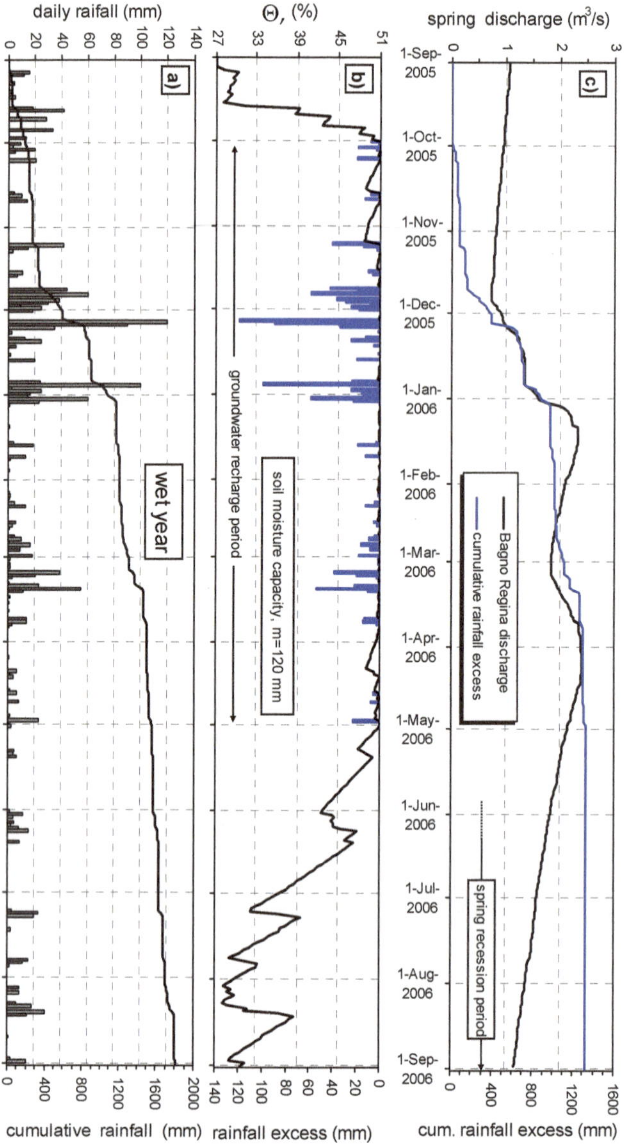

Fig. 8. Hydrological characteristics of the 2005-06 wet year; Serino rain gauge (374 m a.s.l.). a) daily and cumulative rainfall. b) (volumetric) water content, Θ, computed from daily hydrological balance (Θ_{min}=27% and Θ_{max}=51% have been fixed by laboratory test, Fiorillo and Wilson, 2004); rainfall excess (histogram) expresses the rainfall surplus after that maximum soil water content has been reached (Θ_{max}=51%). c) clean positive trend of the cumulative rainfall excess causes the spring discharge increasing; the flood of late March-early April has been connected to snow melting.

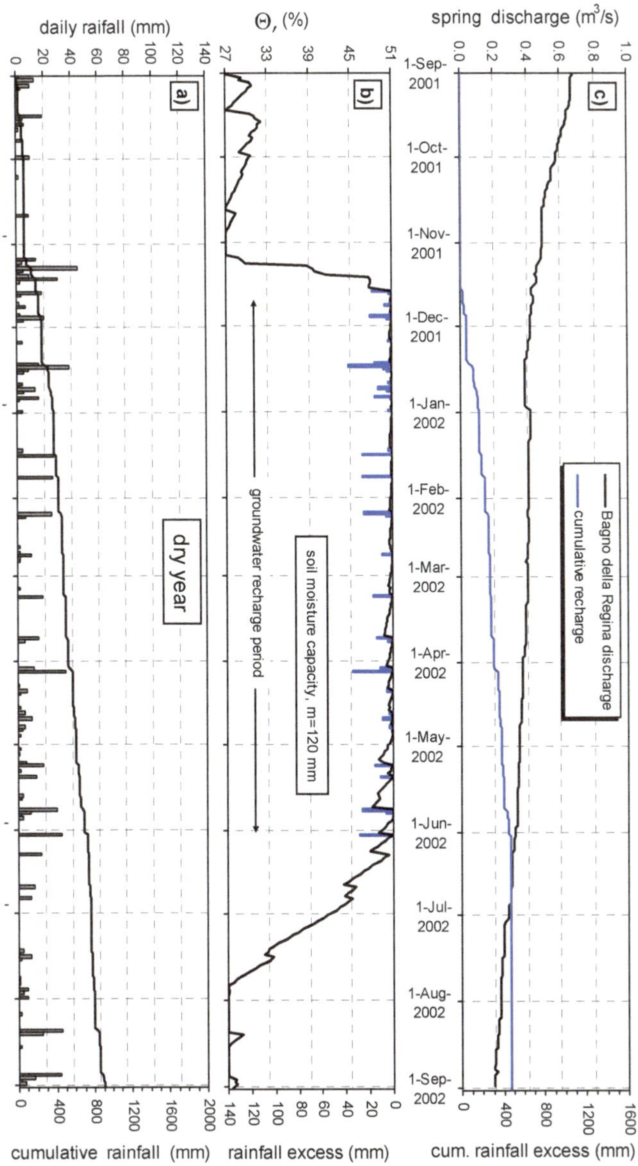

Fig. 9. Hydrological characteristics of the 2001-02 dry year; Serino rain gauge (374 m a.s.l.). a) daily and cumulative rainfall. b) (volumetric) water content, Θ, computed from daily hydrological balance (see figure 8). c) the weak positive trend of the cumulative rainfall excess is unable to increase the spring discharge, which show a continuous decreasing trend.

the linear decrease of the (cumulative) rainfall excess and the linear increase of the evapotranspiration. The trends are linear and parallel, indicating that (i) the differences of the rainfall excess between the two years are almost the same, and (ii) the amount of evapotranspiration, for a fixed soil moisture capacity, dos not depend on the annual rainfall.

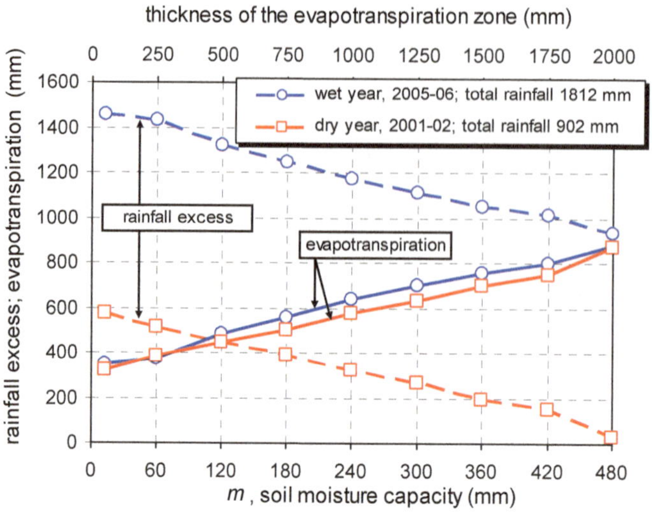

Fig. 10. The dependence of the rainfall excess and evapotranspiration from the soil moisture capacity (data from Serino rain gauge, 374 m a.s.l.). Thickness of evapotranspiration zone is also shown, under hypothesis of homogenous proprieties of the soil and uniform evapotranspiration processes (see text for details).

5. Conclusion

In a typical mountain karst environment of Mediterranean area, the effect of the evapotranspiration processes during the hydrological year cause important consequence on the groundwater recharge processes. During the spring season, due to decrease of the rainfall and increase of the temperature, generally recharge do not occur from late April, and remains almost null up to following autumn. Generally, rainfall occurring during the spring-summer period are completely lost by evapotranspiration processes or adsorbed by the soil mantle as soil moisture, to reduce the water deficit accumulated in previous days. During this period, groundwater is drained by springs, which present a typical recession hydrograph (Fiorillo, 2011). It is very important that recharge occur during autumn-winter period, when water lost by evapotranspiration processes is at minimum level. A different distribution of the rainfall through the hydrological year could reduce the spring discharge, inducing negative consequences for water supply.

A fundamental role has been played by the soil mantle which acts as "filter", regulating the amount of rainfall which reaches the karst substratum, and recharges the aquifer. Higher soil moisture capacity favour higher values of evapotranspiration (Fig.10), but the amount of evapotranspiration seems to be independent from the annual rainfall, as water lost by evapotranspiration processes is almost the same during dry and wet years, as occurred in 2001-02 (dry) and 2005-06 (wet) year. The amount of rainfall which recharge the aquifer decreases rapidly as annual rainfall decreases, and can reaches negligible values during extreme droughts.

A reduction of the rainfall excess can be induced also by warm years. The worldwide records of the temperature increasing suggest that also the evapotranspiration processes are increasing (Cassardo et. al, 2007), widening the period of no-recharge.

The effect of the evapotranspiration increased can be observe in many aquifers, characterized by the drop of groundwater. This drop is observed since the eighteens, coinciding with the marked beginning of the temperature increasing worldwide (Brunetti et al., 2006; MOHC, 2010), and is well-documented in the karst areas of Southern Italy (Fiorillo et al. 2007; Fiorillo & Guadagno, 2010).

6. References

Andreo B., Vías J., Durán J.J., Jiménez P., López-Geta J.A., Carrasco. *Methodology for groundwater recharge assessment in carbonate aquifers: application to pilot sites in southern Spain.* Hydrogeology Journal, 16, 911-925.

Brunetti M.A., Maugeri M., Monti F., Nanni T. (2006). *Temperature and precipitation variability in Italy in the last two centuries from homogenised instrumental time series.* Int. J. Climatol. 26: 345-381.

Cassardo C., Loglishi N., Romani M. (2007). *Preliminary results o fan attempt to provide soil moisture dataset in order to verify weather prediction models.* Nuovo Cimento della Società di Italiana Fisica, 30 (6), pp 653.

Celico P., Civita M. (1976). *Sulla tettonica del massiccio del Cervialto (Campania) e le implicazioni idrogeologiche ad essa connesse.* Boll. Soc. Natur., 85, Naples.

Civita M. (1969). Idrogeologia *del massiccio del Terminio-Tuoro (Campania).* Memorie e Note Istituto di Geologia Applicata. Università di Napoli, Italy.

Fiorillo F. (2009). *Spring hydrographs as indicators of droughts in a karst environment.* Journal of Hydrol 373:290-301.

Fiorillo F. (2011). *Tank reservoir drainage as a simulation of the recession limb of karst spring hydrographs.* Hydrogeology Journal, 19 (5), 1009-1019.

Fiorillo F., Esposito L., Guadagno F.M. (2007). *Analyses and forecast of water resources in an ultra- centenarian spring discharge series from Serino (Southern Italy).* Journal of Hydrol 336:125-138.

Fiorillo F., Guadagno F.M. (2010) - *Karst spring discharges analysis in relation to drought periods, using the SPI* - Water Resources Management, 24, 1864-1884.

Fiorillo F. & Wilson R.C. (2004) - *Rainfall induced debris flows in pyroclastic deposits, Campania (southern Italy)* - Engineering Geology, 75, 263-289.

ISPRA (2009) *Geological map of Italy,* 1:50.000 scale. Istituto Superiore per la Protezione e la Ricerca Ambientale (ISPRA), Rome. http://www.apat.gov.it/Media/carg/.

MOHC (2010). *Global average temperature series 1850-2009.* Met Office Hadley Centre (MOHC), http://www.metoffice.gov.uk/climatechange/science/hadleycentre/.

Parotto M., Praturlon A. (2004). *The southern Apennine Arc.* Special Volume of the Italian Geological Society, 32° Int. Geol. Conf., Florence, Crescenti U., D'Offizi S., Merlino S., Sacchi L. eds; pp. 34-58.

Tallaksen L.M., Van Lanen H.L.J. (2004). *Drought as a Natural Hazard.* In "Hydrological Drought: Processes and Estimation Metods for Stream Flow and Groundwater", Tallaksen and Van Lanen eds., Elsevier, 579 pp.

Thornthwaite, C.W. 1948. *An approach towards a rational classification of climate*. Geograph. Rev., 38: 55-94.

Ward R.C., Robinson M. (2000). *Principle of Hydrology*. McGraw-Hill Publishing Company, England, 450 pp.

Part 5

ET and Climate

20

The Evapotranspiration in Climate Classification

Antonio Ribeiro da Cunha and Edgar Ricardo Schöffel
São Paulo State University - Superior School Agronomic Sciences
Federal University of Pelotas - Superior School Eliseu Maciel Agronomy
Brazil

1. Introduction

1.1 Climate

Climate is the average atmospheric condition of a particular place or region, ranging from months to millions of years, being 30 years the classical period defined by the World Meteorological Organization (WMO). It's represents different weather conditions prevailing at a site or region, considering the analysis of a large amount of data, affecting the majority of human activities, mainly in agriculture.
Several researchers have defined the climate word, such as: Köppen (1936) – "the sum of the atmospheric conditions that make a place the earth's surface more or less habitable for humans, animals and plants"; Trewartha (1937) - "the mean or normal condition over a long period, such as 20, 30 and 100 years"; Blair (1942) - "the summation of weather conditions in historical time" or "climate is the summary of all the manifold weather influences"; Thornthwaite (1948) – "the interaction of meteorological factors that contribute to give the place its character and individuality"; Miller (1959) -"the science that discusses the weather condition of the earth surface"; Oliver (1981) - "climate is the aggregate of weather at a given area for a given time period"; and Anwar (1993) - "the generalized picture of weather is called climate".

1.2 Climatological series

For thousands of years, historians recorded information about the weather, but most were based on reports with no reliability. With the creation of weather records in the first half of the 20th century, these manuscripts had extreme events, which were a cause proud of the observation and recording of such information, because the stamp or the signature of the person which recorded allowed reliability - personal source history. After the establishment of WMO was standardized the forms and procedures regarding records in national archives of meteorology.
Thom (1966) defined climatological series as "a sample of a population of data of one same location within the observation period, in days or months", for example, the average temperatures of 30 January in 30 years of observation. The analysis of climatological data involves verifying the story of a weather station. The displacement of the weather station, a change in the observation period or of instrumentation, discontinuity in the measurements, a change in the environment immediately around the station, such as buildings, can invalidate the climatological series; in addition, tests should be made for consistency data.

Recommendations and criteria are defined by WMO to promote the detection of the usual problems of discontinuity, non-representativeness and data inconsistency.

The data record should be consistent (have no changes in location, instruments, observation practices, etc.; these are identified here as "exposure changes") and have no missing values so a normal will reflect the actual average climatic conditions. This ensures the homogeneity of climatologically series, because that way, the observed variations are caused only by conditions of weather and climate. Another problem is related to the mesh to collect weather data that is still insufficient, affecting the representativeness and damaging the climate studies in a region.

After the end of the 20th century, the weather information has been transmitted digitally, due to advances in computer technology, and with that, it the collection, transmission, processing and storage of meteorological data have been facilitated. This is due to the use of data loggers that reduced the errors of reading, interpreting and typing, providing quality and standardization of data collected.

1.3 Climatological normals

From 1956 to 1989, WMO guided to update the weather data the every 10 years. As of 1989, it's established the general procedures for calculating the monthly and annual averages, which provided an opportunity for the climatological normal standard and provisional (WMO, 2009).

The WMO defines normal as "period averages computed for a relatively long period of at least three consecutive decades", i.e. "mean climatological data computed for the following consecutive periods of 30 years: 1 January 1901 to December 31 , 1930, 1 January 1931 to December 31, 1960, etc" (WMO, 1984).

There is a coordinated international effort to compile data standard - the global standard - every 30 years to monitor climate change that may occur, the latter being obtained for the period 1961-1990.

The use of meteorological allows comparisons between regions through different climate elements, analyzing the changes monthly throughout the year and during the fall-winter and spring-summer, the energy transfer from soil to atmosphere by evapotranspiration and water availability (climatic water balance).

1.4 Climatic classification system

The climate classifications are intended the organize large amount of information to facilitate the rapid retrieval and communication, grouping items according to their similarities to provide an estimate of the climatic resources of a particular place or region, serving for various purposes. They simplify the climatic data of a place or region, provides a concise description of climate factors in terms of real assets (the which creates the local climate), provides a means by which the climatic regions can be accurately identified and can be used in global, local or micro scale, being the starting point for analyzing the causes of climate variations.

The climate classifications provide relevant information to the different types of cultures: allow the interrelationship between the local climate with the productive potential of a given species, the identification of abnormal conditions or agricultural practice, the quantification of water stress levels that the culture is subject, estimate rates of aridity and productivity losses for different locations, identify the conditions conducive or anomalous for the agricultural practices, choose plant varieties most suitable for a region, providing the best sowing time, management recommendations, cultural practices and harvesting, and other information critical to the agroclimatic zoning.

It is difficult to create a suitable climate classification system, because it is defined according to each climatologist and their own criteria, being some more complex systems and other simpler.

According to Trewartha (1968) "classification of climate is a process basic to all sciences, consisting of recognizing individuals with certain important characteristics in common and grouping them into a few classes or types". Griffiths (1978) defined it as, "climatic classification is merely a method of arranging various climatic parameters either singly or grouped into ranks or sets, so, as both to simplify the mass of data and to identify analogies", i.e. "there are no two places in the world that experience exactly the same climate; but it is possible to identify areas with similar climate; this grouping or analogue method is referred to as climate classification".

Classification sistems different are known, such as those of Amend for Supan (1879), Köppen (1936), Kendrew (1941), Blair (1942), Thornthwaite (1948), Flohn (1950), Kazi (1951), Geiger (1953), Miller (1959), Nasrullah (1968), Terjung & Louie (1972), Griffiths (1978), Johnson (1979), Shamshad (1988), Raja & Twidal (1990), F.K. Khan (1991) and J.A. Khan (1993). In which case, we gave priority to two more classifications used throughout the world, Köppen (1936) and Thornthwaite (1948).

The Köppen climate classification is relatively simple and very popular. It associates vegetation types with climate prevailing in the regions, taking into account the temperature and precipitation. Already the Thornthwaite classification is based on two major climate indices, moisture content and the annual potential evapotranspiration.

2. Köppen's versus Thornthwaite's climatic classification

The main difficulty in climate classification is related to the inadequacy of climatic data available, both in terms of surface coverage as in terms of duration and reliability. The climatic elements most frequently used to characterize the climate are the mean values of temperature and precipitation, beyond of other climate elements for a better classification.

The Köppen climate classification contains information more suited to geographical and climatological studies than of agrometeorological, simplifying complex situations of the relationship of climate to the productivity of agricultural crops (Burgos, 1958). Rolim et al. (2007) does not recommend it for the determination of agrometeorological zones because it is efficient only for the macro scale, has a low capacity for the separation of types of climates, and lost in details, since it does not differentiate very well the climatic types (Cunha et al. 2009).

It is inconceivable that Köppen could have produced his original classification and map without using other landscape signals of climate (particularly vegetation) since there would have been so little observed climate data available at that time. In the light of this, the persistence of his scheme of classification is even more remarkable (Peel et al., 2007). The use of Köppen's classification is not confined to teaching, because many researchers routinely use it for their own particular research purposes (McMahon et al., 1992; Peel et al., 2004, 2007). In this sense, Lohmann et al. (1993) state that the Köppen classification is easier to use and is still a useful tool to estimate the ability of climate models to reproduce the present climate and indicate the impact of climate change on the biosphere.

The Köppen classification compensates for its shortcomings because have wide acceptance (Stern et al., 2000), and remains the best known scheme for the climatic zoning, according to Hudson & Brown (2000).

Sunkar (2008) says "the differentiation between the wet and the dry seasons provides adequate information to properly evaluate the potential of agricultural land, when using the Köppen classification system".

Thornthwaite considers the moisture as factor truly active, using it as a basis for identifying the most of its major climatic types, except for three areas of cold in which it analyzes the temperature as active factor or critical. It considers the vegetation is a physical means by which it is possible to transport water from the soil into the atmosphere, defining a type of climate as wet or dry depending on the water requirements of vegetation (Trewartha, 1954). Undoubtedly this is one of the great virtues of Thornthwaite classification, consider the potential evapotranspiration as an essential element in determining of other indexes. Thornthwaite's introduced the concept of potential evapotranspiration which was improved over time on several occasions. The potential evapotranspiration is one opposite process of precipitation it refers to "water used by an extensive vegetated surface in active growth, completely covering the ground, unrestricted and soil water " (Thornthwaite, 1946). The comparison between precipitation and potential evapotranspiration obtained from in the climatic water balance indicate deficiencies and excesses of moisture throughout the year or in season of crop growth (Pereira et al., 1997). The potential evapotranspiration is estimated by the temperature, due to the difficulty in obtaining information from other climatic elements to use a more accurate method (Pereira et al., 2002).

Climatic classification Thornthwaite presents a detailed climate, generating more information a local called of climatic subtype, which favors local studies. According to Balling (1984), this methodology has brought a much greater sensitivity in the definition of climates, and allows separated efficiently climates in topoescala, summarize to efficiently the information generated by normal water balance and the ability to determine agroclimatic zones (Rolim et al., 2007). This classification takes into account temperature, precipitation and evapotranspiration, and presents in detail the period of deficit and hidric surplus of a local (Rolim et al., 2007; Cunha et al., 2009). The Köppen classification associate the vegetation with temperature, while the Thornthwaite classification with the moisture factor (Essenwanger, 2001). For all this, Thorthwaite classification is considered a more refined method than that of Koppen for agricultural applications, since it takes into account the evapotranspiration (Trewartha, 1954).

The most important contribution to the modeling of vegetation and climate is the idea of potential evaporation proposed by Thornthwaite (Hudson & Brown, 2000) and the Thornthwaite map that includes the clear divisions of wetter and drier regions (Logan, 2006). By comparing maps generated by the Köppen and Thornthwaite methods, Doerr (1962) says the Thornthwaite classification is more accurate and helpful to describe the climate of a place. Most geographers readily admit that the Thornthwaite climate classification represents a major step forward in terms of conceptual sophistication about the Köppen system (Malmström, 1969).

3. Climatologic water balance

When relate evapotranspiration with the rain, taking into account the storage of water in the soil, periods of deficit and surplus are revealed, allowing a more critical analysis and appropriate a place or region. This allows quantification of the levels of water stress to which he is subjected a particular culture, and also estimate indexes of drought and reduced productivity.

According to Thornthwaite (1948), deficiencies and surpluses of water over one year affects the climate of a region amending the conditions of humidity. Excess water in the rainy season, in large part does not solve the drought in the dry season, but may soften, especially if the plants have broad and deep root system. Would be better if the rain was always greater than evapotranspiration, which always occurs to small soil water surplus, provided that no surface flow (runoff), thereby ensuring a water storage in soil under conditions of field capacity, this means that actual evapotranspiration is close to potential evapotranspiration.

The climatic water balance calculations depend on average monthly temperature data, monthly precipitation, potential evapotranspiration and water retention characteristics of soil-water site. The depth of soil layer defines the ability of water available for plants, the greater the depth of soil the greater the water holding capacity and the root system tends to be more superficial. When there is low water availability in the soil, the plants develop more their root system.

Calculations of water stress and evapotranspiration potential and real that participate in the water balance are needed to quantify levels of water stress to which that finds a culture, as well as estimating the productive potential of different locations. For this, the climatic water balance should represent the water availability of a place or region and must be made from time series uninterrupted, consistent and homogeneous, at least 30 years.

In 1948, Thornthwaite proposed the climatic water balance monthly as a simple method, using average monthly air temperature and precipitation accumulated, as well as the storage capacity of soil water, assuming that the soil is considered a reservoir, and that all the water available to the soil first will contribute to the evapotranspiration demand. Still, defines the potential evapotranspiration, as the transfer of water to the atmosphere of a large area with dense vegetation, with active growth covering the entire surface (bahiagrass) and under soil conditions without water restriction - more appropriate concept for climatological studies (Varejão-Silva, 2006).

Thornthwaite & Mather (1955) developed a climatic water balance monthly to monitor the variation of soil water storage throughout the year. By accounting for the supply of water to the soil - precipitation, and the atmospheric demand - potential evapotranspiration, and with a maximum storage - available water capacity, the water balance provides estimates of actual evapotranspiration, of deficit and water excess and soil water storage (Pereira et al., 2002).

4. Methodology

This study used climatologically normal data (1961-1990) published by Brazil's National Institute of Meteorology (INMET, 2009): data of monthly mean air temperature and cumulative monthly rainfall. The locations used to evaluate the effect of altitude: Campos do Jordão (SP) and Santos (SP), the effect of latitude: Boa Vista (RR) and Santa Maria (RS), the effect of longitude: João Pessoa (PB) and Porto Velho (RO) and the effect of ocean currents: Angra dos Reis (RJ) and Cabo Frio (RJ). Geographic coordinates of the locations used in the study (Table 1).

District	State	Latitude	Longitude	Altitude
Campos do Jordão	SP	22° 45′ S	45° 36′ W	1,642.0
Santos	SP	23° 56′ S	46° 20′ W	13.5
Boa Vista	RR	02° 49′ N	60° 39′ W	90.0
Santa Maria	RS	29° 42′ S	53° 42′ W	95.0
João Pessoa	PB	07° 06′ S	34° 52′ W	7.4
Porto Velho	RO	08° 46′ S	63° 55′ W	95.0
Angra dos Reis	RJ	23° 01′ S	44° 19′ W	3.0
Cabo Frio	RJ	22° 59′ S	42° 02′ W	7.0

Table 1. Geographic coordinates of the locations used in the study.

With these normal climatologically data were used two classification systems climate, Köppen and Thorntwaite, aiming to evaluate the role of evapotranspiration.

4.1 Description of criteria and symbols by Köppen (1936) method

The "**1st letter**", it was considered P = annual precipitation (cm) and T = average annual temperature (°C); analyzing the conditions:

Condition (1): winter precipitation: 70% of total annual precipitation occurs during the six coldest months of the year;

Condition (2): summer precipitation: 70% of total annual precipitation occurs during the six hottest months of the year;

Condition (3): when does not apply any of the above conditions.

Condition (1) true:

$P > 2T \Rightarrow$ climate is **A**, **C** or **D**

$2Q < P < T \Rightarrow$ climate is **BS** (steppe)

$P < T \Rightarrow$ climate is **BW** (desert)

Condition (2) true:

$P > 2\,(T + 14) \Rightarrow$ climate is **A**, **C** or **D**

$(T + 14) < P < 2\,(T + 14) \Rightarrow$ climate is **BS** (steppe)

$P < (T + 14) \Rightarrow$ climate is **BW** (desert)

Condition (3) true:

$P > 2\,(T + 7) \Rightarrow$ climate is **A**, **C** or **D**

$(T + 7) < P < 2\,(T + 7) \Rightarrow$ climate is **BS** (steppe)

$P < (T + 7) \Rightarrow$ climate is **BW** (desert)

Therefore, according to the average temperature, the "**1st letter**" can be:

A: tropical, temperature of the coldest month is above 18 °C;

B: dry climates, limits determined by the temperature and precipitation;

C: temperate climate, temperature of the coldest month between 18 and -3 °C;

D: cold weather, temperature of the warmest month above 10 °C and temperature of the coldest month below -3 °C;

E: polar climates: temperature of the warmest month is below 10 °C;

F: the hottest month is below 0 °C;

G: mountain weather;

H: high altitude climates.

The "**2nd letter**" is obtained as a function of precipitation:

Af: no dry season - the driest month precipitation is greater than 6 cm; **Am**: the driest month precipitation shows higher or equal to (10 - P/25); **Aw**: when the previous conditions do not apply and the driest period occurs in winter;
BS; **BW**: see the previous conditions (1), (2) and (3);
Cs; **Ds**: when the precipitation is winter and the wettest month of winter precipitation has equal or greater than three times that of the driest month; **Cw**; **Dw**: the wettest month of summer precipitation has greater than or equal to 10 often the driest month; **Cf**; **Df**: moist, do not apply when sewing;
EF: displays all months of the year with average temperatures below 0 °C; **ET**: when the warmest month has temperature between 0 and 10 °C; **EB**: perpetual snow or tundra.
The "**3rd letter**"is obtained as a function of temperature:

a. hot summer, is the hottest month temperature is above 22 °C;
b. moderately warm summer, the temperature of the warmest month is below 22 °C and at least four months have temperatures above 10 °C;
c. short summer and moderately cold, less than 4 months has a temperature higher than 10 °C;
d. very cold winter, the coldest month has temperatures below -38 °C;

OBS: For arid local (**BS** or **BW**):
BSh' or **BWh'**: very warm, with average annual temperatures over 18 °C and warmest month with temperatures above 18 °C;
BSh or **BWh**: warm, with average annual temperatures over 18 °C and warmest month temperatures below 18 °C;
BSk or **BWk**: cold, with average annual temperature below 18 °C and warmest month with temperatures above 18 °C;
BSk' or **BWk'**: very cold, with average annual temperature below 18 °C and warmest month temperatures below 18 °C.

4.2 Description of criteria and symbols by Thornthwaite (1948) method

The calculation of potential evapotranspiration was done according to the Thornthwaite (1948) method:

$$ET = 16\left(\frac{10.Tn}{I}\right)^{a} \text{ to } 0 \le Tn < 26.5°C \quad (1)$$

$$ET = -415.85 + 32.24Tn - 0.43Tn^{2} \text{ to } Tn \ge 26.5°C \quad (2)$$

Where "Tn" is the average temperature of the month "n", in °C, "n" ranges from 1 to 12 (January through December); "I" index that expresses the level of heat available in the region, according to the equation:

$$I = \sum_{n=1}^{12} (0.2Tn)^{1.514} \quad (3)$$

The exponent of the equation (1) is a function of "I", calculated by the equation:

$$a = 6.7510^{-7} I^{3} - 7.7110^{-5} I^{2} + 1.791210^{-2} I + 0.49239 \quad (4)$$

The value of "ETp" is a standard condition: evapotranspiration that occurs in a month for 30 days with a photoperiod of 12 hours, dependent on thermal conditions, and therefore needs correction.

$$ETp = ET\left(\left(\frac{d}{30}\right)\left(\frac{N}{12}\right)\right) \tag{5}$$

Where "d" is the number of days of the month and "N" photoperiod of the month in question.
The calculating the climatic water balance was done by the Thornthwaite & Mather (1955) method, assuming the available water capacity of the soil equal to 100 mm for comparative purposes.
The estimation indexes of humidity (Ih), aridity (Ia) and moisture (Im) were calculated according to Thornthwaite (1948):

$$Ih = \frac{EXC}{ETP}100 \tag{6}$$

$$Ia = \frac{DEF}{ETP}100 \tag{7}$$

$$\mathrm{Im} = Ih - 0{,}6(Ia) \tag{8}$$

Where "EXC" is water excess and "DEF" is water deficit, comings from the climatic water balance (mm); "ETP" is potential evapotranspiration (mm).
The following tables that contains the sort keys: Table 2 (1st key), Table 3 (2nd key), Table 4 (3rd key) and Table 5 (4th key).

Climatic types	Moisture index (Im)
A - perhumid	$100 \leq \mathrm{Im}$
B_4 - humid	$80 \leq \mathrm{Im} < 100$
B_3 - humid	$60 \leq \mathrm{Im} < 80$
B_2 - humid	$40 \leq \mathrm{Im} < 60$
B_1 - humid	$20 \leq \mathrm{Im} < 40$
C_2 - moist subhumid	$0 \leq \mathrm{Im} < 20$
C_1 - dry subhumid	$-20 \leq \mathrm{Im} < 0$
D - semiarid	$-40 \leq \mathrm{Im} < -20$
E - arid	$-60 \leq \mathrm{Im} < -40$

Table 2. Climatic types. 1st key based on the moisture index.

Moist climates (A, B_4, B_3, B_2, B_1 e C_2)	Aridity index (Ia)	Dry climates (C_1, D e E)	Humidity index (Ih)
r - little or no water deficiency	$0 \leq Ia < 16.7$	d - little or no surplus water	$0 \leq Ih < 10$
s - moderate summer water deficiency	$16.7 \leq Ia < 33.3$	s - moderate winter water surplus	$10 \leq Ih < 20$
w - moderate winter water deficiency	$16.7 \leq Ia < 33.3$	w - moderate summer water surplus	$10 \leq Ih < 20$
s_2 - large summer water deficiency	$33.3 \leq Ia$	s_2 - large winter water surplus	$20 \leq Ih$
w_2 - large winter water deficiency	$33.3 \leq Ia$	w_2 - large summer water surplus	$20 \leq Ih$

Table 3. Climatic types. 2nd key based on the aridity indexes and humidity.

Climatic types	Thermal index (It) (ET_P anual)
A′ - megathermal	$1{,}140 \leq ET_P$
B'_4 - mesothermal	$997 \leq ET_P < 1{,}140$
B'_3 - mesothermal	$855 \leq ET_P < 997$
B'_2 - mesothermal	$712 \leq ET_P < 855$
B'_1 - mesothermal	$570 \leq ET_P < 712$
C'_2 - microthermal	$427 \leq ET_P < 570$
C'_1 - microthermal	$285 \leq ET_P < 427$
D′ - tundra	$142 \leq ET_P < 285$
E′ - frost	$142 > ET_P$

Table 4. Climatic types. 3rd key based on the thermal index and annual potential evapotranspiration.

Climatic subtypes	Concentration of ET_P in summer (%)
a′	$ET_s < 48\%$
b'_4	$48.0 \leq ET_s < 51.9$
b'_3	$51.9 \leq ET_s < 56.3$
b'_2	$56.3 \leq ET_s < 61.6$
b'_1	$61.6 \leq ET_s < 68.0$
c'_2	$68.0 \leq ET_s < 76.3$
c'_1	$76.3 \leq ET_s < 88.0$
d'	$88.0 \leq ET_s$

Table 5. Climatic subtypes. 4th key based on the relationship summer/annual potential evapotranspiration in % (ET_s).

5. Examples of application

5.1 Different altitude

In Tables 6 and 7 following information regarding the climatologic water balance (Thornthwaite & Mather, 1955) to Campos do Jordão and Santos, respectively. They were required to Thornthwaite climatic classification.

Months	T (°C)	P (mm)	ET_P (mm)	P-ET_P (mm)	NEG	GW (mm)	ALT (mm)	ET_A (mm)	DEF (mm)	EXC (mm)
J	17.3	306.1	82.5	223.7	0.0	100.0	0.0	82.4	0.0	223.7
F	17.5	265.6	76.4	189.4	0.0	100.0	0.0	76.2	0.0	189.4
M	16.7	193.5	75.7	117.8	0.0	100.0	0.0	75.7	0.0	117.8
A	14.7	98.9	57.7	41.2	0.0	100.0	0.0	57.7	0.0	41.2
M	11.9	79.3	41.8	37.5	0.0	100.0	0.0	41.8	0.0	37.5
J	10.1	51.4	30.7	20.6	0.0	100.0	0.0	30.8	0.0	20.6
J	9.5	42.1	28.9	13.2	0.0	100.0	0.0	28.9	0.0	13.2
A	11.3	58.5	38.1	20.4	0.0	100.0	0.0	38.1	0.0	20.4
S	13.4	91.6	49.5	42.1	0.0	100.0	0.0	49.5	0.0	42.1
O	14.9	159.3	62.9	96.3	0.0	100.0	0.0	63.0	0.0	96.3
N	15.9	205.9	70.4	135.5	0.0	100.0	0.0	70.4	0.0	135.5
D	16.6	300.1	79.9	220.2	0.0	100.0	0.0	79.9	0.0	220.2
Ann	14.2	1,852.3	694.5	1,157.8		1,200.0	0.0	694.5	0.0	1,157.8

Table 6. Climatologic water balance according to Thornthwaite and Mather (1955) for the period 1961 to 1990. Campos do Jordão, SP, Brazil, INMET (2009). (T = temperature; P = precipitation; ETP = potential evapotranspiration; NEG = negative accumulated; GW = ground water; ALT = GWactual -GWprevious; ETA = actual evapotranspiration; DEF = water deficit; EXC = water excess).

To compare two nearby site, with small differences in latitude and longitude, but with a reasonable difference in altitude, and using the methodology of Köppen, Campos do Jordão shows the climatic type **Cfb** (humid temperate climate with moderately warm summer) while Santos shows the climatic type **Afa** (humid tropical climate with hot summers). The two sites are similar with respect to humidity, sites with high rainfall in all months of the year and show no water stress throughout the year. However, Campos do Jordão shows itself as a place colder than Santos due to the influence of altitude on air temperature.

Months	T (°C)	P (mm)	ET_P (mm)	P-ETP (mm)	NEG	GW (mm)	ALT (mm)	ET_A (mm)	DEF (mm)	EXC (mm)
J	28.6	255.9	197.2	58.7	0.0	100.0	0.00	197.2	0.0	58.7
F	28.9	220.3	185.6	34.7	0.0	100.0	0.00	185.6	0.0	34.7
M	28.1	221.1	178.6	42.5	0.0	100.0	0.00	178.6	0.0	42.5
A	26.3	193.6	129.9	63.7	0.0	100.0	0.00	129.9	0.0	63.7
M	24.8	144.3	103.7	40.6	0.0	100.0	0.00	103.7	0.0	40.6
J	23.2	106.2	76.5	29.7	0.0	100.0	0.00	76.5	0.0	29.7
J	22.8	121.6	73.8	47.7	0.0	100.0	0.00	73.9	0.0	47.7
A	22.8	78.4	76.3	2.1	0.0	100.0	0.00	76.3	0.0	2.1
S	22.4	130.2	73.5	56.7	0.0	100.0	0.00	73.5	0.0	56.7
O	24.2	146.0	104.9	41.1	0.0	100.0	0.00	104.9	0.0	41.1
N	25.8	162.0	133.4	28.6	0.0	100.0	0.00	133.4	0.0	28.6
D	27.4	210.9	175.2	35.7	0.0	100.0	0.00	175.2	0.0	35.7
Ann	25.4	1,990.5	1,508.7	481.8		1,200.0	0.00	1,508.7	0.0	481.8

Table 7. Climatologic water balance according to Thornthwaite and Mather (1955) for the period 1961 to 1990. Santos, SP, Brazil, INMET (2009). (T = temperature; P = precipitation; ET_P = potential evapotranspiration; NEG = negative accumulated; GW = ground water; ALT = GWactual - GWprevious; ET_A = actual evapotranspiration; DEF = water deficit; EXC = water excess).

Using the method of Thornthwaite realizes that Campos do Jordão presents the climatic type **ArB'$_1$a'** (perhumid without water deficit throughout the year, with moderate temperatures and annual potential evapotranspiration of 694.5 mm concentrated in 33.7% in the summer) while Santos presents the climatic type **B$_1$rA'a'** (humid without water deficit throughout the year, with high temperatures and annual potential evapotranspiration of 1,508.7 mm concentrated in 37.2% in the summer). The district of Santos presents increased evaporative demand due to their higher temperatures, thus the annual water surplus is smaller than in Campos do Jordão, which has a water surplus much higher compared to Santos due to low temperatures which occur along the year, which reduces evapotranspiration.

Comparing the two methods of classification, it is noted which Campos do Jordão is differs of Santos at altitude and relief (Mantiqueira Mountains), which means it has lower values of temperature, and thus, presents a lower demand evapotranspiration. The rainfall in these regions tends to be higher in the rainy season (October-March) in relation to sea level - effect of atmospheric circulation that brings moisture from the Atlantic Ocean. For agricultural activities have thermal constraints for many crops - risk of frost as a limiting factor, however, the natural landscape of these regions offer excellent conditions for ecotourism activities.

Interestingly, the Köppen classification does not differentiate the two sites for moisture content (function of precipitation), considering them as **f** (wet), while the Thornthwaite classification shows the difference between Campos do Jordão and Santos, characterizes them as **A** (perhumid) and **B$_1$** (humid), respectively, according to the difference in water surplus. Moreover, the Thornthwaite classification of stands out for identifying the absence of water stress (**r**) and potential evaporation concentrated in the summer (**a'**), being the 3rd letter, according to Thornthwaite classification, identifies the difference between the evapotranspiration demand between the two locations, indicating no need for supplemental irrigation.

5.2 Different latitude

Tables 8 and 9 present information regarding the climatologic water balance (Thornthwaite & Mather, 1955), necessary for climatic classification for districts of Boa Vista and Santa Maria respectively.

Months	T (°C)	P (mm)	ET_P (mm)	$P-ET_p$ (mm)	NEG	GW (mm)	ALT (mm)	ET_A (mm)	DEF (mm)	EXC (mm)
J	27.5	25.1	151.0	-125.9	-513.2	0.6	-1.5	26.6	124.4	0.0
F	28.0	18.1	152.2	-134.1	-647.2	0.1	-0.4	18.5	133.6	0.0
M	28.4	30.9	179.4	-148.5	-795.7	0.0	-0.1	31.0	148.4	0.0
A	28.0	88.5	165.0	-76.5	-872.2	0.0	-0.0	88.5	76.5	0.0
M	26.9	213.0	145.8	67.2	-39.8	67.2	67.2	145.8	0.0	0.0
J	25.9	321.3	121.6	199.7	0.0	100.0	32.8	121.6	0.0	166.9
J	25.8	267.8	123.8	144.0	0.0	100.0	0.0	123.8	0.0	144.0
A	26.6	188.0	139.6	48.4	0.0	100.0	0.0	139.6	0.0	48.4
S	27.7	99.4	158.3	-58.9	-58.9	55.5	-44.5	143.9	14.4	0.0
O	28.2	63.5	174.8	-111.3	-170.2	18.2	-37.2	100.7	74.1	0.0
N	28.0	60.8	163.3	-102.5	-272.7	6.5	-11.7	72.5	90.8	0.0
D	27.6	44.0	158.5	-114.5	-387.2	2.1	-4.5	48.5	110.0	0.0
Ann	27.4	1,420.4	1,833.3	-412.9		450.0	0.0	1,061.1	772.2	359.3

Table 8. Climatologic water balance according to Thornthwaite and Mather (1955) for the period 1961 to 1990. Boa Vista, RR, Brazil, INMET (2009). (T = temperature; P = precipitation; ET_P = potential evapotranspiration; NEG = negative accumulated; GW = ground water; ALT = GWactual – GWprevious; ET_A = actual evapotranspiration; DEF = water deficit; EXC = water excess)

Located in the Northern Hemisphere, but very close to the equator, Boa Vista introduces climatic type **Awa** (tropical climate with dry season in winter and warm all year round), second the methodology of Köppen. The low latitude of Boa Vista question influence in the definition of the seasons because the summer no presents the highest temperatures. On the other hand, the district of Santa Maria is located further south of the Tropic of Capricorn, and shows the climatic type **Cfa** (humid temperate and hot summer).

These two sites were similar with respect to high temperatures in summer (**a**), however, the air temperature is higher during the year, which explains the high values of evapo-

transpiration. Regarding precipitation, the two sites have high values, but the monthly distribution is different. Boa Vista has dry and rainy season well defined, while in Santa Maria the precipitation is uniform throughout the year.
Second the method of Thornthwaite, Boa Vista presents the climatic type **C_1dA'a'** (dry subhumid with little excess water during June, July and August, high temperatures and annual potential evapotranspiration of 1,833.3 mm concentrated in 26.3% in the summer). Already, Santa Maria has the climatic type **B_4rB'$_3$a'** (humid without water deficit, with moderate temperatures and annual potential evapotranspiration of 896.5 mm concentrated in 38.7% in the summer).

Months	T (°C)	P (mm)	ET_P (mm)	P-ET_P (mm)	NEG	GW (mm)	ALT (mm)	ET_A (mm)	DEF (mm)	EXC (mm)
J	24.2	163.0	130.7	32.3	0.0	100.0	0.0	130.7	0.0	32.3
F	23.9	127.2	115.0	12.2	0.0	100.0	0.0	115.0	0.0	12.2
M	21.9	136.2	101.1	35.1	0.0	100.0	0.0	101.1	0.0	35.1
A	18.4	121.4	64.0	57.4	0.0	100.0	0.0	64.0	0.0	57.4
M	15.9	127.5	45.8	81.7	0.0	100.0	0.0	45.8	0.0	81.7
J	13.9	139.3	31.9	107.4	0.0	100.0	0.0	31.9	0.0	107.4
J	14.1	144.9	33.6	111.3	0.0	100.0	0.0	33.6	0.0	111.3
A	14.2	142.1	35.5	106.6	0.0	100.0	0.0	35.5	0.0	106.6
S	16.5	124.3	50.0	74.3	0.0	100.0	0.0	50.0	0.0	74.3
O	18.6	128.2	70.8	57.4	0.0	100.0	0.0	70.8	0.0	57.4
N	21.0	120.5	93.6	26.9	0.0	100.0	0.0	93.6	0.0	26.9
D	23.3	142.2	124.4	17.8	0.0	100.0	0.0	124.4	0.0	17.8
Ann	18.8	1,616.8	896.5	720.3		1,200.0	0.0	896.5	0.0	720.3

Table 9. Climatologic water balance according to Thornthwaite and Mather (1955) for the period 1961 to 1990. Santa Maria, RS, Brazil, INMET (2009). (T = temperature; P = precipitation; ET_P = potential evapotranspiration; NEG = negative accumulated; GW = ground water; ALT = GWactual - GWprevious; ET_A = actual evapotranspiration; DEF = water deficit; EXC = water excess).

Analyzing the 1st letter of the climate type, Boa Vista has the highest average monthly temperatures in relation to the humid climate of Santa Maria, which increases the demand for evapotranspiration. The climate of Santa Maria (humid) is associated with lower levels of temperature due to the mountainous regions, which requires less evapotranspiration. These conditions are confirmed by the 2nd letter, which indicates **d** (small excess of water) and **r** (no drought) to Boa Vista and Santa Maria, respectively, and also the 3rd letter **A'** (high temperatures associated with high evapotranspiration). Both classifications show that in monthly scale there is no need for irrigation of agricultural crops in Santa Maria. While in Boa Vista the classification of Thornthwaite specifies more appropriately through of high aridity index (Ia = 42) and of negative moisture index (Im =- 6), showing the need for the use of irrigation during the months September to April.

5.3 Different longitude

The climatologic water balance (Thornthwaite & Mather, 1955) for districts of João Pessoa and Porto Velho can be found in Tables 10 and 11, respectively.
According to the methodology of Köppen, João Pessoa and Porto Velho present the same climatic type **Ama** (tropical climate with a dry season and hot summer).

Months	T (°C)	P (mm)	ET_P (mm)	P-ET_P (mm)	NEG	GW (mm)	ALT (mm)	ET_A (mm)	DEF (mm)	EXC (mm)
J	27.1	75.8	151.1	-75.3	-439.0	1.2	-1.4	77.2	74.0	0.0
F	27.2	108.4	141.8	-33.4	-472.4	0.9	-0.3	108.8	33.1	0.0
M	27.0	252.2	150.9	101.3	0.0	100.0	99.1	150.9	0.0	2.2
A	26.7	349.8	137.9	211.9	0.0	100.0	0.0	137.9	0.0	211.9
M	26.0	307.3	127.5	179.8	0.0	100.0	0.0	127.5	0.0	179.8
J	25.2	346.1	109.1	237.0	0.0	100.0	0.0	109.1	0.0	237.0
J	24.2	346.2	97.3	248.9	0.0	100.0	0.0	97.3	0.0	248.9
A	24.3	183.5	99.5	84.0	0.0	100.0	0.0	99.5	0.0	84.0
S	25.1	87.2	109.8	-22.6	-22.6	79.7	-20.2	107.4	2.4	0.0
O	26.3	35.4	136.4	-101.0	-123.6	29.0	-50.7	86.1	50.3	0.0
N	26.7	24.9	141.6	-116.7	-240.3	9.0	-20.0	44.9	96.7	0.0
D	26.9	28.5	151.8	-123.3	-363.7	2.6	-6.4	34.9	116.9	0.0
Ann	26.1	2,145.3	1,554.9	590.4		723.0	0.0	1,181.6	373.3	963.7

Table 10. Climatologic water balance according to Thornthwaite and Mather (1955) for the period 1961 to 1990. João Pessoa, PB, Brazil, INMET (2009). (T = temperature; P = precipitation; ET_P = potential evapotranspiration; NEG = negative accumulated; GW = ground water; ALT = GWactual - GWprevious; ET_A = actual evapotranspiration; DEF = water deficit; EXC = water excess).

Months	T (°C)	P (mm)	ET_P (mm)	P-ET_P (mm)	NEG	GW (mm)	ALT (mm)	ET_A (mm)	DEF (mm)	EXC (mm)
J	25.5	320.9	123.9	197.0	0.0	100.0	0.0	123.9	0.0	197.0
F	25.5	316.0	114.5	201.5	0.0	100.0	0.0	114.5	0.0	201.5
M	25.6	273.9	126.4	147.5	0.0	100.0	0.0	126.4	0.0	147.5
A	25.7	251.0	121.4	129.6	0.0	100.0	0.0	121.4	0.0	129.6
M	25.3	126.6	116.6	10.0	0.0	100.0	0.0	116.6	0.0	10.0
J	24.7	49.6	102.5	-52.9	-52.9	58.9	-41.0	90.7	11.8	0.0
J	24.6	24.2	104.1	-79.9	-132.8	26.5	-32.4	56.6	47.5	0.0
A	25.9	36.4	125.5	-89.1	-221.8	10.9	-15.6	52.0	73.4	0.0
S	26.2	119.9	128.7	-8.8	-230.6	10.0	-0.9	120.8	7.9	0.0
O	26.1	192.7	133.8	58.7	-37.6	68.7	58.7	134.0	0.0	0.0
N	26.0	225.2	130.5	94.7	0.0	100.0	31.3	130.5	0.0	63.4
D	25.5	319.1	127.8	191.3	0.0	100.0	0.0	127.8	0.0	191.3
Ann	25.6	2,255.5	1,455.9	799.6		875.0	0.0	1,315.3	140.6	940.2

Table 11. Climatologic water balance according to Thornthwaite and Mather (1955) for the period 1961 to 1990. Porto Velho, RO, Brazil, INMET (2009). (T = temperature; P = precipitation; ET_P = potential evapotranspiration; NEG = negative accumulated; GW = ground water; ALT = GWactual - GWprevious; ET_A = actual evapotranspiration; DEF = water deficit; EXC = water excess).

According to the Thornthwaite method, Porto Velho presents the climatic type **B_3rA'a'** (humid with little water deficit in the months from June to September, high temperatures and annual potential evapotranspiration of 1,455.9 mm concentrated in 25.1% in the summer), while João Pessoa presents the climatic type **B_2sA'a'** (humid with moderate water stress during the months of January, February and from September to December, with high temperatures and annual potential evapotranspiration of 1,554.9 mm concentrated in 29.0% in the summer).

Thornthwaite's method shows differences with respect to temperature monthly. In Porto Velho the variation of monthly temperature range is less while in João Pessoa is higher due to the effect of continentality/ocean. It also differentiates the sites with respect to deficiencies and excess water throughout the year. The proximity of the ocean generates less thermoregulatory effect provided by the moisture present in the interior of the continent, which showed higher temperatures in winter in João Pessoa. The rainfalls are concentrated during summer in Porto Velho and autumn-winter in João Pessoa, showing respectively, deficiencies in winter and spring-summer. The irrigation of cultures is recommended from June to September in Porto Velho, while in João Pessoa is necessary from September to February, once the annual aridity index is high (Ia = 24). The use of the climatologic water balance in the classification of Thornthwaite becomes advantageous because it allows identifying the level of disability and the season when occurs water deficit.

5.4 Ocean currents

Information related to climatologic water balance (Thornthwaite & Mather, 1955), for Angra dos Reis and Cabo Frio, necessary to climatic second classification Thornthwaite are presented in Tables 12 and 13, respectively.

Months	T (°C)	P (mm)	ET_P (mm)	P-ET_P (mm)	NEG	GW (mm)	ALT (mm)	ET_A (mm)	DEF (mm)	EXC (mm)
J	25.9	241.0	143.1	97.9	0.0	100.0	0.0	143.1	0.0	97.9
F	26.4	221.9	137.3	84.6	0.0	100.0	0.0	137.3	0.0	84.6
M	25.7	233.4	135.1	98.3	0.0	100.0	0.0	135.1	0.0	98.3
A	23.8	163.8	99.7	64.1	0.0	100.0	0.0	99.7	0.0	64.1
M	22.1	105.3	79.3	26.0	0.0	100.0	0.0	79.3	0.0	26.0
J	20.7	74.3	61.3	13.0	0.0	100.0	0.0	61.3	0.0	13.0
J	20.2	71.9	58.7	13.2	0.0	100.0	0.0	58.7	0.0	13.2
A	20.8	78.9	65.6	13.3	0.0	100.0	0.0	65.6	0.0	13.3
S	21.4	109.0	72.6	36.4	0.0	100.0	0.0	72.6	0.0	36.4
O	22.3	152.1	89.0	63.1	0.0	100.0	0.0	89.0	0.0	63.1
N	23.5	171.0	105.0	66.0	0.0	100.0	0.0	105.0	0.0	66.0
D	24.9	261.1	131.8	129.2	0.0	100.0	0.0	131.9	0.0	129.2
Ann	23.1	1,883.7	1,178.6	705.1		1,200.0	0.0	1,178.6	0.0	705.1

Table 12. Climatologic water balance according to Thornthwaite and Mather (1955) for the period 1961 to 1990. Angra dos Reis, RJ, Brazil, INMET (2009). (T = temperature; P = precipitation; ET_P = potential evapotranspiration; NEG = negative accumulated; GW = ground water; ALT = GWactual – GWprevious; ET_A = actual evapotranspiration; DEF = water deficit; EXC = water excess).

According with the methodology of Köppen, Angra dos Reis presents the climatic type **Afa** (tropical climate with no dry season and hot summer) while Cabo Frio is climatic type **Awa** (tropical climate with dry season in winter and hot summer). In this case, the difference between the climates is the amount and distribution of rainfall throughout the year, which is much smaller in Cabo Frio. Angra dos Reis presents high rainfall, especially in spring and summer.

When the Thornthwaite method is applied, Angra dos Reis presents the climatic type **B_3rA'a'** (humid without water stress, with high temperatures and annual potential evapotranspiration of 1,178.6 mm concentrated in 35.3% in the summer), while Cabo Frio displays the climatic type **C_1dA'a'** (dry subhumid, with drought in nearly all months of the year, with high temperatures and annual potential evapotranspiration of 1,156.6 mm concentrated in 32.9% in summer). Thus, we find that the differences between these two sites are mainly related to precipitation and water deficit, because as the precipitation is higher there will be less water deficiency in soil. In contrast, the water surplus occurs when rainfall exceeds field capacity.

Months	T (°C)	P (mm)	ET_P (mm)	P-ET_P (mm)	NEG	GW (mm)	ALT (mm)	ET_A (mm)	DEF (mm)	EXC (mm)
J	25.0	74.6	129.8	-55.2	-409.5	1.7	-1.2	75.8	54.0	0.0
F	25.2	37.0	120.7	-83.7	-493.2	0.7	-0.9	37.9	82.8	0.0
M	25.3	58.1	129.4	-71.3	-564.6	0.3	-0.4	58.5	71.0	0.0
A	24.1	78.6	103.4	-24.8	-589.4	0.3	-0.1	78.7	24.8	0.0
M	22.6	74.0	84.7	-10.7	-600.1	0.2	-0.0	74.0	10.6	0.0
J	21.6	47.9	69.3	-21.4	-621.5	0.2	-0.0	47.9	21.4	0.0
J	21.1	47.1	66.6	-19.5	-641.0	0.2	-0.0	47.1	19.5	0.0
A	21.0	37.6	67.8	-30.2	-671.2	0.1	-0.0	37.6	30.1	0.0
S	21.2	58.1	71.0	-13.0	-684.1	0.1	-0.0	58.1	12.9	0.0
O	22.0	90.6	86.1	4.5	-307.6	4.6	4.5	86.1	0.0	0.0
N	23.3	92.7	102.8	-10.2	-317.7	4.2	-0.5	93.1	9.7	0.0
D	24.4	88.3	124.8	-36.5	-354.3	2.9	-1.3	89.6	35.2	0.0
Ann	23.1	784.6	1,156.6	-372.0		16.0	0.0	784.6	372.0	0.0

Table 13. Climatologic water balance according to Thornthwaite and Mather (1955) for the period 1961 to 1990. Cabo Frio, RJ, Brazil, INMET (2009). (T = temperature; P = precipitation; ET_P = potential evapotranspiration; NEG = negative accumulated; GW = ground water; ALT = GWactual - GWprevious; ET_A = actual evapotranspiration; DEF = water deficit; EXC = water excess).

These differences between the two local are restricted to the effect of ocean currents on the Atlantic coast, the Brazilian coast, influencing the precipitation regime. As the ocean current is warmer near Angra dos Reis, this presents higher values rainfall throughout the year, especially during the hottest periods of the year.

The climatic classifications of Köppen and Thornthwaite suggest that in Angra dos Reis is not necessary the use of irrigation for local agriculture, but to Cabo Frio, the differences between the two classifications are important. According to Köppen classification the use of irrigation be essential in the winter, however hides the water deficit that occurs in the remainder of the year in Cabo Frio, a fact that is detected by the Thornthwaite classification, which stresses the use of irrigation throughout the years, except in October.

Therefore, in general, to Köppen sees only a greater quantity of rain, but not how it is distributed, nor whether it is sufficient to avoid water deficit. Thus, the Thornthwaite classification is more appropriate because it allows not only see the differences in rainfall, as well as other pertinent differences and coming from the water balance, such as evapotranspiration, water deficiencies and excesses throughout the year, important in the planning of an agricultural region. Thus, Köppen loses in details when the need is for agricultural use, being recommended the Thornthwaite's method.

6. Conclusion

It is difficult to find a climate classification that can be considered perfect, because each classification has its own merits, limitations and failures. Despite the differences between climatic classifications, either for one reason or another, knowledge of the climate of a place (region) allows a better orientation to agriculture.

The Köppen classification is still the most widely used despite its limitations, since it only depends on temperature and precipitation. The Thornthwaite classification seems more appropriate in the scope of its subdivisions and other climatic types according to temperature, precipitation and evapotranspiration (moisture factor), featuring more detail for a place or region, even using the estimated potential evapotranspiration according to the annual temperature variation and photoperiod, is still a more complete and comprehensive. Currently, it is not possible to use bolder methods for estimating evapotranspiration due to failure to obtain sufficient data for this, beyond poor spatial distribution of meteorological stations in several countries, as in Brazil, being recommended using the original method for calculating potential evapotranspiration suggested by Thornthwaite (1948).

The use of the water balance by the classification of Thornthwaite enriches the basis of classification based on the amounts of rainfall and temperature, allowing to identify a place or region by the characteristics: the changes in air temperature and cumulative monthly and annual rainfall, the deficit and water excess throughout the year, the potential evapotranspiration monthly and annual, the seasons drought and rainy - the distribution of rainfall seasonally and the index of aridity. This will allow identifying periods where there is need for irrigation.

Through the classification of Thornthwaite, when analyzing the deficiency or excess moisture, the concentration of thermal efficiency or potential evapotranspiration during the summer, allows climatic information more detailed from location - climate subtype, showing that Thornthwaite (1948) improved the climate classification system when introduced the water factor as a function of evapotranspiration and water balance.

7. Acknowledgements

This study was supported by Fundação de Amparo à Pesquisa do Estado de São Paulo (FAPESP).

8. References

Anwar, M.M. (1993). *A geography of Pakistan*, Book-World Islamabad and Lahore, pp. 20-60.

Balling, R.C. (1984). Classification in climatology, *in* Gaile, G.L. & Willmott, C.J. (ed.), *Spatial statistics and models*, Dordrecht: Reídel Publishing, Hingham, pp. 81-108.

Blair, T.A. (1942). *Climatology, general and regional*, Prentice-Hall, INC., New York. pp. 484.

Burgos, J.J. (1958). *Agroclimatic classifications and representations: report of the applications value of climatic and agroclimatic classification for agricultural purposes*, Comission for Agricultural Meteorology, WMO, Varsovia.

Cunha, A.R. & Martins, D. (2009). Climatic classification for the districts of Botucatu and São Manuel, SP, *Brazilian Journal of Irrigation and Drainage* Vol. 14 (No. 1): 1-11.

Doerr, A.H. (1962). Thornthwaite's original climatic classification and Oklahoma's climate, *Proceedings of the Oklahoma Academy of Science* Vol. 42 (No. 1): 231-235.

Essenwanger, O.M. (2001). Classification of climates, *World Survey of Climatology* Vol. 1C, General Climatology. Elsevier, Amsterdam, pp. 102.

Flohn, H. (1950). Neue auschavgen über die allgemeinen zirkulation der atmosphare und ihre klimatische bedeutung, *Erdkunde* Vol. 4 (No. 1): 141-162.

Geiger, R. (1953). Eine neue Weltkarte der Klimagebiete der Erde, *Erdkunde* Vol. 8 (No.1/2): 58-60.

Griffiths, J.F. (1978). *Applied climatology*. An Introduction, 2nd Edition, Oxford University Press London, pp. 136.

Hudson, J.C. & Brown, D.A. (2000). *Rethinking grassland regionalism*.

URL://www.geog.umn.edu/Faculty/brown/grasslands/RGR1.htm

INMET - National Institute of Meteorology. (2009). *Climatological Normals of Brazil 1961-1990, in* Ramos, A.M., Santos, L.A.R., Fortes, L.T.G. pp. 465.

Johnson, B.L.C. (1979). *Pakistan*, Heinemann, London, pp. 51-59.

Kazi, S.A. (1951). Climatic regions of west Pakistan, *PGR* Vol. 6 (No. 1): 1-39.

Kendrew, W.G. (1941). *Climates of the continents*, Oxford University Press London, pp. 608.

Khan, F.K. (1991). *A geography of Pakistan*, Oxford University Press Karachi, Pakistan, pp. 245.

Khan, J.A. (1993). *The climate of Pakistan*, Rahber Publishers Karachi, pp. 79.

Köppen, W. (1936). Das geographische system der climate, *in* Köppen, W. & Geiger, R. (ed.), *Handbuch der Klimatologie*, CG Borntrager, Berlin. pp. 46.

Logan, B. (2006). *A Statistical Examination of the Climatic Human Expert System, The Sunset Garden Zones for California*, Masters of Science in Geography, Faculty of the Virginia Polytechnic Institute & State University, pp. 144.

Lohmann, U., Sausen, R., Bengtsson, L., Cubasch, U., Perlwitz, J. & Roeckner, E. (1993). The Köppen climate classification as a diagnostic tool for general circulation models, *Climate Research* Vol. 3 (No. 3): 177–193.

Malmström, V.H. (1969). A new approach to the classification of climate, *The Journal of Geography* Vol. 68 (No. 6): 1-12.

McMahon, T.A., Finlayson, B.L., Haines, A.T., & Srikanthan, R. (1992). *Global runoff – continental comparisons of annual flows and peak discharges*, Catena Verlag, Cremlingen, pp. 166.

Miller, A.A. (1959). *Climatology*, Mehtewn London and E.P. Dulton and Co. INC., New York, pp. 313.

Nasrullah, K. (1968). Climate of West Pakistan according to Thornthwaite system of classification of climates, *PGR* Vol. 23 (No. 1): 12-25.

Oliver, J.E. (1981). *Climatology, Selected Applications*, Great Britain, Richard Clay Ltd, Bungay, Suffolk, p. 260.

Peel, M.C., Finlayson, B.L. & McMahon, T.A. (2007). Updated world map of the Köppen-Geiger climate classification, *Hydrology and Earth System Sciences* Vol. 11 (No. 5): 1633–1644.

Peel, M.C., McMahon, T.A. & Finlayson, B.L. (2004). Continental differences in the variability of annual runoff - update and reassessment, *Journal of Hydrology* Vol. 295 (No. 1-4): 185–197.

Pereira, A.P, Villa Nova, N.A. & Sediyama, G.C. (1997). *Evapotranspiration*. Piracicaba: Fealq. pp. 183.

Pereira, A.P, Angelocci, L.R. & Sentelhas, P.C. (2002). *Agrometeorology: fundamentals and practical applications*, Lavras: Agricultural. pp. 478.

Raja, I.A. & Twidal, J.W. (1990). Distribution of global insulation over Pakistan, *Journal of Solar Energy* Vol. 44 (No. 2): 21-29.

Rolim, G.S., Camargo, M.B.P., Lania, D.G. & Moraes, J.F.L. (2007). Climatic classification of Köppen and Thornthwaite sistems and their applicability in the determination of agroclimatic zonning for the state of São Paulo, Brazil, *Bragantia* Vol. 66 (No. 4): 711-720.

Shamshad K.M. (1988). *The meteorology of Pakistan. Climate and Weather of Pakistan,* Royal Book Company Karachi Pakistan, pp. 313.

Stern, H., De Hoedt, G. & Ernst, J. (2000). Objective classification of Australian climates, *Australian Meteorological Magazine* Vol. 49 (No.1): 87–96.

Sunkar, A. (2008). *Sustainability in karst resources management: the case of the Gunung Sewu in Java,* Thesis of Doctor of Philosophy, The University of Auckland, pp. 230.

Supan, A. (1879). Die Temperaturzonen der Erde, *Petermanns Geography Mitt* Vol. 25 (No. 1): 349-358.

Terjung, W.H. & Louie, S. (1972). Energy input output climates of the world: a preliminary attempt, *Archiv für Meteorologie Geophysik und Bicklimatologie* Vol. 2: 129-166.

Thom, H.C.S. (1966). *Some methods of climatological analysis,* Technical note no. 81. WMO, Geneva, Switzerland, pp. 53.

Thornthwaite, C.W. (1946). The moisture factor in climate, *American Geophysical Union Transactions* Vol. 27 (No. 1): 41-48.

Thornthwaite, C.W. (1948). An approach toward a rational classification climate, *Geographical Review* Vol. 38 (No. 1): 55-94.

Thornthwaite, C.W. & Mather, J.R. (1955). The water balance, *Drexel Institute of Tecnology* Vol. 8 (No. 1): 1-14.

Trewartha G.T. (1937,1954). *An introduction to climate,* 3rd Edition, Mc Grawhill Book Company. INC, pp. 395.

Trewartha G.T. (1968). *An introduction to climate,* 4th Edition, Mc Grawhill Kogakusha, LTD, pp. 408.

Varejão-Silva, M.A. (2006). *Meteorology and climatology.* Digital Version 2. Recife, Pernambuco, pp. 463.

WMO - World Meteorological Organization (1984). *Technical Regulations,* Vol. I. WMO Publication No. 49. Geneva, Switzerland.

WMO - World Meteorological Organization No. 100 (2009). *Guide to climatological practices,* 3rd ed. pp. 180.
URL://www.wmo.int/pages/prog/wcp/ccl/guide/guide_climat_practices.html

Permissions

The contributors of this book come from diverse backgrounds, making this book a truly international effort. This book will bring forth new frontiers with its revolutionizing research information and detailed analysis of the nascent developments around the world.

We would like to thank Dr. Giacomo Gerosa, for lending his expertise to make the book truly unique. He has played a crucial role in the development of this book. Without his invaluable contribution this book wouldn't have been possible. He has made vital efforts to compile up to date information on the varied aspects of this subject to make this book a valuable addition to the collection of many professionals and students.

This book was conceptualized with the vision of imparting up-to-date information and advanced data in this field. To ensure the same, a matchless editorial board was set up. Every individual on the board went through rigorous rounds of assessment to prove their worth. After which they invested a large part of their time researching and compiling the most relevant data for our readers. Conferences and sessions were held from time to time between the editorial board and the contributing authors to present the data in the most comprehensible form. The editorial team has worked tirelessly to provide valuable and valid information to help people across the globe.

Every chapter published in this book has been scrutinized by our experts. Their significance has been extensively debated. The topics covered herein carry significant findings which will fuel the growth of the discipline. They may even be implemented as practical applications or may be referred to as a beginning point for another development. Chapters in this book were first published by InTech; hereby published with permission under the Creative Commons Attribution License or equivalent.

The editorial board has been involved in producing this book since its inception. They have spent rigorous hours researching and exploring the diverse topics which have resulted in the successful publishing of this book. They have passed on their knowledge of decades through this book. To expedite this challenging task, the publisher supported the team at every step. A small team of assistant editors was also appointed to further simplify the editing procedure and attain best results for the readers.

Our editorial team has been hand-picked from every corner of the world. Their multi-ethnicity adds dynamic inputs to the discussions which result in innovative outcomes. These outcomes are then further discussed with the researchers and contributors who give their valuable feedback and opinion regarding the same. The feedback is then collaborated with the researches and they are edited in a comprehensive manner to aid the understanding of the subject.

Apart from the editorial board, the designing team has also invested a significant amount of their time in understanding the subject and creating the most relevant covers. They scrutinized every image to scout for the most suitable representation of the subject and create an appropriate cover for the book.

The publishing team has been involved in this book since its early stages. They were actively engaged in every process, be it collecting the data, connecting with the contributors or procuring relevant information. The team has been an ardent support to the editorial, designing and production team. Their endless efforts to recruit the best for this project, has resulted in the accomplishment of this book. They are a veteran in the field of academics and their pool of knowledge is as vast as their experience in printing. Their expertise and guidance has proved useful at every step. Their uncompromising quality standards have made this book an exceptional effort. Their encouragement from time to time has been an inspiration for everyone.

The publisher and the editorial board hope that this book will prove to be a valuable piece of knowledge for researchers, students, practitioners and scholars across the globe.

List of Contributors

Jerry L. Hatfield and John H. Prueger
National Laboratory for Agriculture and the Environment, United States of America

F. Castellví
Dept. Environmental and Soil Sciences, University of Lleida, Lleida, Spain

Simona Consoli
University of Catania, Department of Agri-food and Environmental Systems Management, Hydraulic Section Catania, Italy

Giacomo Gerosa, Antonio Ballarin Denti and Riccardo Marzuoli
Dep. of Matematics and Physics, Università Cattolica del Sacro Cuore BresciaI, Italy

Angelo Finco
Dep. of Matematics and Physics, Università Cattolica del Sacro Cuore BresciaI, Italy
Ecometrics s.r.l., Spin-off Company Università Cattolica del Sacro Cuore, Italy

Simone Mereu
Department of Economics and Woody Plant Systems, University of Sassari, Italy

Adel Zeggaf Tahiri
Manzanar-Mauritania, B-151, F-Nord, Tevragh Zeina, Nouakchott, Mauritania

José O. Payero
The University of Queensland, Queensland Alliance for Agriculture and Food Innovation (QAAFI), 203 Tor St. Toowoomba, QLD 4350, Australia

Suat Irmak
University of Nebraska-Lincoln, 241 L.W. Chase Hall, Lincoln, NE 68583-0726, USA

J. Pivec, V. Brant and K. Hamouzová
University of Life Sciences Prague, the Czech Republic

David R. Bryla
U.S. Department of Agriculture, Agricultural Research Service, USA

André Pereira and Luiz Pires
State University of Ponta Grossa, Brazil

Salah El-Hendawy and Mohamed Alboghdady
Faculty of Agriculture, Suez Canal University, 41522 Ismailia, Egypt

Jun-Ichi Sakagami
Japan International Research Center for Agricultural Sciences, Tsukuba, Japan

Urs Schmidhalter
Department of Plant Sciences, Technische Universität München, Germany

Rutilo López- López, Ignacio Sánchez-Cohen and Victor González-Lauck
Instituto Nacional de Investigaciones Forestales Agrícolas y Pecuarias, Km. 1 Carretera, Huimanguillo-Cárdenas. 86400, Huimanguillo, Tabasco, México

Ramón Arteaga Ramírez
Postgrado en Ingeniería Agrícola y Uso Integral del Agua, Universidad Autónoma, Chapingo. Km 38.5 Carretera México-Texcoco. 56230, Chapingo, Texcoco, México

Waldo Ojeda Bustamante
Instituto Mexicano de Tecnología del Agua, Paseo Cuauhnáhuac 8532, Progreso, Jiutepec, Morelos, 62550, México

Hossein Dehghanisanij
Department of Irrigation Technology, Agricultural Engineering Research Institute, Karaj, Iran

Hanieh Kosari
Department of Irrigation and Reclamation Engineering, Agricultural and Natural Resources Campus, University of Tehran, Karaj, Iran

Ahmed Al-Busaidi
College of Agricultural & Marine Sciences, Department of Soils, Water and Agricultural Engineering, Sultan Qaboos University, Muscat, Oman

Tahei Yamamoto
Arid Land Research Center, Tottori University 1390 Hamasaka, Japan

Laura Bacci, Piero Battista and Bernardo Rapi
Istituto di Biometeorologia (IBIMET) - Consiglio Nazionale delle Ricerche - Firenze, Italia

Mariateresa Cardarelli, Youssef Rouphael and Giuseppe Colla
Dipartimento di Scienze e Tecnologie per l'Agricoltura, le Foreste, la Natura e l'Energia - Università della Tuscia - Viterbo, Italia

Giulia Carmassi, Luca Incrocci, Fernando Malorgio and Alberto Pardossi
Dipartimento di Biologia delle Piante Agrarie - Università di Pisa - Pisa, Italia

Nikolaos Katsoulas
University of Thessaly, Dept of Agriculture, Crop Production and Rural Environment, Volos, Greece

Constantinos Kittas
University of Thessaly, Dept of Agriculture, Crop Production and Rural Environment, Volos, Greece
Centre for Research and Technology Thessaly Institute of Technology and Management of Agricultural Ecosystems, Volos, Greece

Rached Ben Younes
University of Gafsa, Tunisia

Richard C. Beeson
University of Florida, USA

Tanaka K.
Japan Agency for Marine-Earth Science and Technology, Japan

Wakahara T. and Shiraki K.
2Tokyo University of Agriculture and Technology, Japan

Yoshifuji N.
Kyoto University, Japan

Suzuki M.
The University of Tokyo, Japan

Francesco Fiorillo
Dipartimento di Science per la Biologia, Geologia e l'Ambiente, University of Sannio Benevento, Italy

Antonio Ribeiro da Cunha and Edgar Ricardo Schöffel
São Paulo State University - Superior School Agronomic Sciences, Federal University of Pelotas - Superior School Eliseu Maciel Agronomy, Brazil